Euclid Vindicated from Every Blemish

Classic Texts in the Sciences

Classic Texts in the Sciences offers essential readings for anyone interested in the origin and roots of our present-day culture. Considering the fact that the sciences have significantly shaped our contemporary world view, this series not only provides the original texts but also extensive historical as well as scientific commentary, linking the classic texts to current developments. *Classic Texts in the Sciences* presents classic texts and their authors not only for specialists but for anyone interested in the background and the various facets of our civilization.

Gerolamo Saccheri

Euclid Vindicated from Every Blemish

Edited and Annotated by Vincenzo De Risi

Translated by G.B. Halsted and L. Allegri

Birkhäuser

Gerolamo Saccheri

Editor
Vincenzo De Risi
Max Planck Institute for the History of Science
Berlin, Germany

ISBN 978-3-319-37791-9 ISBN 978-3-319-05966-2 (eBook)
DOI 10.1007/978-3-319-05966-2
Springer Cham Heidelberg New York Dordrecht London

Softcover reprint of the hardcover 1st edition 2014

Printed on acid-free paper.

Springer is a part of Springer Science+Business Media
www.springer.com

Foreword to the English Edition

This book is the English translation of my annotated edition of Saccheri's *Euclides vindicatus*, published in Italian for the 'Edizioni della Normale' in 2011. With respect to the Italian edition, I have corrected some minor mistakes and typos, added some bibliographical references and deleted those that may have no interest for the English-speaking reader.

I have not re-translated Saccheri into English (a difficult task that awaits a native English-speaking scholar), and have based the present edition on Halsted's classical translation which first appeared in volume form in 1920 and was subsequently reprinted several times. Halsted's translation received some favorable reviews by R.C. Archibald and A. Emch ("Amer. Math. Monthly", 1921; "Bull. Amer. Math. Soc.", 1922) and harsh criticisms by T. Heath ("Nature", 1922). It seems to me fairly good, and Halsted's geometrical skills allowed him to produce a text that contains no significant mathematical misunderstandings. Today his language may sound old-fashioned, but it mimics Saccheri's own baroque and cumbersome Latin very well, which is almost impossible to translate into modern, elegant prose. Nonetheless, I have taken the liberty of modifying and correcting Halsted's version at some places. Besides correcting a few minor mistakes noticed by Heath, and a pair of mathematical imprecisions, my changes mostly regard the choice of terminology, some of which I have justified in my commentary. The most obvious modification is in the title of Saccheri's work whose translation varies largely depending on the translator or interpreter in question. In the *Introduction*, I have tried to justify the preservation of Euclid 'vindicated' rather than the more common 'emended' or 'corrected' (or Halsted's 'freed'), which do not seem to reflect the author's intentions. I prefer the term 'blemish' to other more abstract terms (such as 'error' or 'flaw'), given that Savile's original text makes use of an explicit corporal metaphor ("two blemishes, two moles in the most beautiful body of geometry").

Halsted's translation, in any case, covered only the First Book of *Euclid Vidicated*. A complete English translation of the Second Book is to be found in the doctoral dissertation of Linda Allegri (Columbia University, 1960) which was intended to complement Halsted's work. I publish Allegri's text here which, however, required a larger number of corrections and revisions. To my knowledge, this is the first complete English edition of Saccheri's masterwork.

My *Introduction* and *Notes* have been translated by Marco Santi and Caterina Benincasa, whose painstaking efforts deserve my deepest gratitude. I also thank Rebecca Rothfeld and Chiara Fabbrizi for the editing of the texts.

I would also like to express my thanks to Massimo Mugnai, Mariano Giaquinta and Paolo Freguglia for their invaluable help with the original Italian edition; to Massimo Mugnai and Massimo Girondino for providing me with their edition of Saccheri's *Logica* when it was still in proofs; to Marvin J. Greenberg, Victor Pambuccian, and Roshdi Rashed for their important suggestions on how to improve the English edition; to the many scholars who commented on my talks on Saccheri in several workshops organized by Arianna Betti, Michael Detlefsen and Roshdi Rashed.

I am grateful to my colleagues and friends at the Max Planck Institute for the History of Science for their support and advice, to the Library of the Institute for providing me with a digitalization of the Latin text of Saccheri and to the Max Planck Society for financial support.

Finally, and most of all, I thank Jürgen Jost for making this edition possible and for proposing that I publish it in this Series. I also thank the 'Edizioni della Normale' for allowing this translation. The present volume will soon be complemented (in the same Series) by a similar English edition of Lambert's *Theorie der Parallellinien.*

Berlin, January 2014

Table of Contents

Au commencement de la géométrie, on dit: "On donne le nom de parallèles à deux lignes qui, prolongées à l'infini, ne se rencontrent jamais." Et, dès le commencement de la *Statique*, cet insigne animal de Louis Monge a mis à peu près ceci: *Deux lignes parallèles peuvent être considérées comme se rencontrant, si on les prolonge à l'infini*. Je crus lire un catéchisme et encore un des plus maladroits. Ce fut en vain que je demandai des explications à M. Chabert. Le monstre, s'approchant de son tableau en toile cirée et traçant deux lignes parallèles et très voisines, me dit: "Vous voyez bien qu'à l'infini on peut dire qu'elles se rencontrent." Je faillis tout quitter. Un confesseur, adroit et bon jésuite, aurait pu me convertir à ce moment en commentant cette maxime: "Vous voyez que tout est erreur, ou plutôt qu'il n'y a rien de faux, rien de vrai, tout est de convention."

(Stendhal, *Vie de Henry Brulard*)

Introduction

1. Euclid and Saccheri

Euclid Vindicated from Every Blemish, by the Jesuit mathematician Gerolamo Saccheri,[1] appeared in print in Milan in 1733. The work enjoyed little success in the eighteenth century, was completely forgotten during the following century, and was rediscovered

[1] Saccheri was born in San Remo (a city under the dominion of Genoa) in 1667. In his youth, he lectured on grammar, philosophy, and theology at the Jesuit colleges in Cremona, Milan, Turin and Pavia. In 1699, he was appointed to the chair of mathematics at the university of Pavia, which he held till his death in 1733. It is known that he displayed outstanding mathematical abilities from an early age, and that he enjoyed the highest esteem in the Italian scientific circles of his time. He had close personal ties to the mathematicians Guido Grandi and Tommaso and Giovanni Ceva; we also possess letters that he exchanged with Vincenzo Viviani. He did not authored many scientific texts, and all sources report that his uncommon laziness as a writer left for later ages a less vivid impression of his intelligence and thought than his acquaintances had of him. Before *Euclid Vindicated*, his major work, Saccheri had published certain *Quaesita Geometrica* in 1693, an important *Logica Demonstrativa*, which appeared anonymously in 1696–1697 and was reprinted several times, and a *Neo-Statica* in 1708; he had also published a few theological works. Many details of Saccheri's life may be found in the biography written (in Italian) by Count Francesco Gambarana between 1733 and 1739. This text, of which we have two slightly different versions, remained manuscript and was recently published as an Appendix to the Italian edition of *Logica Dimostrativa*, ed. by M. Mugnai and M. Girondino, Pisa, Edizioni della Normale 2012. A previous account of Saccheri's life is that of A. Pascal, *Girolamo Saccheri nella vita e nelle opere*, "Giornale di Matematiche di Battaglini", 52, 1914, pp. 229–51; cf. also H. Bosmans, *Le géomètre Jérome Saccheri, S.J. (1667–1733)*, "Revue des Questions Scientifiques", 7, 1925, pp. 401–30.

G. Saccheri, *Euclid Vindicated from Every Blemish*, Classic Texts in the Sciences 1,
DOI 10.1007/978-3-319-05966-2_1,

and circulated only in the early nineteen hundreds. Today it is rather well-known, at least in outline, and is usually considered to be the birthplace of research on non-Euclidean geometry. The strategy of *Euclid Vindicated* is also widely regarded as one of the largest misunderstandings in the whole history of mathematics – and the most felicitous error in eighteenth-century geometry – as Saccheri's intention was in fact to demonstrate Euclid's Fifth Postulate, the parallel axiom, and thus to prove the impossibility of the very non-Euclidean geometries of which he is today regarded as the father. He undertook to prove the Fifth Postulate *per absurdum* and sought to spot a contradiction in the vast geometric theory that he constructed, for the first time in history, on the negation of the Euclidean axiom – a geometric theory that we nowadays identify without doubt as a genuine and well-structured system of hyperbolic geometry. Saccheri did not find the supposed contradiction, as it was nowhere to be found, but he was unable to convince himself that the new geometry he had erected might in fact be a reasonable alternative to Euclid's *Elements* rather than a green-eyed monster: consequently, he pointed to a contradiction of his own making, and thereby proved himself to be nothing more than a Jesuit.[2] This effort notwithstanding, the sacrilege, so to speak, had already been committed, and Saccheri's outstanding achievements towards the construction of hyperbolic geometry, while disowned by their author and relegated to a book printed in quite few copies, sneaked into European mathematical culture and poisoned the minds of certain more acute, unprejudiced, or simply more modern geometers. One century after the Jesuit's death, these scholars eagerly welcomed Saccheri's 'monster' in their writings, thus celebrating the triumph of non-Euclidean geometry. Following this widespread story, Saccheri unwittingly (yet brilliantly) anticipated one of the most momentous conceptual revolutions in the genesis of contemporary mathematics.

Instead of following this approach and treating Saccheri as an unwitting innovator, it will be worthwhile to try and place *Euclid Vindicated* in the history of the mathematics of its age. This is not to deny that its significance extends beyond its time: it is indisputable that Saccheri's results on the theory of parallelism are extraordinarily more advanced than any previous treatment of the subject, and that he was able to manufacture, in almost perfect isolation, a construction of exceptional amplitude and depth. Rather, it is for just this reason that we need to inquire how, where and why a work with intents so classical as to appear reactionary, and contents so novel as to start a revolution, could emerge at the beginning of the eighteenth century.

The structure and aim of Saccheri's book are indeed completely classical, and the work is firmly embedded in a long tradition of commentary and elaboration on Euclid's *Elements*

---

[2] The expression comes from Valéry, who levels the same charge of priestly pusillanimity against Saccheri and Cavalieri, as the latter, with his theory of indivisibles, almost arrived at the genuine theory of infinitesimal calculus – only to fail in the end. Cf. P. Valéry, *Mélanges*, in *Oeuvres*, ed. by J. Hytier, Paris, Gallimard 1957, vol. 1, p. 473. At any rate, Cavalieri belonged to the Jesuates (a mendicant order suppressed in 1668), not the Jesuits.

dating back to the Renaissance. This tradition represents one of the most precious fruits of sixteenth- and seventeenth-century mathematics. The way for Saccheri had been paved by the great sixteenth-century commentaries on Euclid (most notably, that of Clavius), and the tradition had subsequently consolidated through the production of texts for the study and teaching of elementary geometry, especially in Italy. These works began as commentaries on the *Elements*, but tended to transform gradually into autonomous treatises on certain controversial issues in the classical text. Most of these commentaries endeavored to reconstruct Euclid's intentions, and in certain cases did so with the help of philological tools. Other commentators openly distanced themselves from the ancient mathematician's solutions and aimed at producing a clearer or more rigorous foundation for the first elements of geometry. Thus, before Saccheri's *Euclides vindicatus,* we find in Italy alone a *Euclides restitutus* by Borelli (1658), a *Euclides adauctus et methodicus* by Guarini (1671), a *Euclide restituto* by Giordano (1680) and a *Euclides reformatus* by Marchetti (1709), along with countless other works with different titles and similar contents.[3] In fact, Saccheri's *Euclid Vindicated* represents in many ways the highest point and the ripest fruit (almost rotten, to be sure) of that direction of studies: not only because its mathematical content is undoubtedly more relevant than that of the older treatises, but also because it has assimilated their results to such extent that its theory stands on the shoulders of the whole tradition of sixteenth- and seventeenth-century commentaries on Euclid. Even the structure of *Euclid Vindicated* is autonomous from the partitions of the Euclidean text: Saccheri's book presents itself as a monographic treatment of the unconnected problems that still plagued would-be interpreters of the *Elements.*

Saccheri groups such problems under two headings, that of parallel theory (for which *Euclid Vindicated* is still famed) and that of the theory of proportions. His 1733 work is consequently divided into two books addressing these two topics, respectively. This arrangement was not at all unusual, as it was not out of the ordinary for a seventeenth-century geometer to indicate exactly the Fifth Postulate and the definitions of 'ratio' and 'proportion' as the two parts of elementary mathematics most in need of clarification and foundation. In 1621, Henry Savile described these difficult subjects as "two blemishes, two moles" spoiling Euclid's "beautiful body". Indeed, Savile even went so far as to establish a mathematics chair

[3] The works just mentioned, which certainly represent the closest precedents for the *title* of Saccheri's work, differ consistently as to their intrinsic values. Guarini's book, for instance, is a ponderous and verbose commentary on Euclid in the Scholastic style and contains very little mathematics of any import. The work of Borelli, and, building on this, those of Giordano and Marchetti, are of great mathematical significance, and we will discuss them in the next sections. The title of Borelli's book was itself probably derived from the term and concept of 'restitution' as it was employed in a previous edition of Euclid published by F.F. Candale, *Euclidis Megarensis Mathematici clarissimi Elementa Geometrica, libris XV. ad germanam Geometriae intelligentiam è diversis lapsibus temporis iniuria contractis restituta*, Paris, Royer 1566.

in Oxford designed to produce the research needed to wash the ugly blemishes away.[4] Half a century later, in 1663, John Wallis, the Savilian Professor of Geometry, wrote a short treatise on these two topics, which he took to present a solution to all the difficulties associated with them. Wallis concluded that he had accomplished the task assigned to him by his remote benefactor: in his words, he had finally "vindicated" Euclid.[5]

It is therefore with one eye towards the Italian tradition of commentaries on Euclid, and the other towards the English conceptions of the two major foundational problems of elementary geometry, that Saccheri baptizes his *Euclid Vindicated*, which is hence firmly situated in the mathematical research of its age. Let us now briefly outline the state of the theories of the parallels and of proportions in Saccheri's era.

2. The theory of parallels in the seventeenth century

The Fifth Postulate in Book I of Euclid's *Elements* affirms that if two straight lines that are both intersected by a third form with it two interior angles (on the same side) whose sum is less than π (two right angles), then the two lines, if extended, will end up meeting (on that side). We know of early attempts to prove this postulate in Classical Antiquity. In fact, these attempts probably preceded the composition of the *Elements*, suggesting that Euclid perhaps assumed his Fifth Postulate unwillingly, because he could not devise the proof for it that he sought. Nowadays, we ignore the methods, developments, and results of these ancient discussions almost completely; perhaps we even ignore their aim, because the mathemati-

---

[4] Sir Henry Savile (1549–1622), English philologist and humanist most noted for his contribution to the *King James Version* of the Bible. The Oxford professorships for geometry and astronomy that he instituted in 1619 have yielded an incredible scientific progeny and are still active to this day. In his *Praelectiones tresdecim in principium elementorum Euclidis*, Oxford, Lichfield&Short 1621, he wrote: "In pulcherrimo Geometriae corpore duo sunt naevi, duae labes, nec, quod sciam, plures, in quibus cluendis & emaculandis, cùm veterum tùm recentiorum, ut postea ostendam, vigilavit industria. Prior est hoc postulatum [the Fifth Postulate], posterior pertinet ad compositionem rationum" (lect. VII, p. 140).

[5] John Wallis (1613–1703), great English mathematician, was Savilian Chair at Oxford from 1649 until his death. The work in question is *De Postulato Quinto; & Quinta Definitione Lib. 6 Euclidis; Disceptatio Geometrica*, which is the transcription of two lectures given by Wallis on February 1551 and then on July 11th, 1663. It was published in 1693 as an appendix to *De Algebra tractatus cum variis appendicibus*, and in J. Wallis, *Opera mathematica*, Oxford, Theatrum Sheldonianum 1693–1699, vol. 2, pp. 665–78; repr. Olms 1972, ed. by C.J. Scriba (we will always refer to this edition). *De Postulato Quinto* ends with the words: "Atque haec, in Euclidis vindicias, sufficiant". A comprehensive introduction to Wallis' mathematical activity is L. Maierù, *John Wallis. Una vita per un progetto*, Catanzaro, Rubbettino 2007.

cal epistemology underpinning them is quite obscure to us.[6] At any rate, these attempts to prove the postulate went on uninterrupted throughout the following centuries, originating a tradition of foundational studies in geometry that, to be sure, pursued different aims (as well as methods) in different ages. This tradition, which is partially known to us, begins with the immediate heirs of Euclid in the Hellenistic age (Posidonius, Geminus), continues through the Roman Imperial age (Ptolemy) to Late Antiquity (Proclus, Simplicius), and disappears completely in the Christian Middle Ages, along with almost the entirety of geometrical studies. The tradition, however, survives and prospers in the Islamic Middle Ages through scores of studies of a very high quality (Thābit ibn Qurra, An-Nayrīzī, Avicenna, Al-Haytham, Khayyām, Nasīr ad-Dīn), whence it returns to Europe with a handful of Late Medieval and Early Humanistic treatises (Witelo, Gersonides, Alfonso). It finally blooms in the Renaissance, with the mature sixteenth- and seventeenth-century research to which Saccheri's work is immediately tied. A detailed retelling of this story, which has been told very well many times, does not belong here,[7] but some of the results, difficulties, and per-

6 The level of sophistication of these Ancient studies might have indeed rivaled the elegance and rigor of the *Elements*: cf. I. Mueller, *Remarks on Euclid's Elements* I, *32 and the Parallel Postulate*, "Science in Context", 16, 2003, pp. 287–97. The somewhat old but classic – and groundbreaking – study on this subject is M. Dehn, *Beziehungen zwischen der Philosophie und der Grundlegung der Mathematik im Altertum*, "Quellen und Studien zur Geschichte der Mathematik, Astronomie und Physik", B4, 1938, pp. 1–28. I find it problematic, however, to interpret the attempted proofs of the Fifth Postulate in terms of non-Euclidean developments in the geometry of Classical Antiquity, which is the famous thesis put forward in I. Tóth, *Das Parallelenproblem im Corpus Aristotelicum*, "Archive for History of Exact Sciences", 3, 1967, pp. 249–422; also see I. Tóth, *Fragmente und Spuren nicheuklidischer Geometrie bei Aristoteles*, Berlin, DeGruyter 2010; and I. Tóth, *Aristotele e i fondamenti assiomatici della geometria. Prolegomeni alla comprensione dei frammenti non-euclidei nel "Corpus Aristotelicum" nel loro contesto matematico e filosofico*, Milano, Vita e Pensiero 1997.

7 Today, many very valuable texts on the history of non-Euclidean geometries are available. The older classic, somewhat aged but still an excellent read, is R. Bonola, *La geometria non-euclidea. Esposizione storico-critica del suo sviluppo*, Bologna, Zanichelli 1906; translated into English by H.S. Carslaw as *Non-Euclidean Geometry*, Chicago, Open Court 1912; and anticipated by R. Bonola, *Sulla teoria delle parallele e sulle geometrie non-euclidee*, in *Questioni riguardanti le matematiche elementari*, ed. by F. Enriques, Bologna, Zanichelli 1924–1927 (1900¹), Book I, vol. 2, pp. 309–427. The most extensive work on the subject is currently J.-C. Pont, *L'aventure des parallèles. Histoire de la géométrie non euclidienne: précurseurs et attardés*, Berne, Lang 1986. Equally notable is B.A. Rosenfeld, *A History on Non-Euclidean Geometry. Evolution of the Concept of a Geometric Space*, transl. from the Russian by A. Shenitzer, New York, Springer 1988, which presents much interesting material, especially in the part devoted to Arabic-Persian mathematics. On the latter topic, however, the indispensable survey is still K. Jaouiche, *La théorie des parallèles en pays d'Islam*, Paris, Vrin 1986. Many mathematical handbooks of hyperbolic geometry also contain similar, usually shorter, historical overviews of the research on parallels. Finally, I may refer to the relevant section in the well-known bibliography P. Riccardi, *Saggio di una Bibliografia Euclidea*, Hildesheim, Olms 1974 (1887–1892¹), which is now quite dated and very incomplete, but nonetheless provides a good starting point for studying the Early Modern tradition of foundational work in elementary geometry. See also the more specific D.M.Y. Sommerville, *Bibliography of Non-Euclidean Geometry including the Theory of Parallels, the Foundations of Geometry, and Space of n Dimensions*, London, Harrison 1911.

sonalities that contributed in a special way to the genesis of *Euclid Vindicated* are worth mentioning. First, we consider some more remote discussions, which precede and, as it were, prepare the Renaissance studies.

The oldest, most natural, and most important of the misunderstandings interspersed throughout the history of the efforts to prove the Fifth Postulate is directly tied to the definition of parallel straight lines. Euclid's definition characterizes them as straight lines lying on the same plane and not intersecting. The definition had long been a subject of discussion and at times of criticism, since, firstly, the concept of non-intersection, i. e., of two straight lines that do not meet even if in(de)finitely extended, seemed to involve an illegitimate reference to the obscure concept of infinity. Secondly, the definition did not seem to present a distinctive property of parallelism as such, since there are asymptotic curves, which never intersect and which nonetheless no one wanted to qualify as parallels. Already Posidonius (first century B.C.) had therefore defined parallels as equidistant straight lines,[8] which appeared as a more complete and intuitive characterization than non-intersection. It is likely that this was the accepted definition even before the composition of the *Elements*; at any rate, it enjoyed exceptional success in the Medieval and Modern times. The problem, however, is that the existence and possibility of such equidistant parallel straight lines requires a demonstration. Though Euclid had proved the existence of his non-intersecting parallels (in *Elements* I, 31, which is independent of the Fifth Postulate), thereby establishing the possibility of his definition, he had not established the existence of equidistant straight lines, which he never considers. As a matter of fact, the existence of such equidistant straight lines can only be proved by assuming the Fifth Postulate itself: in hyperbolic geometry, the line that is in any point equidistant from a given straight line is not itself straight (but rather a curve, a *hypercycle*). Nevertheless, the clear and intuitive grasp of parallelism as equidistance, and the confusion about Euclid's existential proof (which was expected to remain valid even though the definition had changed), tied the intricate knot. Thus, there were many mathematicians who did not challenge the intrinsic possibility of equidistant straight lines, but rather just defined parallels as equidistant, assumed their existence, and then undertook a (perhaps difficult, but nevertheless possible) demonstration of the Fifth Postulate.

This *petitio principii*, of course, was uncovered early on – though not universally acknowledged, to be sure – so that the first and original paralogism in the history of the demonstrations of the parallel axiom evolved, in the mindset of some geometers, into a proof strategy: to prove the Fifth Postulate, one first had to show that the line equidistant from a straight line is itself straight. This argumentative path was very difficult, anyway, as no good definition of straight lines was available – Euclid's definition, as is well known, is so ambiguous as to be almost unserviceable – nor was it easy to find one. Thus, the history of proofs of the Fifth Postulate went for a long time hand in hand with that of the definitions of straight lines.

[8] See PROCLUS DIADOCHUS, *In primum Euclidis*, 176 (ed. by G. Friedlein, Leipzig, Teubner 1873). Posidonius of Apamea (ca. 135–50 B.C.) is the eminent Stoic philosopher. Cf. also HERO, *Definitiones* 70 (note, however, that Hero is talking about equidistant *lines*, γραμμαί, not about equidistant *straight lines*).

Beginning in the Middle Ages, there was yet another hurdle: the incorrect but very tempting argument involving the employment of motion in geometry. Let the end point of a line segment slide continuously along a straight line, and let the segment remain perpendicular to the line while moving: then the other end point will trace out a line, undoubtedly equidistant from the given straight line; and it was evident to all that the line generated in this way could not be other than a straight line itself. To be sure, the evident certainty attained in this way is not very different from that of the original proposition, that the line equidistant from a straight line is straight; but for many centuries the genetic character of this procedure deceived countless mathematicians, who could not explain *how* such a simple and uniform movement might generate a curve – in fact, it would be hard for anyone lacking an account of curvature, that is, anyone before Gauss, to provide such explanation.[9]

So it happened that most of the attempts to prove the Fifth Postulate in Ancient times (first of all that of Proclus, much celebrated by the Renaissance),[10] in the Middle Ages (virtually all of the Arabic demonstrations), in the Renaissance, and even long after Saccheri, were doomed by their appeal to equidistant lines. This happened either because their authors did not notice that the possibility of such lines was in want of a demonstration, because they

[9] Arguably, the creator of the proof from motion was the Arab physician, mathematician and astronomer Thābit ibn Qurra (836–901). The argument was probably the most difficult to refute in the history of the alleged demonstrations of the Fifth Postulate – unless one banishes the use of kinematic concepts in geometry altogether, indeed a rather common option. Saccheri himself, who was among the few to dispute this proof, ran into a similar difficulty in Scholium 2 to Proposition 37 of *Euclid Vindicated*: he claimed that the line drawn by the end point of the moving segment must be, if not straight, at least of the same length as the underlying line (which is again false in hyperbolic geometry). Thābit's two works on the theory of parallels are translated in French in Jaouiche's above-mentioned edition, as well as in R. Rashed, C. Houzel, *Thābit ibn Qurra et la théorie des parallèles*, "Arabic Science and Philosophy", 15, 2005, pp. 9–55, which also offers a long commentary. An English edition of the text is in A.I. Sabra, *Thābit Ibn Qurra on Euclid's Parallels Postulate*, "Journal of the Warburg and Courtauld Institutes", 31, 1968, pp. 12–32.

[10] Proclus of Lycia (412–485), the great Neoplatonist philosopher, wrote an important commentary on Book I of Euclid's *Elements* that is one of the primary sources of our knowledge of Ancient mathematics. In the Renaissance, the Greek text was edited by Simon Grynaeus, who published it in his *editio princeps* of Euclid (Basel, Herwagen 1533). A famous Latin translation was prepared by Francesco Barozzi (Padova, Percacino 1560), who also began writing a commentary on the text. The demonstration in question is found in Proclus, *In primum Euclidis*, 371–73. The argument is discussed by Saccheri in his Scholium 1 after Proposition 21 of *Euclid Vindicated.*

thought they could rely on some ill-devised definition of straight line, or, finally, because they were convinced of the validity of the rigid motion argument.[11]

The mathematicians of the Islamic Middle Ages formulated another fallacious argument that gained much currency in the Renaissance and beyond. This argument, which may have first appeared in the mathematical work of Nasīr ad-Dīn, consists in splitting the proof of the Fifth Postulate into two steps and claiming that (1) if two straight lines intersected by a perpendicular line form internal angles that sum to less than π, then they approach each other; and (2) if two straight lines approach each other, then they meet. As a matter of fact, however, *each* of these statements is equivalent to the Fifth Postulate.[12] Indeed, it is only locally that statement (1) enjoys absolute validity: in the neighborhood of their point of intersection with the transversal line, the two straight lines do, in fact, approach each other – but nothing prevents that they may subsequently move away from each other (as is the case with *ultraparallel lines* in hyperbolic space). Statement (2), in turn, is falsified in hyperbolic geometry by the existence of *asymptotic* parallels; we should note that the possibility of asymptotic straight lines was already apparent to the thinkers of Classical Antiquity and had provided one of the most important (and, ultimately, correct) reasons to doubt the truth of the Fifth Postulate.[13] The tangle presented by these two results posed great difficulties for

---

[11] This tendency can also be shown in the terminology: sometimes Euclid's 'parallels' are translated into Latin as *rectae aequidistantes*, even when the definition is correctly stated as that of straight lines that do not meet. This is the case with Gerard of Cremona's translation of the *Elements* from an Arabic original, which became the principal Medieval translation of Euclid's text: see P.M. Tummers, *The Latin Translation of Anaritius' Commentary on Euclid's Elements of Geometry, Books I–IV*, Nijmegen, Ingenium 1994, pp. 23–5. Something similar is still to be found in the important and otherwise very accurate sixteenth-century edition of Euclid by Federico Commandino (1509–1575), the last edition before Clavius': F. Commandino, *Euclidis Elementorum libri XV, unà cum Scolijs antiquis*, Pesaro, Camillo Franceschini 1572.

[12] Nasīr ad-Dīn at-Tūsī (1201–1274), Persian scientist. He formulated, and attempted to prove, the two principles in his *Treatise to Cure Doubts Regarding Parallel Lines* (before 1251) and later restated them in his commentary on Euclid's *Elements* (the so-called *Shorter Version*). After his death, a disciple composed another commentary on Euclid that he attributed to his mentor (the so-called *Longer Version* of 1298). The two theorems also played a crucial role in this later work. The *Longer Version*, however, only provides a proof of statement (2) and assumes the more delicate claim (1) as an unproved lemma. Nasīr ad-Dīn's proof is the latest refinement of several attempts to prove the Fifth Postulate that were envisaged in the Islamic world; the first sketch of this kind of proof is probably to be found in an Arabic manuscript published in A.I. Sabra, *Simplicius's Proof of Euclid's Parallels Postulate*, "Journal of the Warburg and Courtauld Institutes", 32, 1969, pp. 1–24, who attributes it to Simplicius. Jaouiche offers a French translation of Nasīr ad-Dīn's works in *La théorie des parallèles*.

[13] The argument was put forward in Antiquity by Geminus and discussed in Proclus' commentary. The main studies on asymptotic lines from the Renaissance, which Saccheri certainly knew (if indirectly), are surveyed in the two articles by L. Maierù, *L'influsso del Narbonense sui commentatori euclidei del Seicento italiano circa il problema delle parallele*, in Atti del Convegno *La Storia delle matematiche in Italia*, Cagliari, Università di Cagliari 1982, pp. 341–49, and L. Maierù, *Il "meraviglioso problema" in Oronce Finé, Girolamo Cardano e Jacques Peletier*, "Bollettino di Storia delle Scienze Matematiche", 4, 1984, pp. 141–70.

Modern mathematicians, who often made one of two mistakes: they either assumed one of the contentions as obvious and set out to prove the other (a task that is eminently feasible, for as soon as one ascribes truth to either (1) or (2), one has already admitted the Fifth Postulate); or they committed a paralogism in the proof of either claim and then exultantly, and correctly, inferred the other. It was especially statement (1) that proved to be a particular source of error, as the difference between global and local properties was difficult to grasp within the Euclidean synthetic and constructive framework. This crucial difference would start to become clearer only from the second half of the seventeenth century onwards, with the first results of the Calculus. Only Saccheri, at any rate, provided a complete clarification of the matter in classical terms. In this case as well, however, well into the eighteenth (and even nineteenth!) century several latecomers kept contriving proofs of the Fifth Postulate based on some version of Nasīr ad-Dīn's demonstration.

Beside those two sources of error, the theory of equidistance and the theory of local approach, a positive accomplishment of Late Antique and Medieval research should be mentioned, as it plays an exceptionally important role in the history of Saccheri's work. This achievement can be described as the elaboration of a new strategy for proving the Fifth Postulate, a strategy that was not based on criteria governing the intersection of straight lines, but rather on the consideration of the sum of the internal angles of certain figures. Nowadays, after Legendre's results, the most famous of these paths to prove Euclidean geometry is the one that takes as its starting point the sum of internal angles of a triangle: a kind of inversion of *Elements* I, 32, where Euclid infers from the Fifth Postulate that the sum of a triangle's internal angles equals π. For a long time, however, the most relevant, and indeed the most obvious, object of study for this direction of research were quadrilaterals. Geometers attempted to prove that the internal angle sum of a quadrilateral is 2π without relying on the Fifth Postulate, and then proceeded to deduce the Postulate from the former result: a perfectly legitimate approach, provided that one handles with due attention the system of principles employed. Although the traces of such procedure are already present in certain late antique attempts, it seems that the Persian Omar Khayyām was the first to extensively explore this path. He was followed by many other Islamic scientists.[14]

[14] In fact, some moves in this direction are already to be found in Thābit ibn Qurra, but the one who made the most extensive use of this strategy is the great Persian poet and scientist Umar Khayyām (circa 1048–1131), whose *Explanation of the Difficulties in the Postulates of Euclid* probably dates from 1077. The manuscript of Khayyām's work first appeared in print in 1936, but many of its results were discussed in the already quoted *Treatise to Cure Doubts* by Nasīr ad-Dīn. Khayyām's *Explanation* consists of three books, devoted to precisely the same three blemishes mentioned by Saccheri in the Preface of *Euclid Vindicated*: Book I discusses the theory of parallels, Book II the definitions of the equality of ratios, and Book III the composition of ratios. On the whole, Khayyām's work can perhaps be considered the principal Medieval contribution on the foundations of geometry. The section devoted to the parallels can nowadays be read in French in Jaouiche's aforementioned book, or in English (partially) in "Scripta mathematica", 24, 1959, pp. 275–303. The part on proportions is available, again in French, edited by A. Djebbar, in "Farhang. Quarterly Journal of Humanities and Cultural Studies", 14, 2002, pp. 83–136. An edition of the entire work is in R. Rashed, B. Vahabzadeh, *Al-Khayyâm mathématicien*, Paris, Blanchard 1999.

In the European Renaissance, this route was taken by all the major mathematicians who attempted to prove the Fifth Postulate. The path, of course, was not viable, and their expectations had to be disappointed, but it was nonetheless a fruitful path: it led naturally to the examination of a geometry with angular 'excess' (spherical geometry) and one with angular 'defect' (hyperbolic geometry) with regard to the expected value of 2π, hence to exact quantitative considerations of the phenomenon of parallelism. In other words, this strategy represented the dawn of the concept of curvature, at which no-one could have arrived by way of mere consideration of the intersection or non-intersection of straight lines.

Let us now take a closer look at the discussion occurring in the seventeenth century. The real center of all the research on the foundations of classical geometry – not only concerning the theory of parallels – is the extensive commentary on Euclid's *Elements* by Christoph Clavius, which appeared in several editions from 1574.[15] This work represents a genuine watershed in foundational studies in geometry, since it offers one of the first (mostly) accurate editions and translations of the *Elements*, a text that had undergone significant corruptions during the Middle Ages, and it supplements the Ancient text with a systematic collection of all scholia and comments that the tradition regarded as required additions, together with relevant considerations authored by Clavius himself. This edition immediately stood out as the reference text for any scientific study of Euclid's work, a position it held for two centuries. Moreover, Clavius was the first and most prominent among Jesuit scientists. It is therefore not hard to understand why Saccheri's work is consistently informed by Clavius' commentary on Euclid.

Clavius' confrontation with the Fifth Postulate begins in the second edition (1589) of his commentary, probably because it is around this time that he comes upon a manuscript

[15] Christoph Clavius (1538–1612), born in Germany but a resident at the Roman College from 1560, was the most important and authoritative Jesuit mathematician of the Renaissance. Beside his monumental commentaries on Euclid and on Sacrobosco's astronomical treatise, he is remembered principally as the architect of the Gregorian calendar reform. His commentary on the *Elements* is C. Clavius, *Euclidis Elementorum Libri* XV, Roma, Vincenzo Accolto 1574: it was subsequently reissued in several editions with variants and additions, the most important being the second edition of 1589. The last edition was also reprinted in the first of five volumes of C. Clavius, *Opera mathematica*, Mainz, Reinhard Eltz (et al.) 1611–1612; see the reprint edited by E. Knobloch, Olms 1999, to which I will consistently refer. A succinct but very precise introduction to Clavius' vast body of work is E. Knobloch, *Sur la vie et l'oeuvre de Christophore Clavius (1538–1612)*, "Revue d'Histoire des Sciences", 41, 1988, pp. 331–356; more recently and extensively, cf. S. Rommevaux, *Clavius, une clé pour Euclide au XVI siècle*, Paris, Vrin 2005. On the intellectual context of Jesuit mathematics, see U. Baldini, *Legem impone subactis. Studi su filosofia e scienza dei gesuiti in Italia, 1540–1632*, Roma, Bulzoni 1992.

version of Nasīr ad-Dīn's demonstration.[16] Thanks to Nasīr ad-Dīn, he realizes that Proclus' proof, which he included in the first edition of his *Euclid*, is gravely inadequate, insofar as it takes the existence of equidistant straight lines for granted. To contrive his own proofs of the Postulate, Clavius follows two distinct routes, namely, the two erroneous strategies outlined above. The first hinges on the claim that the line equidistant from a straight line is itself straight, which, Clavius believes, is established both by his peculiar interpretation of Euclid's definition of a straight line and by the argument from the rigid motion of a line segment. The second route is the same as in Nasīr ad-Dīn's bipartite theorem. In both cases, Clavius does not directly conclude the Fifth Postulate, but rather goes through a demonstration about the sum of the internal angles of a quadrilateral. Thus, Clavius' work contains all of the principal devices – and the principal errors – that later discussions, in particular Saccheri's, were to take as their starting point.

[16] In the second edition of his commentary, Clavius prefaces the new proof of the Fifth Postulate with this remark: "Id quod in Euclide quodam Arabico factum etiam esse accepi, sed nunquam facta mihi est copia demonstrationem illam legendi, etsi obnixe illud iterum atque iterum ab eo, qui eum Euclidem Arabicum poßidet, flagitavi" (*Euclidis*, p. 50). We know, in fact, that a copy of Nasīr ad-Dīn's three works on parallel lines (including the spurious *Longer Version*) was owned (among many other Arabic manuscripts, such as that of Books 5–7 of Apollonius' *Conics*) by the Medici family; their librarian and orientalist Giovan Battista Raimondi was appointed to publish most of these manuscripts, and to this effect he founded and directed the Typographia Medicea in Rome. The Oriental Press was active from 1584 to the death of Raimondi (1614), and published several works in Arabic to foster the evangelization of the Islamic world. The relations between Clavius and Raimondi were not excellent, and Clavius complained about the delay in the publication of the Arabic texts, trying to supersede Raimondi in the editorial work or, at least, to have access to the manuscripts. We have (for instance) a letter of Guidobaldo del Monte to Clavius (from 1590), asking the Jesuit some questions on the *Conics* of Apollonius, if he could have access to Raimondi's treasure; but in 1605 Raimondi was still protecting Apollonius' manuscript, claiming to be almost ready for an edition of it (see U. Baldini, P.D. Napolitani, *Christoph Clavius. Corrispondenza*, Pisa, Department of Mathematics (preprint) 1992, letters 65 and 256). Apollonius was only published by Borelli in 1661. The publication of the Arabic Euclid was more fortunate, however, and Raimondi begun his edition of Nasīr ad-Dīn's *Longer Version* in 1588, publishing it in 1594. In this case, clearly, Clavius was able to look at the text already in 1588, and made good use of it in his 1589 edition of Euclid. What is more, in the apocryphal *Longer Version* the proof for statement (1) mentioned above was omitted, and it is only to be found in the original (unpublished) works of Nasīr ad-Dīn. Since Clavius' *Euclidis* offers such a proof and it contains the same error as Nasīr ad-Dīn's original demonstration, that is, the passage from a local to a global property, it seems plausible to me that Clavius became acquainted, through Raimondi, with the *Treatise to Cure Doubts* and the *Shorter Version* (or had some glimpses thereof). On the history of the Medici's manuscript, see the special issue of the "Cahiers d'histoire des mathématiques de Toulouse", 9, 1986, edited by J. Cassinet. On Clavius' knowledge of Arabic sources, see E. Knobloch, *La connaissance des mathématiques arabes par Clavius*, "Arabic Science and Philosophy", 12, 2002, pp. 257–84. On Borelli's important edition of Apollonius, see L. Guerrini, *Matematica ed erudizione. Giovanni Alfonso Borelli e l'edizione fiorentina dei libri* V, *VI e VII delle Coniche di Apollonio di Perga*, "Nuncius", 14, 1999, pp. 505–68.

The later developments, for the most part, relied passively on Clavius' proofs. It must be stressed that Clavius' commentary represented a masterpiece of rigor for the study of the foundations of elementary geometry – the standard it set was seldom met in the following decades. Clavius' work became, for instance, the reference text for all of mathematical teaching in Jesuit colleges. Yet mere decades later, it had already become necessary to compose and compile new treatises in elementary geometry, shorter and more accessible to students than the original Clavian work. The works by the Order's great professors, such as André Tacquet and Milliet Dechales, were composed for this purpose.[17] These texts aimed chiefly at a simplification of the original Euclidean edifice and were less oriented towards scientific study than Clavius' commentary. The consequence, at least insofar as it concerns us, is that all these authors simply followed the old master: they mostly restated Clavius' proof appealing to equidistant straight lines, though in a less rigorous version; on the other hand, Nasīr ad-Dīn's bipartite proof must have appeared unnecessarily complicated to them. It is evident that Saccheri was well acquainted with these works, and he might have even made use of them as textbooks for his courses; it is equally evident that he could not hold them in any scientific esteem, and, in fact, *Euclid Vindicated* represents a lively protest against the turn taken by mathematical teaching in Jesuit colleges.[18]

[17] Andreas Tacquet (1612–1660) was a Flemish mathematician and a disciple of the Jesuit geometer Grégoire de Saint-Vincent (who was, in turn, a student of Clavius'), and is nowadays mostly remembered for some studies that were instrumental in the formulation of the Fundamental Theorem of Calculus. In his age, he was also renowned for his edition of Euclid's *Elements*, which was translated into several languages and enjoyed a very wide circulation for educational purposes: A. Tacquet, *Elementa Geometriae planae ac solidae*, Antwerp, Jacob van Meurs 1654. Claude François Milliet Dechales (1621–1678) held a professorship for mathematics in Turin, which explains Saccheri's familiarity with his work, as Saccheri also spent many years in that city. In 1660, Dechales wrote a commentary on some of the books of the *Elements*, which was translated into various languages. First and foremost, however, he is the author of a monumental mathematics textbook in four volumes that treats all sorts of subjects, but whose parts on elementary geometry depend heavily on Tacquet's treatise: C.F.M. Dechales, *Cursus seu Mundus Mathematicus*, Lyon, Posuel&Rigaud 1690 (1674[1]). Honoré Fabri (1607–1688) was a mathematician, physicist and theologian who came under attack from Rome on charges of Cartesianism. It is very likely that Saccheri was acquainted with Fabri's works and was afraid of meeting the same fate (see the notes on this point in the aforementioned edition of the *Logica Demonstrativa* by Mugnai and Girondino). Fabri wrote a short handbook of geometry for students, which enjoyed great popularity: H. Fabri, *Synopsis Geometrica*, Lyon, Molin 1669.

[18] There is a vast body of literature on mathematical teaching in Jesuit colleges, and specifically on the genesis of the Society's first *ratio studiorum*, with which Clavius was involved. Cf. for instance G. Cosentino, *L'insegnamento delle matematiche nei collegi gesuitici nell'Italia settentrionale*, "Physis", 13, 1971, pp. 205–217; D.C. Smolarski, S.J., *The Jesuit Ratio Studiorum, Christopher Clavius, and the Study of Mathematical Sciences in Universities*, "Science in Context", 15, 2002, pp. 447–64; the most comprehensive treatment is probably A. Romano, *La contre-réforme mathématique. Constitution et diffusion d'une culture mathématique jésuite à la Renaissance (1540–1640)*, Roma, École Française 1999. More generally on the subject cf. the two volumes edited by Feingold: *The New Science and Jesuit Science: Seventeenth Century Perspectives*, ed. by M. Feingold, Dordrecht, Kluwer 2003; *Jesuit Science and the Republic of Letters*, ed. by M. Feingold, Cambridge, MIT 2003.

A comparable development also took place outside the Jesuit milieu. Jansenist and Cartesian circles, which were farthest removed from the Society, asserted the intuitive validity of the first principles. For this reason, they rebuked Euclid's – to say nothing of Clavius' – excessive rigor on the charge of obscuring with its subtleties what is manifest to common understanding. In this context, the issue of providing a demonstration of an axiom would not cause much concern. Indeed, the great Arnauld accepted one of the versions of the Fifth Postulate as self-evident, while composing his *New Elements of Geometry*, where he aimed to present elementary mathematics in a more natural and simple way than Euclid had. Nonetheless, he also restates Nasīr ad-Dīn's proof.[19]

The need for rigor in proofs was felt more urgently in the Italian school of geometry, which placed as great an emphasis on foundational work as it did on performing its role as an educational institution. Borelli's *Euclides restitutus* (1658)[20] does not contain impor-

[19] A. ARNAULD, *Nouveaux Elémens de géometrie*, Paris, Savreux 1667, with a second, extensively revised edition in 1683; both editions (now very rare) can be read in the recent *Géométries de Port-Royal*, ed. by D. Descotes, Paris, Champion 2009, to which I shall consistently refer. This geometrical work by Arnauld (1612–1694), which was probably conceived after certain discussions with Pascal in Port-Royal, enjoyed considerable success and also represents in many ways an important example of the Cartesian epistemology of mathematics, beside Arnauld's more famous *Logic* (A. ARNAULD, P. NICOLE, *La logique ou l'art de penser*, Paris, Desprez 1683 (1662^1); reprint ed. by P. Clair and F. Girbal, Paris, PUF 1965). If Arnauld often writes in the *Logic* that it is a mistake to try to prove what is evident in itself (cf. for instance *Logique*, IV, 9; pp. 326–27), the same principle is applied in the *Nouveaux Elémens* to the parallel axiom: "Sixième Axiome. Deux lignes droites qui étant prolongées vers un même côté s'approchent peu à peu, se couperont à la fin. *Euclide prend cette proposition pour un principe et avec raison: car elle a assez de clarté pour s'en contenter, et ce serait perdre le temps inutilement que de se rompre la tête pour le prouver par un long circuit*" (p. 361). Although this position could hardly be more distant from Saccheri's or Clavius', it was common in the whole Cartesian tradition, as well as in the 'modern' Jesuit circles. One can read, as a further example, Malebranche's laconic remarks on the two Euclidean blemishes discussed in Savile's *Praelectiones*: Savile was, in the eyes of the French philosopher, nothing but a pedantic Englishman who wasted his time on trifles (*Recherche de la Vérité*, II, II, 6; in N. MALEBRANCHE, *Oeuvres complètes*, vol. 1, ed. by G. Rodis-Lewis, Paris, Vrin 1991^3, pp. 297–301).

[20] Giovanni Alfonso Borelli (1608–1679) was perhaps the most important Italian scientist around the half of the seventeenth century. He was born in Naples and was related to philosopher Tommaso Campanella. After joining the Galilean school, he taught in Pisa, where he was the mentor of Alessandro Marchetti and was involved in a lively polemic with Vincenzo Viviani, after which he moved to Messina. Eventually, he left the city and sought refuge in Rome, as he was implicated in an anti-Spanish conspiracy, and it was in Rome that he met and befriended Vitale Giordano. He is chiefly known for his studies of mechanics and physiology, but he also wrote an important commentary on Euclid, which is one of Saccheri's primary sources: G.A. BORELLI, *Euclides restitutus, sive priscae Geometriae Elementa brevius et facilius contexta, in quibus praecipue proportionum theoriae nova firmiorique methodo proponantur*, Pisa, Francesco Onofri 1658. The book had an important third edition (Roma, Mascardi 1679), which is the text Saccheri read (and from which I shall quote), and an Italian translation as *Euclide Rinnovato*, transl. from the Latin by D. Magni and corrected by the Author, Bologna, Giovan Battista Ferroni 1663. On the figure of Borelli and his ideas in the epistemology of mathematics, so important for Saccheri, cf. C. VASOLI, *Fondamento e metodo logico della geometria nell'Euclides restitutus del Borelli*, "Physis", 11, 1969, pp. 571–98; then U. BALDINI, *Giovanni Alfonso Borelli e la rivoluzione scientifica*, "Physis", 16, 1974, pp. 97–128.

tant innovations in parallel theory, but it ventures a novel definition of parallelism, which Saccheri will starkly criticize. Most importantly, Borelli seems quite aware that Clavius' demonstrations fail to achieve their goal. In the end, however, he simply accepts the claim that an equidistant line is straight as an axiom. Vitale Giordano's *Euclide restituto* (1680),[21] on the other hand, puts forward another demonstration of the Fifth Postulate that rejects Clavius' proofs as insufficiently rigorous, and explores an autonomous path. Giordano's central argument is itself vitiated by a rather subtle error, but its principal feature is that he treats the theory of the quadrilaterals' internal angles at great length, producing new results that precipitate the direction later followed by Saccheri.

Finally, we cannot avoid mentioning Wallis' 1663 proof, which bore in many ways a prime theoretical and historical importance.[22] First of all, it is noteworthy that Wallis, in the course of his efforts, published for the first time Nasīr ad-Dīn's text,[23] which was thus made available to scholars without Clavius' mediation. Equally noteworthy is Wallis' explicit rejection of Nasīr ad-Dīn's conclusions. But most important of all is Wallis' original attempt to prove the Fifth Postulate itself, which demonstrates that the postulate is equivalent to the possibility of constructing triangles similar, though not congruent, to a given triangle: in other words, (non-trivial) transformations by similarity are only possible in Euclidean geometry, as opposed to hyperbolic or spherical geometry. Wallis reasoned that this amounted to a proof of the postulate, since it was derived from a principle that he held to be indisputable, namely, that of the possibility of similarity. Wallis also sketches a metaphysical proof, claiming that the possibility of transformations by similarity depends on the very nature of quantity, that is, on the fact that the category of *quality* (under which falls the *shape* of a geometric figure, construed as invariant by similarity) differs from that of *quantity* (which determines that figure's *magnitude*). Indeed, quantity and quality represent two distinct highest genera of being, such that it is always possible to vary the properties of the former while preserving those of the latter, and *vice versa*.

---

[21] Vitale Giordano da Bitonto (1633–1711) retired in Rome to teach mathematics after a very turbulent youth as a soldier, during which he faced a couple of charges of assault and murder. His Italian commentary on Euclid is V. Giordano, *Euclide restituto, ovvero gli antichi elementi geometrici ristaurati e facilitati*, Roma, Angelo Bernabò 1686 (1680[1]); although Giordano criticizes Borelli in several passages, his text was certainly written as a homage to the Neapolitan mathematician, and it was published the year after Borelli's death.

[22] It is the aforementioned proof in *De Postulato Quinto*, first edited in 1693.

[23] Nasīr ad-Dīn's *Longer Version*, which had been published in Rome in 1594, in Arabic, was translated into Latin in 1651, on the occasion of Wallis' first lecture in Oxford on the Fifth Postulate. This Latin version, by the English orientalist Edward Pocock, was published as an appendix to Wallis' *De Algebra Tractatus* in 1693. We know that Pocock had access to the aforementioned Medicean manuscript, and translated in Latin both the original work of Nasīr ad-Dīn and the spurious essay that was published in Rome. Wallis' *Algebra* only contains the latter, but Wallis himself had access to Pocock's translation of the former, which he added in manuscript to his own copy of the *Opera mathematica*. Pocock's unpublished translation is still available in manuscript at the Bodleian Library, and is reprinted in the edition of Cassinet (see above).

Wallis' demonstration bore great historical significance: not only was it very elegant and deep mathematically speaking, but it also inaugurated the metaphysical discussions of the parallel postulate that were to bloom fully in later centuries. Let me stress that the intense philosophical discussion on the *significance* of non-Euclidean geometries that agitated the cultural and scientific scene for a good half of the nineteenth century and even continued, though in an evolved form, after the discovery of the general theory of relativity – and which constitutes one of the most interesting aspects of a study of the history of parallelism – was a unique product of the nineteenth century. In fact, it seems that, in the whole course of Antiquity and the Early Modern Age, despite the attention lavished on the Fifth Postulate, not a single philosopher detected in the problem of parallels any special metaphysical difficulty (which means, a difficulty concerning the nature of space) – at least, no one before Wallis. If general philosophical debate on the topic arose only much later, after Gauss' discoveries and Kant's philosophy had gained sway, there is no doubt that some metaphysical minds of the eighteenth century had already grappled with the issue. This is true, for instance, of Leibniz, Thomas Reid, and Johann Lambert: all of them tackled the problem from the viewpoint inaugurated by Wallis. These discussions remained quite isolated, however, as it is confirmed by the work of many other geometers of the same century, who were completely insensitive to any philosophical implications of the subject. This is also true of Saccheri: although he was rather gifted in philosophy, a subtle logician, and a professor of theology, he never seemed to discern in the mathematical problem that most strained his intellectual forces any philosophical theme, nor anything motivating a project broader than the systemization of Euclidean axiomatics. On the contrary, he overtly dismissed Wallis' demonstration as too metaphysical. While the rebuttal of Wallis' proof was certainly a gain for the progress of science, it also typifies the epoch's struggle with the idea of a structural or functional interpretation of physical space, as opposed to a merely 'substantial' one.

3. The theory of proportions in the seventeenth century

Early Modern discussions on the theory of proportions were much broader and livelier than those on parallel theory. The problem of parallels, in fact, was nothing more than a byproduct of Renaissance research on the foundations of geometry, inherited from antique discussions since lost. The scientific reach of this issue was limited to interpreting classical texts or to teaching elementary geometry, and its scientific efficacy gradually decreased as modern mathematics moved on towards territories untouched by Euclid. On the contrary, the theory of proportions, thanks to its greater generality and its immediately operational aspects, was the classical mathematical device most suited to facilitating a transition towards the new seventeenth-century discoveries. Indeed, Fermat's and Descartes' wide-ranging algebraic reform of geometry, which represents one of the chief landmarks of Modern mathematics, was not applicable to mechanics, a discipline whose methods have little in common with the elegant purity of Algebra. The mathematization of natural philosophy – its metamorphosis into a scientific, quantitative mechanics – could not make use of Descartes' algebraic equations. Moreover, even within pure mathematics, the new Calculus, like the various studies of the indivisibles before it, was rendered possible by a reformed theory of proportions and rational numbers. Therefore, Book V of Euclid's *Elements*, where the theory of ratios receives its most extensive treatment, was one of the few vestiges of Greek mathematics that still had an active role to play in the age of the Scientific Revolution.

To be sure, there had been numerous foundational discussions of the theory of proportions in older times, and these discussions were quite akin in spirit to those concerning the Fifth Postulate. In particular, Euclid's outstanding definition of the equality of ratios, which in some sense represents the peak of abstraction of Classical mathematics (and which is probably to be ascribed to Eudoxus), was gravely misunderstood in the Middle Ages, mainly due to an intricate tangle of textual problems.[24] Yet, nothing could save the whole edifice of Euclid's theory from collapse, if this definition (as well as a few others) underwent textual corruption. A few attempts were made to fix the problem: Arabic mathematics explored an alternative theory, the so-called anthyphairetic theory of ratios,[25] whereas the European tradition resorted to numeric interpretations of proportions through the concept of the denomination of a ratio.[26] The fundamental issue of the treatment of incommensurable magnitudes, however, resisted all solution. Only in the sixteenth century, eventually, new editions of the *Elements* were sufficiently cleansed of textual corruption to allow scholars to

[24] Let me note that Greek mathematics, on the other hand, never seemed to find anything problematic about the theory of proportions after Eudoxus' systematization. On the Ancient theory see for instance the classic, and controversial, F. Beckmann, *Neue Gesichtpunkte zum 5. Buch Euklids*, "Archive for History of Exact Sciences", 3, 1967, pp. 1–144.

[25] This theory consists basically in defining the equality of two ratios as follows: by first applying an algorithm of successive divisions to each of them and then ascertaining the identity of the successions of factors in the two procedures. In the case of irrational magnitudes, this amounted essentially to a definition of the latter with reference to what we today call continued fractions. It should be noted that the procedure could be executed geometrically, and it was not necessary to work with numbers instead of segments. Whether such a theory could have already been present in Classical Antiquity and in pre-Euclidean mathematics (a thesis first put forward by Zeuthen and Becker, then followed by many others) is a question that is much discussed among historians of science. In any event, anthyphairesis was certainly employed in the Islamic world, where it was often ascribed to Greek mathematicians. Umar Khayyām was one of its greatest theorists, in the aforementioned Book II of his *Explanation*; cf. B. Vahabzadeh, *Al-Khayyām's conception of ratio and proportionality*, "Arabic Science and Philosophy", 7, 1997, pp. 247–63; B. Vitrac, *Umar al Khayyam et l'anthyphérèse: Étude du deuxième Livre de son commentaire "Sur certaines prémisses problématiques du Livre d'Euclide"*, "Farhang. Quarterly Journal of Humanities and Cultural Studies", 14, 2002, pp. 137–92; and more in general E.B. Plooij, *Euclid's Conception of Ratio and his Definition of Proportional Magnitudes as Criticised by Arabian Commentators*, Rotterdam, Hengel 1950.

[26] In theories of this kind, every ratio between magnitudes is expressed by a numeric fraction associated with it. Such approaches proceed to define (for instance) the equality of ratios through the identity of fractions, and apply this method to the other Euclidean definitions. It is to be noted, nevertheless, that this did not entail the identification of a ratio with its fraction, as would occur as part of the algebraization of mathematics in the Modern Age. The two approaches also differed in some other details.

make sense of Euclid's original strategy. The point of reference is once again Clavius' commentary, with tens of pages devoted to scholia that expound the correct Eudoxian theory and amend the misunderstandings that had plagued geometry for centuries.

Clavius' commentary also inaugurated a more critical reception of the Euclidean theory. For the first time, scholars began to focus not only on the strengths but also on some flaws and difficulties (or blemishes) in the classical theory of proportions. One of the major issues called into question concerned the so-called principle of the existence of the fourth proportional: in some theorems, Euclid simply and tacitly accepts that, given any three magnitudes A, B, and C, there always exists a fourth magnitude D such that A:B::C:D. It appeared, however, that this assumption should be considered as an additional axiom, and, as such, it was re-classified and counted among the Euclidean axioms in the work of Clavius, followed in this by many later interpreters. One of the problems with this strategy was that the axiom appeared to be existential and not constructive: it warranted the generic assumption of the fourth element, but gave no clue for producing the fourth magnitude concretely. In sum, the dispute was between mathematical constructivism and non-constructivism – one that is certainly not easy to settle. The birth of a new mechanics made it even bitterer. Although the non-constructive axiom is (perhaps) not indispensable to the *Elements*, and alternative demonstrations for the theorems that Euclid proves by tacit recourse to the non-constructive principle can be provided, Clavius' explicit axiomatization induced a series of mathematicians, first and foremost Galileo, to undertake a radical reform of the Euclidean theory of proportions that went so far as to be *based* on the existence of a fourth proportional. With this new system they hoped to sidestep some of the weakest points of the Greek theory – which had originated chiefly, or maybe solely, for geometric purposes – and to widen the scope of the doctrine of ratios to include the non-geometric magnitudes that were now employed in the new natural science. This new theory of proportions, whose final form Galileo had only outlined, was later developed in order to provide a justification for those 'functional' procedures that were required by, and had in fact already been employed in, the mathematization of the physical world.[27]

This reform of the theory of proportions, required by mechanics, demanded in turn the theoretical transvaluation of an almost certainly apocryphal Euclidean definition, namely, that of the composition of ratios. Today we could say that such composition consisted in nothing more than the product of two fractions, but such an operation would have been hard to justify within the classical Euclidean context, where ratios were not seen as numbers or quantities, which can be multiplied by one another, but rather as mere relations. This spurious definition of the composition of ratios, which may have been introduced by some Late Antique commentator as a means of clarifying the terms employed in a couple of theo-

[27] The standard reference work on the theory of proportions in seventeenth-century Italy is E. Giusti, *Euclides reformatus. La teoria delle proporzioni nella scuola galileiana*, Torino, Bollati Boringhieri 1993. For a comparison between the Galilean reform and Clavius' theory, which Saccheri will adopt with little alteration, cf. P. Palmieri, *The Obscurity of the Equimultiples. Clavius' and Galileo's Foundational Studies of Euclid's Theory of Proportions*, "Archive for History of Exact Sciences", 55, 2001, pp. 555–97.

rems of the *Elements*, represented initially only a problem for the interpreters of the Greek text. However, it ended up becoming, in the hands of Galileo and his followers, an essential instrument enabling those operations and calculations that were indispensable to the new sciences. So it happened that a very peripheral locus of Euclid's vast work rapidly came to the fore of the scientific debate, even though the correct strategy to deal with it was unclear. Savile certainly had such developments in mind when he came to regard the definition of the composition of ratios as one of the ugly blemishes on Euclid's body.

The transformation that the theory of proportions in the *Elements* underwent in the Modern Age was therefore radical: ultimately, it also affected its very meaning – which, to be sure, had been a matter of debate since the early Renaissance. The controversy was mainly about the object of the theory of ratios – the enigmatic concept of 'magnitude' around which all the theorems in Book V of the *Elements* revolve and which Euclid does not define elsewhere. If by that concept one means the three different (that is, non-homogeneous, according to the definition) geometrical magnitudes, i. e., lengths, areas, and volumes, and maybe, in addition to these, angular width or numbers, then the theory of proportions applies perfectly, and solely, to the objects with which the other twelve books of the *Elements* are concerned. On the other hand, one could see in the concept of magnitude a generality that is absent from the rest of Euclid's work and allows us to apply proportions and ratios, under the same laws expressed in the *Elements*, to time, motion, speed, musical intervals, and potentially many other things – in a word, to anything that falls under the more general concept of 'magnitude' as quantity. In this second case, a way was opened towards a general mathematical theory of magnitudes as a *mathesis universalis*, and indeed towards the application of Classical mathematics to the wide world of natural science.[28] Yet this broader move required a discussion of the very foundational questions (the Euclidean blemishes) associated with the definition of the equality of ratios, the existence of a fourth proportional, and the exact characterization of the operation of the composition of ratios.

At stake here was the complete reworking of a Classical theory that was very well structured and that could be adapted to novel applications only with difficulty. This exceptional enterprise spanned for about the entire first half of the seventeenth century and engaged chiefly the Galilean school – and, of course, the commentators of Euclid. The highest points of this endeavor are, first, the composition of the *Fifth Day* of Galileo's *Two New Sciences*, second, the analogous Galilean treatises by Torricelli and Viviani, and, finally and most importantly, the most advanced work of the Italian school on this subject, Borelli's *Euclides*

[28] Among the general reference works on this point, see the classic G. Crapulli, *Mathesis Universalis. Genesi di un'idea nel XVI secolo*, Roma, Edizioni dell'Ateneo 1969; and more recently D. Rabouin, *Mathesis Universalis. L'idée de "mathématique universelle" d'Aristote à Descartes*, Paris, PUF 2009.

restitutus.[29] These works represent a genuine hand-to-hand combat with the Euclidean text, and combine inseparably attentive reading, loose interpretation, emendation, reformation, and explicit rebuttal.

The discussion on the theory of proportions, however, was not exclusively Italian, although in Italy it developed in constant proximity with the original Euclidean text and bore the most important results of geometrical significance.

In France, on the other hand, the most productive research in mathematics was moving in a completely different direction, and mostly consisted in a development of Descartes' *Géométrie*. Consequently, few of the great French geometers of the time would be concerned with the Euclidean theory of proportions, which had little to do with the new algebra. Nonetheless, we can find significant exceptions to this general direction if we examine French mathematical textbooks, where Euclid's *Elements* still played an essential role. Thus it happens that the didactic works of the mathematicians mentioned above, be they Jesuits or Jansenists, devote plenty of space to the interpretation and reformation of the theory of proportions. The motives that prompted the first and most widely followed of these geometers, André Tacquet, to undertake a revision of the Euclidean theory are hard to determine. He may have had educational concerns: perhaps he deemed Eudoxus' complex system of definitions too obscure for his students, and intended to simplify the theory of proportions; other parts of his revision of the *Elements*, in fact, had been informed by this procedure – at times to the detriment of the work's rigor. Alternatively, he may have been motivated by worries similar to those of the Galilean school, since he certainly had a strong interest in mechanics and was very probably acquainted with the works of the Italian scientists.[30] Finally, it is also possible that he was motivated by foundational concerns of older origin

---

[29] This *Fifth Day* of the *Discourses and Mathematical Demonstrations Concerning Two New Sciences* (published in 1638) is devoted wholly to the theory of proportions. It was conceived by Galileo (1564–1642) as a later addition to the work and dictated – it seems – to Torricelli in 1641. The *Fifth Day* was only published in 1674, in the treatise on proportions by V. Viviani, *Quinto Libro degli Elementi di Euclide, ovvero Scienza Universale delle Proporzioni spiegata colla dottrina del Galileo*, Firenze, Condotta 1674; nowadays, it is to be found in G. Galilei, *Opere*, Firenze, Barbera 1968, vol. 8, pp. 347–62. Evangelista Torricelli (1608–1647) also wrote a *De Proportionibus Liber* (1647), which, despite being first printed in the twentieth century, was distributed in universities in manuscript form. Indeed, Borelli certainly relies on it. It can be read in E. Torricelli, *Opere*, ed. by G. Loria and G. Vassura, Faenza, Montanari 1919–1944, vol. 1, pp. 293–327; and a critical edition is in appendix to Giusti, *Euclides reformatus*, pp. 299–340. Vincenzo Viviani (1622–1703), 'Galileo's last disciple', offered a probably mistaken or at least misleading, but very influential, reading of Galileo's work on proportions, in the text mentioned above, and added his own considerations on the subject. Viviani also treated the topic in an Italian edition of the *Elements* (Firenze, Carlieri 1690) that enjoyed a wide circulation and was certainly known to Saccheri.

[30] Tacquet's theory of proportions is probably autonomous and original, but on several points it undoubtedly displays similarities with those developed in Italy. Cf. F. Palladino, *Sulla teoria delle proporzioni nel Seicento. Due "macchinazioni" notevoli: Le sezioni dei razionali del galileiano G.A. Borelli; Le classi di misura del gesuita A. Tacquet*, "Nuncius", 6, 1991, pp. 38–81.

and was not persuaded that Clavius' edition, with its amended Greek text and its diligent scholia detailing how all the Medieval and Renaissance mathematicians were mistaken, had itself solved the interpretive problems associated with the *Elements* – as this work offered (he may have thought) a structurally confused and wrong theory that called for a rather deep revision. Nowadays, it is not easy to share Tacquet's concerns. Eudoxus' theory strikes us as quite masterful, whereas Tacquet's own reform may appear weakly conceived; in any event, it seems undeniable that these concerns must have prompted the Flemish Jesuit's work. The other mathematicians mentioned above, from Dechales to Arnauld, followed his approach all the way.[31]

In England, the situation was different still, and the most significant trend was probably a gradual movement away from the Euclidean methods. Instead of following Euclid's approach, English scholars attempted to completely re-ground the theory of proportions with arithmetical or algebraic methods. The point of departure was probably the Medieval anthyphairetic theory, which allowed the first moves towards the introduction of the concept of an irrational number; a theory that surfaced in the European Renaissance and probably informed some important works on arithmetic of the sixteenth century (such as Stevin's *Arithmétique* from 1585).[32] This approach had a great success in Britain, and the works of the two greatest mid-century mathematicians, Wallis and Barrow, both point in this direc-

[31] The French Jesuit mathematicians often refer to Tacquet explicitly, even going so far as to copy his definitions and demonstrations. Dechales, for instance, engages in this practice. Arnauld's case is more complex, but the first edition of his *Nouveaux Elémens* (1667) includes a theory of proportions that is very close to Tacquet's. Since we know with a good deal of certainty that Arnauld began elaborating his reform and systematization of the Euclidean *Elements* in 1655, and Tacquet's *Elementa Geometriae*, which explicitly pursue the same goal, appeared in 1654, it is even possible that the motivation for Arnauld's work (besides Pascal's suggestion to this effect) was Tacquet's edition, although Arnauld, to be sure, never mentions him. On the other hand, by the second edition of the *Nouveaux Elémens* (1683) Arnauld has completely reworked his treatment of proportions. In this later text, he employs a positively 'modern' method, which rests on arithmetization rather than on Euclid's or Tacquet's synthetic geometry: it seems that Arnauld had Nonancourt's early work in mind, rather than English writings on the subject. Cf. F. de Nonancourt, *Euclides Logisticus sive de ratione euclidea*, Louvain, Bouvet 1652, also reprinted in the aforementioned *Géométries de Port-Royal*, pp. 801–20. The mathematician François de Nonancourt (1624–1686) was close to the Port-Royal school from 1669 and was in frequent personal contact with Arnauld in the years 1679–1680.

[32] The anthyphairetic theory, in fact, allowed one to take a continuous fraction as the expression of a ratio, and thus may represent the first introduction of irrational numbers; it arrived to the West through An-Nayrizi's commentary. Stevin's *Arithmétique* is to be found in his *Oeuvres mathematiques*, ed. A. Girard, Leyden 1634 (vol. 1). On Clavius' (and thus Saccheri's) connection with Stevin and this tradition of studies, see A. Malet, *Renaissance notions of number and magnitude*, "Historia Mathematica", 33, 2006, pp. 63–81.

tion: they no longer use ratios, but rather fractions and numbers.[33] This strategy avoided confrontation with the authentic foundational difficulties: Wallis and Barrow escaped, as it were, rather than engaging in battle. But it was the most victorious escape of all, because the Euclidean theory of proportions was progressively abandoned, in France and Italy as well, in favor of the infinitesimal calculus: in short time, this new theory produced the modern concept of a function, which would in the end supersede all the discussions on the homogeneity of magnitudes, on a composition of magnitudes that was not a true multiplication, or on the non-constructive existence of a quotient.[34]

Whatever judgment one may pass on the historical developments in the second half of the seventeenth century, it is certain that the new mathematics forcefully swept away all the ancient foundational difficulties, and the discipline transformed so completely as to make Euclid quite useless to the professional mathematician. Borelli's and Tacquet's studies towards a reform of the classical theory of proportions, dating from the 1650s, are therefore the swan's song of an epoch in decline. There were certainly continuators and latecomers to the Euclidean scene, but the later works of the Italian school, such as the two volumes evidently inspired by Borelli's work, Giordano's *Euclide restituto* and Marchetti's *Euclides reformatus*, had to come to terms with the numeric and algebraic theories of proportions

---

[33] The algebraization and symbolic formalization – which, nevertheless, is not an arithmetization – of the Euclidean theory of proportions undertaken by Isaac Barrow (1630–1677) is to be found in Book V of his edition of the *Elements* (Cambridge, Nealand 1655). Two years later, Wallis attempted a genuine arithmetization of the theory, against which Barrow took a stand: cf. especially his *Euclidis Elementum quintum Arithmetice demonstratum*, which is Chapter 35 of the 1657 *Mathesis universalis* (Wallis, *Opera mathematica*, vol. 1, pp. 183–93). On the relationship between Barrow's and Wallis' very different projects, see C. Sasaki, *The Acceptance of the Theory of Proportion in the Sixteenth and Seventeenth Centuries. Barrow's Reaction to the Analytic Mathematics*, "Historia Scientiarum", 29, 1985, pp. 83–116. On the treatment of the theory of proportions in Britain, see J.A. Goldstein, *A Matter of Great Magnitude: The Conflict over Arithmetization in 16th-, 17th-, and 18th-Century English Editions of Euclid's Elements Books I through VI (1561–1795)*, "Historia Mathematica", 27, 2000, pp. 36–53.

[34] To gain an impression of the complete de-geometrization of the theory of proportions at the end of the seventeenth century, one may look at Leibniz's 1695 *Mathesis Universalis*. In this work, Leibniz takes a characteristically radical stance from the epistemological point of view: "Sed ego reprehendi regulariter non esse opus in calculo peculiaribus signis pro rationibus et proportionibus, earumque analogiis seu proportionalitatibus, sed pro ratione sive proportione sufficere signum divisioni, et pro analogia seu proportionum coincidentia sufficere signum aequalitatis. Itaque rationem seu proportionem ipsius a ad ipsum b sic scribo: a:b seu a/b, quasi de divisione ipsius a per b ageretur" (GM vii, p. 56). See also Leibniz's discussion of the subject from a more philosophical point of view in his exchange with the Newtonian 'classicist' Samuel Clarke: at § 54 of Leibniz's *Fifth Paper* to Clarke (GP vii, p. 404) and in Clarke's indignant *Fifth Reply* (GP vii, pp. 429–30). On the dissolution of the Classical theory of proportion in Early Modern mathematics, see A. Malet, *Changing notions of proportionality in pre-modern mathematics*, "Asclepio", 1, 1990, pp. 183–211; and A. Malet, *Euclid's Swan Song: Euclid's Elements in Early Modern Europe*, in *Greek Science in the Long Run*, ed. P. Olmos, Newcastle, Cambridge Publishing 2012, pp. 205–34.

that were gaining traction everywhere.[35] Euclid's blemishes, vindicated or not, were soon forgotten.

4. Sources and composition of Euclid Vindicated

This was, generally speaking, the state of the question of the theory of parallels and proportions at the time when Saccheri undertook the composition of *Euclid Vindicated*. We must then inquire as to which of those authors Saccheri actually knew, and how his endeavor came about. Regarding both questions, however, we only possess very scant clues.

For one thing, all evidence indicates that Saccheri was not a very well-read scientist. Although the fact that his mathematical activity was performed in isolation surely accounts for the absolute novelty of his book, it also means that we cannot know the details of his activity and we are left with mere conjectures. Notwithstanding, he was thoroughly acquainted with Clavius, whose commentary on the *Elements* Saccheri always keeps in mind and treats as the true touchstone, the starting point, and, so to say, the holy book of his geometrical endeavors. Even on the rare occasions on which he raises objections against Clavius, Saccheri takes pains to undermine his criticisms' importance, thereby confirming his reverence for the old Jesuit master. He is certainly acquainted with more recent Jesuit commentators, but he is critical of the scandalous lack of rigor to which the new course of the *ratio studiorum* appears to condemn geometrical teaching. Among these newer mathematicians, Milliet Dechales is the only one to appear in *Euclid Vindicated*; though he is never mentioned by name, Saccheri quotes his text verbatim, eliminating doubt as to the identity of his polemic target. Moreover, it is very likely that Saccheri was familiar with the work of André Tacquet, who was in many ways the model for Dechales, but this is not certain. The same holds for Honoré Fabri.

Of the Italians, Saccheri names only Giovanni Alfonso Borelli, whose influence on *Euclid Vindicated* is second only to Clavius' – though Borelli features in Saccheri's text almost

[35] Angelo Marchetti (1674–1753) was the son of the scholar and mathematician Alessandro Marchetti, who had studied under Borelli in Pisa and is remembered today for the first Italian translation of Lucretius' *De rerum natura*. Angelo Marchetti continued his father's quarrel against Viviani and in favor of Borelli in the controversy over Galileo's legacy. He wrote, in Italian, *La natura della proporzione e della proporzionalità con nuovo, facile, e sicuro metodo*, Pistoia, Gatti 1695. The work was later included in A. Marchetti, *Euclides reformatus*, Livorno, Celsi 1709, which in addition contains a short Latin treatise on the same subject (pp. 101–34), followed by a complete rephrasing of Book V of the *Elements*. Both this work by Marchetti and Giordano's *Euclide* still follow the classical (not algebraic) approach to proportions. It cannot be denied, however, that they display – alongside a vaguely outmoded feel – a certain openness towards the new developments. For instance, Giordano defines the composition of ratios simply as a multiplication, without tackling such foundational issues as the homogeneity of the terms (*Euclide restituto*, p. 181). For a brief presentation of Giordano's work on compound proportion cf. § 12 in G. Vailati, *Sulla teoria delle proporzioni*, in *Questioni riguardanti le matematiche elementari*, Parte Prima, vol. 1, pp. 143–91. We also know that Giordano wrote a lengthy scientific letter (yet to be published) against some aspects of Marchetti's theory of proportions.

exclusively as his main antagonist. Saccheri must have found in Borelli a sort of kindred spirit, a fellow practitioner of the classical geometrical style uncorrupted by the seductions of algebra: that is, a classicist – but a classicist who nonchalantly amended and rectified Euclid instead of explaining and vindicating him. Saccheri must also have seen Borelli as the ideal target, for his work represented the peak of the reform of elementary geometry undertaken by the Galilean school. Hence, a refutation of Borelli's position would represent a triumph over a century of Italian research. Moreover, criticizing Borelli was a more prudent move than directly criticizing the great Galileo or his disciple Vincenzo Viviani, with whom the young Saccheri had been on good terms.[36] Furthermore, both Borelli and his pupil Marchetti held materialistic positions in natural philosophy that met with the disapproval of the Society of Jesus and Saccheri himself.[37]

Saccheri is also undoubtedly acquainted with Wallis' *De Postulato Quinto*, a text he repeatedly cites. It does not seem, however, that the presence of *De Postulato Quinto* in *Euclid Vindicated* evidences any actual openness towards English mathematics (not even elementary mathematics) on Saccheri's part. On the contrary, Saccheri appears poorly acquainted with Wallis' other works, even the most important among them (or those concerning proportions). Had he known them, he would probably have despised their algebraic rather than synthetic method. The same holds true of Saccheri's hypothetical stance on Isaac Barrow, a sophisticated interpreter of Euclid whom some scholars have tried to connect to the genesis of *Euclid Vindicated*. It is true that Saccheri would later advocate for some of the same causes as Barrow did, and it is also true that the English mathematician had rejected Tacquet's criti-

[36] Of the young Saccheri's correspondence with the elderly Viviani, who appreciated his early mathematical efforts, only two letters are extant: cf. A. Favaro, *Due lettere inedite del P. Gerolamo Saccheri d.C.d.G. a Vincenzio Viviani*, "Rivista di Fisica, Matematica e Scienze Naturali", 4, 1903, pp. 426–30. We should remember that Galileo's theory of proportions had been first published by Viviani, hence the two men's doctrines were often believed to be one and the same. Moreover, both Borelli and Marchetti fought violent disputes against Viviani. To criticize Borelli was thus in some sense to choose an affiliation. Saccheri also criticizes Borelli's approach to the theory of parallelism, although the Neapolitan mathematician's ideas on the subject were not very original: hence, either Saccheri's criticism is the fruit of his biases and obstinacy, or perhaps Saccheri had initially composed the part of *Euclid Vindicated* devoted to proportions, to include a severe rebuttal of Borelli's ideas, and had then continued criticizing him throughout the work with some inertia.

[37] We almost completely ignore Saccheri' metaphysical convictions. He must have shared many of the positions held by late Scholastic Aristotelians, but we can also find in Guido Grandi an assessment of him as a seven-eights Cartesian (being Aristotelian for the remaining eight). This is, however, a biased judgment, as Grandi himself took a Cartesian view, tempered with a pinch of Aristotelianism. All in all, it is likely that Saccheri's actual position was analogous to the one Grandi ascribes to him, although in the reversed percentages: "Godo che il suddetto padre sia per sette ottavi Cartesiano, anzi non credo sia d'uopo avanzarsi quell'altro ottavo che rimane verso Cartesio, per accostarsi alla verità, perché in qualche cosa credo abbia mostrato d'esser uomo ancora Cartesio, né bisogna sciogliersi da Aristotele per legarsi con un altro filosofo …" (Grandi to Tommaso Ceva, May 8th, 1700; in L. Tenca, *Relazioni fra Gerolamo Saccheri e il suo allievo Guido Grandi*, "Studia Ghisleriana", 1, 1952, pp. 19–45). Be as that may be, Saccheri certainly did not espouse Borelli's and Alessandro Marchetti's 'Epicurean' atomism.

cisms of Euclid with the same spirit – and indeed very similar arguments – that the Italian 'avenger' would later employ. I think, however, that these facts point towards Saccheri's ignorance of Barrow, rather than his acquaintance with the English mathematician: first because there is no explicit mention of Barrow in *Euclid Vindicated*, as there is of Wallis, and second because Barrow's arguments against Tacquet are much better than Saccheri's corresponding refutations of Dechales. I believe that Saccheri would have considerably improved his essay had he been aware of Barrow's work. In any case, he would not have neglected to mention explicitly a renowned and very respected ally in his battle for Euclid's honor.

Through Wallis, Saccheri is certainly acquainted with Nasīr ad-Dīn, whose geometric ideas are also explicitly discussed in *Euclid Vindicated*. The question of other Arabic sources is more complicated. Some scholars have ventured the hypothesis that Saccheri had direct knowledge of some works of Medieval Islamic geometry, particularly of Khayyām's book on parallels, since some of the theorems that figure in the ancient work are quite similar to the first two or three Propositions in the First Part of Book One of *Euclid Vindicated*.[38] It is true that Nasīr ad-Dīn knew and cited Khayyām, but these citations do not appear in the portion of Nasīr ad-Dīn's work translated into Latin by Pocock and published by Wallis, on which Saccheri apparently relied. Of course, we cannot categorically rule out the possibility that Saccheri, like Clavius before him, may have come into contact with some Arabic manuscript (and with an able translator), and in particular with Khayyām's, but the hypothesis seems quite speculative. Moreover, the results that we find in Khayyām's text, and that, according to this hypothesis, Saccheri must have taken from there, are also found, with few variations, in Vitale Giordano. This, of course, only shifts the problem: was Giordano acquainted with Khayyām's work? and was Saccheri acquainted with Giordano's work? A negative answer to the first question seems more plausible to me: after all, Giordano was in a position to derive these same theorems himself, building on Nasīr ad-Dīn's work. Indeed, Giordano read in Clavius' commentary a succinct account of Nasīr ad-Dīn's contributions, whose unabridged text Saccheri would later read in Wallis' publication.[39] It is equally unnecessary to imagine that Saccheri needed Giordano or Khayyām to furnish

---

[38] See the two articles, now somewhat dated, by D.E. Smith, *Euclid, Omar Khayyâm, and Saccheri*, "Scripta Mathematica", 3, 1935, pp. 5–10, and by K. Jaouiche, *De la fécondité mathématique: d'Omar Khayyam à G. Saccheri*, "Diogène", 57, 1967, pp. 97–113. None of these authors takes notice of the fact, which I take to be much more significant, that the three books in Khayyām's work are devoted to the solution of precisely the three Euclidean blemishes mentioned by Saccheri. Although the book on parallel lines bears some embryonic resemblance to Saccheri's later studies, Khayyām's solutions to the problems of Euclid's theory of proportions are altogether different from those presented in *Euclid Vindicated*.

[39] The difficulties, therefore, were not of a technical nature, but consisted solely in determining the best route by which to tackle the issue of parallelism – and this route was already excellently suggested in the works of Clavius and Nasīr ad-Dīn. It seems unlikely, but is nonetheless possible, that the question of Giordano's sources will be further clarified by the publication of some of his many unpublished papers. For a summary overview of these materials, cf. M.T. Borgato, *Una presentazione di opere inedite di Vitale Giordani (1633–1711)*, as well as *Scritti inediti di Vitale Giordani*, both in *Giornate di storia della matematica*, ed. by M. Galuzzi, Cetraro, EditEl 1991, pp. 1–56.

him with the first – and quite simple – theorems appearing in *Euclid Vindicated*. Nonetheless, it is hard to imagine that, as Saccheri confronted the issue of parallelism, he did not at least skim through Giordano's *Euclide restituto*: a quite famous Italian book that already revealed its subject in the title and comprised so many efforts at proving the Fifth Postulate. On the other hand, if Saccheri read Giordano, I cannot see why he did not mention him. Giordano's demonstration was ingenious but seriously flawed, and would have fit well alongside Nasīr ad-Dīn's or Wallis' proofs, which Saccheri explicitly and instructively criticizes in a pair of scholia.[40]

The last author whom Saccheri explicitly names is the German philosopher and mathematician Christian Wolff,[41] whom Saccheri mentions only in connection with a very marginal issue regarding the definition of ratio. Wolff was most likely known to Saccheri as metaphysician, and, given Wolff's Cartesian and Scholastic sympathies and his interest in elementary geometry, he must have appeared to Euclid's vindicator as a sort of kindred spirit. Wolff's mathematical epistemology, however, could not be farther removed from Saccheri's, and his stature as a mathematician is not even comparable to that of the Italian Jesuit. At any rate, the short and almost irrelevant mention of Wolff in *Euclid Vindicated* is quite possibly the only reason that Saccheri's work has survived through the centuries.

Saccheri also mentions some other Classical or Renaissance authors, from Proclus to Oronce Fine, but the context in which these allusions occur and the kind of remarks in which they consist suggest that they are second-hand references – which is to say that Saccheri himself had no direct knowledge of the authors' works. It is unclear whether Saccheri was acquainted with other mathematical studies that would have been relevant to *Euclid Vindicated*. How much he knew about the numerous French studies of elementary geometry, for instance, remains unclear. The question is especially relevant with respect to Arnauld's aforementioned *Nouveaux Elémens de géometrie*, which had become a very

---

[40] It is possible that Vitale Giordano had led too deplorable a life, by any Jesuit moral standard, to receive an approving mention by a member of the Society. There is no doubt that Saccheri was very cautious on these matters, but I am not sure whether this explanation may be reasonable enough. Another possibility is that Saccheri's silence was motivated by *esprit de corps*, since at the end of the seventeenth century Giordano had defended Galilean mechanics in quite a violent mathematical dispute with the Jesuit Giovanni Francesco Vanni; Giordano's contempt for the Jesuits is manifest in his letter to Viviani of May 14th, 1689, which can be read in M. Torrini, *Dopo Galileo. Una polemica scientifica (1684–1711)*, Firenze, Olschki 1979, pp. 120–21. In any case, the question of Giordano's influence on Saccheri divides interpreters: for instance, Klügel, Stäckel, Engel and Bonola believe that Saccheri was not acquainted with *Euclide restituto*, whereas Halsted, Allegri and Maierù think that it is quite likely that Saccheri was directly influenced by the text.

[41] The philosopher Christian Wolff (1679–1754) lectured on mathematics at Halle for many years and wrote a monumental *Elementa Matheseos universae* in five volumes (Halle, 1713–1741); however, he was not an original mathematician, and his positions on the Euclidean issues relevant to Saccheri were definitely traditional. He proved the Fifth Postulate with appeal to the notion of equidistance, which he assumed completely uncritically, and steered a middle course between a classical and an algebraic approach to the theory of proportions. We will discuss his role in the reception of *Euclid Vindicated* later.

important text in France, and which represented one of the most relevant mathematical fruits of Port Royal. Since Arnauld's work is in many ways similar to Dechales', it is possible that Saccheri was acquainted with both texts – and that by attacking his Jesuit colleague he also intended to criticize the great Jansenist theologian without even deigning to write his name. In any case, it does not seem to me that there is conclusive evidence of the dependence of *Euclid Vindicated* on the *Nouveaux Elémens*, nor of the hypothesis that Saccheri read Arnauld's work and then passed it over in silence.

Finally, we conclude with a question about Saccheri's familiarity with 'modern' mathematics, that is, first with algebra and second with the developments of the Calculus. It seems that he knew quite some algebra – but did not love it – and little or no infinitesimal calculus. We shall soon see that *Euclid Vindicated* contains a couple of very big blunders due to Saccheri's poor skills in dealing with very small terms. In any case, Saccheri never mentions any mathematician active in the field of the Calculus, nor any significant result of this direction of research, in *Euclid Vindicated*; and the references to the Calculus found in his other works are totally generic.[42] A potential source for Saccheri in the domain of Calculus was Guido Grandi,[43] who was briefly a student of the Jesuit's in Cremona and with whom Saccheri remained in contact all his life. Other unwritten sources of which Saccheri may have made use in his mathematical work were Pietro Paolo Caravaggio and especially

[42] Compare the passage from the *Neo-Statica*: "Rursum adhibemus infinitesimas, seu temporis, seu spatij, hoc est particulas infinitè parvas, tanquam compendium à Geometris recentioribus inventum ad evitandas operosas prolixasque demonstrationes circa curvas, quas prisci per inscriptionem, & circumscriptionem examinare consueverunt. Eorum usum tutissimum, rationesque certissimas invenias passim apud Clarissimos Geometras Leibnitium, Vallisium, Guidonem Grandum, Marchionem Hospitalium, Gabrielem Manfredum, atque alios, à quibus supersedeo, ne longiori digressione benigno lectori morem injciam" (p. 97). As this passage demonstrates, Saccheri regarded the Calculus as nothing more than a way to obtain more rapidly the same results yielded by the Classical method of exhaustion – a position that was not exceptional at the beginning of the eighteenth century. Among the authors mentioned by Saccheri, we know that Leibniz read the *Neo-Statica*, which he criticized in some respects: cf. Leibniz's letter to Des Bosses of August 23rd, 1713, in GP II, p. 482. In fact, Saccheri espoused in this work some rather questionable tenets of the so-called 'geostatic' theory, which had enjoyed a brief bout of popularity in France in the 1630s before it was abandoned by virtually everyone. The theory had also circulated in Italy, and Giordano had also adhered to it. The only modern analysis of Saccheri's book is the one offered by P. DUHEM, *Les origines de la statique*, Paris, Hermann 1906, vol. 2, pp. 261–65.

[43] Guido Grandi (1671–1742), professor in Pisa from 1700, was one of the major promoters of the Calculus in Italy and interacted with the best European mathematicians of his time. A useful sketch of Grandi in his historical context can be found in S. GIUNTINI, *Gabriele Manfredi – Guido Grandi: Carteggio (1701–1732)*, "Bollettino di storia delle scienze matematiche", 13, 1993, pp. 5–144; on Grandi's relationships with European mathematical culture, cf. A. ROBINET, *Leibniz et les mathématiciens italiens*, "Symposia Mathematica", 27, 1986, pp. 101–21. On Grandi's relations with Ceva, see L. SIMONUTTI, *Guido Grandi, scienziato e polemista, e la sua controversia con Tommaso Ceva*, "Annali della Scuola Normale Superiore di Pisa. Classe di Lettere", III 19, 1989, pp. 1001–26.

the Ceva brothers,[44] who enjoyed Saccheri's close friendship. But in none of these cases is it possible to identify a clear influence on any specific passage of *Euclid Vindicated.*

Saccheri's personal relations with the Ceva brothers and with Grandi, however, justify some hypotheses regarding the composition or at least the development of *Euclid Vindicated.* We know from their correspondence that, in 1713, Saccheri discussed with Tommaso Ceva Wallis', Borelli's and Marchetti's definitions and demonstrations of the Fifth Postulate – and easily refuted them. Saccheri's correspondence with Ceva also reveals that, as of the same date, Saccheri believed himself to be in a position to prove the Postulate from the properties of straight lines: this is the route he eventually took in the First Part of Book One of *Euclid Vindicated* (Propositions 1–33), which is by far the most extensive and important section of the whole work. It is hence very likely that Saccheri's best results, those for which he is still remembered to this day, were the fruit of twenty years of reflection; indeed, he may have worked on them for his whole life, if it is true that he had already come up with the bulk of his theorems on this subject in 1713.[45]

On the other hand, there are no external clues that help us establish the date of composition of the Second Part of Book One (i. e., Propositions 34–39), where Saccheri attempts

---

[44] Giovanni Ceva (1647–1734) studied in Pisa under Marchetti and lectured for his whole life in Mantua. He wrote a famous short work (*De Lineis rectis*, 1678), in which he proved an important theorem of elementary geometry that in some ways makes him one of the forerunners of projective geometry. Saccheri himself later published some of his friend Ceva's results, along with his own, in his first work, *Quaesita Geometrica*, where he tackled some mathematical difficulties proposed by Count Ruggero Ventimiglia. The complete title is G. Saccheri, *Quaesita Geometrica a Comite Rugerio de Vigintimillis omnibus proposita*, Milano, Malatesta 1693. This booklet enjoyed some circulation and was also mentioned by G. de l'Hospital, *Traité analytique des sections coniques*, Paris, Montalant 1720, p. 259. For a mathematical study of Saccheri's relationship to Ceva in this early work, see A. Brigaglia, P. Nastasi, *Le soluzioni di Girolamo Saccheri e Giovanni Ceva al "Geometram Quaero" di Ruggero Ventimiglia: Geometria proiettiva italiana nel tardo seicento*, "Archive for History of Exact Sciences", 30, 1984, pp. 7–44. Tommaso Ceva (1648–1737), Jesuit, lived and taught in Milan his whole life, and was a mathematician, an Arcadian poet and a philosopher, though he did not excel in any of these fields. He was probably the scientist who had the closest personal relationship to Saccheri: friendship and familiarity in everyday life. Pietro Paolo Caravaggio (1658–1723), son of a Milan mathematics professor of the same name, succeeded his father on the same chair. At the age of twelve (!), he published a translation of the *Primi sei libri d'Euclide tratti in volgare* (Milano, Lodovico Monza 1671).

[45] This information can be obtained from the Ceva-Grandi correspondence quoted in the aforementioned article by Tenca, *Relazioni fra Gerolamo Saccheri e il suo allievo Guido Grandi.* In 1695, an edition of Euclid was published in Turin, edited by Carlo Edoardo Filippa, who had been a student of the young Saccheri and dedicated the book to his teacher: C.E. Filippa, *Euclidis priora elementa sex, auctore Carolo Eduardo Taurinense, reverendo patri Hieronymo Saccherio Societatis Iesu dicata*, Torino, Zappata 1695. The fact that this work stresses the role of the Fifth Postulate in elementary geometry, though it is accepted as a postulate without attempting a proof, may be taken to attest to the atmosphere surrounding Saccheri's teaching activity. On the other hand, however, the rest of the work takes a standard Clavian route and contains no evidence suggesting that Saccheri had already developed any of his mature theories concerning parallels and proportions. A contrary interpretation has been proposed by L. Brusotti, *Gli "Elementa" di Carlo Edoardo Filippa allievo di Girolamo Saccheri*, "Atti dell'Accademia Ligure di Scienze e Lettere", 9, 1952, pp. 155–164.

to provide a different proof of the Fifth Postulate. Instead of building on the properties of straight lines, this alternative demonstration is based on an application of *consequentia mirabilis*. What concerns us now, however, is that the proof carried out in the Second Part, is not only much shorter but also much more classical than the one presented in the First Part. It rests entirely on the concept of equidistance and on the motion of line segments, that is, on the usual proofs of the Fifth Postulate from Proclus onwards, although it attempts to rectify and improve them. Moreover, in the whole of the Second Part, Saccheri exploits only Propositions 2 and 3 of the previous part, that is, the simple theorems that were already found in the Arabic-Persian tradition, in Vitale Giordano, and to a great extent in Clavius himself – no use is made of the impressive results that make up the true mathematical body of the First Part. In a word: it is quite possible that this Second Part, which is shorter, more classical, simpler (and weaker) than the First, had been written long before the latter, or at least independently of it.

As regards the theory of proportions, the subject of Book Two, we know that Saccheri was engaged in the topic since his youth, to the point where he refers in *Euclid Vindicated* to some results already obtained in the *Demonstrative Logic* from 1697. And if it is true that those references are quite unspecific, and that the *Logic* does not contain any relevant mathematical proof in the theory of ratios, it is equally true that the strengths of Book Two of *Euclid Vindicated* are not those few and poorly executed demonstrations, but precisely its logical structure and its epistemological considerations. In the end, it seems as though Saccheri could very well have held the contents of Book Two in his hands from the beginning of his mathematical career, although the actual composition must have taken place much later (the text of Wolff mentioned by Saccheri is also from 1713).

The book as a whole must hence have had a very long, and probably very composite, gestation. *Euclid Vindicated* was granted the *imprimatur* on August 16th, 1733, and on October 25th of the same year Saccheri died "after a long illness with fever and oppression of the head" (in the words of his biographer, Count Gambarana); we do not know whether he ever saw the printed volume. Saccheri's innate laziness as a writer, the unfinished state of many logical threads in *Euclid Vindicated*, and the account of Saccheri's long illness all suggest that he must have put off the completion of the book from year to year during his life and hurried with its publication as his death was nearing. The First Part of Book One is nearly perfect from the formal viewpoint, and it must have long been ready and polished: it could easily have been an independent essay, and had perhaps been conceived with this aim. On the other hand, the Second Part of Book One is on the whole quite unevenly balanced, and it ends with such an incredible mathematical error that it is hard to imagine how someone, if not Saccheri at least Ceva or Grandi, may have failed to notice it;[46] the same holds for one very clumsy proof in the following Book on proportions. Therefore, it seems likely that Saccheri, wishing to preserve his important researches on the Fifth Postulate (the First Part of Book One) from oblivion, resolved to seize the opportunity represented by that publication to combine in one volume other subjects that he valued and which had probably constituted the core of his university teaching. It appears certain, moreover, that some of Saccheri's works had an academic origin as lecture notes or hand-outs.[47] Therefore, *Euclid Vindicated* may have crystallized around a well-polished core of geometrical texts on the problem of parallels, to which old papers, materials for research and teaching, maybe even student notes were then added. It is probable that Saccheri revised all of the material, if in a hurry and with difficulty, and took the trouble of writing (or maybe dictating) the work's general Preface; but in the end it is not even certain that all the parts of *Euclid Vindicated* come from his hand. The original manuscript appears to be lost.

5. Aims and epistemological structure of Euclid Vindicated

Such were, hence, the sources and influences that occasioned the publication of Saccheri's book. We now must inquire as to its aim – as to the goals that Saccheri must have set for himself in writing *Euclid Vindicated*. At first glance, the question seems very simple, as Saccheri himself is unambiguous on this point: he wants to prove the Fifth Postulate and improve the Euclidean theory of proportions. Despite this explicit statement, an interpreter quickly notices an incongruity, that is, that Saccheri would like to defend Euclid from the

[46] Saccheri and Grandi did at times discuss mathematical subjects, and Grandi provided detailed criticism of some of the results of Saccheri's *Neo-Statica*. There is evidence of their discussions in some letters edited by A. Agostini, *Due lettere inedite di Girolamo Saccheri*, "Memorie della Reale Accademia d'Italia. Classe di Scienze Fisiche, Matematiche e Naturali", 2, 1931, pp. 3–20. It is nonetheless true that this close personal and scientific relation seems to fade in Saccheri's last years: most notably, Grandi published in 1731 his own *Elementi geometrici piani e solidi di Euclide posti brevemente in volgare* without the slightest mention of Saccheri's foundational researches. Grandi even repeated Clavius' proof of the Fifth Postulate to the letter, though he undoubtedly knew that Saccheri regarded it as fallacious. It seems, therefore, that all scientific communication between the two had died out. On the other hand, it may be that in 1731 Saccheri was already afflicted by the illness that ultimately caused his death, and in those same years Grandi was certainly entering into a state of intellectual decline that would soon lead to dementia. Perhaps the composition of *Euclid Vindicated* took place when it was already too late for a fruitful scientific exchange.

[47] This thesis was already put forward by Arnold Emch, with respect to the *Demonstrative Logic*, in his doctoral dissertation (Harvard, 1934). The main contents of the dissertation are presented in a three-part article: A. Emch, *The Logica Demonstrativa of Girolamo Saccheri*, "Scripta Mathematica", 3, 1935, pp. 51–60, 143–52, 221–33. On the composition of the *Logic* and its various editions, also see P. Pagli, *Two unnoticed Editions of Saccheri's Logica Demonstrativa*, "History and Philosophy of Logic", 30, 2009, pp. 331–40.

criticisms raised against him, but as he undertakes to do so, he employs a seemingly anti-Euclidean strategy. Should Saccheri actually prove the Fifth Postulate (to leave the theory of proportions aside for the moment), he would thereby deny exactly what Euclid had affirmed, that is, that the principle concerning parallel lines must be regarded as an axiom rather than a theorem. So it happened that some readers of Saccheri considered his *Euclides Vindicatus* a *Euclid amended*, or *rectified*, or *freed* from the errors that the ancient geometer had committed, the most prominent of which had been to view as a postulate what can, as a matter of fact, be demonstrated. And yet, this interpretation certainly involves a misunderstanding of the aims of the work in question, since Saccheri, on many occasions, voices his deep contempt for this attitude towards Euclid, which is common among mathematicians of his age (first and foremost Borelli). Saccheri does not long to write another *Euclid Renewed*, but rather to show precisely that the original text and strategy of the *Elements* suffice to answer all criticisms, so that Euclid can be *vindicated* from the errors with which he was unjustly charged. Thus, what is to be investigated is how Saccheri understands his own celebrated (and unsuccessful) proof of the Fifth Postulate.

In fact, we need to leave aside the opposition between axioms and theorems that is obvious in modern mathematical epistemology, and try to penetrate deeper into the mathematical mindset of the seventeenth century in general and of Saccheri in particular. There, the fundamental character of an axiom is not to be indemonstrable, but rather to be evident: that is, an axiom is not in want of (not: incapable of) a proof – which does not mean that a proof cannot be offered. It costs us some effort, of course, to raise these statements from the mere psychological level, on which the reference to the axiom's evidence seems to place them, and towards a more abstract epistemological formulation. However, a logical and epistemological understanding of the notion of evidence is at work in the mathematical practice of the time and in the works of Saccheri in particular.

In the *Demonstrative Logic*, Saccheri distinguishes, in fact, two kinds of axioms.[48] Those of the first class are most evident, though unprovable, principles that it is necessary to assume before any demonstration at all can take place – such as the Principle of Contradiction. The other class of axioms – the majority of them – contains evident propositions that are found to be true once the terms that compose their statement are understood; however, their truth can and must be exhibited explicitly, instead of remaining implicit in their definitional structures. Their evident status must, in other words, be gained and explicated in some way: for axioms of this kind, *proofs* can certainly be provided. Thus, for instance, that given two

[48] Here is the passage: "Axiomatis nomine censetur, non modò propositiones per se immediatè certae, nec omininò habentes medium, unde demonstrentur, ut *idem non potest simul esse, & non esse*, & fortasse *quodlibet est, vel non est*, sed etiam propositiones aliae immediatè patentes ex sola terminorum intellectione, licet quaepiam de eisdem confici possit demonstratio, ut *totum est maius sua parte*" (*Logica demonstrativa*, p. 127 of the 1701 edition, the first one explicitly authored by Saccheri, and reprinted in the 2012 Italian edition by Mugnai and Girondino; p. 200 of the first, anonymous edition from 1697, reprinted by Risse in 1980).

points one straight line always passes through them (Euclid's First Postulate) is an axiom:[49] however, this fact is not axiomatic in virtue of its being a strictly indemonstrable proposition, nor (much less!) because, as it were, it has been assumed as an arbitrary stipulation in order to ground a particular formal system. Rather, it is an axiom because it proves immediately true as soon as the definitions of 'straight line', 'point', 'two' etc. are grasped; its truth can hence be exhibited demonstratively by going back to the definitions. It is apparent that in this way the entire logical weight of the axiom is shifted onto definitions – in this case, that of a straight line. For this reason, Saccheri, like many others at the time, takes pains to provide a theory considering definitions as the *principles* of demonstration – an epistemological move that, to some extent, had also been make by Aristotle and was by no means a novelty. Theories of this sort hinge on a basic distinction between *real definitions* (that is, definitions *quid rei*) and *nominal definitions* (*quid nominis*).[50] A real definition is in fact a genuine proposition that joins certain concepts together and baptizes their union – concepts that have already been defined or that are already known through other sources (from experience, for instance). If I define a parallel line as a straight line equidistant from a given straight, I have provided a real definition: in this definition of a parallel, I join together two properties already given, being a straight line and being equidistant from another straight line. Therefore, it would not be implausible, even from a present-day perspective, to try to found a mathematical theory on such real definitions: they are in fact nothing other than 'axioms in disguise' (and the one just mentioned is in fact equivalent to Euclid's Fifth Postulate). This, however, does not exhaust the strategy that Saccheri and his fellow seventeenth-century mathematicians and logicians intended to follow, since this would amount to simply calling 'real definition' what others would call 'axiom'. On the contrary, these real definitions, insofar as they are propositions, must be capable of proof, just like any other proposition. Saccheri will discuss this matter more than once in *Euclid Vindicated*, where he aims to show that many mathematicians who wanted to rectify Euclid's *Elements* ended up accepting uncritically and without proof certain real definitions (such as that of equidistant parallels just mentioned, or others concerning proportions and the equality of ratios) that called for demonstration. In addition, we have nominal definitions: these are

[49] The criteria for distinguishing between axioms and postulates were discussed extensively in the Renaissance and the early Modern Age. On this subject, Saccheri seems to adopt for the most part Clavius' denomination and divisions, without scrutinizing them in much detail (cf. my *Notes* for further remarks). In Clavius, the first three Euclidean postulates maintain their label, whereas the latter two (including the parallel postulate) are re-classified as axioms, as they are not constructive.

[50] For Saccheri, cf. *Logica demonstrativa*, ed. 1701, pp. 119–26 [ed. 1697, pp. 189–200]. But this was a rather common distinction, even though the terminology and the details of the definitions were not always homogeneous. The most famous seventeenth-century locus is probably the *Logique de Port Royal*, in Chapter 12 of the First Part (p. 86). The broadest eighteenth-century exploration of this issue was probably accomplished by Lambert, another 'non-Euclidean' thinker. His discussion is part of a fierce polemic against Wolff, whom Lambert charged, perhaps unfairly, with accepting real definitions without proof. On Saccheri and nominal definitions, see G. Lolli, *Saccheri e le definizioni "filiae plurium demonstrationum"*, in G. Lolli, *Le ragioni fisiche e le dimostrazioni matematiche*, Bologna, Il Mulino 1985.

linguistic stipulations about the use of terms, and assign a name to a certain simple concept. Saccheri's project for proving axioms (and real definitions) consists therefore in an attempt to reduce all the principles in question to the nominal definitions of the terms involved.[51]

This is, ostensibly, a sort of 'logicist' program: the theorems must be reduced to the axioms, and the latter to the mere definitions of the terms of which they consist. This is not the place to discuss the radical impossibility of such an operation, nor the inescapable difficulties it encountered, nor the means by which some of these difficulties were overcome (or not), largely via recourse to views on the meaning of 'demonstration', or, more generally, the meaning of 'logic', which differed much from ours; let us for now deal with projects rather than with outcomes. It is very well known that this kind of logicist program in geometry was championed by Leibniz, who is sometimes regarded as its originator and inventor. Although Leibniz carried this program out further, with greater rigor and perceptiveness than any other mathematician of his age, he was not the only one to undertake it.[52] It suffices to open any commented edition of the *Elements* from Clavius' onwards to find the statements of the 'Axioms' and 'Postulates', labeled as such, followed each by a 'Proof'. Although this appears disconcerting to contemporary epistemology, it was the norm for the mathematical mindset of the Early Modern Age. An important point, however, needs to be clarified: the question of why these axioms and postulates, once proved, are not immediately re-classified as theorems so as to yield a theory that is completely demonstrated, hence devoid of principles; we have to grasp, that is, what prevents the straightforward collapse of the notion of an axiom into that of a theorem.

There is no doubt that in some instances the identification of principles and theorems was avoided through a simple appeal to the authority of tradition. Many of the Renaissance mathematicians who produced editions of Euclid conceived of mathematics as an unhypothetical science, free of genuinely unprovable principles, but at the same time they hoped to provide a correct edition of the Classical text: therefore, they did not rename the propositions that Euclid had labeled 'axioms' and 'postulates'. Instead, they added demonstrations of their own – which turned the 'axioms' and 'postulates' (in fact, if not by name) into explicitly proved theorems. We can, however, also discern that some geometers of that age, Saccheri among them, applied more rigorous epistemological criteria. They construe axioms as *immediately* demonstrable propositions, which is to say that an axiom can be proved with reference to the nominal definitions of its terms alone, without appealing to

[51] Cf. *Logica demonstrativa*: "Omnia axiomata (exceptis duobus universalissimis) famulari debent definitioni *quid nominis*, & ex ea confirmari. Sit axioma *totum est maius sua parte*. Dico admittendum non esse, nisi post definitiones *quid nominis* terminorum, ex quibus demonstrativè elici possit" (ed. 1701, p. 128 [ed. 1697, p. 201]).

[52] Therefore, the peculiarity of the logicist program of Leibniz (1646–1716) does not consist in the philosopher's aspiration for an unhypothetical mathematics. One could instead consider his broader conception of logic, which encompasses a rich theory of relations alongside the classical doctrine of syllogism. This theoretical model justifies Leibniz's hope for purely logical proofs of axioms and theorems, proofs that do not require the actual construction of the figures, that is, the demonstrative techniques characteristic of synthetic geometry.

other axioms or theorems. Thus, we have, first, nominal definitions; then some propositions that follow from them 'immediately', by purely logical entailment, hence 'by the mere meaning of the terms' (these are the axioms); and finally a multitude of theorems, which are not proved immediately, but rather with reference to the axioms and other theorems. Propositions belonging to this last category do not derive their truth solely from the meaning of the terms *in question*, but also from a body of other definitions – and, possibly, from auxiliary constructions with compass and straightedge. We do not push the inquiry so far as to ask whether or not these axioms or theorems should be considered analytically true, or whether the distinction between axioms and theorems that I have sketched here could withstand serious logical critique: such complicated issues cannot be settled so broadly, and we would need to consider the work of each author specifically. For our purposes, it will be enough to remark that Saccheri is firmly embedded in this logicist or pseudo-logicist tradition, and that he intends to prove all of Euclid's axioms without calling their axiomatic status into question.

Saccheri thus hopes to provide a demonstration of Euclid's Fifth Postulate that will also show that it is *an axiom* – this is to say, that it is a demonstrable principle, but nevertheless an evident and primitive one. For Saccheri, the latter features are not a psychological marks of a proposition, consisting, for instance, in its being obvious or self-evident: rather, it consists primarily in its being capable of an immediate proof – that is, a proof that yields the truth of the proposition in question without appealing to anything but the terms that compose it. It is thus only in the context of this broader epistemological program that it becomes apparent how it was possible for Saccheri to intend to *vindicate* Euclid and his Axiom by way of demonstration.

In fact, it should be expressly noted that this logicist program differs in essential ways from the agenda that had inspired similar attempts to prove the Fifth Postulate in Antiquity, the Middle Ages and even as late as the Renaissance. In these earlier cases, the evidence of truths was often construed as a psychological feature. The first four Euclidean postulates appeared self-evident and unproblematic and, for this reason, many authors believed them not to be in need of a demonstration. Only the Fifth was obscure, and it was hoped that it could be turned into a *theorem* (in the proper sense).[53] Saccheri, on the other hand, believed that all axioms *as axioms* call for a demonstration, including those that appeared self-evident

[53] The positions were of course extraordinarily varied, and before the Modern Age, geometers may have conceived projects that were also 'logicist' in some other sense. It is not at all clear that Euclid conceived of his work as a genuine axiomatic system, nor is it clear that pre-Euclidean mathematics should be understood in this way. On this point see A. Seidenberg, *Did Euclid's Elements, Book I, Develop Geometry Axiomatically?*, "Archive for History of Exact Sciences", 10, 1975, pp. 263–295, which goes so far as to claim that the only principle of the *Elements* that Euclid holds to be indemonstrable is perhaps the Fifth Postulate. For a much more nuanced position see I. Mueller, *Sur les principes des mathématiques chez Aristote et Euclide*, in *Mathématique et Philosophie de l'antiquité à l'âge classique*, ed. by R. Rashed, Paris, CNRS 1991, pp. 101–13; and I. Mueller, *On the Notion of a Mathematical Starting Point in Plato, Aristotle, and Euclid*, in *Science and Philosophy in Classical Greece*, ed. A.C. Bowen, London, Garland 1991, pp. 59–97.

to his predecessors. If he does not contrive demonstrations as complete and exhaustive for these other axioms in *Euclid Vindicated*, it is simply because he is satisfied with the proofs of them provided by Clavius and others:[54] the Fifth Postulate only had resisted unconquered every logicist assault as late as 1733.

However, the issue is not yet settled. We need to ask how Saccheri could possibly regard his demonstration of the Fifth Postulate as an 'immediate' proof, one conducted solely on the basis of the meaning of the relevant terms – when it comprises a hundred pages of proofs, thirty-nine propositions and innumerable lemmas, scholia and corollaries, and, moreover, when it explicitly relies on some other twenty theorems from the *Elements,* none of which appeals to the postulate in question. The issue is clearly an important one, since it seems that, according to Saccheri's own epistemological framework, he managed to prove (insofar as he did prove) only a theorem about parallels – not the parallel axiom *as an axiom*. Saccheri may have carried out an impressive reform of Euclid, but he failed to vindicate him. In fact, this corresponds to the stance on the Fifth Postulate taken by Clavius, Saccheri's illustrious model. Clavius proved all the axioms from the *Elements*, without coming to regard them as anything but genuine axioms for this reason – with one exception. This was the single axiom that Clavius had not been able to prove via straightforward and immediate inference, but only with the recourse to five difficult Lemmata, many constructions, and the first twenty-eight propositions of Euclid: this axiom was of course Euclid's Fifth Postulate, and Clavius was forced to concede that it was after all a *theorem*.[55] Hence, we must try to determine how Saccheri intended to amend not only Clavius' demonstration, but also its epistemological implications.

It appears that the answer to this question can be found in the *consequentia mirabilis*. This is a deductive technique that had received a detailed theoretical examination in Saccheri's *Demonstrative Logic*, and that he now applies *in concreto* in a couple of crucial propositions of *Euclid Vindicated*. A proof by *consequentia mirabilis* (which he calls *redargutio*, 'redargution') consists in deriving a proposition from its own negation, thereby establishing the unhypothetical validity of the proposition in question. The logical details are rather uncertain and not easy to formalize, as it is not apparent which role the first derivation is supposed to play. Undoubtedly, it cannot be a simple material (Philonian) conditional of

[54] Saccheri himself proves many Euclidean axioms and postulates (or other principles that the later tradition had attributed to Euclid) in the body of *Euclid Vindicated*; most of these foundational discussions are compressed into five Lemmata that are proved in the course of Proposition 33. All of Saccheri's proofs depend heavily on their Clavian counterparts, which Saccheri from time to time seeks to refine, but which he regards as basically acceptable. In fact, the whole seventeenth-century mathematical community deemed Clavius' proofs valid: they were constantly presented with only slight modifications in the later commentaries on Euclid. The only exception, of course, was the Fifth Postulate: on this point, Clavius had convinced no one (or only very few).

[55] At the conclusion of his proof, Clavius explicitly affirms that Axiom 13, that is, the Fifth Postulate, has now become a theorem, and that he only labeled it an axiom so as to avoid deviating from the Greek text: "… ut iure optimo inter theoremata, & non inter principia poßit connumerari; tamen ne ordinem Euclidis in quoquam immutemus, utumur eo in omnibus propositionibus, quarum demonstrationes ex ipso pendent, tamquam pronunciato" (*Euclidis*, p. 53). Note that Clavius never writes anything of the sort about his proofs of the remaining axioms: they preserve the status of principles.

the kind employed in modern extensional logic, but it is not easy to determine whether and how the derivation can be construed as a 'strict' or 'relevant' implication, in modern terms.[56] Accordingly, the history of the *consequentia mirabilis* is also quite complex: some believe it can be found already in Plato, or in Stoic logic, others trace its operational use to certain Euclidean demonstrations.[57] In the Modern Age, it was regarded by Cardano as a novel and original deductive technique, though Clavius later maintained that the procedure had already been employed by Euclid and Theodosius.[58] Clavius' mention drew the attention of several scholars, mostly Jesuits, to the *consequentia mirabilis*. Tacquet discussed the mathematical meaning of this logical procedure at length in the second edition of his *Elementa geometriae* (which Saccheri probably knew), and he engaged in a lively debate about its demonstrative power with Huygens.[59] In those years, the *consequentia mirabilis* left for the first time the narrow field of geometrical studies and was employed in several theological

[56] In extensional terms, this would simply be the tautology $(\neg\alpha\rightarrow\alpha)\rightarrow\alpha$, or the analogous rule of inference. It is unlikely, however, that Saccheri and his contemporaries would interpret it this way, as the first implication is probably intended in a non-extensional way. On Saccheri's conception of consequence, see Mugnai's Introduction to the Italian edition of *Demonstrative Logic*.

[57] We owe the first historical studies on the subject to Giovanni Vailati, who conducted the research in question after his rediscovery of Saccheri's *Demonstrative Logic*. Cf. G. Vailati, *Sur une classe remarquable de raisonnements par réduction à l'absurde*, "Révue de Métaphysique et de Morale", 12, 1904, pp. 799–809.

[58] Gerolamo Cardano (1501–1576), the famous mathematician and astrologer, wrote a treatise on proportions in 1570. In this work, he proved one theorem (proposition 201) through a redargutive argument by *consequentia mirabilis*. He thought it necessary to emphasize the peculiarity of the proof procedure: "… & hoc numquam fuit factum ab aliquo, imò videtur plane impossibile. Et est res admirabilior quae inventa sit ab orbe condito, scilicet ostendere aliquid ex suo opposito, demonstratione non ducente ad impossibile & ita, ut non possit demonstrari ea demonstratione nisi per illud suppositum quod est contrarium conclusioni, velut si quis demonstraret quòd Socrates est albus quia est niger, & non posset demonstrare aliter, & ideo est longè maius Chrysippeo Syllogismo" (G. Cardano, *Opus novum de proportionibus numerorum*, Basel, Heinrich Petri 1570, p. 227). Clavius discussed Cardano's *admirabilis demonstratio*, as he was the first to call it, three years later, that is, in the first edition of his *Euclidis*, where he claimed that it had already been employed by Euclid in *Elements* IX, 12, and by Theodosius in *Sphaerica* I, 12. The latter assertion is historically false: it was Clavius himself who, following Maurolycus, had rewritten Theodosius' proof in that form. As for Euclid, interpreters are divided. The locus of Clavius' commentary is the Scholium to the aforementioned Euclidean demonstration (*Euclidis*, p. 365).

[59] The debate between Tacquet and Christiaan Huygens (1629–1695) was only the beginning of a dispute that spread rapidly throughout the Netherlands: cf. G. Nuchelmans, *A 17th-Century Debate on the Consequentia Mirabilis*, "History and Philosophy of Logic", 13, 1992, pp. 43–58. We should note, however, that Saccheri does not mention Tacquet's work in *Demonstrative Logic*, where he begins his discussion of the *consequentia mirabilis*, and it seems that his only source is Clavius (cf. *Logica demonstrativa*, ed. 1701, p. 82 [ed. 1697, p. 130]). It is still possible, however, that Saccheri read Tacquet at any time between 1696 and 1733.

treatises of the Society of Jesus.[60] Through all these channels it came to Saccheri's attention, and Euclid's avenger became one of its most fervid proponents. A critical point is that some geometers and logicians took arguments by redargution, or *consequentia mirabilis*, to be a form of *reductio ad absurdum*; moreover, this latter sort of argument had a bad reputation among certain epistemologists – chiefly, but not only, those in the Cartesian camp. These critics conceded that a *reductio* proof of a proposition was capable of leading the soul to assent to it, but not of illuminating the mind: the proposition did not stand out as positively evident.[61] Other logicians, on the other hand, denied that this procedure was a genuine *reductio ad absurdum*, and regarded it as straightforwardly positive.[62]

For his part, Saccheri was not concerned with the Cartesian issue of a proposition's evidence but rather with the logical correctness of inferential rules: therefore, not only he accepted without hesitation the usual *reductio* procedures in mathematics, but he even deemed the *consequentia mirabilis* the best demonstrative technique available. Even more than the mathematical use of this technique in Euclid, Clavius, or Tacquet, Saccheri must have had

---

[60] For a history of the *consequentia mirabilis* see F. Bellissima, P. Pagli, *Consequentia Mirabilis. Una regola logica fra matematica e filosofia*, Firenze, Olschki 1996; a technical exposition is in C.F.A. Hoormann, *A further Examination of Saccheri's Use of the "Consequentia Mirabilis"*, "Notre Dame Journal of Formal Logic", 17, 1976, pp. 239–47.

[61] The liveliest protest came from the Cartesian camp, and Cartesian objections were to be fond in the logic and geometry of Port Royal. See this clear passage from the *Logique*: "Démonstration par l'impossible. Ces sortes de démonstrations qui montrent qu'une chose est telle, non par ses principes, mais par quelque absurdité qui s'ensuivroit si elle étoit autrement, sont très-ordinaires dans Euclide. Cependant il est visible qu'elles peuvent convaincre l'esprit, mais qu'elles ne l'éclairent point, ce qui doit être le principal fruit de la science. Car notre esprit n'est point satisfait, s'il ne sait non seulement que la chose est, mais pourquoi elle est; ce qui ne s'apprend point par une démonstration qui réduit à l'impossible. Ce n'est pas que ces démonstrations soient tout-à-fait à rejetter ..." (p. 328). It is even possible that Arnauld was motivated to write the *Nouveaux Elémens de géometrie* at least in part because he hoped to expunge arguments *per absurdum* from classical elementary geometry. Let me note that Aristotelians also raised objections against *reductio* arguments. Their disapproval stemmed from a common interpretation of the concept of scientific demonstration in mathematics found in the *Posterior Analytics*: from this point of view, the issue belonged to the broader debate *de certitudine mathematicarum* that had animated the Italian Cinquecento. For a good discussion of the general state of this debate in the seventeenth century, see P. Mancosu, *On the Status of Proofs by Contradiction in the Seventeenth Century*, "Synthese", 88, 1991, pp. 15–41.

[62] Should we express the *reductio* and the *consequentia mirabilis* as inference rules, we get respectively:

$$\frac{\neg\alpha \to \bot}{\alpha} \quad \text{and} \quad \frac{\neg\alpha \to \alpha}{\alpha}.$$

These are extensionally equivalent, but if the inference ($\to$) is intended as an intensional operator expressing mathematical provability in classical sense (whatever this may mean), then the idea is that the passage ($\neg\alpha\to\alpha$) in *consequentia mirabilis* actually shows a proof of α. Thus, in concluding α through *consequentia*, we also have an explicit, direct proof of the statement, and therefore the conclusion is not indirect as in *reductio* arguments. At any rate, Saccheri regarded the *consequentia mirabilis* as a direct proof: cf. *Demonstrative Logic*, ed. 1701, p. 82 [ed. 1697, p. 130]. On the topic also see I. Angelelli, *Saccheri's Postulate*, "Vivarium", 33, 1995, pp. 98–111.

in mind its logical interpretation as an essential and foundational element of Aristotelian ontology. It is apparent that Saccheri recognized the first, grand employment of *consequentia mirabilis*, the archetype for every subsequent application of it, in the famous ἔλεγχος of the Fourth Book of Aristotle's *Metaphysics*, where a proof 'by redargution' establishes the Principle of Contradiction itself: should an opponent of the Principle set about to deny it, he would thereby be forced to accept it in his very act of denying, hence the Principle withstands the attacks of the skeptic.[63] Clearly, this reasoning leads us back to the very definition of an axiom, which can either be proved immediately on the basis of the mere meaning of the terms that figure in it (that is, of their nominal definitions), or be unprovable and, as it were, a condition of demonstration in general. And yet, even for this second kind of axiom, which Saccheri always illustrates with the Principle of Contradiction, a kind of demonstration is still possible: not an immediate proof, but an oblique (yet not indirect) proof that is unassailable. In a word, the sort of proof found in Aristotle's celebrated ἔλεγχος.

Following this approach, it does not matter how long our demonstration is, how many propositions have to be proved, how many diagrams drawn, or how many Euclidean lemmata, corollaries, scholia, arguments and theorems put to work. If we succeed in showing that anyone who denies the Fifth Postulate is in the end forced to admit it, and to derive it from the heart of his very hypothsis, then (at least in Saccheri's view) we have not only proved the Postulate, but have shown that it is a genuine axiom – an unhypothetical and undeniable proposition like the *principium contradictionis* itself. Just as Saccheri's epistemological convictions led him to distinguish two ways of proving an axiom as such (the immediate proof from the meaning of the terms, and the oblique proof by redargution), they also led him to believe that his proof of the Fifth Postulate had to proceed by *consequentia mirabilis*. We can now appreciate that this particular proof strategy was by no means motivated by stylistic concerns, nor by a predilection for the deductive tool so dear to the scientists and theologians of the Society of Jesus, nor by Saccheri's desire to pay belated homage to his own juvenile discussion in the *Demonstrative Logic*. On the contrary, the choice of the *consequentia mirabilis* was first and foremost motivated by an imperative epistemological need, which is to say, by Saccheri's need to preserve the axiomatic status of a proposition whose demonstration, as opposed to Clavius' proof, was meant to *vindicate* Euclid. Still, one quite significant fact cannot be denied: Saccheri accepts immediate demonstrations for all the other Euclidean axioms. He feels compelled to complete a proof by redargution of the parallel axiom alone, thereby endowing it with a decidedly special epistemological status among the foundations of geometry.

Since Saccheri does not take as his starting point a single negation of the entire Fifth Postulate, but rather divides the negation into two distinct hypotheses, that of the obtuse

[63] The locus of the Aristotelian discussion is, of course, *Metaph.* Γ 4. The source of Aristotle's argument may be a similar passage in *Protrepticus*: see W. Kneale, *Aristotle and the Consequentia Mirabilis*, "The Journal of Hellenic Studies", 77, 1957, pp. 62–6. It is certainly possible (but maybe unlikely) that Saccheri associated Aristotle's famous refutation with the equally well-known Cartesian claim about the evidence of the *cogito*, which also proceeds (in a broad sense) by redargution.

angle (elliptic geometry) and that of the acute angle (hyperbolic geometry), the task he undertakes is that of refuting via redargution both of these rival hypotheses. This twofold endeavor serves as the basis for the whole of Book One of *Euclid Vindicated*, which takes the form of a monumental *reductio* of these two hypotheses. Moreover, Saccheri aims not only to reveal that both of these non-Euclidean geometries yield contradictions, but also to prove, under each of the two hypotheses, the validity of Euclid's parallel postulate. It is in the gap between the declaration of intent and the actual (fallacious) redargutions that Saccheri builds all the vast and most beautiful constructions in elliptic and hyperbolic geometry, and it is in light of these that *Euclid Vindicated* is regarded to this day as a masterpiece of eighteenth-century geometry.

The first redargution takes place at Proposition 14 in the First Part of Book One and concerns elliptic geometry. At the end of the same part, at Proposition 33, Saccheri refutes the hypothesis of the acute angle with a simple *reductio ad absurdum*, that is to say, with reference to a contradiction that Saccheri believes he has derived from his theorems and the definition of a straight line. It is clear, however, that this proof *per absurdum* cannot accomplish the goal that Saccheri set for himself, as it falls short of a *consequentia mirabilis*. If Saccheri stopped there (and his proofs were correct), he would have established the Fifth Postulate as a theorem, but failed to vindicate Euclid. Therefore, Saccheri adds a whole Second Part to Book One of *Euclid Vindicated*, the stated aim of which is to prove hyperbolic geometry false for the second time, this time by showing that it leads to the truth of Euclidean geometry. Saccheri believes he has accomplished this at Proposition 39, which ends the Book devoted to the theory of parallels and concludes the demonstration that the Fifth Postulate is really an axiom (in the sense of the word discussed above).

Yet, if we analyze in more detail the deeper logical form of Propositions 14 and 39, we discover that none of them actually achieves a *consequentia mirabilis*. Saccheri must have run into more serious hurdles than he expected in the realization of his declared epistemological program.[64] From this viewpoint, *Euclid Vindicated* appears to be a complete failure not only with respect to its stated mathematical goals concerning the theory of the parallels – what appears obvious and inevitable from a modern standpoint – but even more importantly with respect to its primary logical aim, the justification of an axiom. The Second Part of Book One, in which Saccheri recasts his refutation of the acute angle hypothesis as a *consequentia mirabilis*, is notoriously weak, and Saccheri commits some blatant mathematical errors therein. Moreover, as we have already noted, this Second Part draws heavily on Clavius' research, and is likely to have been composed by Saccheri before the 'direct' refutation of hyperbolic geometry that he accomplishes in Proposition 33 of the First Part and that is far more sophisticated. The weakness of the proof of Proposition 39 of the Second Part led some interpreters to hypothesize that Saccheri was aware of not having proved the Fifth Postulate and consequently delayed the publication of his book because he was not satisfied with his own results. Other interpreters even suggest that the Jesuit was by now uncertain as to very provability of the Fifth Postulate, and tell the (rather fanciful) story of Saccheri's conversion

---

[64] The question will receive a detailed treatment in the *Notes*.

to non-Euclidean geometry. I am of the opinion that Saccheri could not be satisfied with the shaky conclusion of Proposition 39 – if he had been, he would have suppressed the by then useless, and logically weaker, Proposition 33; but he must have been convinced of his Propositions 14 (refutation of elliptic geometry by *consequentia mirabilis*) and 33 (refutation of hyperbolic geometry by simple *reductio*), about which he never betrays any doubt or hesitation. Thus, if Saccheri died dissatisfied with his own work, it was not because he doubted the validity of his proof of the Fifth Postulate, but rather because he worried that the task may not have been achieved via perfect redargution – a failing that would have made the Jesuit Euclid's unwitting adversary instead of his willing vindicator.

6. Results and errors of Euclid Vindicated

It will be fruitful to take a closer look at Saccheri's three pseudo-refutations, that of elliptic geometry and the two of hyperbolic geometry, since the failure of the method of redargution can shed some light on the significance of the – more momentous – failure of Saccheri's mathematical proofs in general.[65]

Any refutation of elliptic geometry, the one appearing in the work of Saccheri (in Proposition 14 of *Euclid Vindicated*) as well as those by the authors immediately following him, always hinges on some strategic application of the Exterior Angle Theorem (*Elements* I, 16), the first theorem in which Euclid either implicitly or explicitly, but in an essential way, makes use of the property of straight lines of being unbounded.[66] The task of justifying this refutation hence amounts to that of determining whether the unboundedness of straight lines is explicit in classical Euclidean axiomatics. If it is not, *Elements* I, 16, is guilty of a paralogism, along with all of the refutations of elliptic geometry that depend on it (Saccheri's among them), as has been maintained. The problem is that Euclid's Second Postulate, which states that a straight line may be extended indefinitely, is ambiguous: it can be read as asserting that straight lines are complete and unbounded (be they either infinite or closed), or

[65] I do not aspire to present all the results of *Euclid Vindicated* in the field of non-Euclidean geometries here: I provide surveys of specific results in the *Notes* to individual propositions or to the bigger sections of the work. Several reliable summaries of Saccheri's main theorems are available, and some of these offer an overview of the Jesuit's contributions to hyperbolic geometry. The first such study to appear – after the essay of Beltrami, who first rediscovered *Euclid Vindicated* – was the one by M.P. Mansion, *Analyse des recherches du R.P. Saccheri, S.J. sur le postulatum d'Euclide*, "Annales de la société scientifique de Bruxelles", 14.2, 1889–1890, pp. 46–59; soon followed by the superior treatment in Bonola, *La geometria non-euclidea*, pp. 18–27 (English ed., pp. 22–44). There are a good number of such surveys today: the English reader may see for instance J. Gray, *Ideas of Space: Euclidean, Non-Euclidean, and Relativistic*, Oxford, Clarendon 1989[2], pp. 60–69; R. Hartshorne, *Geometry: Euclid and Beyond*, New York, Springer 2000, pp. 306–11; M.J. Greenberg, *Euclidean and Non-Euclidean Geometries. Development and History*, New York, Freeman 2008[4], pp. 176–207, 218–9.

[66] Further complications arise from the assumption of Archimedes' Axiom in this connection, as well as from Saccheri's commitment to restricting his use of *Elements* I, 16 in the proofs in *Euclid Vindicated*. I will discuss the technical facets of this problem in the *Notes*.

as straightforwardly asserting that they are infinite (i. e., lines of infinite measure). Without a doubt, all geometers of the Modern Age, including Saccheri, interpret the postulate the second way. This is manifest, for instance, in Clavius' 'proof' of the principle in question, which, however, is too tautological to have much explicative value.[67] Thus, these geometers' mistake seems only to consist in carelessly assuming a postulate as true without noticing that a highly interesting geometry could be developed from its negation – or rather, in all historical fairness, the error consisted in their blind confidence in a proof intended to show the Second Postulate's absolute logical necessity.[68]

Granted the interpretation of the Second Postulate that states that straight lines are unbounded, the proof of the impossibility of elliptic geometry is unquestionably correct.[69] Indeed, this supposed impossibility depends *solely* on the Second Postulate: the parallel axiom in Euclid's formulation is perfectly consistent with a metric of positive curvature, and the other axioms with which later commentators intended to supplement the text of the *Elements*, such as the axiom holding that through two points only one straight line can pass, were by themselves insufficient to conclusively rule out spaces of positive curvature. In fact, it is widely known that a projective plane with a metric of constant positive curvature (which is the model of the so-called simple elliptic geometry) satisfies all the axioms of classical and Renaissance geometry except the Second Postulate as it is interpreted above. The fact that such a model displays strange properties, foreign to Saccheri's epoch (for one, it is not orientable), would not have represented a hindrance to an axiomatic treatment of elliptic geometry – both because global properties went completely unnoticed at the time (and thus could not possibly disconcert a geometer) and because the very concept of a model was altogether missing from the Early Modern geometrical mindset. Be that as it may be, nowhere in *Euclid Vindicated* does Saccheri try to use the principle that only one straight line passes through two points to prove the impossibility of spherical geometry, despite his

[67] Clavius, *Euclidis*, p. 22. Note, however, that an interpretation of unboundedness in the sense of "closed" was already hinted by Simplicius *apud* An-Nayrīzī, in commenting *Elements* I, def. 3. See Tummers, *The Latin Translation of Anaritius*, p. 4.

[68] Geometers were of course well aware that elliptic geometry is fertile and interesting, and treatises on spherical geometry have existed since Classical Antiquity. What mathematicians before the late eighteenth century found particularly troublesome was the task of providing an axiomatization of spherical geometry independent from the system put forward in the *Elements* – namely, an axiomatization that no longer views the sphere as a surface embedded in a Euclidean space. The reason for this difficulty, of course, lies in the geometrical logicism that is implicit in all these early researches. It is evident, at any rate, that the elaboration of a geometry where the Second Postulate does not hold, an idea ordinarily attributed to Riemann, was a direct product of investigations into a geometry where the Fifth Postulate does not hold (this is also true of all other non-classical geometrical systems). As for Riemann's celebrated *Habilitationsschrift* from 1855, see the recent annotated edition, B. Riemann, *Über die Hypothesen, welche der Geometrie zu Grunde liegen*, ed. by Jürgen Jost, Berlin, Springer 2013.

[69] The present-day equivalent of this proof is a simple and direct consequence (for instance) of the Bonnet Theorem, stating that if the curvature of a manifold M is everywhere lower bounded by some fixed $\varepsilon > 0$, then M is compact.

recurrent use of that principle elsewhere: therefore, if we wanted to provide a present-day model for his 'hypothesis of the obtuse angle', it would probably be the projective one. As a consequence, Saccheri was unable to explicitly reconnect this hypothesis to the model of a sphere, or to acknowledge that the geometry of the obtuse angle has in fact a (local) realization in a sphere embedded in Euclidean space.[70]

The issue of the dependence of the elliptic hypothesis on the Second Postulate is important in that it shows why investigations on spherical geometry should not have, in principle, any role in the research on the parallel axiom. In fact, a long and important line of attempts to prove the Fifth Postulate were never concerned with spaces of positive curvature. For instance, a discussion of spherical geometry is notably absent from the work of Legendre, who famously proved, independently from the Fifth Postulate, that the sum of internal angles in a triangle is less than or equal to two right angles – that is, space is either Euclidean or hyperbolic. Spherical geometry is equally absent from the works of Lobachevsky and Bolyai, who consistently discussed the possibility of *two* geometries, Euclidean and non-Euclidean (hyperbolic), depending on whether the Fifth Postulate or its negation holds. In this regard, Saccheri is part of an altogether different intellectual genealogy: his proof is rooted in a tradition of investigations of the parallel axiom that do not take a direct route to their object – that is, a route that begins with a discussion of the intersection of straight lines. Rather, they follow a more serpentine path, beginning with a *primitive* consideration of the sum of the angles of a triangle or quadrilateral. In a word, Saccheri belongs to the line of mathematicians descended from Khayyām, a line that includes Clavius and Vitale Giordano and which remained unknown to – or overlooked by – the age after Saccheri. However, the scant remarks on angular excess and defect of quadrilaterals found in the works of Saccheri's predecessors develop, in *Euclid Vindicated*, into a grand system of *three* (not just two) hypotheses, which are dealt with and expanded on for countless pages. It is important to stress that Saccheri displays a genuine geometrical interest in the development of these three hypotheses, which is not just a function of his ultimate goal of proving the Fifth Postulate. To this (quite limited) extent, he thus betrays a sort of unconfessed inclination towards the formal foundation of the non-Euclidean geometries.[71]

[70] The first mathematician to reconnect the hypothesis of the obtuse angle to the geometry on a sphere was Lambert. This advancement was especially due to Lambert's studies on spherical trigonometry (a "modern" field of research that Saccheri do not employ in *Euclid Vindicated*), which in turn mostly relied on Euler's *Principes de la trigonométrie sphérique* from 1755 (now in L. Euler, *Opera omnia*, vol. I, 27, pp. 309–39).

[71] For instance, when proving Propositions 15, 16, 18, and 19, Saccheri takes into account the obtuse angle hypothesis as well, although he refuted this same hypothesis as early as Proposition 14. Unless one wants to maintain that editorial haste alone accounts for this redundancy, and for the insertion of those theorems in a position in which they prove superfluous, one has to admit that Saccheri retained an interest in elliptic geometry even after having shown its impossibility. The same can be said (*a fortiori*) of hyperbolic geometry: the most advanced theorems in Saccheri's whole work (Propositions 29–32 on asymptotic hyperbolic straight lines) are never employed in his two attempts to refute the hypothesis of the acute angle.

The crucial point, at any rate, is that Saccheri was the first thinker to regard the discussion of elliptic geometry as essentially bound up with the issue of parallelism, and did it systematically. This approach culminates and is (allegedly) justified precisely in Proposition 14, where Saccheri intends to refute the hypothesis of the obtuse angle by *consequentia mirabilis*. It is with the aim of constructing a redargutive proof that Saccheri tries to move from the obtuse angle hypothesis to the Fifth Postulate, and from there to the truth of Euclidean geometry. As a result, the obtuse angle hypothesis would destroy itself. The proof of Proposition 14 is perfectly correct given the axiomatic premises of *Euclid Vindicated*, and still it does not really, nor could ever consist in a genuine redargution: in the course of the proof, Saccheri makes a crucial use of the usual *Elements* I, 16, whereas his many references to the Fifth Postulate are unnecessary and ultimately irrelevant. To this extent, it can be maintained that Legendre's theorem on the angular defect of triangles is already present in Saccheri's work, so that it is appropriate to credit the Jesuit with its discovery; on the other hand, however, Saccheri evidently failed to grasp its real significance, that is, its complete independence from the theory of parallels.

It is quite interesting, however, that the error committed by Saccheri in his refutation of the hypothesis of the obtuse angle, which is not mathematical but rather epistemological in nature (it amounts to a confusion as to which principles he is really employing), ultimately provides the reason for his inclusion of elliptic geometry among the various geometrical possibilities that are opened up by the discussion of parallelism. This idea was definitely mistaken, in the framework Euclid's axiomatic approach, and Legendre and Lobachevsky were right in ignoring the issue of spherical geometry. However, this error proved seminal to later treatments of the parallel problem, which dealt with the issue in analytic terms of curvature, rather than in terms of synthetic axiomatics. Of course, a connection of the notion of parallelism with that of metric required very complex mathematical innovations that would arise only after Gauss, in Riemann's work and in the subsequent research on parallel transport and affine connections. There is no doubt, however, that the starting point for these discussions was the consideration of the angular defect of a quadrilateral (appropriately re-read with the Calculus). In a word: future progress with respect to the theory of parallelism was achieved via Saccheri's tripartite, rather than Legendre's bipartite, route, and the Jesuit certainly deserves credit for revealing for the first time that the difficulties associated with the Fifth Postulate also pointed towards other mathematical horizons.[72] That Saccheri should come to show this only thanks to an mistake, or rather, only thanks to his classicist conviction that a refutation of the obtuse angle hypothesis required the *consequentia mirabilis*, cannot but strengthen his reputation as a man of genius *malgré lui*.

[72] If we reverse the perspective adopted above, we may claim that Saccheri's work hints at the insight that Euclid's Second Postulate is independent from the Fifth Postulate, but it is not independent from a full-fledged theory of parallelism – and in this consists the deep meaning of the above-mentioned Bonnet theorem. Let me note that Lambert refutes the obtuse angle hypothesis (elliptic geometry) without employing *Elements* I, 16. Instead, he *directly* applies the principle holding that straight lines are unbounded (*Theorie der Parallellinien*, §§ 62–64, pp. 341–3): this tactic reveals a greater awareness on Lambert's part.

The influence of philosophy and epistemology on mathematical research is also apparent, to a lesser degree, in Saccheri's other two famous refutations, which deal with hyperbolic geometry. The deductive chain ending with Proposition 33, which is supposed to disprove the acute angle hypothesis with a simple *reductio ad absurdum*, does not in itself contain any significant mathematical error. On the contrary, that argument constitutes, as a whole, that First Part of Book One of *Euclid Vindicated* that enjoys the highest respect among scholars, as it offers a consistent and rather advanced picture of the principal elementary properties of hyperbolic geodesics; it represents, in fact, Saccheri's mathematical masterpiece. After erecting this splendid construction of non-Euclidean geometry, Saccheri arrives at the contradiction he seeks, via the simple, unproven *assertion* that the results he has obtained so far are repugnant to the 'nature' of straight lines. This 'nature', according to Saccheri's demonstrations, is such that two straight lines cannot be tangent at one point (short of complete coincidence). Because there are certain hyperbolic straight lines that are (loosely speaking) tangent at a point at infinity, Saccheri concludes that hyperbolic geometry is untenable. Evidently, the error resides in Saccheri's attribution of a property that holds only for the finite (and that Saccheri claims to have established for that domain) to a point at infinity. That is to say, the error consists in understanding the point at infinity as an ordinary mathematical object, rather than the limit of a convergence. It should be stressed, however, that there are no other proofs in *Euclid Vindicated* in which Saccheri handles the points at infinity so carelessly – nor does Proposition 33 contain any discussion of the subject, nor any actual mathematical fallacy in its demonstration. Saccheri's wanted conclusion is claimed in a short philosophical, or metaphysical passage, that attempts to disprove hyperbolic geometry with a single sentence lacking any preparation or further conclusion. Ultimately, this refutation has the force of a particularly emphatic 'No!', an exclamation for which no argument is given. As such, it was received by Saccheri's readers, whom it could not convince for an instant.

At least, Saccheri's quasi-metaphysical appeal to the infinite has the advantage of sparing him the shame attendant on real mathematical error, like that he ultimately makes in the Second Part of Book One of *Euclid Vindicated*, where he endeavors to prove the truth of the Fifth Postulate via redargution of the acute angle hypothesis. Once again, the proof fails to take the genuine form of a *consequentia mirabilis*, of which Saccheri must have been aware, since he devotes a lengthy Scholium to convincing the reader (and perhaps himself) of the contrary, with little success. Saccheri's fervid zeal in this Scholium is motivated by the central role of the *consequentia mirabilis* in his epistemological program, as we have seen. His failure, on the other hand, is due to the fact that this second proof is not likely to have been originally constructed with an eye to providing the sort of formal redargution that Saccheri desired at this point in his argument. Rather, it represents Saccheri's attempt to modify Clavius' and Giordano's proofs to his satisfaction, an attempt which he hastily decided to insert in the book as a means of legitimating his own ends.

In this second demonstration, at any rate, the error is real, and it is very significant (it occurs at Proposition 37). Once again, the problem lies in Saccheri's poor grasp of passages to the infinite – in this case, of a limit procedure in which he uses infinitesimals (which he also treats as indivisibles) to prove a completely false result. Saccheri reaches his conclusions by way of an incredible argumentative muddle that entails, more or less, that all the curves between two given extremes have equal length. Some interpreters have generously tried to argue that Saccheri's reasoning, though seriously misguided, can be explained by the lack of rigorous foundations (at the Age) for the Calculus and for handling infinitesimals in general. Yet a closer look reveals that Saccheri's error is far too grave to be ascribed to the confusion of an epoch. Rather, it seems undeniable that not only any skilled analyst in 1733, but even Cavalieri a century before, who was working with even less subtle theoretical instruments, would have immediately detected the error. The lack of formal rigor in Calculus was perhaps a foundational problem, but it did not present a real issue for the mathematician – at least, not at this elementary level. Most mathematicians were perfectly capable of distinguishing a correct argument concerning infinitesimals from a faulty one. In fact, Saccheri's second attempt at a refutation resulted as unconvincing as his first, and the first writer to comment on it, thirty years after its publication (Simon Klügel, whom we will discuss later), not only recognized it immediately but also rebuked its author for his naiveté.

In conclusion, the error in Proposition 37 is so serious that I believe that only some biographical or textual incident could possibly account for it – though this is admittedly a poor explanation.[73] The error, however, is deeply rooted in Saccheri's general attitude towards the Calculus and towards symbolic and algebraic manipulation at large, which is of some interest to the historian and is therefore worth mentioning here.

In fact, given that the goals of *Euclid Vindicated* were hyper-classicistic and conservative, it comes as no surprise that the *methods* employed by Saccheri were equally classicistic and conservative. On the one hand, he hoped to restore the text of the *Elements* in its original formulation – so much so that he set out to prove Euclid's axiom only after preparing a precautionary epistemological apparatus capable of showing that the proved proposition was nonetheless an axiom, as in Euclid's text. On the other hand, it was in the same spirit that he relied exclusively on the synthetic geometry of Greek mathematics, even in a century in which algebra and the Calculus were celebrating their greatest triumphs.

To be sure, this route was also taken by some other (great and celebrated) mathematicians of the age, who maintained that the modern algebraic and analytic techniques could not produce elegant and well-formed mathematical reasoning, although they were perhaps suitable as an *ars inveniendi*. Saccheri does not count among these scholars, as his knowledge of algebra was solid but modest and his familiarity with Calculus nonexistent. Consequently, none of the results in *Euclid Vindicated* go beyond elementary synthetic geometry in terms

[73] The same holds for another demonstration concerning proportions, which appears in Book Two of *Euclid Vindicated* and which interpreters also recognized as patently fallacious, around the same years as Klügel's criticisms. I will discuss this briefly in the *Notes*.

of either their technique or outcomes.[74] It may well be that the roots of this methodological choice are to be found both in Saccheri's personal mathematical abilities and in the generally underdeveloped state of Italian mathematics in the early eighteenth century. Be as that may be, Saccheri chose to develop his incompetence into a system, hence his explicit theorizing about the superiority of classical methods over their algebraic and analytic rivals. Statements to this effect are virtually a constant in his works. In his oldest mathematical text, the *Quaesita Geometrica* from 1693, Saccheri declares that solving the problems raised in the work with the resources of algebra would have been undemanding: his booklet is important precisely because it shows that these questions can be explained in a synthetic fashion, and that this is in fact the only difficulty and merit of the work.[75] Forty years later, in *Euclid Vindicated*, Saccheri still feels compelled to end the book with an Appendix where he explains that the symbolic manipulation characteristic of algebra is blind to the Euclidean structure of geometry. In other words, algebra is indifferent to the truth of the Fifth Postulate, and while this represents its strength from a present-day perspective, for Saccheri it means that algebra remains in want of a foundation – unless it is supplemented by synthetic geometry, which determines the true structure of spatial relations.

It is undeniable that such methodological classicism represents both one of the salient features and one of the most serious weaknesses of *Euclid Vindicated*, all the more as it could easily have been avoided in 1733, a time when the evolution of European mathematics had moved so far beyond the study of compass-and-straightedge constructions. The book's anachronistic method accounts not only for its errors but also for its failure to fulfill its potential. If we consider the future evolution of the subject matter of Book One, that is, the development of eighteenth- and nineteenth-century research on parallelism, we will notice that such studies are so important precisely in virtue of their algebraic

---

[74] There are other places where Saccheri resorts to passages to the limit – indeed, it is difficult to explore hyperbolic geometry in any depth without them. The most sophisticated geometric results in *Euclid Vindicated* concern asymptotic hyperbolic straight lines, and Saccheri is forced to manipulate the concepts of convergence and limit in the course of these more advanced studies. He insists, however, on dealing with such notions in a classical synthetic manner. The fact that he does not make any big mistakes is thus a sign that the mathematical common sense that had prevented so many analysts before Cauchy and Weierstrass from committing errors did not fail him either. His confused terminology and shaky arguments, however, betray his very imperfect knowledge of analytic tools.

[75] Truth to be told, on the second page of the Preface to *Quaesita Geometrica*, Saccheri affirms that he took this synthetic route mostly because Ventimiglia had formulated in a classical form the geometric problems that he undertakes to solve: "Neque enim reor satis esse ad investigationem eius loci, more Carthesij genus aequationis (quod facile esset) expendere; sed aut lineas omnes praecipui nominis singillatim excludere aut aliquam ex ijs certam statuere, & quidem geometricè, iuxta mentem proponentis necessarium existimo". It should also be noted that Viviani took a similar position in his 1676 *Diporto Geometrico*, with great success. Viviani later attempted to embark upon the same synthetic and classical route in his 1692 *Aenigma geometricum*. The latter treatise, however, had to face Leibniz's infinitesimal methods in solving one and the same problem, and comparing the results no one could have any doubt that the elderly Viviani had finally been defeated by modern times.

and quantitative dimension – the very dimension that Saccheri so thoroughly expunged from his classicist treatise. Lambert's research surpasses *Euclid Vindicated* only in its use of the calculation of hyperbolic trigonometric functions, that is, in its employment of quantitative (rather than qualitative) estimation of spatial curvature. Similarly, one of the most important parts of Lobachevsky's and Bolyai's later studies, for which they are held in such great esteem that they are often regarded as the real founders of non-Euclidean geometries, is the formula for calculating the angle of parallelism. Now, the concept of an angle of parallelism is clearly formulated in Saccheri's work, but the Jesuit never deigns to associate it with an analytic function, nor is he in a position to do so. The same point can be made with respect to the – even more consequential – Gaussian concept of curvature. In sum, the only route that could lead beyond *Euclid Vindicated* was the route that our Jesuit's classicist whims did not allow him to take: that of giving his synthetic results an algebraic interpretation.[76]

Saccheri's resistance to algebra is especially deleterious in Book Two of *Euclid Vindicated*: new developments in algebra and analysis rendered the classical Euclidean theory of proportions useless, and, although the Galilean school had considerably reformed the theory, the traditional approach remained completely unserviceable when it came to solving the most intricate problems posed by eighteenth-century mechanics. Thus, in this case, Saccheri's choice of an anti-algebraic method ends up, so to speak, determining the contents themselves, because in those years the Euclidean theory of proportions was metamorphosing completely into modern Calculus. This makes Saccheri's failure complete, as there was no point in reviving the classical theory of ratios in an outdated form that was – at best – adequate to Galileo's mechanical theory: Saccheri's construction could perhaps have an archaeologic interest, but was of no consequence to Modern mathematicians, who relied on algebra and infinitesimals. As we have already mentioned, the classical, non-algebraic theory of proportions had been virtually abandoned in France and England by the second half of the seventeenth century. It persisted mainly as an antiquarian curiosity in Italy, where it was explored in the works of Galileo's and Borelli's epigones. To ignore developments in the field of the Calculus at the turn of the new century was therefore either to commit oneself to a severe methodology – or just sheer foolishness. To employ Euclidean synthetic methods in a discussion of proportions as late as 1733, as Saccheri did, appeared without a doubt as a baffling oddity, and it is understandable that European mathematicians looked at *Euclid Vindicated* as a strange relic or a curiosity for a *Wunderkammer*.

---

[76] Saccheri's stance can be contrasted with the position later held by Lobachevsky, who nonchalantly maintained that his *pangeometry* (Euclidean and non-Euclidean) was a branch of the Calculus. On the application of infinitesimal analysis to the theory of parallelism as the real turning point in the history of non-Euclidean geometry, see J. Gray, *Non-Euclidean Geometry – A Re-interpretation*, "Historia mathematica" 6, 1979, pp. 236–58.

7. The reception of Euclid Vindicated

Since *Euclid Vindicated* was in many ways outdated at the time of its publication, it is easy to understand why it had to be so poorly received.

Eighteenth-century mathematical research in analysis, algebra, and mechanics was rapidly moving in new and promising directions, and scholars were loath to interrupt its progress to scrutinize the foundations of elementary geometry. Modern mathematicians even laid aside some of the central foundational issues in the Calculus, which they did not regard as immediately pressing. In nations where mathematical research was more advanced, especially France, only a few geometers considered an investigation of the Fifth Postulate or the theory of proportions worth pursuing. To be sure, some of the greatest French mathematicians devoted some time to these issues, but the works in which their investigations on the subject appear are marginal and often even unpublished (Lagrange, Fourier).[77] Moreover, in the long run, the approach to teaching inaugurated by the Port-Royal geometers and the seventeenth-century Jesuits won out over the rigorous style for which Saccheri had pleaded. As a consequence, the degree of demonstrative rigor found in textbooks of synthetic geometry was considerably relaxed: the handbooks authored by Tacquet and Dechales were substituted with the even simpler ones authored by Clairaut.[78] The only significant exception to this trend was Legendre, who was charged with the task of writing a new handbook of geometry for the age of the Revolution, and who used this undertaking as an opportunity to revisit the problem of parallelism from the foundations. His various attempts at a proof do not, however, indicate that he had any acquaintance with the work of Saccheri, whom he never mentions. It is thus completely reasonable to suppose that the sixty years intervening between the publication of *Euclid Vindicated* and the Élé-

[77] The attempt of Lagrange (1736–1813) at proving the Fifth Postulate dates back to 1806, but was first published in M.T. Borgato, L. Pepe, *Una memoria inedita di Lagrange sulla teoria delle parallele*, "Bollettino di Storia delle Scienze Matematiche", 8, 1988, pp. 307–335. It had already been mentioned and discussed two years before by Jean-Claude Pont, who was also responsible for the appearance of Fourier's unpublished writings on the subject (see below). On Lagrange's attempt, see also J.V. Grabiner, *Why Did Lagrange "Prove" the Parallel Postulate?*, "The American Mathematical Monthly", 116, 2009, p. 3–18; V. Pambuccian, *On the Equivalence of Lagrange's Axiom to the Lotschnittaxiom*, "Journal of Geometry", 95, 2009, pp. 165–71.

[78] Indeed, the successful Éléments de géométrie (Paris, Lambert&Durand 1741) written by the mathematician and astronomer Alexis Claude Clairaut (1713–1765) is more of a handbook of practical geometry than an edition of the *Elements*. It often leaves out the definitions of the terms it employs; demonstrations of main propositions are also often omitted, as the author relies on their immediately evident character. The Port-Royal epistemology of clear and distinct ideas has taken such a radical form in Clairaut's book that, if we were to read it as a genuine textbook of pure mathematics, we would have to regard it as in fact accomplishing a *reductio ad absurdum* of its epistemological assumptions. At any rate, it is interesting that Clairaut seems to reduce the Fifth Postulate to the assumption of the existence of a rectangle (*Éléments*, pp. 10–11), which essentially corresponds to Saccheri's hypothesis of the right angle. Nonetheless, there is no reason to suppose that Clairaut had read *Euclid Vindicated*.

ments de géométrie were sufficient to consign to oblivion the former work, which Legendre would have found highly interesting.[79]

Scientific studies in elementary geometry continued to thrive in some mathematically backward states, universities and intellectual circles, which still regarded Euclid's *Elements* as a scientifically fruitful text. It is undoubtedly due to the peculiar state of research on the foundations of mathematics that non-Euclidean geometries first bloomed in Russia (with Lobachevsky), in Hungary (with the two Bolyai's), then in Germany. This last case is the most interesting for our purposes, since questions concerning the Fifth Postulate were so widely discussed in Germany that by the end of the century it was hard to keep track of the vast literature on the subject: it is difficult to locate a German mathematician who did not write at least an article, if not entire volumes, on the proof or the unprovability of the famous parallel axiom. Indeed, the few works on which Saccheri's ideas exerted some influence all belong to the German tradition (be this the cause or, more likely, the effect of the German interest in parallel theory).[80] The failure of *Euclid Vindicated* was so great that the text went unmentioned even in Italy, although Saccheri's country saw a marked increase in the number of editions of the *Elements* printed – and with them, a marked increase in the

[79] Adrien-Marie Legendre (1752–1833) worked in diverse fields of mathematics and nourished an especial interest in the problems of elementary geometry. His *Éléments de géometrie* (Paris, Didot 1794) saw countless reissues and enjoyed great success for about a century. In the work, Legendre attempted to arrive at a proof of the Fifth Postulate by many different routes: in his first edition, he built on Wallis' proof; in the second, on the principle that there is always one circle passing through three points, which is equivalent to the Fifth Postulate; in the third through the eighth editions, he devised an original proof that ran into difficulties because of an illicit transition to the limit; from the ninth through the eleventh, somewhat disheartened, he reinstated the proposition as a postulate; but from the twelfth onwards, he attempted a different analytic proof, which also contained a fallacy concerning the convergence of a limit. His attempts, along with other material, were collected in the year of his death in A.-M. Legendre, *Réflexions sur différentes manières de démontrer la théorie des parallèles ou le théorème sur la somme des trois angles du triangle*, "Mémoires de l'Académie Royale des Sciences de l'Institut de France", 12, 1833, pp. 367–410. We should remark that in the twelfth edition of the *Éléments* (1823) Legendre concludes his proof of the Fifth Postulate stating that a certain consequence of its denial "répugne à la nature de la ligne droite" (p. 279), which is the very expression employed by Saccheri in Proposition 33. As Legendre's proof is completely different from Saccheri's, however, it doesn't seem that this phrasing should be taken as an evidence of a direct reading.

[80] L. Allegri, *The Mathematical Works of Girolamo Saccheri, S.J. (1667–1733)*, Ph.D. diss., Columbia University, 1960, p. 160, also mentions some documents belonging to the University of Pavia, where Saccheri taught. These documents demonstrate that a significant contingent of the University's students came from Germany, and that academic relations between that university and those in German-speaking regions were quite close.

number of attempts at proving the Fifth Postulate. Virtually none of these even mentions *Euclid Vindicated*.[81]

Let us then briefly consider the history of the reception of Saccheri's work. Right after its publication, *Euclid Vindicated* was mentioned in several scientific journals across Europe. This was, however, not particularly noteworthy, as almost any work received this treatment, and it did not necessarily lead to recognition or widespread circulation. Strange as it may seem, the altogether marginal reference to Wolff, which appears in Saccheri's discussion of proportions, was possibly the main reason behind the eighteen-century diffusion of the book. *Euclid Vindicated* received its first review (not a mere mention) in the *Nova acta eruditorum* of 1736. This was a simple summary of the work's contents, and dealt mainly with Saccheri's Preface, not even taking a stance as to the correctness of the proofs in question: nonetheless, it was a start.[82] Some years later, in 1741, Wolff himself mentioned *Euclid Vindicated* (most likely without having read the parts of the work not discussing his theories) in the fifth volume of his *Elementa Matheseos universae*,[83] which was widely circulated in Germany. This was probably the beginning of the (meager, but still observable) critical reaction to Saccheri's work in that nation. Wolff was, however, the least suitable scholar to discuss Saccheri's work, since he maintained that the Fifth Postulate could be easily proved from the notion of equidistance. *Euclid Vindicated* had unambiguously condemned this proof as a fallacy, but Wolff did not take notice of this refutation in later editions of the *Elementa Matheseos*.

In the following years, *Euclid Vindicated* received citations in two of the earliest histories of mathematics: it is mentioned, though not discussed, in the books by Heilbronner (1742) and Montucla (1758).[84] The first mathematical discussion of the contents of Saccheri's book – extensive, acute, and harshly critical – was to appear in the very successful edition of

---

[81] Here we could mention G.M. Pagnini, *Theoria rectarum parallelarum ab omni scrupolo vindicata*, Parma, Rossi&Ubaldi 1783: in spite of this title, the author is not acquainted with *Euclid Vindicated*, and he commits the very errors that Saccheri appropriately criticized. Pagnini explicitly mentions Saccheri in a 1794 supplement to this work, without any benefit to the solidity of his proofs, however. We could also point to the following essays: G. Venturi, *Memoria intorno alle linee parallele*, Modena, Società Tipografica 1784; F.M. Franceschinis, *Teoria delle parallele rigorosamente dimostrata*, Bassano, Remondini 1787; G. Saladini, *Nuovo trattato delle parallele*, Bologna, S. Tommaso 1805; none of these authors read Saccheri. Nineteenth-century Italian books and some minor foreign works mentioning Saccheri are listed in Pascal, *Girolamo Saccheri*, pp. 250–51.

[82] *Nova acta eruditorum*, 1736, pp. 277–81.

[83] *De praecipuis scriptis mathematicis brevis commentatio* in C. Wolff, *Elementa Matheseos Universae*, vol. 5, p. 28.

[84] Histories of mathematics had already been written in the previous centuries, and Dechales' edition of the *Elements*, for instance, opened with a massive introduction on the history of geometry: *De Progressu Matheseos & illustribus mathematicis*, in *Cursus*, vol. 1, pp. 1–108. However, J.C. Heilbronner, *Historia Matheseos universae*, Leipzig, Gleditsch 1742 and E. Montucla, *L'Histoire des Mathématiques*, Paris, Jombert 1758 were among the first monographic works on the topic. Heilbronner's source for the mention of Saccheri was probably Wolff himself.

the *Elements* by Simson (1756). Simson, however, does not consider Saccheri's important theory of parallels, but concentrates all his critical energy on examining Saccheri's theory of proportions, which he regards as seriously flawed – and not without reason.[85]

Meanwhile, in the second half of the eighteenth century, Abraham Kästner was establishing his reputation as the best teacher of mathematics in Germany,[86] and he was indeed a far better geometer than Wolff. Kästner regarded the Fifth Postulate as indemonstrable, and encouraged his pupil Simon Klügel to write a dissertation reviewing all the (failed) attempts to prove the axiom.[87] Klügel's *Recensio* from 1763 represented a turning point for *Euclid Vindicated*, which was the work most extensively discussed therein. Klügel provides a summary of the general structure and an abridgement of several theorems from Saccheri's work, then proceeds to level all sorts of objections against it. The mathematical objections are correct, and succeed in uncovering the fallacies in Saccheri's arguments. The epistemological objections, on the other hand (that one should never consider a proposition requiring so painstaking a proof to be an axiom; and one should rather rely on the axiom's clear evidence rather than on such a labyrinthine proof) are somewhat naive, though natural, and they betray Klügel's lack of familiarity with the *Demonstrative Logic*. Most importantly, Klügel laments that Saccheri devotes so much time in proving theorems in the acute angle hypothesis – that is, in hyperbolic geometry – which are not strictly necessary for the Jesuit's conclusions: a staunch Euclidean, Klügel thus overlooked what we (along with Saccheri, perhaps) view as the most beautiful fruits of *Euclid Vindicated*.

In any case, it was Klügel's dissertation that precipitated the general discussion of the parallel axiom in Germany, a discussion that would yield legions of essays and inconclusive demonstrations over the next fifty years, and that would eventually culminate in Gauss'

---

[85] Robert Simson (1687–1768) was a Scottish mathematician who devoted his whole life to editing and commenting on numerous Classical mathematical texts. His work enjoyed significant success. The volume in question is R. SIMSON, *Euclidis Elementorum libri priores sex item undecimus et duodecimus ex versione latina Federici Commandini*, Glasgow, Foulis 1756, with an English version appearing in the same year and with the same publisher. It enjoyed a particularly wide circulation in Germany, where Simson's discussion of proportions, containing an explicit reference to Saccheri, was restated by C.F. PFLEIDERER, *Deduction der Euclidischen Definitionen 3,4,5,7 des fünftes Buchs der Elemente*, "Archiv für reine und angewandte Mathematik", 2, 1797–1798, pp. 257–87 and 440–47. We should note that Simson almost completely ignores Saccheri's theory on the parallels, and even believes he can prove the Fifth Postulate (pp. 345–7) on the basis of Proclus' ancient proof, though *Euclid Vindicated* contains a convincing refutation of the latter. This proves that Simson had not seriously studied Saccheri's text.

[86] Abraham Gotthelf Kästner (1719–1800) was an esteemed mathematician and poet, but primarily a great scientific popularizer and organizer. The extraordinary vitality of the mathematical faculty at the university of Göttingen is due in large part to his efforts. On Kästner's stance on the Fifth Postulate, cf. W.S. PETERS, *Das Parallelenproblem bei A. G. Kästner*, "Archive for History of Exact Sciences", 1, 1962, pp. 480–87.

[87] G.S. KLÜGEL, *Conatuum praecipuorum theoriam parallelarum demonstrandi recensio*, Göttingen, Schultz 1763. Georg Simon Klügel (1739–1812), the author of a famous mathematical dictionary, enjoyed particular renown in Germany in his time.

work. It would be useless to discuss all these minor works, which sometimes mention Saccheri and his theorems but tend to rely on Klügel's interpretations thereof.[88]

We should, however, at least consider that Lambert had certainly read the *Recensio*, on which he depended considerably in his 1766 *Theorie der Parallellinien* – a work that restates the distinction between the three hypotheses of the right, the obtuse and the acute angle and which proves a variety of theorems about quadrilaterals and asymptotic straight lines that also figure in *Euclid Vindicated*.[89] It is unclear whether Lambert had first-hand acquaintance with Saccheri's book, but this hypothesis is not really necessary to explain his achievement, given Lambert's genius and the abundance of details provided by Klügel's dissertation. The peculiarity of Lambert's work, however, lies in its considerable openness to algebra and the theory of trigonometric functions, of which Euler had recently given a wide-ranging treatment. This theory allows him to move beyond Saccheri's results in synthetic geometry and offer a model of hyperbolic geometry (the sphere of imaginary radius) that would later prove an excellent heuristic tool for research in the field. At any rate, it is in virtue of Lambert's algebraic approach that the deductive course of his work differs so significantly from Saccheri's synthetic course. If *Euclid Vindicated* indeed influenced *Parallellinien* directly – beside its obvious indirect influence through Klügel's work – this was limited to the realm of general ideas, and it certainly did not motivate the particular theorems appearing in Lambert's work.

Lambert's essay was so excellent that it played an instrumental role in facilitating the circulation of Saccheri's ideas about parallelism throughout Germany and Europe. At the same time, however, *Parallellinien* obviated the need for consultation of the 1733 work itself. Another indirect route to Saccheri's work became available in 1824, when Camerer published an edition of the *Elements* that included another summary of Saccheri's chief results on parallelism. This work was less critical than Klügel's, and it was certainly written independently from it.[90]

---

[88] Among the many works on the topic, one merits particular mention: C.F. Hindenburg, *Anmerkungen über das neue System der Parallellinien*, "Leipziger Magazin zur Naturkunde, Mathematik und Oekonomie", 1781, pp. 342–71. Hindenburg affirms that he has offered a proof of the Fifth Postulate by *consequentia mirabilis*, but also believes that he has employed an altogether new demonstrative procedure: "Da dieser Beweis etwas Eignes hat, und, so viel ich mich erinnere, ein ähnliches Beyspiel in der Geometrie nirgends vorkommt" (p. 348). Hindenburg will mention Saccheri cursorily in a later article on Lambert from 1786.

[89] The philosopher and mathematician Johann Heinrich Lambert (1728–1777) attempted to prove the Fifth Postulate in 1766, but he did not publish his efforts, as he was aware of their failure. In a later work, he accepted Wallis' similarity principle as an axiom. Lambert's work on parallels was published posthumously by Bernoulli and Hindenburg twenty years later: J.H. Lambert, *Theorie der Parallellinien*, "Magazin für reine und angewandte Mathematik", 1786, pp. 13–64 and 325–58; today it can be read in the anthology on non-Euclidean geometries by Engel and Stäckel. An English edition of the work is forthcoming in the present Series.

[90] J.G. Camerer, *Euclidis Elementorum libri sex priores*, Berlin, Reimer 1824; Camerer is of the opinion that Lambert was not directly acquainted with Saccheri. We may remark that J.J.I. Hoffmann, *Critik der Parallel-Theorie*, Jena, Erdker 1807, has a long section discussing previous failed attempts to prove the Fifth Postulate (sometimes updating Klügel's *Recensio*), but does not mention Saccheri.

In addition, interpreters have debated the question of whether Gauss, Lobachevsky and Bolyai were acquainted with *Euclid Vindicated*.[91] We know that a copy of the book existed in Göttingen, and it is therefore possible – even likely – that Gauss read it. In any event, Gauss was probably more interested in Lambert's algebraic approach than in Saccheri's synthetic one. Saccheri's influence on Gauss, then, must have been at best indirect. Lobachevsky and Bolyai, on the one hand, do not discuss elliptic geometry at all: this aligns them with Legendre, whom they certainly know, rather than with Clavius and Saccheri, who distinguished three hypotheses instead of two (Euclidean versus non-Euclidean). Lobachevsky and Bolyai's interest in hyperbolic trigonometry, on the other hand, can be traced back to Lambert's work. In any case, it does not seem necessary to conjecture that these thinkers drew directly on a work as obscure as Saccheri's. We should also note that these new developments in hyperbolic geometry, which in some sense mark its official origin, take as their starting point an absolute definition of parallel lines (valid for both Euclidean and hyperbolic space), which hinges on the concept of an asymptotic limit straight line. Such a line is also the subject of the last theorem proved by Saccheri under the acute angle hypothesis. Hence, Saccheri's last word represents the basis for and beginning of Gauss', Lobachevsky's and Bolyai's researches. *Euclid Vindicated*, thereby, fulfilled in some sense the function of a ladder that one must throw away after he has climbed it: a new era of mathematics had begun.

It is also worth considering the influence of *Euclid Vindicated* on the philosophical discussions on the structure of space and the nature of geometry in the years that preceded the actual discovery of non-Euclidean geometries. Such influence was, of course, exiguous, for many reasons: Saccheri's book had few readers, Saccheri himself was very skeptical of the metaphysical applications of his own geometrical studies, and, finally, few eighteenth-century thinkers paid much attention to the subject. It should be noted, however, that Thomas Reid, who developed a theory of visual space as a spherical space and provided one of the earliest axiomatizations of elliptic geometry (unsophisticated as it may have been), read and annotated Saccheri's work. Reid even attempted several proofs of the Fifth Postulate, and it is clear that his geometric interests went hand in hand with his philosophical research on the nature of space. It seems, however, that Reid's interest in the physiology and metaphysics of visual space predated his first encounter with *Euclid Vindicated*, which therefore provided him with better mathematical devices, but not with the original idea of inquiring

[91] Three mathematicians are usually credited with the discovery of hyperbolic geometry: Carl Friedrich Gauss (1777–1855), who, however, never published on the subject during his lifetime, Nikolai Ivanovich Lobachevsky (1792–1856), who published his results in 1829–30, and János Bolyai (1802–1860), who published his results independently in 1832–33. The first article on Saccheri's influence on these authors is C. Segre, *Congetture intorno all'influenza di Girolamo Saccheri sulla formazione della geometria non euclidea*, "Atti della Regia Accademia di Scienze di Torino", 38, 1902–1903, pp. 351–63. Segre hypothesizes that these scholars were very likely acquainted with *Euclid Vindicated*.

into spaces with non-Euclidean structure.[92] The same, more or less, can be said of the other scientist who spelled out the philosophical consequences of his mathematical reflections on the theory of parallels, that is, Lambert. His philosophical reflections on space, in fact, are not grounded on Saccheri's results, and they rather build on Wallis' thesis that the Euclidean axiom is equivalent to the possibility of similarity transformations.[93]

Yet Saccheri himself hints at a philosophical development of *Euclid Vindicated* in a Scholium where he explicitly considers the possibility of providing a 'physical-geometrical' proof of the Fifth Postulate by empirically measuring certain angles between real objects. This should by no means suggest that Saccheri envisaged an empiricist foundation for Euclidean geometry, or for the metric of space – a position later held by Gauss and Lobachevsky.[94] Saccheri indeed regards these demonstrations as simply confirming a result that he expects to derive from pure logic. If he mentions such measurements, it is with the goal of stressing the enormous power of his geometric results, which state that the curvature of space can be determined by an analysis of a single object, as the measurement of the angular sum of a single triangle suffices to establish the validity of Euclidean, elliptic or hyperbolic geometry. At any rate, it is very significant that Saccheri realizes that this theorem represents the condition of possibility for an empirical evaluation of the metric of space (given the hypothesis of isotropy). Even though he finally judges that such an experimental measurement cannot be conceptually relevant, *Euclid Vindicated* will offer support to thinkers willing to put forward a program in geometrical empiricism.[95] Equally significant is the fact that Saccheri identi-

---

[92] The main philosophical work of Thomas Reid (1710–1796) is the 1764 *Inquiry into the Human Mind*, which provides a first tentative axiomatization of spherical geometry: cf. T. Reid, *Philosophical Works. With notes and supplementary dissertations by sir William Hamilton*, Edinburgh, Maclachlan&Stewart 1895[8], vol. 1, pp. 147–52. On Reid's annotations on *Euclid Vindicated* cf. G.B. Grandi, *Thomas Reid's geometry of visibles and the parallel postulate*, "Studies in History and Philosophy of Science", 36, 2005, pp. 79–103.

[93] Cf. J.H. Lambert, *Anlage zur Architectonic*, Riga, Hartknoch 1771, § 79, vol. 1, pp. 61–62.

[94] It seems, however, that the story in which Gauss uses geodetic measurements to establish the curvature of space is in fact a legend. For an account, see A.I. Miller,*The Myth of Gauss' Experiment on the Euclidean Nature of Physical Space*, "Isis", 63, 1972, pp. 345–8; or E. Breitenberger, *Gauss's Geodesy and the Axiom of Parallels*, "Archive for History of Exact Sciences", 31, 1984, pp. 273–89. On Lobachevsky's physical experiments, see his *Pangeometry* from 1855 (recently edited by A. Papadopoulos). At any rate, we can notice that Mach was a very attentive reader of Saccheri's reflections, and quoted them in his work aimed to a sensualistic foundation of the metric of space; cf. E. Mach, *Erkenntnis und Irrtum. Skizzen zur Psychologie der Forschung*, Leipzig, Barth 1905, pp. 397–99.

[95] This result is also found in Legendre, but the French mathematician did not include it in any of the editions of his *Éléments*. It was first presented in the late edition of his reflections on parallel lines, published in 1833 (exactly a century after *Euclid Vindicated*). In this work, Legendre envisions a sort of physical-geometric proof, to be carried out with real instruments – a compass and a straightedge (Legendre, *Réflexions*, pp. 378–79). Saccheri's work was thus the only precedent available to Gauss and Lobachevsky, although we have already discussed the complex circumstances surrounding its reception and availability.

fies the structure of geometric space with that of physical space,[96] even though his "most accurate physical experimentations" fall short of a detailed and adequate epistemological reflection in the end.[97]

We may also remark that even before proceeding to geodetic or astronomical measurements of spatial curvature, the late eighteenth- and early nineteenth-century scientific world could count on a couple of very well-devised theorems that showed the equivalence between the parallel axiom and certain principles of statics, thereby establishing that it suffices to consult a scale to determine the curvature of space.[98] It is regrettable that Saccheri did not notice this important relation between statics and parallelism, especially when he was open to the possibility of a physical-geometrical experiment on space. His 1708 *Neo-Statica* does not contain the foundation for an elliptic or hyperbolic static theory, which was, however, well within his reach – and which would have clarified some theorems of *Euclid Vindicated*. This is not even chief among Saccheri's oversights, however, and blaming him for failing to draw all the consequences from his insight is, of course, historically unfair.[99]

To return to our history of the reception of *Euclid Vindicated* after the official discovery of non-Euclidean geometries, we must note the extent to which Saccheri's work had faded into oblivion during the nineteenth century. In the years between 1830 and 1860, research on hyperbolic geometry remained very marginal in European mathematical culture, and Saccheri's book was at any rate too old – and too naive – to make significant contributions, at least compared with the more recent studies undertaken by Lobachevsky and Bolyai. The

[96] In his identification of physical and geometrical space (or better extension), Saccheri eventually reveals a Cartesian tendency. We should note that the cosmological theses put forward in the *Demonstrative Logic* (ed. 1697, pp. 277–8), on the other hand, are firmly embedded in an Aristotelian conception of cosmic space. It is apparent, however, that no one in a Catholic university in 1697 could have openly espoused Cartesian positions in an academic discussion: the condemnation of Honoré Fabri, the Jesuit with Cartesian sympathies, served as an effective deterrent. On the figure of Fabri, see E. Caruso, *Honoré Fabri, gesuita e scienziato*, in *Miscellanea Secentesca*, Dipartimento di Filosofia dell'Università di Milano, Cisalpino 1987, pp. 85–126; and several remarks in E. De Angelis, *Il metodo geometrico nella filosofia del Seicento*, Firenze, Le Monnier 1964.

[97] In fact, Saccheri appears to completely lack well-developed epistemological views on scientific experiments. As Russell correctly argues vis-à-vis comparable nineteenth-century positions, no experiment whatsoever, however 'accurate', can ever determine the Euclidean structure of space (that is, that the sum of the angles of a triangle amounts to *exactly* two right angles). Although this observation was certainly within Saccheri's reach, he appears to have something altogether different from a real experiment in mind, as every real experiment has a margin of error.

[98] For instance, there is a vast discussion of this topic authored by Fourier (1768–1830), which, however, remained unpublished. See the extensive account in Pont, *L'aventure des parallèles*, pp. 554–62; on this subject, also see Bonola, *La geometria non euclidea*, pp. 100–11 (English ed., pp. 181–99).

[99] Should one want to continue in this line, one could point to Saccheri's failure to draw connections between the theory of parallels and the first inquiries into projective geometry, a discipline that Saccheri had cultivated in his youth in the *Quaesita Geometrica*. One could, for instance, mention that Ceva's theorem, very familiar to Saccheri, admits a projective and thus hyperbolic formulation, which is relevant to the study of the foundations of that geometry; cf. O. Perron, *Nichteuklidische Elementargeometrie der Ebene*, Stuttgart, Teubner 1962, § 32.

next thirty years, following the appearance of Gauss' unpublished writings and Klein's and Poincaré's new studies, witnessed the explosion of a wide mathematical (and philosophical) debate over non-Euclidean geometries. The exceptional speed at which research on the subject accelerated forced mathematicians to focus on nascent theories, rather than engage on the archaeological reconstruction of their origins. Although Saccheri was occasionally mentioned, references to his work resembled the typical eighteenth-century allusions motivated by erudition or politeness rather than by any real interest in Saccheri's thought. We do not know of any extensive discussions of his results.[100]

The rediscovery of *Euclid Vindicated* occurred only in in 1889, when another Jesuit, Father Manganotti, revealed the existence of Saccheri's book to Eugenio Beltrami. Beltrami studied the work and proceeded to write an important article, in which he presented Saccheri as a precursor of Legendre and Lobachevsky.[101] By then, hyperbolic geometry was ripe enough to serve as the subject of historical inquiry, and Beltrami's article inaugurated a series of studies dedicated to exploring *Euclid Vindicated* and unearthing Saccheri's other works. This research aimed to evaluate the geometrical significance of Saccheri's work and to investigate its legacy and its historical influence. This is not the appropriate place to analyze the reception of Saccheri in the twentieth century, which was rather good, however; it suffices to note that, today, one rarely reads about non-Euclidean geometries without finding some reference to *Euclid Vindicated.*

It will be appropriate, however, to mention the modern editions of Saccheri's text. Immediately after Beltrami's announcement, a German edition of Book One (on parallel lines) was prepared by Friedrich Engel and Paul Stäckel. This edition appeared in 1895 in an important anthology of texts on non-Euclidean geometries, edited by both authors.[102] The translation provides a few explanatory notes and an introductory essay, which concentrates on Saccheri's (meager, according to the authors) impact on the later protagonists of the non-Euclidean revolution. In the same year, a Saccheri scholar, the Jesuit Paul Mansion, brought a copy of *Euclid Vindicated* to the United States and entrusted it to Bruce Halsted, an energetic mathematician and enthusiastic advocate of the new geometries who would later

[100] On the development of non-Euclidean geometry in the late nineteenth century, see J.-D. VOELKE, *Renaissance de la géométrie non euclidienne entre 1860 et 1900*, Bern, Lang 2005; and K. VOLKERT, *Das Undenkbare denken. Die Rezeption der nichteuklidischen Geometrie im deutschsprachigen Raum (1860–1900)*, Berlin, Springer 2013.

[101] Eugenio Beltrami (1836–1900) is renowned for devising the first Euclidean model of hyperbolic geometry in 1868. He thereby proved the independence of the Fifth Postulate from Euclid's other assumptions. The work quoted here is E. BELTRAMI, *Un precursore italiano di Legendre e di Lobatschewsky*, "Rendiconti dell' Accademia dei Lincei", 5, 1889, p. 441–48. Beltrami's essay precipitated the rediscovery of the *Demonstrative Logic*, which was analyzed for the first time by G. VAILATI, *Di un'opera dimenticata del P. Girolamo Saccheri ("Logica Demonstrativa" 1697)*, "Rivista Filosofica", 6, 1903. Among the immediate reactions to Beltrami's article in Italy is G. VERONESE, *Fondamenti di Geometria*, Padova, Seminario 1891, pp. 569–70.

[102] F. ENGEL, P. STÄCKEL, *Die Theorie der Parallellinien von Euklid bis auf Gauss*, Leipzig, Teubner 1895, pp. 33–135.

translate the works of Lobachevsky and Bolyai into English. Independent of the German editors, Halsted resolved to publish an English version of Book One of Saccheri's work. This translation appeared in five issues of the *American Mathematical Monthly* starting in 1894 and was collected into a volume in 1920, with a new introduction (but no commentary) by Halsted himself; this translation enjoyed many reprints in the twentieth century and is reproduced in the present volume.[103] The first Italian translation of the entire book appeared in 1903, edited by Giovanni Boccardini; it is an extraordinarily free translation, where the editor summarizes certain propositions, omits others, and handles yet others as he sees fit. Although it is rich in remarks on elementary geometry, it does not offer a sound picture of what Saccheri strove to accomplish with the composition of *Euclid Vindicated*.[104] A more accurate edition, by Alberto Pascal, was almost completed in 1914 but, due to its editor's untimely death, it never appeared in print.[105] Saccheri's entire work, in fact, has only been recently translated into Italian.[106] Finally, there exists a partial edition of *Euclid Vindicated* in Interlingua.[107]

[103] Halsted's scientific romanticism led him to suppose that Saccheri had perceived the possibility of non-Euclidean geometries, but had refuted them with a very weak argument to obtain the *imprimatur*. This legend, without a doubt historically unfounded – it does not seem that the Church could have anything to fear from hyperbolic space – was repeated in several popular works. It is true, however, that a religious component seems to creep into the discussions on Euclidean and non-Euclidean geometry at Halsted's times: see the above-mentioned VOELKE, *Renaissance de la géométrie non euclidienne*, which also offers some biographical notes on Bruce Halsted (1853–1922) and Paul Mansion (1844–1919). On the figure of Halsted as a scholar and popularizer of mathematics, see the obituary by F. Cajori in "The American Mathematical Monthly", 29, 1922, pp. 338–40.

[104] *L'Euclide emendato del p. Gerolamo Saccheri*, transl. and annotated by G. Boccardini, Milano, Hoepli 1904.

[105] The edition is referred to in PASCAL, *Girolamo Saccheri*, p. 229. Halsted reports on Pascal's premature death in his *Introduction* to the English translation.

[106] Before my annotated edition for the Edizioni della Normale (2011), Saccheri's complete book had already been published as G. SACCHERI, *Euclide liberato da ogni macchia*, ed. by P. Frigerio, with an introductory essay by I. Tóth and E. Cattanei, Milano, Bompiani 2001.

[107] It is included in the anthology C.E. SJOSTEDT, *Le axiome de paralleles de Euclide a Hilbert: un probleme cardinal in le evolution del geometrie*, Uppsala, Interlingue-Fundation 1968.

Euclides
Ab Omni Naevo Vindicatus:
Sive
Conatus Geometricus
Quo Stabiliuntur
Prima ipsa universae Geometriae Principia.
Auctore
Hieronymo Saccherio
Societatis Jesu
In Ticinensi Universitate Matheseos Professore.
Opusculum
Ex▲[Mo] Senatui
Mediolanensi
Ab Auctore Dicatum.
Mediolani, MDCCXXXIII.
Ex Typographia Pauli Antonii Montani.
Superiorum permissu.

Euclid Vindicated from Every Blemish or A Geometric Endeavor in which are Established the Foundation Principles of Universal Geometry

Procemium ad lectorem

[IX] Quanta sit Elementorum Euclidis praestantia, ac dignitas, nemo omnium, qui Mathematicas disciplinas noverint, ignorare potest. Lectissimos hanc in rem testes adhibeo Archimedem, Apollonium, Theodosium, aliosque pene innumeros, ad haec usque nostra tempora rerum Mathematicarum Scriptores, qui non aliter haec Euclidis Elementa usurpant, nisi ut principia jam diu stabilita, ac penitus inconcussa. Verum tanta haec nominis celebritas vetare non potuit, quin multi ex Antiquis pariter, ac Recentioribus, iique Magni Geometrae naevos quosdam a se depraehensos censuerint in his ipsis pulcherrimis, nec unquam satis laudatis Elementis. Tres autem hujusmodi naevos designant, quos statim subnecto.

Primus respicit definitionem parallelarum, & sub ea Axioma, quod apud Clavium est decimumtertium Libri primi, ubi Euclides sic pronunciat: *Si in duas rectas lineas, in eodem plano existentes recta incidens linea duos ad easdem partes internos angulos minores duobus rectis cum eisdem efficiat, duae illae rectae lineae ad eas partes in infinitum protractae inter se mutuo incident.* Porro nemo est, qui dubitet de veritate expositi Pronunciati; sed in eo unice Euclidem accusant, quod nomine Axiomatis usus fuerit, quasi nempe ex solis terminis rite perspectis sibi ipsi faceret fidem. Inde autem non pauci (retenta caeteroquin Euclidaea parallelarum definitione) illius demonstrationem aggressi sunt ex iis solis Propositionibus Libri primi Euclidaei, quae praecedunt vigesimam nonam, ad quam scilicet usui esse incipit
[X] controversum Pronunciatum.

Sed rursum; quoniam antiquorum in hanc rem conatus visi non sunt adamussim scopum attingere; factum idcirco est, ut multi proximiorum temporum eximii Geometrae, idem pensum aggressi, necessariam censuerint novam quandam parallelarum definitionem. Itaque; cum Euclides parallelas rectas lineas definiat, *quae in eodem plano existentes, si in utramque partem in infinitum producantur, nunquam inter se mutuo incidunt*; postremis expositae definitionis vocibus has alias substituunt: *Semper inter se aequidistant*; adeo ut nempe singulae perpendiculares ab uno quolibet unius illarum puncto ad alteram demissae aequales inter se sint.

At nova rursum hinc oritur scissura. Nam aliqui, & ii sane acutiores, demonstrare conantur parallelas rectas lineas prout sic definitas, unde utique gradum faciant ad demonstrandum sub ipsis Euclidaeis vocibus controversum Pronunciatum, cui nimirum ab ea vigesima nona Libri primi Euclidaei (pauculis quibusdam exceptis) universa innititur Geometria. Alii vero (non sine magno in rigidam Logicam peccato) eas tales rectas lineas parallelas, nimirum *aequidistantes*, assumunt tanquam datas, ut inde gradum faciant ad reliqua demonstranda.

Preface to the Reader

Of all who have learned mathematics,[1] none can fail to know how great is the excellence and worth of Euclid's *Elements*. As erudite witnesses here I summon Archimedes, Apollonius, Theodosius, and others almost innumerable, writers on mathematics even to our times, who use Euclid's *Elements* as foundation long established and wholly unshaken. But this so great celebrity has not prevented many, ancients as well as moderns, and among them distinguished geometers, maintaining they had found certain blemishes[2] in these most beauteous nor ever sufficiently praised *Elements*. Three such flecks they designate, which now I name.

The first pertains to the definition of parallels and with it the Axiom which in Clavius[3] is the thirteenth of the First Book of the *Elements*, where Euclid says: *If a straight line falling on two straight lines, lying in the same plane, make with them two internal angles toward the same parts less than two right angles, these two straight lines infinitely produced toward those parts will meet each other*. No one doubts the truth of this Assertion;[4] but solely they accuse Euclid as to it, because he has used for it the name Axiom, as if obviously from the right understanding of its terms alone came conviction.[5] Whence not a few (withal retaining Euclid's definition of parallels) have attempted its demonstration from those propositions of Euclid's First Book alone which precede the 29th, wherein begins the use of the controverted Assertion.[6]

But again, since the endeavors of the ancients in this matter do not seem to attain the goal, so it has happened that many distinguished geometers of ensuing times, attacking the same idea, have thought necessary a new definition of parallels. Thus, while Euclid defines parallels' as straight lines *lying in the same plane, which, if infinitely produced toward both sides, nowhere meet,* they substitute for the last words of the given definition these others: *always equidistant from each* other;[7] so that all perpendiculars from any points on one of them let fall upon the other are equal to one another.

But again here arises a new fissure. For some, and these surely the keenest, endeavor to demonstrate the existence of parallel straight lines as so defined, whence they go up to the proof of the debated Assertion as stated in Euclid's terms, upon which truly from that *Elements* I, 29 (with some very few exceptions) all geometry rests. But others (not without gross sin against rigorous logic) assume such parallel straight lines, forsooth *equidistant,* as if given, that thence they may go up to what remains to be proved.

Et haec quidem satis sunt ad praemonendum Lectorem super iis, quae materiam exhibebunt Libro priori hujus mei Opusculi: Nam uberior praedictorum omnium explicatio habebitur in Scholiis post Prop. vigesimam primam enunciati Libri, quem dividam in duas veluti partes. In priore imitabor antiquiores illos Geometras, nihil propterea sollicitus de natura, aut nomine illius lineae, quae omnibus suis punctis a quadam supposita recta linea aequidistet: Sed unice in id incumbam, ut controversum Euclidaeum Axioma citra omnem petitionem principii clare demonstrem; nunquam idcirco adhibens ex ipsis prioribus Libri primi Euclidaei Propositionibus, non modo vigesimam septimam, aut vigesimam octavam, sed nec ipsas quidem decimam sextam, aut decimam septimam, nisi ubi clare agatur de tri-
[XI] angulo omni ex parte circumscripto. Tum in posteriore parte, ad novam ejusdem Axiomatis confirmationem demonstrabo non nisi rectam lineam esse posse, quae omnibus suis punctis a quadam supposita recta linea aequidistet. Horum autem occasione prima ipsa universae Geometriae Principia rigido examini subjicienda hic esse nullus est, qui non videat.

Transeo ad alios duos naevos Euclidi objectos. Prior respicit definitionem sextam Libri quinti super aeque proportionalibus: Posterior Definitionem quintam Libri sexti super compositione rationum. Hic autem erit secundi mei Libri unicus scopus, ut dilucide explicem praefatas Euclidaeas Definitiones, simulque ostendam non aequo jure hac in parte Euclidis nomen vexatum fuisse.

Prodest tamen rursum praemonere, demonstratum a me iri hac occasione unum quoddam Axioma, quod tutissime per omnem Geometriam versetur, sine indigentia illius *Postulati*, sub nomine Axiomatis ab interpretibus (ut reor) intrusi, cujus usus incipit ad 18. quinti.

And this is enough to indicate to the reader what will be the material of the First Book of this booklet of mine: for a more complete explication of all that has been said will be given in the Scholia after Proposition 21 of this First Book. I divide this Book into two parts. In the First Part I will imitate the antique geometers, and not trouble myself about the nature or the name of that line which at all its points is equidistant from a given straight line; but merely undertake without any *petitio principii* clearly to demonstrate the disputed Euclidean Axiom. Therefore never will I use from those prior propositions of *Elements* I, 27 and 28, but not even *Elements* I, 16 and 17, except where clearly it is question of a triangle every way restricted.[8] Then in the Second Part for a new confirmation of the same Axiom, I shall demonstrate that the line which at all its points is equidistant from a given straight line can only be a straight line. But every one sees that on this occasion the very first principles of all geometry are to be subjected to a rigid examination.

I go on to the other two blemishes charged against Euclid. The first pertains to Definition 6 of the Fifth Book of the *Elements* about equiproportionals; the second to Definition 5 of the Sixth Book about the composition of ratios.[9] It will be the sole aim of my Second Book to clearly expound the Euclidean definitions mentioned, and at the same time to show that Euclid's fame is here unjustly attacked.

Yet again it is well to state that on this occasion I shall prove a certain Axiom that may safely be applied throughout the whole of geometry, without need of that *Postulate*, put in (as I believe) by commentators under the name of Axiom,[10] whose use begins at *Elements* V, 18.

Indicis loco addenda censeo, quae sequuntur

[XII] 1. In I. & II. Propos. Lib. primi duo jaciuntur principia, ex quibus in III. & IV. demonstratur, angulos interiores ad rectam jungentem extremitates aequalium perpendiculorum, quae ex duobus punctis alterius rectae, veluti basis, versus easdem partes (in eodem plano) erigantur, non modo fore inter se aequales, sed praeterea aut rectos, aut obtusos, aut acutos, prout illa jungens aequalis fuerit, aut minor, aut major praedicta basi: Atque ita vicissim. *a pag. 1*

2. Hinc sumitur occasio secernendi tres diversas hypotheses, unam anguli recti, alteram obtusi, tertiam acuti: circa quas in V. VI. & VII. demonstratur, unam quamlibet harum hypothesium fore semper unice veram, si nimirum depraehendatur vera in uno quolibet casu particulari. *a pag. 5*

3. Tum vero; post interpositas tres alias necessarias Propositiones; demonstratur in XI. XII. ac XIII. universalis veritas noti Axiomatis, respectu habito ad priores duas hypotheses, unam anguli recti, & alteram obtusi; ac tandem in XIV. ostenditur absoluta falsitas hypothesis anguli obtusi. Atque hinc incipit diuturnum proelium adversus hypothesin anguli acuti, quae sola renuit veritatem illius Axiomatis. *a pag. 10*

[XIII] 4. Itaque in XV. ac XVI. demonstratur stabilitum iri hypotheses aut anguli recti, aut obtusi, aut acuti, ex quolibet triangulo rectilineo, cujus tres simul anguli aequales sint, aut majores, aut minores duobus rectis; ac similiter ex quolibet quadrilatero rectilineo, cujus quatuor simul anguli aequales sint, aut majores, aut minores quatuor rectis. *a pag. 20*

5. Sequuntur quinque aliae Propositiones, in quibus demonstrantur alia indicia pro secernenda vera hypothesi a falsis. *a pag. 23*

6. Accedunt quatuor principalia Scholia; in quorum postremo exhibetur figura quaedam geometrica, ad quam fortasse respexit Euclides, ut suum illud Pronunciatum assumeret tanquam per se notum. In tribus prioribus ostenditur non valuisse ad intentum praecedentes insignium Geometrarum conatus. Sed quia controversum Axioma exactissime demonstratur ex duabus praesuppositis rectis lineis *aequidistantibus*; monet ibi Auctor contineri in eo praesupposito manifestam petitionem *Principii*. Quod si provocari hic velit ad communem persuasionem, atque item exploratissimam *praxim*; rursum monet provocari non debere ad experientiam, quae respiciat puncta infinita, cum satis esse possit unica experientia uni cuivis puncto affixa. Quo loco tres ab ipso afferuntur invictissimae Demonstrationes Physico-Geometricæ. *a pag. 29*

[XIV] 7. Supersunt duodecim aliae Propositiones, quae primae Parti hujus Libri finem imponunt. Non expono particularia assumpta, quia nimis implexa. Solum dico ibi tandem manifestae falsitatis redargui inimicam hypothesim anguli acuti, utpote quae duas rectas agnoscere deberet, quae in uno eodemque puncto commune reciperent in eodem plano perpendiculum: Quod quidem naturae lineae rectae repugnans esse demonstratur per quinque Lemmata, in quibus concluduntur quinque principalia Geometriae Axiomata, quae respiciunt lineam rectam, ac circulum, cum suis correlativis Postulatis. *a pag. 43*

In place of an Index should be added, I think, what follows

1. In Propositions 1 and 2 of the First Book two principles are established, from which in Propositions 3 and 4 is proved, that interior angles at the straight joining the extremities of equal perpendiculars erected toward the same parts (in the same plane) from two points of another straight, as base, not merely are equal to each other, but besides are either right or obtuse or acute according as that join is equal to, or less, or greater than the aforesaid base: and inversely. *From page 1 on.*
2. Hence occasion is taken to distinguish three different hypotheses, one of right angle, another of obtuse, a third of acute: about which in Propositions 5, 6, and 7 is proved, that any one of these hypotheses is always alone true if it is found true in any one particular case. *From page* 5 *on.*
3. Then after the interposition of three other necessary Propositions, is proved in Propositions 11, 12, and 13, the universal truth of the famous Axiom, respect being had to the first two hypotheses, one of right angle, and the other of obtuse; and at length in Proposition 14 is shown the absolute falsity of the hypothesis of obtuse angle. And here begins a lengthy battle against the hypothesis of acute angle, which alone opposes the truth of that Axiom. *From page 10 on.*
4. And so in Propositions 15 and 16 is proved that the hypothesis either of right angle, or obtuse, or acute is established from any rectilineal triangle whose three angles together are equal to, or greater, or less than two right angles; and in like way from any rectilineal quadrilateral, whose four angles are together equal to, or greater, or less than four right angles. *From page 20 on.*
5. Five other Propositions follow, in which are proved other indications for distinguishing the true hypothesis from the false. *From page* 23 *on.*
6. Now come four fundamental Scholia. In the last is exhibited a certain geometric diagram, of which Euclid perhaps thought, in order that his Assertion might assume self-evidence. In the preceding three is shown that the prior endeavors of distinguished geometers have not reached their aim. Since however the debated Axiom can be exactly proved from two straight lines presupposed *equidistant,* the author here shows a manifest *petitio principii* to be contained in that presupposition. If one wishes here to appeal to common persuasion, and surest *experience,* again he shows appeal should not be taken to an experience involving an infinity of points, when a single experiment pertaining to any one point can suffice. In this place are set forth by him three invincible physico-geometric demonstrations. *From page* 29 *on.*
7. To the end of the First Part of this Book there remain twelve other Propositions. I do not state the particular assumptions, because they are too complex. I only say here at length I have disproved the hostile hypothesis of acute angle by a manifest falsity, since it must lead to the recognition of two straight lines which at one and the same point have in the same plane a common perpendicular. That this is contrary to the nature of the straight line is proved by five Lemmata, in which are contained five fundamental Axioms relating to the straight line and circle, with their correlative postulates. *From page 43 on.*

8. Secunda pars continet sex Propositiones. Ibi autem; post expensam (juxta hypothesim anguli acuti) naturam illius lineae, quae omnibus suis punctis a quadam praesupposita recta linea aequidistet; multis modis ostenditur, eam fore aequalem contrapositae basi, unde infertur praenunciatae hypothesis certissima falsitas. Quare tandem in ultima Propos. quae est XXXIX. exactissime demonstratur celebre illud Euclidaeum Axioma, cui nempe (ut omnes sciunt) universa Geometria innititur. *a pag. 87*
9. Secundus Liber digeri commode non potuit in Propositiones, etiamsi locis opportunis plura intermista sint utilissima Theoremata, ac Problemata. Meretur nihilominus expresse notari unum quoddam Axioma, cujus ibi demonstratur non modo veritas, [XV] verum etiam universalis utilitas pro omni Geometria, sine indigentia alterius parum decori Postulati, quod ab interpretibus censeri potest intrusum sub nomine Axiomatis, cujus nempe usus incipit ad 18. quinti. Et id quidem pro prima Parte hujus Libri, in qua vindicatur Def. sexta quinti Euclidaei. *a pag. 102*
10. Tum in secunda Parte; praeter nonnulla alia opportune addita, ad tuendas reliquas Definitiones ejusdem Quinti circa magnitudines proportionales; demonstratur priore loco (respectu habito ad magnitudines commensurabiles) quinta Definitio Sexti, etiamsi recipi ea deberet in *quid rei*, veluti Axioma: Sed rursum multis exemplis, ex ipso Euclide petitis, ostenditur nullius demonstrationis indigam eam esse, quia Definitionem *puri nominis*. Atque ita, post opportunam additam Appendicem, huic Operi finis imponitur. *a pag. 132*

8. The Second Part contains six Propositions. Here, after investigating the nature (assuming the hypothesis of acute angle) of that line which at all its points is equidistant from a given straight line, it is shown in many ways that it equals the base opposite, whence is inferred the certain falsity of the aforesaid hypothesis. Wherefore at length in the last Proposition 39 is exactly proved that famous Axiom of Euclid, upon which (as everybody knows) the whole of geometry rests. *From page* 87 *on.*
9. The Second Book cannot conveniently be divided into Propositions, although at opportune places are intercalated many most useful theorems and problems. Nevertheless is worthy of express mention a certain Axiom, of which not merely the truth is there demonstrated but also the universal utility for all geometry, without need of the other inelegant postulate supposedly inserted by commentators under the name of Axiom, whose use begins at *Elements* V, 18. So much for the First Part of this Book, in which is defended *Elements* V, Def. 6. *From page 102 on.*
10. Then in the Second Part, besides some other things opportunely added for the purpose of maintaining other Definitions of *Elements* V about proportional magnitudes, is demonstrated in the first place (with respect to commensurable magnitudes) *Elements* VI, Def. 5, even if it ought to be taken as a *real* definition, i. e. as an axiom. But on the contrary is shown by many examples drawn from Euclid himself that this needs no demonstration, because it is a *purely nominal* definition. And so after an Appendix opportunely added, an end is put to this work. *From page 132 on.*

Euclidis ab omni naevo vindicati Liber Primus

In quo demonstratur:

duas quaslibet in eodem plano existentes rectas lineas, in quas recta quaepiam incidens duos ad easdem partes internos angulos efficiat duobus rectis minores, ad eas partes aliquando invicem coituras, si in infinitum producantur.

Pars Prima

[1] **Propositio I.**

Si duae aequales rectae (Fig. 1) *AC, BD, aequales ad easdem partes efficiant angulos cum recta AB: Dico angulos ad junctam CD aequales invicem fore.*

Demonstratur. Jungantur AD, CB. Tum considerentur triangula CAB, DBA. Constat (ex quarta primi) aequales fore bases CB, AD. Deinde considerentur triangula ACD, BDC. Constat (ex octava primi) aequales fore angulos ACD, BDC. Quod erat demonstrandum.

Propositio II.

Manente uniformi quadrilatero ABDC, latera AB, CD, bifariam dividantur (Fig. 2) *in punctis*
[2] *M, & H. Dico angulos ad junctam MH fore hinc inde rectos.*

Demonstratur. Jungantur AH, BH, atque item CM, DM. Quoniam in eo quadrilatero anguli A, & B positi sunt aequales, atque item (ex praecedente) aequales sunt anguli C, & D; constat ex quarta primi (cum alias nota sit aequalitas laterum) aequales fore in triangulis CAM, DBM, bases CM, DM; atque item, in triangulis ACH, BDH, bases AH, BH. Quare; collatis inter se triangulis CHM, DHM, ac rursum inter se triangulis AMH, BMH; constabit (ex octava primi) aequales invicem fore, atque ideo rectos angulos hinc inde ad puncta M, & H. Quod erat demonstrandum.

G. Saccheri, *Euclid Vindicated from Every Blemish,* Classic Texts in the Sciences 1,
DOI 10.1007/978-3-319-05966-2_2,

Euclid Vindicated from every Blemish
Book One

In which is proved:

any two coplanar straight lines, falling upon which any straight makes toward the same parts two internal angles less than two right angles, at length meet each other toward those parts, if infinitely produced.

First Part

Proposition 1.

If two equal straights (Fig. 1) *AC, BD, make with the straight AB angles equal toward the same parts: I say that the angles at the join* CD *will be mutually equal.*

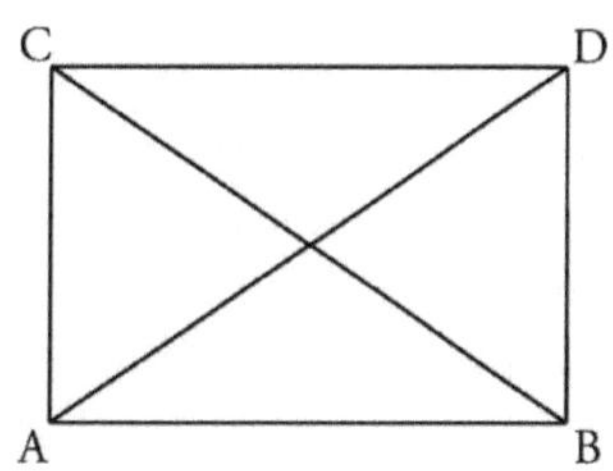

Fig. 1

Proof. Join AD, CB. Then consider the triangles CAB, DBA. It follows (*Elements* I, 4) that the bases CB, AD will be equal. Then consider the triangles ACD, BDC. It follows (*Elements* I, 8) that the angles ACD, BDC will be equal. This is what was to be demonstrated.

Proposition 2.

Retaining the uniform quadrilateral ABCD, bisect the sides AB, CD (Fig. 2) *in the points M and H. I say the angles at the join MH will then be right.*

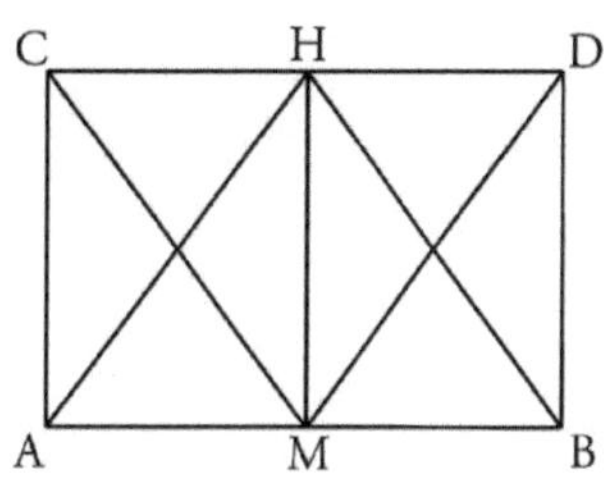

Fig. 2

Proof. Join AH, BH, and likewise CM, DM. Because in this quadrilateral the angles A and B are taken equal and likewise (Proposition 1) the angles C, and D are equal; it follows (*Elements* I, 4) (noting the equality of the sides) that in the triangles CAM, DBM, the bases CM, DM will be equal; and likewise, in the triangles ACH, BDH, the bases AH, BH. Therefore; comparing the triangles CHM, DHM, and in turn the triangles AMH, BMH; it follows (*Elements* I, 8) that we have mutually equal, and therefore right, the angles at the points M, and H. This is what was to be demonstrated.

Propositio III.

Si duae aequales rectae (Fig. 3) *AC, BD, perpendiculariter insistant cuivis rectae AB: Dico junctam CD aequalem fore, aut minorem, aut majorem ipsa AB, prout anguli ad eandem CD fuerint aut recti, aut obtusi, aut acuti.*

Demonstratur prima pars. Existente recto utroque angulo C, & D; sit, si fieri potest, alterutra ipsarum, ut DC, major altera BA. Sumatur in DC portio DK aequalis ipsi BA, jungaturque AK. Quoniam igitur super BD perpendiculariter insistunt aequales rectae BA, DK, aequales erunt (ex prima hujus) anguli BAK, DKA. Hoc autem absurdum est; cum angulus BAK sit ex constructione minor supposito recto BAC; & angulus DKA sit ex constructione externus, atque ideo (ex decimasexta primi) major interno, & opposito DCA, qui supponitur rectus. Non ergo alterutra praedictarum rectarum, DC, BA, est altera major, dum anguli ad junctam CD sint recti; ac propterea aequales invicem sunt. Quod erat primo
[3] loco demonstrandum.

Demonstratur secunda pars. Si autem obtusi fuerint anguli ad junctam CD, dividantur bifariam AB, & CD, in punctis M, & H, jungaturque MH. Quoniam ergo super recta MH perpendiculariter insistunt (ex praecedente) duae rectae AM, CH, poniturque ad junctam AC angulus rectus in A, non erit (ex prima hujus) recta CH aequalis ipsi AM, cum desit angulus rectus in C. Sed neque erit major: caeterum sumpta in HC portione KH aequali ipsi AM, aequales forent (ex prima hujus) anguli ad junctam AK. Hoc autem absurdum est, ut supra. Nam angulus MAK est minor recto; & angulus HKA est (ex decimasexta primi) major obtuso, qualis supponitur internus, & oppositus HCA. Restat igitur, ut CH, dum anguli ad junctam CD ponantur obtusi, minor sit ipsa AM; ac propterea prioris dupla CD minor sit posterioris dupla AB. Quod erat secundo loco demonstrandum.

Demonstratur tertia pars. Tandem vero, si acuti fuerint anguli ad junctam CD, ducta pariformiter (ex praecedente) perpendiculari MH, sic proceditur. Quoniam super recta MH perpendiculariter insistunt duae rectae AM, CH, poniturque ad junctam AC angulus rectus in A, non erit (ut supra) recta CH aequalis ipsi AM, cum desit angulus rectus in C. Sed neque erit minor: caeterum; si in HC protracta sumatur HL aequalis ipsi AM; aequales forent (ut supra) anguli ad junctam AL. Hoc autem absurdum est. Nam angulus MAL est ex constructione major supposito recto MAC; & angulus HLA est ex constructione internus, & oppositus, atque ideo minor (ex decimasexta primi) externo HCA, qui supponitur acutus. Restat igitur, ut CH, dum anguli ad junctam CD sint acuti, major sit ipsa AM, atque ideo prioris dupla CD major sit posterioris dupla AB. Quod erat tertio loco demonstrandum.

[4] Itaque constat junctam CD aequalem fore, aut minorem, aut majorem ipsa AB, prout anguli ad eandem CD fuerint aut recti, aut obtusi, aut acuti. Quae erant demonstranda.

Proposition 3.

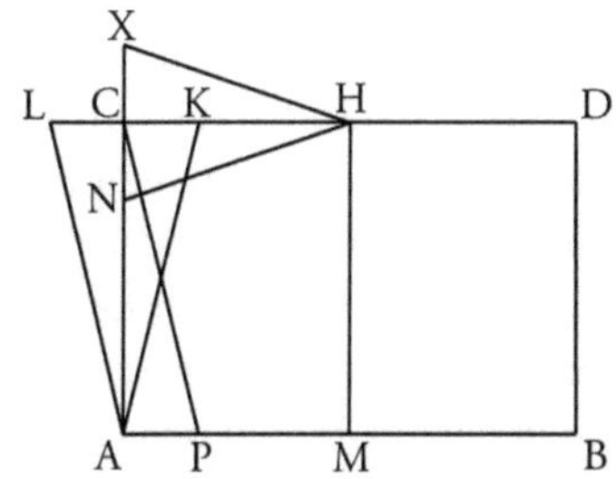

Fig. 3

If two equal straights (Fig. 3) *AC, BD, stand perpendicular to any straight AB: I say the join CD will be equal to, or less, or greater than, AB, according as the angles at CD are right, or obtuse, or acute.*

Proof of the first part. Each angle C, and D, being right; suppose, if it were possible, either one of those, as DC, greater than the other BA. Take in DC the piece DK equal to BA, and join AK. Since therefore on BD stand perpendicular the equal straights BA, DK, the angles BAK, DKA will be equal (Proposition 1). But this is absurd; since the angle BAK is by construction less than the assumed right angle BAC; and the angle DKA is by construction external, and therefore (*Elements* I, 16) greater than the internal and opposite DCA, which is supposed right. Therefore neither of the aforesaid straights, DC, BA, is greater than the other, whilst the angles at the join CD are right; and therefore they are mutually equal. This is what was to be demonstrated in the first part. Proof of the second part. But if the angles at the join CD are obtuse, bisect AB, and CD, in the points M, and H, and join MH. Since therefore on the straight MH stand perpendicular (Proposition 2) the two straights AM, CH, and at the join AC is a right angle at A, the straight CH will not be (Proposition 1) equal to this AM, since a right angle is lacking at C. But neither will it be greater: otherwise in HC the piece KH being assumed equal to this AM, the angles at the join AK will be (Proposition 1) equal. But this is absurd, as above. For the angle MAK is less than a right; and the angle HKA is (*Elements* I, 16) greater than an obtuse, such as the internal and opposite HCA is supposed.[1] It remains therefore, that CH, whilst the angles at the join CD are taken obtuse, is less than this AM; and therefore CD double the former is less than AB double the latter. This is what was to be demonstrated in the second part.

Proof of the third part. Finally, however, if the angles at the join CD are acute, MH being constructed as before perpendicular (Proposition 2), we proceed thus. Since on the straight MH stand perpendicular two straights AM, CH, and at the join AC is a right angle at A, the straight CH will not be equal to this AM (as above), since the angle at C is not right. But neither will it be less: otherwise, if in HC produced HL is taken equal to this AM, the angles at the join AL will be (as above) equal. But this is absurd. For the angle MAL is by construction greater than the assumed right MAC; and the angle HLA is by construction internal, and opposite, and therefore less than (*Elements* I, 16) the external HCA, which is assumed acute. It remains therefore, that CH, whilst the angles at the join CD are acute, is greater than this AM, and therefore CD the double of the former is greater than AB the double of the latter. This is what was to be demonstrated in the third part.

Therefore it is established that the join CD will be equal to, or less, or greater than this AB, according as the angles at the same CD are right, or obtuse, or acute. This is what was to be demonstrated.

Corollarium I.

Hinc in omni quadrilatero continente tres quidem angulos rectos, & unum obtusum, aut acutum, latera adjacentia illi angulo non recto minora sunt, alterum altero, lateribus contrapositis, si ille angulus sit obtusus, majora autem, si sit acutus. Id enim demonstratum jam est de latere CH relate ad contrapositum latus AM; similique modo ostenditur de latere AC relate ad contrapositum latus MH. Cum enim rectae AC, MH, perpendiculares sint ipsi AM, nequeunt (ex prima hujus) esse invicem aequales, propter inaequales angulos ad junctam CH. Sed neque (in hypothesi anguli obtusi in C) potest quaedam AN, portio ipsius AC, aequalis esse ipsi MH, qua nimirum major sit praedicta AC: caeterum (ex eadem prima) aequales forent anguli ad junctam HN; quod est absurdum, ut supra. Rursum vero (in hypothesi anguli acuti in eo puncto C) si velis quandam AX, sumptam in AC protracta, aequalem ipsi MH, qua nimirum minor sit modo dicta AC; jam eodem titulo aequales erunt anguli ad HX; quod utique absurdum itidem est, ut supra. Restat igitur, ut in hypothesi quidem anguli obtusi in eo puncto C, latus AC minus sit contraposito latere MH; in hypothesi autem anguli acuti sit eodem majus. Quod erat intentum.

Corollarium II.

Multo autem magis erit CH major portione qualibet ipsius AM, ut puta PM, ad quam
[5] nempe juncta CP acutiorem adhuc angulum efficiat cum ipsa CH versus partes puncti H, & obtusum (ex decimasexta primi) cum ea PM versus partes puncti M.

Corollarium III.

Rursum constat praedicta omnia aeque procedere, sive assumpta perpendicula AC, & BD, fuerint certae cujusdam apud nos longitudinis, sive sint, aut supponantur infinite parva. Quod quidem notari opportune debet in reliquis sequentibus Propositionibus.

Propositio IV.

Vicissim autem (manente figura praecedentis Propositionis) anguli ad junctam CD erunt aut recti, aut obtusi, aut acuti, prout recta CD aequalis fuerit, aut minor, aut major contraposita AB.

Demonstratur. Si enim recta CD aequalis sit contrapositae AB, & nihilominus anguli ad eandem sint aut obtusi, aut acuti; jam ipsi tales anguli eam probabunt (ex praecedente) non aequalem, sed minorem, aut majorem contraposita AB; quod est absurdum contra hypothesim. Idem uniformiter valet circa reliquos casus. Stat igitur angulos ad junctam CD esse aut rectos, aut obtusos, aut acutos, prout recta CD aequalis fuerit, aut minor, aut major contraposita AB. Quod erat demonstrandum.

Corollary 1.

Hence in every quadrilateral containing three right angles, and one obtuse, or acute,[2] the sides adjacent to this oblique angle are less respectively than the opposite sides if this angle is obtuse, but greater if it is acute. For this has just now been demonstrated of the side CH relatively to the opposite side AM; in the same way it is demonstrated of the side AC relatively to the opposite side MH. For since the straights AC, MH, are perpendicular to this AM, they cannot (Proposition 1) be mutually equal, on account of the unequal angles at the join CH. But neither (in the hypothesis of an obtuse angle at C) can a certain AN, a piece of this AC, be equal to this MH (of which certainly the aforesaid AC would then be greater): otherwise (Proposition 1) the angles at the join HN would be equal; which is absurd, as above. Again however (in the hypothesis of an acute angle at this point C), if you take a certain AX, assumed on AC produced, equal to this MH (of which certainly the just mentioned AC would be less,); now by this same title the angles at HX will be equal; which assuredly is absurd in the same way, as above. It remains therefore, that indeed in the hypothesis of an obtuse angle at this point C, the side AC is less than the opposite side MH; but in the hypothesis of an acute angle is greater than it. This is what we wanted.

Corollary 2.

But by much more will CH be greater than any piece of this AM, as for instance PM, since of course the join CP makes an angle still more acute with this CH toward the parts of the point H, and obtuse (*Elements* I, 16) with this PM toward the parts of the point M.

Corollary 3.

Again it abides that all things aforesaid equally result, whether the assumed perpendiculars AC, and BD are of some length fixed by us, are, or are supposed infinitesimal. This indeed ought opportunely to be noted in remaining subsequent Propositions.[3]

Proposition 4.

But inversely (the figure of the preceding Proposition remaining) the angles at the join CD will be right, or obtuse, or acute, according as the straight CD is equal, or less, or greater than the opposite AB.

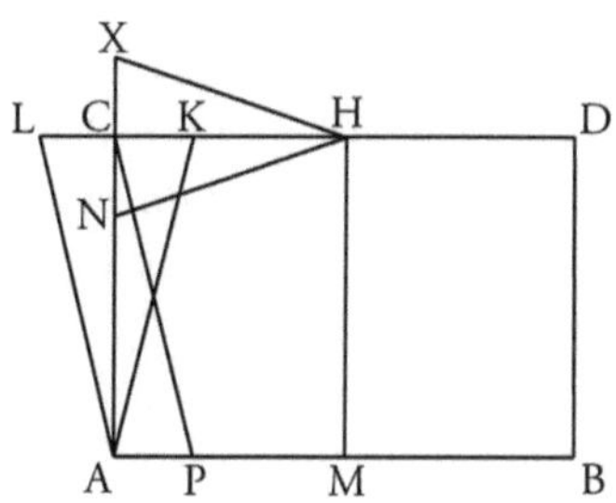

Fig. 3

Proof. For if the straight CD is equal to the opposite AB, and nevertheless the angles at it are either obtuse, or acute; now these such angles prove it (Proposition 3) not equal, but less, or greater than the opposite AB; which is absurd against the hypothesis. The same uniformly avails in regard to the remaining cases. It holds therefore that the angles at the join CD are either right, or obtuse, or acute, according as the straight CD is equal to, or less, or greater than the opposite AB. This is what was to be demonstrated.

Definitiones.

Quandoquidem (ex prima hujus) recta jungens extremitates aequalium perpendiculorum
[6] eidem rectae (quam vocabimus basim) insistentium, aequales efficit angulos cum ipsis perpendiculis; tres idcirco distinguendae sunt hypotheses circa speciem horum angulorum. Et primam quidem appellabo hypothesim anguli recti; secundam vero, & tertiam appellabo hypothesim anguli obtusi, & hypothesim anguli acuti.

Propositio V.

Hypothesis anguli recti, si vel in uno casu est vera, semper in omni casu illa sola est vera.

Demonstratur. Efficiat juncta CD (Fig. 4) angulos rectos cum duobus quibusvis aequalibus perpendiculis AC, BD, uni cuivis AB insistentibus. Erit CD (ex tertia hujus) aequalis ipsi AB. Sumantur in AC, & BD protractis duae CR, DX, aequales ipsis AC, BD; jungaturque RX. Facile ostendemus junctam RX aequalem fore ipsi AB, & angulos ad eandem rectos. Et primo quidem per superpositionem quadrilateri ABDC super quadrilaterum CDXR, adhibita communi basi CD. Deinde elegantius sic proceditur. Jungantur AD, RD. Constat (ex quarta primi) aequales fore in triangulis ACD, RCD, bases AD, RD, atque item angulos CDA, CDR, ac propterea aequales reliquos ad unum rectum, nimirum ADB, RDX. Quare rursum (ex eadem quarta primi) aequalis erit, in triangulis ADB, RDX, basis AB, basi RX. Igitur (ex praecedente) anguli ad junctam RX erunt recti, ac propterea persistemus in eadem hypothesi anguli recti.

Quoniam vero augeri similiter potest longitudo perpendiculorum in infinitum, sub eadem basi AB, consistente semper hypothesi anguli recti, demonstrandum est eandem hypothesim semper mansuram in casu cujusvis imminutionis eorundem perpendiculorum; quod quidem ita evincitur.

[7] Sumantur in AR, & BX duo quaelibet aequalia perpendicula AL, BK, jungaturque LK. Si anguli ad junctam LK recti non sint, erunt tamen (ex prima hujus) invicem aequales. Erunt igitur ex una parte, ut puta versus AB obtusi, & versus RX acuti, ut nimirum anguli hinc inde ad utrunque illorum punctorum aequales sint (ex decimatertia primi) duobus rectis. Constat autem aequalia etiam invicem esse perpendicula LR, KX, ipsi RX insistentia. Igitur (ex tertia hujus) erit LK major quidem contraposita RX, & minor contraposita AB.

Hoc autem absurdum est; cum AB, & RX ostensae sint aequales. Non ergo mutabitur hypothesis anguli recti sub quacunque imminutione perpendiculorum, dum consistat semel posita basis AB.

Sed neque immutabitur hypothesis anguli recti, sub quacunque imminutione, aut majori amplitudine basis; cum manifestum sit considerari posse ut basim quodvis perpendiculum BK, aut BX, atque ideo considerari vicissim ut perpendicula ipsam AB, & rectam aequalem contrapositam KL, aut XR.

Constat igitur hypothesim anguli recti, si vel in uno casu sit vera, semper in omni casu illam solam esse veram. Quod erat demonstrandum.

Definitions.

Since (Proposition 1) the straight joining the extremities of equal perpendiculars standing upon the same straight (which we call base), makes equal angles with these perpendiculars; therefore there are three hypotheses to be distinguished according to the species of these angles. And the first indeed I will call hypothesis of right angle: the second however, and the third I will call hypothesis of obtuse angle, and hypothesis of acute angle.

Proposition 5.

If even in a single case the hypothesis of right angle is true, always in every case it alone is true.

Proof. Let the join CD (Fig. 4) make right angles with any two equal perpendiculars AC, BD, standing upon any straight AB. Then CD will be (Proposition 3) equal to this AB. Assume in AC, and BD produced two straights CR, DX, equal to these AC, BD; and join RX. We may easily show that the join RX will be equal to this AB, and the angles at it right. And first indeed by superposition[1] of the quadrilateral ABDC upon the quadrilateral CDXR, applied to the common base CD. Also we may proceed more elegantly thus. Join AD, RD. It follows (*Elements* I, 4) in the triangles ACD, RCD, the bases AD, RD will be equal and likewise the angles CDA, CDR, and certainly ADB, RDX because equal remainders from a right angle. Whereby in turn (*Elements* I, 4) in the triangles ADB, RDX, the base AB will be equal to the base RX. Therefore (Proposition 4) the angles at the join RX will be right, and so we abide in the same hypothesis of right angle.

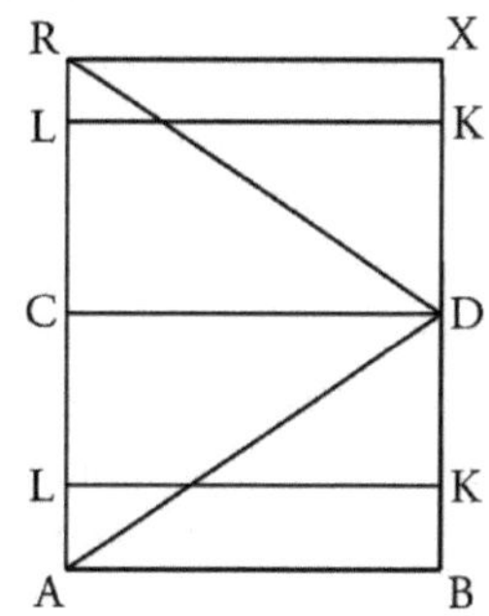

Fig. 4

Since now the length of the perpendiculars can be similarly increased infinitely, under the same base AB, the hypothesis of right angle always subsisting, it only remains to be proved that the same hypothesis will always abide in any case of diminution of those perpendiculars; which indeed is thus evinced.[2]

Assume in AR, and BX any two equal perpendiculars AL, BK, and join LK. If the angles at the join LK are not right, nevertheless (Proposition 1) they will be equal to each other. Therefore they will be toward one part, as suppose toward AB obtuse, and toward RX acute, since certainly the angles here at each of those points are (*Elements* I, 13) equal to two rights. But it also holds that the perpendiculars LR, KX, those standing upon RX, will be mutually equal. Therefore (Proposition 3) LK will be greater indeed than the opposite RX, and less than the opposite AB.

But this is absurd; because AB, and RX have been shown equal. Therefore the hypothesis of right angle is not changed by any diminution of the perpendiculars, whilst abides the once posited base AB.

But neither is the hypothesis of right angle changed for any diminution, or greater amplitude of the base; since manifestly may be considered as base any perpendicular BK, or BX, and therefore may be considered in turn as perpendiculars that AB, and the equal opposite sect KL, or XR.

Therefore is established that if even in a single case the hypothesis of right angle be true, always in every case it alone is true. This is what was to be demonstrated.

Propositio VI.

Hypothesis anguli obtusi, si vel in uno casu est vera, semper in omni casu illa sola est vera.

Demonstratur. Efficiat juncta CD (Fig. 5) angulos obtusos cum duobus quibusvis aequalibus perpendiculis AC, BD, uni cuivis rectae AB insistentibus. Erit CD (ex tertia hujus) minor ipsa AB. Sumantur in AC, BD protractis duae quaelibet invicem aequales portiones
[8] CR, DX; jungaturque RX. Jam quaero de angulis ad junctam RX, qui utique (ex prima hujus) aequales invicem erunt. Si obtusi sunt, habemus intentum. At recti non sunt; quia sic unum haberemus casum pro hypothesi anguli recti, qui nullum (ex praecedente) relinqueret locum pro hypothesi anguli obtusi. Sed neque acuti sunt. Nam sic esset RX (ex tertia hujus) major ipsa AB; ac propterea multo major ipsa CD. Hoc autem subsistere non posse sic ostenditur. Si quadrilaterum CDXR intelligatur impleri rectis abscindentibus ab ipsis CR, DX, portiones invicem aequales, implicat transiri a recta CD, quae minor est ipsa AB, ad RX eadem majorem, quin transeatur per quandam ST ipsi AB aequalem. Hoc autem absurdum esse in hac hypothesi ex eo constat; quia sic (ex quarta hujus) unus haberetur casus pro hypothesi anguli recti, qui nullum (ex praecedente) relinqueret locum hypothesi anguli obtusi. Igitur anguli ad junctam RX debent esse obtusi.

Deinde, sumptis in AC, BD, aequalibus portionibus AL, BK; simili modo ostendemus angulos ad junctam LK nequire esse acutos versus ipsam AB; quia sic illa foret major, quam AB, ac propterea multo major recta CD. Hinc autem reperiri deberet, ut supra, quaedam intermedia inter CD minorem, & LK majorem ipsa AB; intermedia, inquam, aequalis ipsi AB, quae utique, ex jam notis, omnem locum auferret hypothesi anguli obtusi. Tandem propter hanc ipsam causam recti esse nequeunt anguli ad junctam LK; ergo erunt obtusi. Igitur sub eadem basi A B, auctis, aut imminutis ad libitum perpendiculis, manebit semper hypothesis anguli obtusi.

Proposition 6.

If in even a single case the hypothesis of obtuse angle is true, always in every case it alone is true.

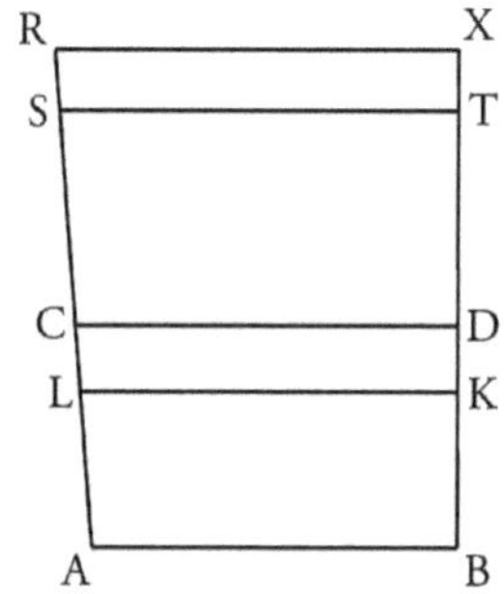

Fig. 5

Proof. Let the join CD (Fig. 5) make obtuse angles with any two equal perpendiculars AC, BD, standing upon any straight AB. Then CD will be (Proposition 3) less than this AB. Assume in AC and BD produced any two mutually equal portions CR and DX; and join RX. Now I investigate the angles at the join RX, which certainly (Proposition 1) will be mutually equal. If they are obtuse we have our assertion. But they are not right; because thus we would have a case for the hypothesis of right angle, which (Proposition 5) would leave no place for the hypothesis of obtuse angle. But neither are they acute. For thus RX would be (Proposition 3) greater than this AB; and still more therefore greater than CD itself. But that this cannot be is thus shown. If the quadrilateral CDXR is taken to be filled up by straights cutting off from these CR, DX, portions mutually equal, this implies transition from the straight CD, which is less than AB itself, to RX greater than it, verily transition[3] through a certain ST equal to this AB. But that this is absurd in the present hypothesis follows so; because thus (Proposition 4) we have a case for the hypothesis of right angle, which (Proposition 5) would leave no place for the hypothesis of obtuse angle. Therefore the angles at the join RX must be obtuse.

Then, equal portions AL, BK being assumed in AC, BD; in a similar manner we show the angles at the join LK cannot be acute toward this AB; because thus LK would be greater than AB, and still more therefore greater than the straight CD. But here would be found, as above, a certain intermediate between CD less, and LK greater than this AB; an intermediate, I say, equal to AB itself, which certainly, from what was just now observed, would take away every place for the hypothesis of obtuse angle. Finally from this very cause the angles at the join LK cannot be right; therefore they will be obtuse. Therefore with the same base AB, the perpendiculars being increased or diminished at will, the hypothesis of obtuse angle will always persist.

Sed debet idem demonstrari sub assumpta qualibet basi. Eligatur (Fig. 6) pro basi quod-
[9] libet ex praedictis perpendiculis, ut puta BX. Dividantur bifariam in punctis M, & H ipsae AB, RX; jungaturque MH. Erit MH (ex secunda hujus) perpendicularis ipsis AB, RX. Est autem angulus ad punctum B rectus ex hypothesi; & obtusus, ex jam demonstratis, ad punctum X. Fat igitur angulus rectus BXP versus partes ipsius MH. Occurret XP ipsi MH in quodam puncto P inter puncta M, & H constituto; cum ex una parte angulus BXH sit obtusus; & ex altera, si jungatur XM, angulus BXM (ex decimaseptima primi) sit acutus. Tum vero; quoniam quadrilaterum XBMP tres continet angulos rectos ex jam notis, & unum obtusum (ex decimasexta primi) in puncto P, quia est externus relate ad internum, & oppositum rectum angulum in puncto H trianguli PHX, erit latus XP (ex Cor. I. post tertiam hujus) minus contraposito BM. Quare; assumpta in BM portione BF aequali ipsi XP; erunt (ex prima hujus) anguli ad junctam PF invicem aequales, nimirum obtusi, cum angulus BFP (ex decimasexta primi) sit obtusus propter rectum angulum internum, & oppositum FMP. Igitur sub qualibet basi BX consistit hypothesis anguli obtusi.

Consistet autem, ut supra, eadem hypothesis sub eadem basi BX, quamvis aequalia perpendicula ad libitum augeantur, aut minuantur. Itaque constat hypothesim anguli obtusi, si vel in uno casu sit vera, semper in omni casu illam solam esse veram. Quod erat demonstrandum.

Propositio VII.

Hypothesis anguli acuti, si vel in uno casu est vera, semper in omni casu illa sola est vera.

Demonstratur facillime. Si enim hypothesis anguli acuti permittat aliquem casum alterutrius hypothesis aut anguli recti, aut anguli obtusi, jam (ex duabus praecedentibus)
[10] nullus relinquetur locus ipsi hypothesi anguli acuti; quod est absurdum. Itaque hypothesis anguli acuti, si vel in uno casu est vera, semper in omni casu illa sola est vera. Quod erat demonstrandum.

Propositio VIII.

Dato quovis triangulo (Fig. 7) *ABD, rectangulo in B, protrahatur DA usque ad aliquod punctum X, & per A erigatur ipsi AB perpendicularis HAC, existente puncto H ad partes anguli XAB. Dico angulum externum XAH aequalem fore, aut minorem, aut majorem interno, & opposito ADB, prout vera sit hypothesis anguli recti, aut anguli obtusi, aut anguli acuti: Et vicissim.*

But the same ought to be demonstrated for any assumed base. Let there be chosen (Fig. 6) for base any one of the aforesaid perpendiculars, as BX suppose. Let AB, RX be bisected in the points M and H; and MH joined. MH will be (Proposition 2) perpendicular to AB, RX. But the angle at the point B is right by hypothesis; and at the point X obtuse, from what has just now been demonstrated. Make therefore the right angle BXP toward the parts of this MH. XP will meet MH itself in some point P situated between the points M and H; since on the one hand the angle BXH is obtuse; and, on the other, if XM be joined, the angle BXM (*Elements* I, 17) is acute. Then however, since the quadrilateral XBMP contains three right angles, from what has just now been noted, and one obtuse (*Elements* I, 16) at the point P, because it is external[4] in relation to the internal and opposite right angle at the point H of the triangle PHX; the side XP will be (Corollary 1 to Proposition 3) less than the opposite BM. Wherefore, assuming in BM the portion BF equal to this XP, the angles at the join PF will be (Proposition 1) mutually equal, certainly obtuse, since the angle BFP (*Elements* I, 16) is obtuse because of the right angle interior and opposite FMP. Therefore the hypothesis of obtuse angle abides for any base BX.

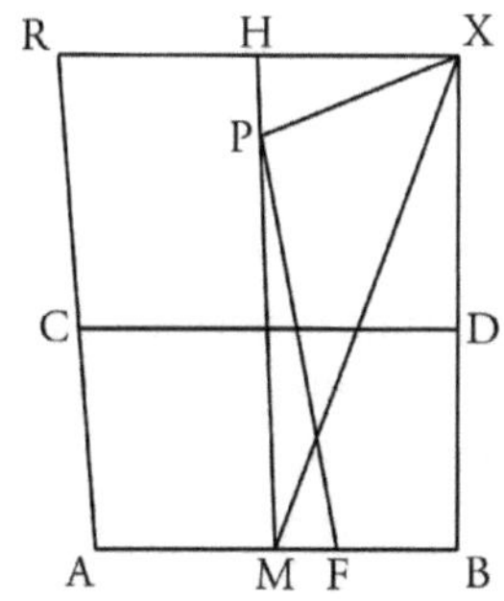

Fig. 6

But, as above, this hypothesis abides for this base BX, however much the equal perpendiculars are augmented or diminished at will. Therefore it holds, that if even in a single case the hypothesis of obtuse angle is true, always in every case it alone is true. This is what was to be demonstrated.

Proposition 7.

If even in a single case the hypothesis of acute angle is true, always in every case it alone is true.

Proof is very easily given. For if the hypothesis of acute angle should permit any case of either other hypothesis, either of right angle, or of obtuse angle, now (Propositions 5 and 6) no place would be left for the hypothesis of acute angle; which is absurd. Therefore if even in a single case the hypothesis of acute angle is true, always in every case it alone is true. This is what was to be demonstrated.

Proposition 8.

Given any triangle (Fig. 7) *ABD, right-angled at B; prolong DA to any point X, and through A erect HAC perpendicular to AB, the point H being within the angle XAB. I say the external angle XAH will be equal to, or less, or greater than the internal and opposite ADB, according as is true the hypothesis of right angle, or obtuse angle, or acute angle: and inversely.*

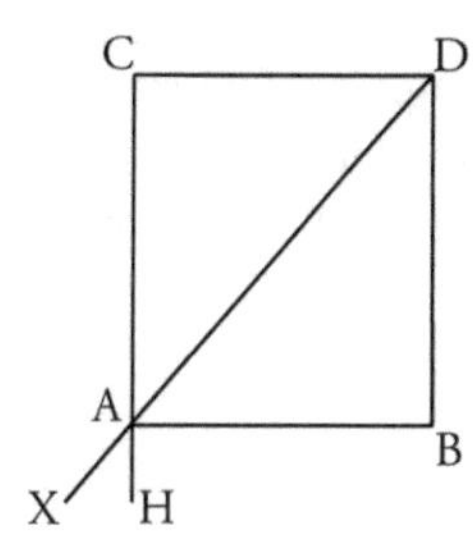

Fig. 7

Demonstratur. Sumatur in HC portio AC aequalis ipsi BD, jungaturque CD. Erit CD, in hypothesi anguli recti, aequalis (ex tertia hujus) ipsi AB. Quare angulus ADB aequalis erit (ex octava primi) angulo DAC, sive ejus aequali (ex decimaquinta primi) angulo XAH. Quod erat primo loco demonstrandum.

Tum, in hypothesi anguli obtusi, erit CD (ex eadem tertia hujus) minor ipsa AB. Quare in triangulis ADB, DAC erit (ex vigesimaquinta primi) angulus DAC, sive (ipsi ad verticem) XAH, minor angulo ADB. Quod erat secundo loco demonstrandum.

Tandem, in hypothesi anguli acuti, erit CD (ex eadem tertia hujus) major contraposita AB. Quare in praedictis triangulis, erit (ex eadem vigesimaquinta primi) angulus DAC, sive (ipsi ad verticem) XAH, major angulo ADB. Quod erat tertio loco demonstrandum.

Vicissim autem: si angulus CAD, sive ejus ad verticem XAH, aequalis sit interno, &
[11] opposito ADB; erit (ex quarta primi) juncta CD aequalis ipsi AB, ac propterea (ex quarta hujus) vera erit hypothesis anguli recti.

Sin vero angulus CAD, sive ejus ad verticem XAH, minor sit, aut major interno, & opposito ADB; erit etiam (ex vigesimaquarta primi) juncta CD minor, aut major ipsa AB; ac propterea (ex quarta hujus) vera erit respective hypothesis aut anguli obtusi, aut anguli acuti. Quae omnia erant demonstranda.

Propositio IX.

Cuiusvis trianguli rectanguli reliqui duo acuti anguli simul sumpti aequales sunt uni recto, in hypothesi anguli recti; majores uno recto, in hypothesi anguli obtusi; minores autem in hypothesi anguli acuti.

Demonstratur. Si enim angulus XAH (manente figura superioris Propositionis) aequalis est (nimirum, ex praecedente, in hypothesi anguli recti) angulo ADB; jam angulus ADB duos rectos efficiet cum angulo HAD, prout eos efficit (ex decimatertia primi) praedictus angulus XAH cum eodem angulo HAD. Quare, dempto recto angulo HAB, aequales manebunt uni recto duo simul anguli ADB, & BAD. Quod erat primum.

Tum vero; si angulus XAH minor est (nimirum, ex praecedente, in hypothesi anguli obtusi) angulo ADB, jam angulus ADB plusquam duos rectos efficiet cum angulo HAD, cum quo duos efficit rectos (ex praedicta decimatertia primi) angulus XAH. Quare, dempto angulo HAB, majores erunt uno recto duo simul anguli ADB, & BAD. Quod erat secundum.

Tandem, si angulus XAH major sit (nimirum, ex praecedente, in hypothesi anguli acuti) angulo ADB; jam angulus ADB minus quam duos rectos efficiet cum angulo HAD, cum
[12] quo duos efficit rectos (ex eadem decimatertia primi) angulus XAH. Quare, dempto angulo recto HAB, minores erunt uno recto duo simul anguli ADB, & BAD. Quod erat tertium.

Proof. Assume in HC the portion AC equal to BD, and join CD. CD will be, in the hypothesis of right angle (Proposition 3) equal to AB. Wherefore the angle ADB will be equal (*Elements* I, 8) to the angle DAC, or to its equal (*Elements* I, 15) the angle XAH. This is what was to be demonstrated in the first part.

Then, in the hypothesis of obtuse angle, CD will be (Proposition 3) less than AB. Wherefore in the triangles ADB, DAC the angle DAC, or its vertical XAH, will be (*Elements* I, 25) less than the angle ADB. This is what was to be demonstrated in the second part.

Finally, in the hypothesis of acute angle, CD will be (Proposition 3) greater than the opposite AB. Wherefore in the said triangles the angle DAC, or its vertical XAH, will be (*Elements* I, 25) greater than the angle ADB. This is what was to be demonstrated in the third part.

But inversely: if the angle CAD, or its vertical XAH, be equal to the internal and opposite ADB; the join CD will be (*Elements* I, 4) equal to AB, and therefore the hypothesis of right angle will be (Proposition 4) true.

But if however the angle CAD, or its vertical XAH, be less, or greater than the internal or opposite ADB; also the join CD will be (*Elements* I, 24) less or greater than AB; and therefore (Proposition 4) will be true respectively the hypothesis of obtuse angle, or acute angle. This is all was needed to be demonstrated.

Proposition 9.

In any right-angled triangle the two acute angles remaining are, taken together, equal to one right angle, in the hypothesis of right angle; greater than one right angle, in the hypothesis of obtuse angle; but less in the hypothesis of acute angle.

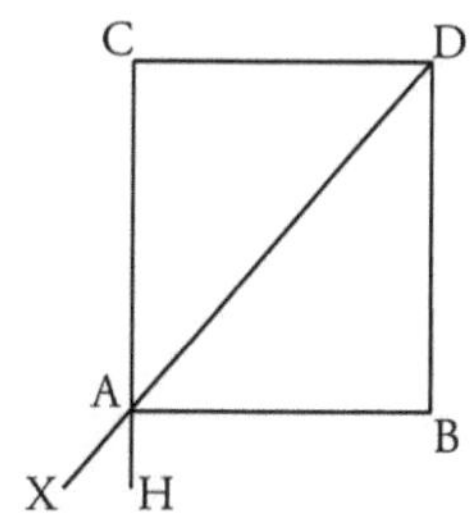

Fig. 7

Proof. For if the angle XAH (Fig. 7) is equal to the angle ADB (certainly, for Proposition 8, in the hypothesis of right angle), then the angle ADB makes up with the angle HAD two right angles, as (*Elements* I, 13) the aforesaid angle XAH makes them up with this angle HAD. Wherefore, the right angle HAB being subtracted, the two angles ADB and BAD remain together equal to one right angle. This is what was to be demonstrated in the first part.

However, if the angle XAH is less than the angle ADB (certainly, for Proposition 8, in the hypothesis of obtuse angle) then the angle ADB makes up with the angle HAD more than two right angles, since with it (*Elements* I, 13) the angle XAH makes up two. Wherefore, the angle HAB being subtracted, the two angles ADB and BAD will be together greater than one right angle. This is what was to be demonstrated in the second part.

Finally, if the angle XAH be greater than the angle ADB (certainly, for Proposition 8, in the hypothesis of acute angle) then the angle ADB will make up less than two right angles with the angle HAD, since with this (*Elements* I, 13) the angle XAH makes up two. Wherefore, subtracting the right angle HAB, the angles ADB and BAD will be together less than one right angle. This is what was to be demonstrated in the third part.

Propositio X.

Si recta DB (Fig. 8) *perpendiculariter insistat cuidam ABM, sitque juncta DM major juncta DA, etiam basis BM major erit basi BA. Et vicissim.*

Demonstratur. Et primo quidem non erunt illae bases invicem aequales. Caeterum (ex quarta primi) aequales forent, contra hypothesim, ipsae AD, DM. Sed neque erit BA major quam BM. Caeterum, sumpta in BA portione BS aequali ipsi BM, junctaque SD, aequales forent (ex eadem quarta primi) anguli BSD, BMD: Est autem angulus BSD (ex decimasexta primi) major angulo BAD. Ergo eodem major foret angulus BMD. Hoc autem est contra decimamoctavam primi; cum latus DM in triangulo MDA supponatur majus latere DA. Restat igitur, ut basis BM major sit basi BA. Quod erat primo loco demonstrandum.

Deinde si alterutra basis, ut puta BA (ne immutetur figura) fingatur major altera BM; tunc juncta DS, quae ex BA abscindat portionem SB aequalem ipsi BM, aequalis erit (ex quarta primi) junctæ DM. Rursum obtusus erit (ex decimasexta primi) angulus DSA, & acutus (ex decimaseptima ejusdem primi) angulus DAS. Quare (ex decimaoctava ejusdem) erit juncta DA major juncta DS, ejusque supposita aequali juncta DM. Quod erat secundo loco demonstrandum. Itaque constant proposita.

[13] **Propositio XI.**

Recta AP (quantaelibet longitudinis) secet duas rectas PL, AD (Fig. 9) *priorem quidem sub recto angulo in P, posteriorem vero in A sub quovis acuto angulo convergente ad partes ipsius PL. Dico rectas AD, PL (in hypothesi anguli recti) in aliquo puncto, & quidem ad finitam, seu terminatam distantiam, tandem coituras, si protrahantur versus illas partes, ad quas cum subjecta AP duos angulos efficiunt duobus rectis minores.*

Demonstratur. Protrahatur DA versus alias partes usque ad aliquod punctum X, & per A erigatur ipsi AP perpendicularis HAC, existente puncto H ad partes anguli XAP. Tum in AD protracta versus partes ipsius PL sumantur duo aequalia intervalla AP, DF, demittanturque ad subjectam AP perpendiculares DB, FM, quae utique cadent, propter decimamseptimam primi, ad partes anguli acuti DAP; jungaturque DM. Ostendere debeo junctam DM aequalem fore ipsi DF, sive DA.

Et primo quidem nequit DM major esse ipsa DF. Caeterum enim angulus DMF minor foret (ex decimaoctava primi) angulo DFM, sive ejus aequali (ex octava hujus, in hypothesi anguli recti) angulo XAH, sive ejus ad verticem CAD. Quare (cum anguli CAM, FMA ponantur aequales, utpote recti) reliquus angulus DMA major foret reliquo angulo DAM. Hoc autem absurdum est (contra decimamoctavam primi) si nempe DM ponatur major ipsa DF, sive DA.

Proposition 10.

If the straight DB (Fig. 8) *stand perpendicular to* a *straight ABM, and the join DM be greater than the join DA, then also the base BM will be greater than the base BA. And inversely.*

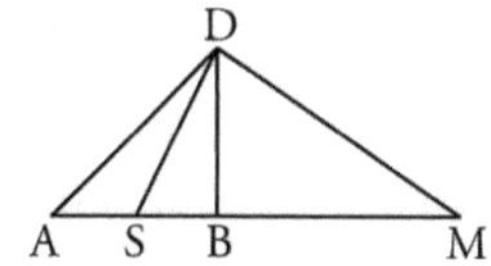

Fig. 8

Proof. And in the first place assuredly these bases will not be mutually equal. Otherwise (*Elements* I, 4) AD and DM would be equal, contrary to the hypothesis. But neither will BA be greater than BM. Otherwise, in BA the portion BS being taken equal to BM, and SD joined, the angles BSD, BMD (*Elements* I, 4) would be equal. But angle BSD is (*Elements* I, 16) greater than angle BAD. Therefore angle BMD would be greater than angle BAD. But this is contrary to *Elements* I, 18; since side DM in triangle MDA is supposed greater than side DA. It remains therefore, that the base BM is greater than the base BA. This is what was to be demonstrated in the first part.

Next if either base, as BA suppose (the figure need not be changed) is conceived as greater than the other BM; then the join DS, which cuts off from BA the portion SB equal to BM, will be equal (*Elements* I, 4) to the join DM. Again angle DSA will be obtuse (*Elements* I, 16) and angle DAS acute (*Elements* I, 17). Wherefore (*Elements* I, 18)[1] the join DA will be greater than the join DS, and the join supposed equal to it DM. This is what was to be demonstrated in the second part.

This completes the proof.

Proposition 11.

Let the straight AP (of any given length) cut the two straights PL, AD (Fig. 9), *the first indeed at the right angles in P, but the latter at A in any acute angle converging toward the parts PL. I say the straights AD, PL (in the hypothesis of right angle) will at length meet in some point, and indeed at a finite, or bounded distance, if they are prolonged toward those parts on which they make with the transversal AP two angles together less than two right angles.*

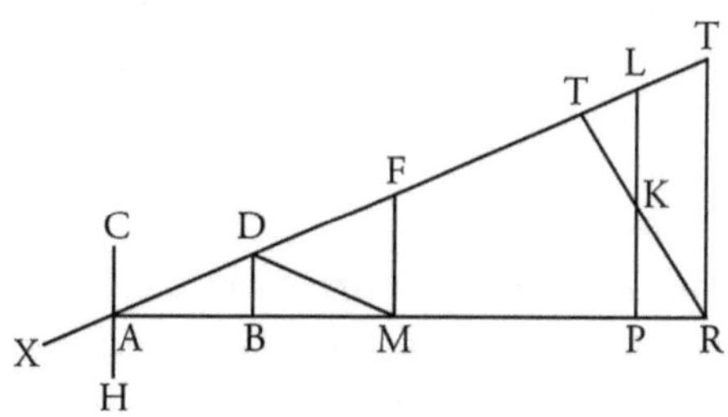

Fig. 9

Proof. Prolong DA toward the other parts even to some point X, and through A erect to AP the perpendicular HAC, the point H being toward the parts of the angle XAP. Then in AD produced toward the parts of PL assume two equal intervals AD,[1] DF, and let fall upon the transversal AP the perpendiculars DB, FM, which certainly (*Elements* I, 17) fall toward the parts of the acute angle DAP; and join DM. I should show that the join DM will be equal to DF, or DA.

And in the first place indeed DM cannot be greater than DF. For otherwise the angle DMF would be less (*Elements* I, 18) than the angle DFM, or its equal (Proposition 8, in the hypothesis of right angle) the angle XAH, or its vertical CAD. Wherefore (since the angles CAM, FMA are assumed equal, as being right)[2] the remaining angle DMA would be greater than the remaining angle DAM. But this is absurd (against *Elements* I, 18) if indeed DM is taken greater than DF or DA.

Sed neque erit DM minor ipsa DF. Caeterum angulus DMF major foret (ex eadem decimaoctava primi) angulo DFM, sive ejus aequali (ex praedicta octava hujus, in hypothesi anguli recti) angulo XAH, sive ejus ad verticem CAD. Quare rursum, ut supra, reliquus
[14] angulus DMA non major, sed minor foret reliquo angulo DAM. Hoc autem absurdum est (contra eandem decimamoctavam primi) si nempe DM ponatur minor ipsa DF, sive DA.

Restat igitur, ut juncta DM aequalis sit ipsi DF, sive DA. Quare in triangulo DAM aequales erunt (ex quinta primi) anguli ad puncta A, & M; atque ideo in triangulis DBA, DBM, rectangulis in B, aequales erunt (ex vigesimasexta primi) bases AB, BM. Quod quidem hoc loco intendebatur.

Quoniam igitur (assumpto in AD continuata intervallo AF duplo intervalli AD) perpendicularis FM ad subjectam AP demissa abscindit ex AP versus P basim AM duplam illius AB, quam abscindit perpendicularis demissa ex puncto D; manifestum est tot vicibus fieri posse hanc praecedentis intervalli duplicationem, ut sic in ipsa AD continuata deveniatur ad quoddam punctum T, ex quo perpendicularis demissa ad continuatam AP abscindat quandam AR majorem ipsa quantalibet finita AP. Constat autem evenire id non posse, nisi post occursum ipsius continuatae AD in quoddam punctum L ipsius PL. Si enim punctum T consisteret ante illum occursum, deberet ipsa perpendicularis TR secare eandem PL in quodam puncto K. Tunc autem in triangulo KPR invenirentur duo anguli recti in punctis P, & R; quod est absurdum contra decimamseptimam primi. Itaque constat rectas AD, PL sibi invicem (in hypothesi anguli recti) in aliquo puncto occursuras (& quidem ad finitam, seu terminatam distantiam) si protrahantur versus illas partes, ad quas cum subjecta AP (quantaelibet finitae longitudinis) duos angulos efficiunt duobus rectis minores. Quod erat demonstrandum.

[15] **Propositio XII.**

Rursum dico rectam AD alicubi ad eas partes occursuram rectae PL (& quidem ad finitam, seu terminatam distantiam) etiam in hypothesi anguli obtusi.

Demonstratur. Nam sumpta, ut in superiore Propositione, DF aequali ipsi AD, demissisque jam notis perpendicularibus, ostendere debeo junctam DM majorem fore ipsa DF, sive DA, atque ideo (ex decima hujus) rectam BM majorem fore ipsa AB. Et primo non erit DM aequalis ipsi DF. Caeterum angulus DMF aequalis foret (ex quarta primi) angulo DFM, atque ideo major (ex octava hujus in hypothesi anguli obtusi) angulo externo XAH, sive ejus ad verticem CAF. Quare (cum anguli CAM, FMA ponantur aequales utpote recti) reliquus angulus DMA minor foret reliquo angulo DAM. Quod est absurdum contra quintam primi, si nempe DM aequalis sit ipsi DF, sive DA.

Sed neque ipsa DM minor est altera DF, sive DA Caeterum (ex decimaoctava primi) angulus DMF major foret angulo DFM, atque ideo (in hac hypothesi anguli obtusi) multo major angulo externo XAH, sive ejus ad verticem CAD. Quare rursum, ut supra, reliquus angulus DMA multo minor foret reliquo angulo DAM. Hoc autem absurdum est, contra eandem decimamoctavam primi, si nempe DM minor sit ipsa DF, sive DA.

But neither will DM be less than this DF. Otherwise the angle DMF would be greater (*Elements* I, 18) than the angle DFM, or its equal (Proposition 8, in hypothesis of right angle) the angle XAH, or its vertical CAD. Wherefore again, as above, the remaining angle DMA will not be greater, but less than the remaining angle DAM. But this is absurd (against *Elements* I, 18) if indeed DM is taken less than DF, or DA.

It remains therefore, that the join DM is equal to DF, or DA. Wherefore in the triangle DAM (*Elements* I, 5) the angles at the points A, and M will be equal; and therefore in the triangles DBA, DBM, right-angled at B, the bases AB, BM will be equal (*Elements* I, 26). This indeed was here our aim.

Since therefore (assuming in AD produced the interval AF double the interval AD) the perpendicular FM let fall on the transversal AP cuts off from AP toward P a base AM double AB, which the perpendicular let fall from the point D cuts off; it is manifest that this duplication of the preceding interval can be so many times repeated, that thus in AD continued we attain to a certain point T, from which the perpendicular let fall upon AP prolonged cuts off a certain AR greater than the finite AP however great.[3]

But it is evident this cannot happen, except after the meeting of the prolonged AD with PL in some point L. For if the point T occurred before that meeting, the perpendicular TR must cut PL in some point K.[4] But then in the triangle KPR would be found two right angles at the points P and R; which is absurd (against *Elements* I, 17). Therefore it holds that the straights AD, PL meet each other mutually (in the hypothesis of right angle) in some point (and indeed at a finite or bounded distance)[5] if they be produced toward that side, on which with the transversal AP (of finite length as great as you choose) they make two angles together less than two right angles. This is what was to be demonstrated.

Proposition 12.

Again I say also in the hypothesis of obtuse angle the straight AD will meet the straight PL somewhere toward those parts (and indeed at a finite, or bounded distance).

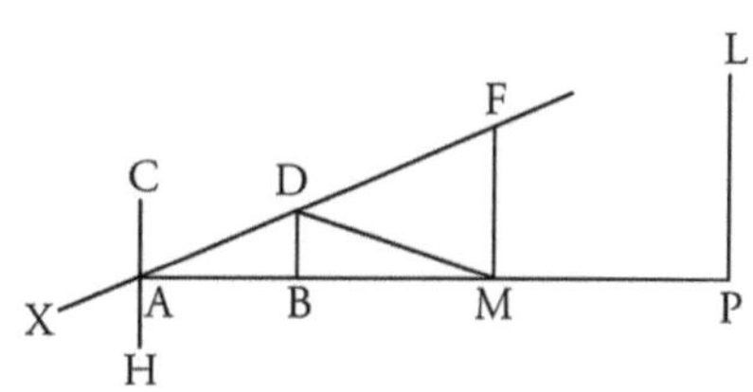

Fig. 10

Proof. For, as in Proposition 11, DF being assumed equal to AD (Fig. 10),[6] and the just noted perpendiculars[7] let fall, I must show the join DM will be greater than DF, or DA, and therefore (Proposition 10) the straight BM will be greater than AB.

And in the first place DM will not be equal to DF. Otherwise the angle DMF would be equal (*Elements* I, 4)[8] to the angle DFM, and therefore greater (Proposition 8, in the hypothesis of obtuse angle)[9] than the external angle XAH, or its vertical CAF. Wherefore (since the angles CAM, FMA are taken equal, as being right) the remaining angle DMA would be less than the remaining angle DAM. This is absurd (against *Elements* I, 15), if indeed DM be equal to DF, or DA.

Restat igitur, ut juncta DM major sit ipsa DF, sive DA, atque ideo (ex decima hujus) ipsa BM major sit altera AB. Quod erat hoc loco intentum.

Quoniam igitur, assumpto in AD continuata intervallo AF duplo intervalli AD, perpendicularis FM ad subjectam AP demissa plus duplo ex eadem abscindit, quam abscindatur a
[16] perpendiculari demissa ex puncto D; multo citius in hac hypothesi anguli obtusi, quam in superiore hypothesi anguli recti, devenietur ad tantum intervallum, ex quo perpendicularis demissa abscindat basim majorem ipsa quantalibet designata AP. Hoc autem, ut in superiore Propositione, contingere nequit, nisi post occursum continuatae AD in aliquod punctum ipsius P L; & quidem ad finitam, seu terminatam distantiam. Quod erat &c.

Propositio XIII.

Si recta XA (quantaelibet designatae longitudinis) incidens in duas rectas AD, XL efficiat cum eisdem ad easdem partes (Fig. 11) *angulos internos XAD, AXL minores duobus rectis: dico, illas duas (etiamsi neuter illorum angulorum sit rectus) tandem in aliquo puncto ad partes illorum angulorum invicem coituras, & quidem ad finitam, seu terminatam distantiam, dum consistat alterutra hypothesis aut anguli recti, aut anguli obtusi.*

Demonstratur. Nam unus praedictorum angulorum, ut puta AXL, erit acutus. Itaque ex apice alterius anguli demittatur ad XL perpendicularis AP, quae utique (propter decimam septimam primi) cadet ad partes anguli acuti AXL. Quoniam igitur in triangulo APX, rectangulo in P, duo simul anguli acuti PAX, PXA, minores non sunt (ex nona hujus) uno recto, in utraque hypothesi aut anguli recti, aut anguli obtusi; si duo isti anguli auferantur a summa angulorum propositorum jam reliquus angulus PAD minor erit recto. Itaque erimus in casu duarum praecedentium Propositionum, dum scilicet alterutra hypothesis consistat aut anguli recti, aut anguli obtusi. Quare (ex eisdem) rectae AD, & PL, sive XL, in aliquo
[17] puncto finitae, seu terminatae distantiae ad notas partes concurrent, tam sub una, quam sub altera praedictarum hypothesium. Quod erat demonstrandum.

But neither is DM less than DF, or DA. Otherwise (*Elements* I, 18) the angle DMF would be greater than the angle DFM, and therefore still greater (in the hypothesis of obtuse angle) than the external angle XAH, or its vertical CAD. Wherefore again, as above, the remaining angle DMA would be still less than the remaining angle DAM. But this is absurd (against *Elements* I, 18) if indeed DM be less than DF, or DA.

It remains therefore, that the join DM is greater than DF, or DA, and therefore (Proposition 10) BM is greater than AB. This indeed was here our aim.

Since therefore, assuming in AD produced the interval AF double the interval AD, the perpendicular FM let fall on the transversal AP cuts off from it more than double what is cut off by the perpendicular let fall from the point D: more quickly by far in this hypothesis of obtuse angle, than in the preceding hypothesis of right angle, we attain to an interval so great, that from it the perpendicular let fall cuts off a base greater than the designated AP however great.

But this, as in Proposition 11, could not happen, unless after the meeting of the produced AD with PL in some point; and indeed at a finite, or bounded distance.

This is what was to be demonstrated.

Proposition 13.

If the straight XA (of designated length however great) meeting two straights AD, XL, makes with them toward the same parts (Fig. 11) *internal angles XAD, AXL less than two right angles: I say, these two (even if neither of those angles be a right angle) at length will mutually meet some point on the side toward those angles, and indeed at a finite, or bounded distance, if either hypothesis holds, of right angle or of obtuse angle.*

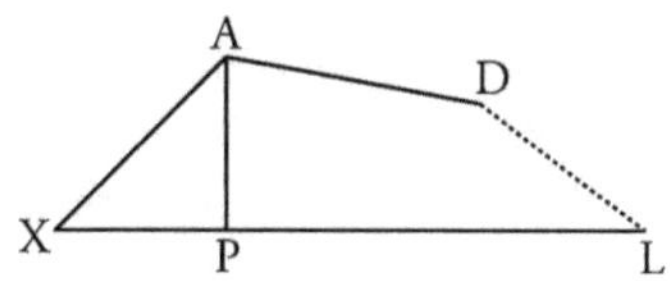

Fig. 11

Proof. For one of the said angles, as AXL suppose, will be acute. Accordingly from the vertex of the other angle is dropped the perpendicular AP on XL, which certainly (because of *Elements* I, 17) falls on the side of the acute angle AXL. Since therefore in the triangle APX, right-angled at P, the two acute angles PAX, PXA, together are not less (Proposition 9) than a right angle, in either hypothesis, of right angle, or of obtuse angle; if these two angles are taken away from the sum of the given angles (XAD and AXL) the then remaining angle PAD will be less than a right angle. Consequently we will be in the case of the two preceding Propositions, since it is obvious that one or the other hypothesis holds, either of right angle, or of obtuse angle. Wherefore (Proposition 11 and 12) the straights AD, and PL, or XL, meet in some point at a finite, or bounded distance on the side noted, as well under the one as under the other mentioned hypothesis. This is what was to be demonstrated.

Scholion I.

Ubi observare licet notabile discrimen ab hypothesi anguli acuti. Nam in ista demonstrari nequiret generalis hujusmodi rectarum concursus, quoties recta aliqua in duas incidens, duos ad easdem partes efficiat internos angulos duobus rectis minores; nequiret, inquam, directe demonstrari, etiamsi in eadem hypothesi admitteretur praedictus generalis concursus, quoties unus duorum angulorum est rectus. Quamvis enim recta AD perpendicularis & ipsa foret rectae AP; quo casu nequiret certe, propter 17. primi, concurrere cum altera perpendiculari PL; nihilominus duo simul anguli DAX, PXA, minores forent duobus rectis, juxta hypothesim praedictam, cum in ea duo simul anguli PAX, PXA minores sint (ex nona hujus) uno recto. Id autem observasse operae pretium fuit.

Qualiter vero ex eo solo admisso generali concursu, dum unus angulorum est rectus, & quidem sub assignata quantumlibet parva incidente, destrui possit hypothesis anguli acuti; docebimus post tres sequentes Propositiones.

Scholion II.

In tribus ante jactis theorematis studiose apposui illam conditionem, quod recta incidens AP, sive XA, intelligatur esse *quantaelibet designatae longitudinis*. Si enim, citra omnem rectae incidentis determinatam mensuram, praecise agatur de exhibendo, ac demonstrando
[18] duarum rectarum concursu in apicem cujusdam trianguli, cujus anguli ad basim sint dati (minores utique duobus rectis) ut puta unus rectus, & altet duobus tantum gradibus, vel, ut libet, minus deficiens a recto; quis est tam expers Geometriae, quin statim rem ipsam demonstrative exhibeat? Nam supponatur (Fig. 12) datus quilibet angulus BAP, ut puta 88. graduum. Si ergo ex quolibet puncto B ipsius AB, demittatur ad subjectam AP (juxta duodecimam primi) perpendicularis BP, constat enim vero in eo triangulo ABP exhibitum fore demonstrative concursum optatum in eo puncto B.

Quod si alter angulus ad basim postuletur & ipse minor recto, ut puta 84. graduum, quem nempe exhibeat datus angulus K: tunc (juxta 23. primi) efficere poteris versus partes rectae AB aequalem angulum APD, occurrente PD ipsi AB in quodam ejus intermedio puncto D. Quare habebitur rursum demonstrative concursus optatus in eo puncto D.

Scholium 1.

Here may be observed a notable difference from the hypothesis of acute angle. For in this the general concurrence of such straights cannot be demonstrated, as often as any straight falling upon two, makes two internal angles toward the same parts less than two right angles; cannot, I say, be directly demonstrated, even if in this hypothesis the aforesaid general concurrence be admitted, as often as one of the two angles is right. For although the straight AD be perpendicular even to the straight AP; in which case it certainly could not concur with another perpendicular PL (*Elements* I, 17); nevertheless the two angles together DAX, PXA, could be less than two right angles, in accordance with the aforesaid hypothesis, since in it the two angles together PAX, PXA may be less (Proposition 9) than one right angle. But it was worth while to have observed this.[10]

But how, solely from the general admission of concurrence when one of the angles is right, and with an assigned indent however small, the hypothesis of acute angle can be demolished; this we shall show after the next three Propositions.[11]

Scholium 2.

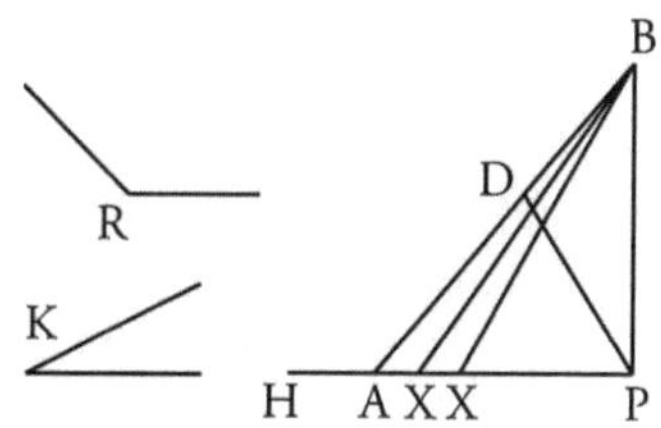

Fig. 12

In the three preceding theorems I have studiously set down this condition, that the cutting straight AP, or XA, is understood to be of *a designated length as great as you choose.* For if, without any determinate extent of the cutting straight, it be discussed precisely concerning the exhibiting and demonstrating of the concurrence of two straights at the apex of a certain triangle, whose angles at the base are given (less indeed than two right angles) as, suppose one right, and the other less than a right by as much as two degrees, or, if you please, by less; who is so devoid of geometry that he could not immediately show the thing demonstratively? For suppose (Fig. 12) given any angle BAP, as, say, 88 degrees. If therefore from any point B of this AB, is let fall on the base AP (*Elements* I, 12) the perpendicular BP, it certainly that in this triangle ABP would be exhibited demonstratively the desired concurrence at this point B.

But if the other angle at the base is postulated, and is less than a right, as, suppose, 84 degrees, which indeed the given angle K represents: then (*Elements* I, 23) one would be able to make toward the parts of the straight AB an equal angle APD, PD meeting this AB in D, some intermediate point of it. Wherefore the desired concourse is again obtained demonstratively in this point D.

Tandem vero: si alter angulus postuletur obtusus, sed minor tamen 92. gradibus, ne cum alio dato angulo BAP compleantur duo recti: exhibitus hic sit in quodam angulo R 91. graduum. Ostendendum est, unum aliquod esse punctum X in ipsa AP, ad quod juncta BX efficiat angulum BXA aequalem dato angulo R 91. graduum; adeo ut propterea sub quadam recta incidente AX habeatur concursus optatus in praedicto puncto B. Sic autem proceditur. Quandoquidem (protracta PA usque in aliquod punctum H) angulus externus BAH & est (propter decimamtertiam primi) 92. graduum, cum angulus interior BAP positus sit 88. graduum; ac rursum, propter decimamsextam primi, major est non solum angulo recto BPA, verum etiam quibusvis eodem titulo obtusis angulis BXA, sumpto puncto X ubilibet intra ipsam PA, & quidem, propter eandem decimamsextam primi, semper majoribus, dum
[19] punctum X assumitur propius puncto A: consequens plane est, ut inter istos angulos, unum 90. graduum in puncto P, & alterum 92. graduum in puncto A, unus reperiatur angulus BXA, qui sit 91. graduum, nimirum aequalis dato angulo R.

Nihilominus, omissa postrema hac observatione circa angulum obtusum, cavere diligentissime oportet, in eo positam esse difficultatem illius pronunciati Euclidaei, quod velit occursum duarum rectarum; in illam utique partem, ad quam cum recta incidente duos angulos efficiant duobus rectis minores; atque ita quidem praedictum occursum velit, *quantaecunque longitudinis sit incidens assignata*. Caeterum enim (ut jam monui in praecedente Scholio) demonstrabo generalem istum occursum ex solo admisso occursu ejusmodi, dum unus angulorum sit rectus; & quidem, etiamsi admisso non pro qualibet assignabili finita incidente, sed solum admisso intra limites cujusdam assignatae parvissimae incidentis.

Propositio XIV.

Hypothesis anguli obtusi est absolute falsa, quia se ipsam destruit.

Demonstratur. Ex hypothesi anguli obtusi, assumpta ut vera, jam elicuimus veritatem illius Pronunciati Euclidaei; quod duae rectae sibi invicem in aliquo puncto ad eas partes occursurae sint, ad quas recta quaedam, easdem secans, duos qualescunque effecerit internos angulos, duobus rectis minores. Stante autem hoc Pronunciato, cui innititur Euclides post vigesimamoctavam sui Libri primi, manifestum est omnibus Geometris, solam hypothesim anguli recti esse veram, nec ullum relinqui locum hypothesi anguli obtusi. Igitur hypothesis anguli obtusi est absolute falsa, quia se ipsam destruit. Quod erat demonstrandum.

[20] Aliter, ac magis immediate. Quandoquidem ex hypothesi anguli obtusi demonstravimus (in nona hujus) duos (Fig. 11) acutos angulos trianguli APX, rectanguli in P, majores esse uno recto; constat talem assumi posse acutum angulum PAD, qui simul cum praedictis duobus acutis angulis duos rectos efficiat. Tunc autem recta AD deberet (ex praecedente, juxta hypothesim anguli obtusi) aliquando concurrere cum ipsa PL, sive XL, respectu habito ad secantem, sive incidentem AP; quod est manifestum absurdum contra decimamseptimam primi, si respicias ad secantem, sive incidentem AX.

But finally: if the other angle is postulated obtuse, but yet less than 92 degrees, lest with the other given angle BAP it should make up two rights: this may be represented in a given angle R of 91 degrees. It is to be shown, that there is some point X of this AP, to which the join BX makes an angle BXA equal to the given angle R of 91 degrees: so that therefore under a certain cutting straight AX the desired meeting in the point B may be obtained. Now we may proceed thus. PA being produced to any point H, since the external angle BAH is (*Elements* I, 13) 92 degrees, because the interior angle BAP is by hypothesis 88 degrees; and again (*Elements* I, 16) is greater not alone than the right angle BPA but also, for the same reason, than any obtuse angle BXA, the point X being assumed wherever you choose within this PA, and indeed always greater as the point X is assumed nearer to the point A (*Elements* I, 16): it is an evident consequence, that between those angles, one of 90 degrees at the point P, and the other of 92 degrees at the point A, one angle BXA is found, which is 91 degrees, truly equal to the given angle R.

None the less, omitting here the last observation about the obtuse angle, it is necessary most diligently to take care that the difficulty of this Assertion of Euclid be fixed in this, that it asserts the meeting of two straights; in particular in that part toward which they make with the cutting straight two angles together less than two right angles; and assuredly that it asserts the aforesaid meeting thus, *of whatever length be the assigned transversal.*[12] However (as I have already mentioned in Scholium 1) I shall demonstrate the general meeting solely from the admitted meeting of this sort when one of the angles is right; and indeed even if it be admitted not for any assignable finite transversal, but alone admitted within the limits of any assigned very small transversal.

Proposition 14.

The hypothesis of obtuse angle is absolutely false, because it destroys itself.

Proof. From the hypothesis of obtuse angle, assumed as true, we have now deduced the truth of Euclid's Assertion: that two straights will meet each other in same point toward those parts, toward which a certain straight, cutting them, makes two internal angles, of whatever kind, less than two right angles. But this Assertion holding good, on which Euclid supports himself after *Elements* I, 28, it is manifest to all Geometers that the hypothesis of right angle alone is true, nor any place left for the hypothesis of obtuse angle. Therefore the hypothesis of obtuse angle is absolutely false, because it destroys itself. This is what was to be demonstrated.

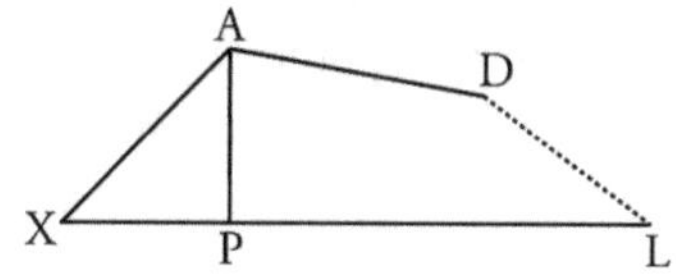

Fig. 11

Otherwise, and more immediately. Since from the hypothesis of obtuse angle we have proved (Proposition 9) that two (Fig. 11) acute angles of the triangle APX, right-angled at P, are greater than one right angle; it follows that an acute angle PAD may be assumed such, that together with the aforesaid two acute angles it makes up two right angles. But then the straight AD must (by Proposition 13, in the hypothesis of obtuse angle) at length meet with this PL, or XL, regard being had to the secant, or incident AP; which is manifestly absurd (against *Elements* I, 17) if we regard the secant, or incident AX.

Propositio XV.

Ex quolibet triangulo ABC, cujus tres simul anguli (Fig. 13) *aequales sint, aut majores, aut minores duobus rectis, stabilitur respective hypothesis aut anguli recti, aut anguli obtusi, aut anguli acuti.*

Demonstratur. Nam duo saltem illius trianguli anguli, ut puta ad puncta A, & C, acuti erunt, propter decimamseptimam primi. Quare perpendicularis, ex apice reliqui anguli B ad ipsam AC demissa, secabit ipsam AC (propter eandem decimamseptimam primi) in aliquo puncto intermedio D. Si ergo tres anguli ipsius trianguli ABC supponantur aequales duobus rectis, constat aequales fore quatuor rectis omnes simul angulos triangulorum ADB, CDB, propter duos additos rectos angulos ad punctum D. Hoc stante: neutrius modo dictorum triangulorum, ut puta ADB, tres simul anguli minores erunt, aut majores duobus rectis; nam sic viceversa alterius trianguli tres simul anguli majores forent, aut minores duobus rectis. Quare (ex nona hujus) ab uno quidem triangulo stabiliretur hypothesis
[21] anguli acuti, & ab altero hypothesis anguli obtusi; quod repugnat sextae, & septimae hujus. Igitur tres simul anguli utriusque praedictorum triangulorum aequales erunt duobus rectis; ac propterea (ex nona hujus) stabilietur hypothesis anguli recti. Quod erat primo loco demonstrandum.

Sin autem tres anguli propositi trianguli ABC ponantur majores duobus rectis; jam duorum triangulorum ADB, CDB omnes simul anguli majores erunt quatuor rectis, propter duos additos rectos angulos ad punctum D. Hoc stante: neutrius modo dictorum triangulorum tres simul anguli aequales praecise erunt, aut minores duobus rectis; nam sic viceversa alterius trianguli tres simul anguli majores forent duobus rectis. Quare (ex nona hujus) ab uno quidem triangulo stabiliretur hypothesis aut anguli recti, aut anguli acuti, & ab altero hypothesis anguli obtusi, quod repugnat quintae, sextae, & septimae hujus. Igitur tres simul anguli utriusque praedictorum triangulorum majores erunt duobus rectis; ac propterea (ex nona hujus) stabilietur hypothesis anguli obtusi. Quod erat secundo loco demonstrandum.

Tandem vero. Si tres anguli propositi trianguli ABC ponantur minores duobus rectis, jam duorum triangulorum ADB, CDB, omnes simul anguli minores erunt quatuor rectis, propter duos additos rectos angulos ad punctum D. Hoc stante: neutrius modo dictorum triangulorum tres simul anguli aequales erunt, aut majores duobus rectis; nam sic viceversa alterius trianguli tres simul anguli minores forent duobus rectis. Quare (ex nona hujus) ab uno quidem triangulo stabiliretur hypothesis aut anguli recti, aut anguli obtusi, & ab altero hypothesis anguli acuti; quod repugnat quintae, sextae, & septimae hujus. Igitur tres simul anguli utriusque praedictorum triangulorum minores erunt duobus rectis;
[22] ac propterea (ex nona hujus) stabilietur hypothesis anguli acuti. Quod erat tertio loco demonstrandum.

Proposition 15.

By any triangle ABC, of which the three angles (Fig. 13) *are equal to, or greater, or less than two right angles, is established respectively the hypothesis of right angle, or obtuse angle, or acute angle.*

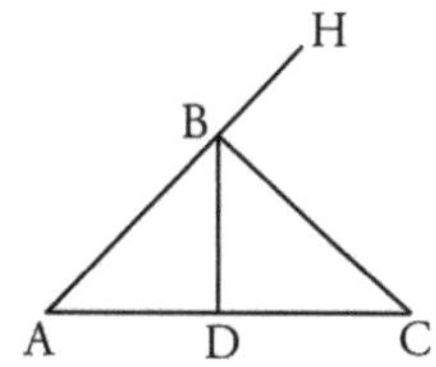

Fig. 13

Proof. For anyhow two angles of this triangle, as suppose at the points A and C, will be acute (*Elements* I, 17). Wherefore the perpendicular, let fall from the apex of the remaining angle B upon AC, will cut AC (*Elements* 1. 17) in some intermediate point D. If therefore the three angles of this triangle ABC are supposed equal to two right angles, it follows that all the angles of the triangles ADB, CDB will together equal to four right angles, because of the two additional right angles at the point D. This holding good, now of neither of the said triangles, as suppose ADB, will the three angles together be less, or greater than two right angles; for thus the three angles together of the other triangle would be (respectively) greater, or less than two right angles. Wherefore (Proposition 9) from one triangle would indeed be established the hypothesis of acute angle, and from the other the hypothesis of obtuse angle;[1] which is contrary to Propositions 6 and 7. Therefore the three angles together of either of the aforesaid triangles will be equal to two right angles; and thereby (Proposition 9) is established the hypothesis of right angle. This is what was to be demonstrated in the first part.

But if however the three angles of the proposed triangle ABC are taken greater than two right angles; now of the two triangles ADB, CDB all the angles together will be greater than four right angles, because of the two additional right angles at the point D. This holding good: now of neither of the said triangles will the three angles together be precisely equal to, or less than two right angles: for thus the three angles of the other triangle would be together greater than two right angles. Wherefore (Proposition 9) from one triangle indeed would be established the hypothesis either of right angle or of acute angle, and from the other the hypothesis of obtuse angle, which is contrary to Propositions 5, 6, and 7. Therefore the three angles together of either of the aforesaid triangles will be greater than two right angles; and therefore is established the hypothesis of obtuse angle. This is what was to be demonstrated in the second part.

But, finally, if the three angles of the proposed ABC are taken less than two right angles, now of the two triangles ADB, CDB, all the angles together will be less than four right angles, because of the two additional right angles at the point D. This holding good: now of neither of the said triangles will the three angles together be equal to, or greater than two right angles; for thus of the other triangle the three angles together would be less than two right angles. Wherefore (Proposition 9) from one triangle indeed would be established the hypothesis either of right angle or of obtuse angle, and from the other the hypothesis of acute angle; which is contrary to Propositions 5, 6, and 7. Therefore the three angles together of either of the aforesaid triangles will be less than two right angles; and therefore (Proposition 9) is established the hypothesis of acute angle. This is what was to be demonstrated in the third part.

Itaque ex quolibet triangulo ABC, cujus tres simul anguli aequales sint, aut majores, aut minores duobus rectis, stabilitur respective hypothesis aut anguli recti, aut anguli obtusi, aut anguli acuti. Quod erat propositum.

Corollarium.

Hinc; protracto uno quolibet cujusvis propositi trianguli latere, ut puta AB in H, erit (ex 13. primi) externus angulus HBC aut aequalis, aut minor, aut major reliquis simul internis, & oppositis angulis ad puncta A, & C, prout vera fuerit hypothesis aut anguli recti, aut anguli obtusi, aut anguli acuti. Et vicissim.

Propositio XVI.

Ex quolibet quadrilatero ABCD, cujus quatuor simul anguli aequales sint, aut majores, aut minores quatuor rectis, stabilitur respective hypothesis aut anguli recti, aut anguli obtusi, aut anguli acuti.

Demonstratur. Jungatur AC. Non erunt (Fig. 14) tres simul anguli trianguli ABC aequales, aut majores, aut minores duobus rectis, quin tres simul anguli trianguli ADC sint ipsi etiam respective aequales, aut majores, aut minores duobus rectis; ne scilicet (ex praecedente) ab uno illorum triangulorum stabiliatur una hypothesis, & ab altero altera, contra quintam, sextam, & septimam hujus. Hoc stante: Si quatuor simul anguli propositi quadrilateri aequales sint quatuor rectis, constat utriusque modo dictorum triangulorum
[23] tres simul angulos aequales fore duobus rectis, atque ideo (ex praecedente) stabilitum iri hypothesim anguli recti.

Sin vero ejusdem quadrilateri quatuor simul anguli majores sint, aut minores quatuor rectis, debebunt similiter illorum triangulorum tres simul anguli respective esse aut una majores, aut una minores duobus rectis. Quare ab illis triangulis stabilietur respective (ex praecedente) aut hypothesis anguli obtusi, aut hypothesis anguli acuti.

Itaque ex quolibet quadrilatero, cujus quatuor simul anguli aequales sint, aut majores, aut minores quatuor rectis, stabilitur respective hypothesis aut anguli recti, aut anguli obtusi, aut anguli acuti. Quod erat demonstrandum.

Corollarium.

Hinc: protractis versus easdem partes duobus quibusvis propositi quadrilateri contrapositis lateribus, ut puta AD in H, & BC in M; erunt (ex 13. primi) duo simul externi anguli HDC, MCD aut aequales, aut minores, aut majores duobus simul internis, & oppositis angulis ad puncta A, & B, prout vera fuerit hypothesis aut anguli recti, aut anguli obtusi, aut anguli acuti.

Accordingly by any triangle ABC, of which the three angles are together equal to, or greater, or less than two right angles, is established respectively the hypothesis of right angle, or obtuse angle, or acute angle. This is what was to be demonstrated.

Corollary.

Hence, any one side of any proposed triangle being produced, as suppose AB to H; the external angle HBC will be (*Elements* I, 13) equal to, or less, or greater than the remaining internal and opposite angles together at the points A, and C, according as is true the hypothesis of right angle, or obtuse angle, or acute angle.[2] And inversely.

Proposition 16.

By any quadrilateral ABCD, of which the four angles together are equal to, or greater, or less than four right angles, is established respectively the hypothesis of right angle, or obtuse angle, or acute angle.

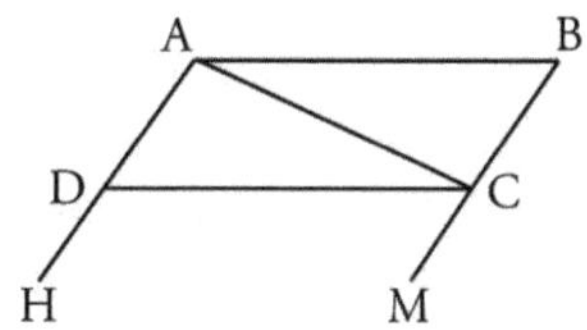

Fig. 14

Proof. Join AC. The three angles of the triangle ABC (Fig. 14) will not be together equal to, or greater, or less than two right angles without the three angles of the triangle ADC being themselves also together respectively equal to, or greater, or less than two right angles, lest obviously (Proposition 15) from one of those triangles be established one hypothesis, and another from the other, against Propositions 5, 6 and 7. This holding good: if the four angles together of the premised quadrilateral are equal to four right angles, it follows that the three angles together of either of the just mentioned triangles will be equal to two right angles, and therefore (Proposition 15) the hypothesis of right angle will be established.

But if indeed the four angles of this quadrilateral be together greater, or less than four right angles, similarly the three angles together of those triangles should be respectively either at the same time greater, or at the same time less than two right angles. Wherefore from these triangles would be established respectively (Proposition 15) either the hypothesis of obtuse angle, or the hypothesis of acute angle.

Therefore by any quadrilateral, of which the four angles together are equal to, or greater, or less than four right angles, is established respectively the hypothesis of right angle, or obtuse angle, or acute angle. This is what was to be demonstrated.

Corollary.

Hence, any two opposite sides of the premised quadrilateral being produced toward the same parts, as suppose AD to H, and BC to M; the two external angles HDC, MCD will be (*Elements* I, 13) either equal to, or less, or greater than the two internal and opposite angles together at the points A, and B, according as is true the hypothesis of right angle, or obtuse angle, or acute angle.

Propositio XVII.

Si uni, ut libet, cuidam parvae rectae AB insistat (Fig. 15) *ad rectos angulos recta AH: Dico subsistere non posse in hypothesi anguli acuti, ut quaevis BD, efficiens cum AB quemlibet angulum acutum versus partes ipsius AH, occursura tandem sit ad finitam, seu terminatam distantiam ipsi AH productae.*

Demonstratur. Jungatur HB. Erit (ex 17. primi) acutus angulus ABH, propter angulum
[24] rectum ad punctum A. Jam (ex 23. primi) ducatur quaedam HD versus partes puncti B, quae non secans angulum AHB efficiat cum ipsa HB angulum acutum aequalem ipsi acuto ABH. Deinde ex puncto B demittatur ad HD perpendicularis BD, quae cadet ad partes praedicti anguli acuti ad punctum H. Quoniam igitur latus HB opponitur in triangulo HDB angulo recto in D, atque item in triangulo BAH angulo recto in A; ac rursum in duobus illis triangulis adjacent eidem lateri HB aequales anguli, qui sunt in priore quidem triangulo angulus BHD, & in posteriore angulus HBA; erit etiam (ex 26. primi) reliquus angulus HBD in priore triangulo aequalis reliquo angulo BHA in posteriore triangulo. Quare integer angulus DBA aequalis erit integro angulo AHD.

Jam vero: non erit uterque praedictorum aequalium angulorum obtusus, ne incidamus (ex praecedente) in unum casum jam reprobatae hypothesis anguli obtusi. Sed neque erit rectus, ne incidamus (ex eadem praecedente) in unum casum pro hypothesi anguli recti, qui nullum (ex 5. hujus) relinqueret locum hypothesi anguli acuti. Uterque igitur illorum angulorum erit acutus. Hoc stante: Quod recta BD protracta occurrere nequeat in quodam puncto K ipsi AH ad easdem partes productae, ex eo demonstratur; quia in triangulo KDH, praeter angulum rectum in D, adesset angulus obtusus in H, cum angulus AHD, in praedicta hypothesi anguli acuti, demonstratus sit acutus. Hoc autem absurdum est, contra 17. primi. Non ergo subsistere potest in ea hypothesi, ut quaevis BD, efficiens cum una, ut libet parva recta AB, quemlibet angulum acutum versus partes ipsius AH, occursura tandem sit ad finitam, seu terminatam distantiam, ipsi AH productae. Quod erat demonstrandum.

Aliter idem, ac facilius. Insistant uni cuidam quantumlibet parvae rectae AB (Fig. 16)
[25] duae perpendiculares AK, BM. Demittatur ad AK ex aliquo puncto M ipsius BM perpendicularis MH, jungaturque BH. Constat acutum fore angulum BHM. Est etiam (ex praecedente) acutus angulus BMH, in hypothesi anguli acuti. Ergo perpendicularis BDX, ex puncto B ad ipsam HM demissa, secabit (ex 17. primi) eam HM in quodam puncto intermedio D. Ergo angulus XBA erit acutus. Constat autem (ex eadem 17. primi) non posse invicem concurrere (saltem ad finitam, seu terminatam distantiam) duas illas utcunque productas AHK, BDX, propter angulos rectos in punctis H, & D. Itaque nequit subsistere in hypothesi anguli acuti, ut quaevis BD, efficiens cum una, ut libet, parva recta AB quemlibet angulum acutum versus partes ipsius AH, eidem AB perpendicularis, occursura tandem sit (ad finitam, seu terminatam distantiam) ipsi AH productae. Quod erat propositum.

Proposition 17.

If straight AH stands (Fig. 15) *at right angles to any certain arbitrarily small straight AB: I say that in the hypothesis of acute angle it cannot hold good, that every straight BD, making with AB toward the parts of this AH any acute angle you choose, will at length meet this AH produced at a finite, or bounded distance.*

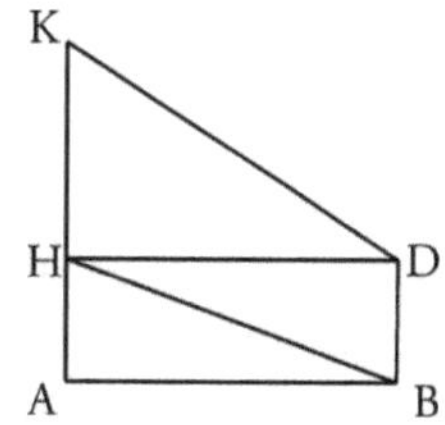

Fig. 15

Proof. Join HE. The angle ABH will be acute (*Elements* I, 17) because of the right angle at the point A. Now draw (*Elements* 1. 23) HD toward the parts of the point B, which not cutting the angle AHB makes with this HB an acute angle equal to this acute angle ABH. Then from the point B is let fall to HD the perpendicular BD, which will fall toward the parts of the aforesaid acute angle at the point H. Since therefore the side HB is opposite in the triangle HDB to the right angle at D, and likewise in the triangle BAH to the right angle at A; and again in those two triangles equal angles are adjacent to this side HB, which are in the first triangle indeed the angle BHD, and in the latter the angle HBA; also (*Elements* I, 26) the remaining angle HBD in the former triangle will be equal to the remaining angle BHA in the latter triangle. Wherefore the entire angle DBA will be equal to the entire angle AHD.

Now however, neither of the aforesaid equal angles will be obtuse, lest we meet (Proposition 16) a case of the now rejected hypothesis of obtuse angle. Nor will either be right, lest we meet (from the same Proposition 16) a case of the hypothesis of right angle, which (Proposition 5) will leave no place for the hypothesis of acute angle. Therefore each one of those angles will be acute. This being the case: that the straight BD produced cannot meet in a certain point K this AH produced toward the same parts, is demonstrated thus; because in the triangle KDH, besides the right angle at D, is present the obtuse angle at H, since the angle AHD in the aforesaid hypothesis of acute angle is proved acute. But this is absurd, against *Elements* I, 17. Therefore it cannot hold good in this hypothesis, that any BD, making with an arbitrarily small straight AB any acute angle toward the parts of this AH, will at length at a finite, or bounded distance, meet this AH produced. This is what was to be demonstrated.

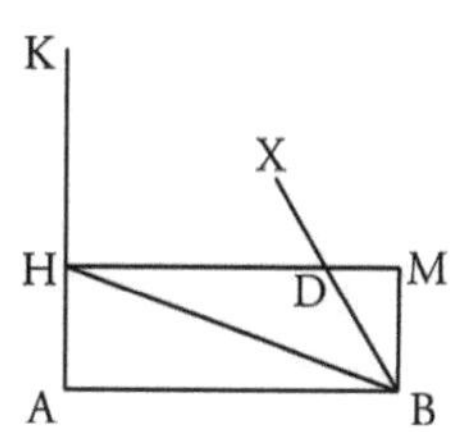

Fig. 16

The same otherwise and more easily. Two perpendiculars AK, BM stand on a certain straight AB, as small as you choose (Fig. 16). From any point M of this BM let fall to AK the perpendicular MH, and join BH. It follows that the angle BHM will be acute. In the hypothesis of acute angle, the angle BMH is also (Proposition 16) acute. Therefore the perpendicular BDX, let fall from the point B to this HM, will cut (by *Elements* I, 17) this HM in some intermediate point D. Therefore the angle XBA will be acute. But it follows (from the same *Elements* I, 17) that those two straights AHK, BDX however produced cannot meet (anyhow at a finite or bounded distance) on account of the right angles at the points H and D. Therefore in the hypothesis of acute angle it cannot hold good, that any BD, making with a straight AB, however small, any acute angle toward the parts of this AH, perpendicular to this same AB, will at length meet (at a finite or bounded distance) this AH produced. This is what was to be demonstrated.

Scholion I.

Atque id est, quod spopondi in Scholiis post XIII. hujus, nimirum destructum iri hypothesim anguli acuti (quae sola obesse jam potest generali illi Pronunciato Euclidaeo) ex solo admisso generali duarum rectarum concursu ad eas partes, versus quas recta quaepiam, quantumlibet parva, in easdem incidens, duos efficiat internos angulos minores duobus rectis; atque ita quidem, etiamsi alteruter illorum angulorum supponi debeat rectus.

Scholion II.

Sed rursum meliore loco, post XXV. hujus, ostendam destructum pariter iri hypothesim
[26] anguli acuti, dum unus aliquis tenuissimus, ut libet, angulus acutus designari possit; sub quo, si recta quaepiam in alteram incidat, debeat haec producta (ad finitam, seu terminatam distantiam) aliquando occurrere cuivis ad quantamlibet finitam distantiam excitatae super ea incidente perpendiculari.

Propositio XVIII.

Ex quolibet triangulo ABC, cujus angulus (Fig. 17) *ad punctum B in uno quovis semicirculo existat, cujus diameter AC, stabilitur hypothesis aut anguli recti, aut anguli obtusi, aut anguli acuti, prout nempe angulus ad punctum B fuerit aut rectus, aut obtusus, aut acutus.*

Demonstratur. Ex centro D jungatur DB. Erunt (ex quinta primi) aequales anguli ad basim AB, atque item ad basim BC, in triangulis ADB, CDB. Quare, in triangulo ABC, duo simul anguli ad basim AC aequales erunt toti angulo ABC. Igitur tres simul anguli trianguli ABC aequales erunt, aut majores, aut minores duobus rectis, prout angulus ad punctum B fuerit aut rectus, aut obtusus, aut acutus. Itaque ex quolibet triangulo ABC, cujus angulus ad punctum B in uno quovis semicirculo existat, cujus diameter AC, stabilitur (ex 15. hujus) hypothesis aut anguli recti, aut anguli obtusi, aut anguli acuti, prout nempe angulus ad punctum B fuerit aut rectus, aut obtusus, aut acutus. Quod erat &c.

Propositio XIX.

Esto quodvis triangulum AHD (Fig. 18) *rectangulum in H. Tum in AD continuata sumatur portio DC aequalis ipsi AD; demittaturque ad AH productam perpendicularis CB. Dico stabilitum hinc iri hypothesim aut anguli recti, aut anguli obtusi, aut anguli acuti, prout portio*
[27] *HB aequalis fuerit, aut major, aut minor ipsa AH.*

Demonstratur. Nam juncta DB erit (ex 4. primi, & ex 10. hujus) aut aequalis, aut major, aut minor ipsa AD, sive DC, pro ut illa portio HB aequalis fuerit, aut major, aut minor ipsa AH.

Scholium 1.

And this is what I promised in the Scholia after Proposition 13, that the hypothesis of acute angle (which alone is able now to stand against that general Euclidean Assertion) will certainly be destroyed by the sole admission of a universal meeting of two straights toward those parts toward which any straight, as small as you choose, meeting them, makes two internal angles less than two right angles; and just so, even if either of those angles is to be supposed right.

Scholium 2.

But again in a better place, after Proposition 25,[1] I shall show that the hypothesis of acute angle will be equally destroyed, provided that any one acute angle as small as you choose can be designated, under which if any straight line meets another, this produced must (at a finite or bounded distance) finally meet any perpendicular erected upon this incident straight at whatever finite distance.

Proposition 18.

From any triangle ABC, of which (Fig. 17) *the angle at the point B is inscribed in any semicircle of diameter AC, is established the hypothesis of right angle, or obtuse angle, or acute angle, according as indeed the angle at the point B is right, or obtuse, or acute.*

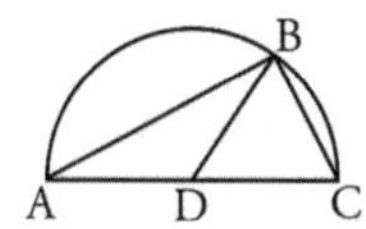

Fig. 17

Proof. From the center D join DB. The angles at the base AB in the triangle ADB will be (*Elements* I, 5) equal, and likewise at the base BC in the triangle CDB. Wherefore, in the triangle ABC the two angles at the base AC will be together equal to the whole angle ABC. Therefore the three angles of the triangle ABC will be together equal to, or greater, or less than two right angles, according as the angle at the point B is right or obtuse, or acute. Therefore from any triangle ABC, of which the angle at the point B is inscribed in any semicircle of diameter AC, is established (Proposition 15) the hypothesis of right angle, or obtuse angle, or acute angle, according as indeed the angle at the point B is right, or obtuse, or acute. This is what was to be demonstrated.

Proposition 19.

Let there be any triangle AHD (Fig. 18) *right-angled at H. Then in AD produced the portion DC is assumed equal to this AD; and the perpendicular CB is let fall to AH produced. I say hence will be established the hypothesis of right angle, or obtuse angle, or acute angle, according as the portion HB is equal to, or greater, or less than AH.*

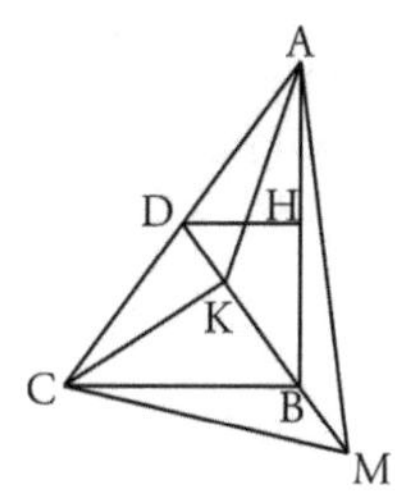

Fig. 18

Proof. For the join DB will be (*Elements* I, 4, and Proposition 10) either equal to, or greater, or less than AD, or DC, according as the portion HB is equal to, or greater, or less than AH.

Et primo quidem sit HB aequalis ipsi AH, ita ut propterea juncta DB aequalis sit ipsi AD, sive DC. Constat circumferentiam circuli, qui centro D, & intervallo DB describatur, transituram per puncta A, & C. Igitur angulus ABC, qui ponitur rectus, existet in eo semicirculo cujus diameter AC. Quare (ex praecedente) stabilietur hypothesis anguli recti. Quod erat primo loco demonstrandum.

Sit secundo HB major ipsa AH, ita ut propterea juncta DB major sit ipsa AD, sive DC. Constat circumferentiam circuli, qui centro D, & intervallo DA, sive DC describatur, occursuram ipsi DB in aliquo puncto intermedio K. Igitur, junctis AK, & CK, erit angulus AKC obtusus, quia major (ex 21. primi) angulo ABC, qui ponitur rectus. Quare (ex praecedente) stabilietur hypothesis anguli obtusi. Quod erat secundo loco demonstrandum.

Sit tertio HB minor ipsa AH, ita ut propterea juncta DB minor sit ipsa AD, sive DC. Constat circumferentiam circuli, qui centro D, & intervallo DA, sive DC describatur, occursuram in aliquo puncto M ipsius DB ulterius protractae. Igitur junctis AM, & CM, erit angulus AMC acutus, quia minor (ex eadem 21. primi) illo angulo ABC, qui ponitur rectus. Quare (ex praecedente) stabilietur hypothesis anguli acuti. Quod erat tertio loco demonstrandum. Itaque constant omnia proposita.

[28] **Propositio XX.**

Esto triangulum ACM (Fig. 19) *rectangulum in C. Tum ex puncto B dividente bifariam ipsam AM demittatur ad AC perpendicularis BD. Dico hanc perpendicularem majorem non fore (in hypothesi anguli acuti) medietate perpendicularis MC.*

Demonstratur. Continuetur enim DB usque ad DH duplam ipsius DB. Foret igitur DH (si DB major sit praedicta medietate) major ipsa CM, ac propterea aequalis cuidam continuatae CMK. Jungantur AH, HK, HM, MD. Jam sic progredimur. Quoniam in triangulis HBA, DBM, aequalia ponuntur latera HB, BA, lateribus DB, BM; suntque (ex 15. primi) aequales anguli ad punctum B; erit etiam (ex quarta ejusdem primi) basis HA aequalis basi MD. Deinde, propter eandem rationem, aequales erunt in triangulis HBM, DBA, bases HM, DA. Quare in triangulis MHA, ADM, aequales erunt (ex 8. primi) anguli MHA, ADM. Rursum in triangulis AHB, MDB, aequalis manebit angulus residuus MHB residuo recto angulo ADB. Igitur rectus erit angulus MHB. At hoc absurdum est, in hypothesi anguli acuti; cum recta KH jungens aequalia perpendicula KC, HD, acutos angulos efficiat (ex tertia hujus) cum eisdem perpendiculis. Non ergo perpendicularis BD major est (in hypothesi anguli acuti) medietate perpendicularis MC. Quod erat demonstrandum.

And first indeed let HB be equal to AH, so that therefore the join DB may be equal to AD, or DC. It follows that circumference of the circle, which is described with the center D and the radius DB, will go through points A and C. Therefore the angle ABC, which is assumed right, is in this semicircle, whose diameter is AC. Wherefore (Proposition 18) is established the hypothesis of right angle. This is what was to be demonstrated in the first part.

Secondly let BH be greater than AH, so that therefore the join DB is greater than AD or DC. It follows that the circumference of the circle which is described with center D, and radius DA, or DC, will meet DB in some intermediate point K. Therefore, AK, and CK being joined, the angle AKC will be obtuse, because greater (*Elements* I, 21) than the angle ABC, which is assumed right. Wherefore (Proposition 18) is established the hypothesis of obtuse angle. This is what was to be demonstrated in the second part.

Thirdly let BH be less than AH, so that therefore the join DB is less than AD, or DC. It follows that the circumference of the circle, which is described with center D, and radius DA, or DC, will meet in some point M this DB produced outwardly. Therefore AM and CM being joined, the angle AMC will be acute, because less (*Elements* I, 21) than the angle ABC, which is assumed right. Therefore (Proposition 18) is established the hypothesis of acute angle. This is what was to be demonstrated in the first part. This completes the demonstration.

Proposition 20.

Let there be a triangle ACM (Fig. 19) *right-angled at C. Then from the point B bisecting this AM let fall the perpendicular BD to AC. I say this perpendicular will not be (in the hypothesis of acute angle) greater than half the perpendicular MC.*

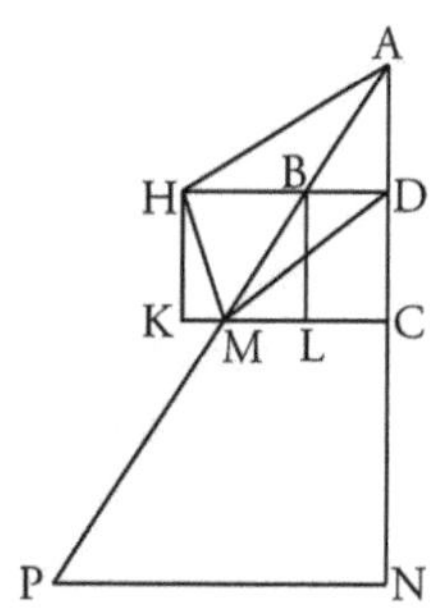

Fig. 19

Proof. For let DB be produced to DH double DB. Therefore DH would be (if DB be greater than the aforesaid half) greater than CM, and therefore equal to a certain continuation CMK.

Join AH, HK, HM, MD. Now we proceed thus. Since in the triangles HBA, DBM, the sides HB, BA are assumed equal to the sides DB, BM; and (*Elements* I, 15) the angles at the point B are equal; the base HA also (*Elements* I, 4) will be equal to the base MD. Then, by the same reasoning, in the triangles HBM, DBA, the bases HM, DA will be equal. Wherefore in the triangles MHA, ADM, the angles MHA, ADM (*Elements* I, 8) will be equal. Again in the triangles AHB, MDB, the residual angle MHB will remain equal to the residual right angle ADB. Therefore the angle MHB will be right. But this is absurd in the hypothesis of acute angle; since the straight KH joining equal perpendiculars KC, HD, makes acute angles with these perpendiculars. Therefore the perpendicular BD is not (in the hypothesis of acute angle) greater than the half of the perpendicular MC. This is what was to be demonstrated.

Propositio XXI.

Iisdem manentibus: Intelligantur in infinitum produci ipsae AM, & AC. Dico earundem distantiam majorem fore (in utraque hypothesi aut anguli recti, aut anguli acuti) qualibet assignabili finita longitudine.

[29] Demonstratur. In AM continuata sumatur AP dupla ipsius AM, demittaturque ad AC continuatam perpendicularis PN. Non erit (ex praecedente) in utravis praedicta hypothesi perpendicularis MC major medietate perpendicularis PN. Igitur PN saltem erit dupla ipsius MC, prout MC saltem est dupla alterius BD. Atque ita semper, si in continuata AM sumatur dupla ipsius AP, ex ejusque termino demittatur perpendicularis ad continuatam AC. Scilicet perpendicularis, quae ex AM semper magis continuata demittetur ad continuatam AC, multiplex erit determinatae BD supra quemlibet finitum assignabilem numerum. Igitur praedictarum rectarum distantia major erit (in utraque praedicta hypothesi) qualibet assignabili finita longitudine. Quod erat demonstrandum.

Corollarium.

Quoniam vero hypothesis anguli obtusi, quae unice obesse hic posset, demonstrata jam est absolute falsa; consequitur sane absolute verum esse, quod distantia unius ab altera praedictarum rectarum, si in infinitum producantur, major sit qualibet finita assignabili longitudine.

Scholion I.

In quo expenditur conatus Procli.

Post Theoremata a me huc usque demonstrata sine ulla dependentia ab illo Pronunciato Euclidaeo, ad cujus nempe exactissimam demonstrationem omnia conspirant; operae pretium facturum me judico, si quorundam etiam celebriorum Geometrarum labores in ean-
[30] dem metam contendentium diligenter expendam. Incipio a Proclo, cujus est apud Clavium in Elementis post XXVIII. Libri primi sequens assumptum: *Si ab uno puncto duae rectae lineae angulum facientes infinite producantur, ipsarum distantia omnem finitam magnitudinem excedet.* At Proclus demonstrat quidem (ut ibi optime advertit Clavius) duas rectas (Fig. 20) ut puta AH, AD ab eodem puncto A exeuntes versus easdem partes, semper magis, in majore distantia ab eo puncto A, inter se distare, sed non etiam ita ut ea distantia crescat ultra omnem finitum designabilem limitem, prout opus foret ad ipsius intentum. Quo loco praefatus Clavius affert exemplum Conchoidis Nicomedeae, quae cum recta AH ex eodem puncto A versus easdem partes exiens, ita semper magis ab eadem recedit, ut tamen ipsarum distantia non nisi ad infinitam earundem productionem, aequalis sit cuidam finitae rectae AB perpendiculariter insistenti ipsis AH, BC, versus easdem partes in infinitum protractis. Quid ni ergo, nisi specialis ratio in contrarium cogat, dici idem possit de duabus suppositis rectis lineis AH, AD?

Proposition 21.

The same remaining: If AM and AC are understood as produced in infinitum I say their distance (in either the hypothesis of right angle, or of acute angle) will be greater than any assignable finite length.

Proof. In AM produced assume AP double of AM, and let fall to AC produced the perpendicular PN. The perpendicular MC will not be (Proposition 20) in either hypothesis aforesaid greater than half the perpendicular PN. Therefore PN will be at least double MC, just as MC is at least double BD. And so always, if in AM produced is assumed double AP, and from the terminus of this a perpendicular is let fall to AC produced. It is obvious that the perpendicular, which from AM ever more produced is let fall to AC produced, will be a multiple of the determinate BD beyond any finite assignable number. Therefore the distance of the aforesaid straights will be (in either aforesaid hypothesis) greater than any assignable finite length. This is what was to be demonstrated.

Corollary.

But since the hypothesis of obtuse angle, which alone could hinder here, is already proved absolutely false; so of course follows as absolutely true, that the distance of one from the other of the aforesaid straights, if they be produced *in infinitum*, is greater than any finite assignable length.

Scholium 1.

In which is weighed the endeavor of Proclus.

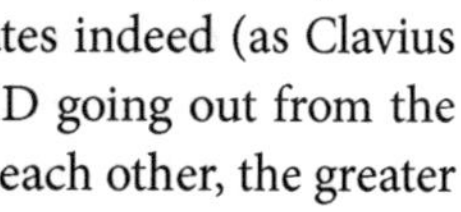

Fig. 20

After the theorems so far demonstrated by me, independently of the Euclidean Assertion, toward an exact proof of which they all conspire; in my judgment it is well if I diligently weigh the labors of my certain well-known Geometers in the same endeavor. I begin from Proclus, of whom Clavius gives the following assumption: *If from a point two straight lines making an angle are produced infinitely, their distance will exceed every finite magnitude.*[1] But Proclus demonstrates indeed (as Clavius there well remarks)[2] that two straights (Fig. 20) as suppose AH, AD going out from the same point A toward the same parts, always diverge the more from each other, the greater the distance from the point A, but not also that this distance increases beyond every finite limit that may be designated, as was requisite for his purpose. In which place the aforesaid Clavius cites the example of the Conchoid of Nicomedes,[3] which going out from the same point A as the straight AH toward the same parts, so recedes always more from it, that nevertheless only at an infinite production is their distance equal to a certain finite sect AB standing perpendicular to AH and BC produced *in infinitum* toward the same parts. Why may not the same be said of the two assumed straight lines, AH, AD, unless a special reason constrains to the contrary?

Neque hic accusari potest Clavius, quod Proclo opponat eam Conchoidis proprietatem, quae nempe demonstrari non potest sine adjumento plurium Theorematum, Pronunciato hic controverso innixorum. Nam dico ex hoc ipso confirmari vim redargutionis Clavianae; quia scilicet ex illo Pronunciato assumpto ut vero manifeste consequitur, duas lineas in infinitum protractas, unam rectam, & alteram inflexam, posse unam ab altera semper magis recedere intra quendam finitum determinatum limitem; unde utique oriri potest suspicio, ne simile quidpiam contingere possit in duabus lineis rectis, nisi aliter demonstretur.

[31] Sed non idcirco; postquam ego in Cor. praecedentis Propositionis manifestam jam feci absolutam veritatem praecitati assumpti; transiri statim potest ad asserendum Pronunciatum illud Euclidaeum. Nam antea demonstrari etiam oporteret, quod duae illae rectae AH, BC, quae cum incidente AB duos ad easdem partes angulos efficiant duobus rectis aequales, ut puta utrunque rectum, non etiam ipsae, ad eas partes in infinitum protractae, semper magis invicem dissiliant ultra omnem finitam assignabilem distantiam. Quatenus enim partem affirmativam praesumere quis velit; quae utique verissima est in hypothesi anguli acuti; non erit sane legitimum consequens, quod recta AD quomodolibet secans angulum HAB; unde nempe minores fiant duobus rectis duo simul ad easdem partes interni anguli DAB, CBA; quod, inquam, ea recta AD, in infinitum producta coire tandem debeat cum producta BC; etiamsi alias demonstratum sit, quod distantia duarum AH, AD in infinitum productarum major semper evadat ultra omnem finitum designabilem limitem.

Quod autem praefatus Clavius satis esse judicaverit veritatem illius assumpti ad demonstrandum Pronunciatum hic controversum; condonari id debet praeconceptae ab ipso Clavio opinioni circa rectas lineas aequidistantes, de quibus in sequente Scholio commodius agemus.

Scholion II.

In quo expenditur idea Clarissimi Viri Joannis Alphonsi Borellii in suo Euclide Restituto.

Accusat doctissimus hic Auctor Euclidem, quod rectas lineas parallelas eas esse definiverit, *quae in eodem plano existentes non concurrunt ad utrasque partes, licet in infinitum*
[32] *producantur.* Rationem accusationis affert, quod talis passio ignota sit: *tum quia*, inquit, *ignoramus, an tales linea infinitae non concurrentes reperiri possint in natura: tum etiam quia infiniti proprietates percipere non possumus, & proinde non est evidenter cognita passio ejusmodi.*

Nor here can Clavius be blamed that he opposes to Proclus this property of the Conchoid, which cannot be demonstrated except with the aid of many Theorems resting upon the here controverted Assertion. For I say from this itself the force of the Clavian redargution is confirmed; because it is certain, this Assertion being assumed, it manifestly follows, that two lines *in infinitum* protracted, one straight and the other curved, can recede one from the other ever more within a certain finite determinate limit; whence at any rate may arise a suspicion lest the same may happen for two straight lines, unless otherwise demonstrated.[4]

But it is not therefore possible, when I now have made manifest in the Corollary to Proposition 21 the absolute truth of the aforesaid assumption, immediately to go over to the assertion of the Euclidean Assertion. For previously must also be demonstrated, that those two straights AH, BC, which with the transversal AB make two angles two toward the same parts equal to two right angles, as for example each a right angle, do not also, protracted toward these parts *in infinitum*, always separate more from one another beyond all finite assignable distance. For if one chooses to presume the affirmative, which is indeed entirely true in the hypothesis of acute angle; it certainly will not be a legitimate consequence, that a straight AD in any way cutting the angle HAB, hence of course making at the same time two internal angles DAB, CBA toward the same parts less than two right angles; that, I say, this straight AD produced *in infinitum*, must at length meet with BC produced; even if it were at another time demonstrated, that the distance of the two AH, AD produced *infinitum* goes out ever greater beyond all finite assignable limit.

But since the aforesaid Clavius judged the truth of this assumption sufficient for demonstrating the Assertion here in question; one should forgive Clavius' preconceived opinion about equidistant straight lines, which we may discuss more conveniently in the next Scholium.

Scholium 2.

In which is weighed an idea of that brilliant man Giovanni Alfonso Borelli in his Euclides Restitutus.

This most learned Author[1] blames Euclid, because he defines parallel straight lines to be those, *which being in the same plane do not meet on either side, even if produced in infinitum.*[2] He offers as grounds for his accusation, that such relation is unknown: *first,* he says, *because we are ignorant, whether such infinite non-concurrent lines can be found in Nature; then also because we cannot perceive the properties of the infinite, and hence a relation of this sort is not clearly cognized.*[3]

Sed pace tanti Viri dictum sit: Numquid reprehendi potest Euclides, quod *quadratum* (ut unum inter innumera exemplum proferam) definiverit esse *figuram quadrilateram, aequilateram, rectangulam*; cum dubitari possit, an figura ejusmodi locum habeat in natura? Repraehendi, inquam, aequissime posset; si, ante omnem Problematicam demonstrativam constructionem, figuram praedictam assumpsisset tanquam datam. Hujus autem vitii immunem esse Euclidem ex eo manifeste liquet, quod nusquam praesumit quadratum a se definitum, nisi post Prop. 46. Libri primi, in qua problematice docet, ac demonstrat *quadrati prout ab ipso definiti, a data recta linea descriptionem*. Simili igitur modo reprehendi nequit Euclides, quod rectas lineas parallelas eo tali modo definiverit, cum eas nusquam ad constructionem ullius Problematis assumat tanquam datas, nisi post Prop. 31. lib. primi, in qua Problematice demonstrat, quo pacto *a dato extra datam rectam lineam puncto duci debeat recta linea eidem parallela*, & quidem juxta definitionem ab eo traditam parallelarum, *ita ut nempe in infinitum protractae in neutram partem sibi invicem occurrant*: Quodque amplius est; id ipsum demonstrat sine ulla dependentia a Pronunciato hic controverso. Itaque Euclides sine ulla petitione principii demonstrat *reperiri posse in natura duas tales lineas rectas, quae* (in eodem plano consistentes) *in utramque partem in infinitum protractae nunquam concurrant*; ac propterea *cognitam nobis evidenter facit eam passionem*, per quam rectas lineas parallelas definit.

Pergamus porro, quo nos invitat diligens Euclidis accusator. Parallelas rectas lineas ap-
[33] pellat duas quaslibet rectas AC, BD, quae perpendiculariter ad easdem partes (Fig. apud me 21) insistant uni cuidam rectae AB. Nihil moror, quin definitio ejusmodi exposita sit *per passionem* (ut ipse ait) possibilem, & evidentissimam; cum (ex undecima primi) a quolibet in data recta puncto excitari possit perpendicularis.

Verum hanc ipsam & possibilitatem, & evidentiam jam demonstravi circa definitionem traditam ab Euclide. Quare unice restat, ut conferatur notum illud Pronunciatum Euclidaeum cum altero itidem Pronunciato, quod usui esse debeat ad ulteriorem progressum post novam istam parallelarum definitionem. Ecce autem alterum istud Pronunciatum apud Clavium (ad quem diserte provocat ipse Borellius) in Scholio post Prop. 28. lib. primi: Si recta linea, ut puta BD super aliam rectam, ut putae BA, in transversum moveatur constituens cum ea in suo extremo B angulos semper rectos, describet alterum illius extremum D lineam quoque rectam DC, dum nempe ipsa BD pervenerit ad congruendum alteri aequali AC.

But with the reverence for so great a man it may be said: Can Euclid be blamed, because (to bring forward one among innumerable examples) he defines *a square* to be *a figure quadrilateral, equilateral, rectangular*, when it may be doubted, whether a figure of this sort has place in Nature. He could, say I, most justly have been blamed, if, before as a problem demonstrating the construction, he had assumed the aforesaid figure as given. But that Euclid is free from this fault follows manifestly from this, that he nowhere assumes the square defined by him, except after *Elements* I, 46, in which in form of a Problem he teaches, and demonstrates *the description from a given straight line, of the square as defined by him*,[4] In the same way therefore Euclid ought not to be blamed, because he defined parallel straight lines in this manner, since he nowhere assumes them as given for the construction of any Problem, except after *Elements* I, 31, in which as a problem he demonstrates, how *should be drawn from a given point without a given straight line a straight line parallel to this*, and indeed according to the definition of parallels given by him, *so that produced indeed into the infinite on neither side do they meet one another*. And what is more; he demonstrates this without any dependence from the Assertion here controverted.[5] Thus Euclid demonstrates without any *petitio principii* that *there can be found in Nature two such straight lines, which* (lying in the same plane) *protracted on each side into the infinite never meet*, and therefore *makes clearly known to us that relation* by which he defines parallel straight lines.

Let us continue onward, whither the scrupulous accuser of Euclid invites us. Parallel straight lines he calls any two straights AC, BD, which toward the same parts stand at right angles to a certain straight AB (Fig. 21). I admit that such a definition is set forth *by a property* (as he says) possible and most evident; since (*Elements* I, 11) from any point in a given straight a perpendicular can be erected.[6]

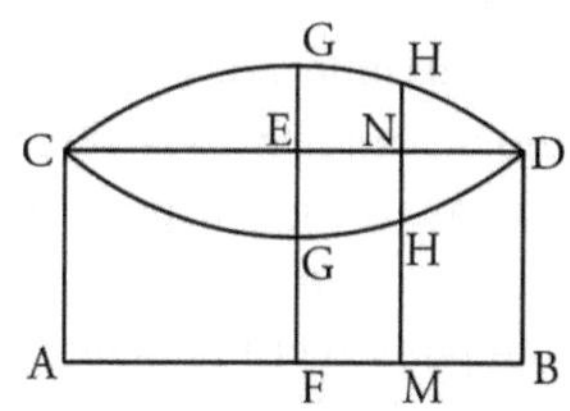

Fig. 21

But precisely both this possibility and clearness I have just now demonstrated about the definition propounded by Euclid. Wherefore remains only to compare that known Assertion of Euclid with the other like Assertion, which must e used for farther progress after the new definition of parallels. But behold this other Assertion in Clavius (to whom Borelli himself expressly refers) in the Scholium after *Elements* I, 28: if a straight line, as suppose BD upon another straight, as suppose BA, moves transversely making with it at its extremity B always right angles, its other extremity D describes a line also straight DC, until this BD shall have come to congruence with the other equal sect AC.[7]

Agnosco opportunitatem Pronunciati, ut inde transitus fiat ad demonstrandum illud alterum Euclidaeum, quo nempe fulciri tandem debet reliqua omnis Geometria. Nam antea proposuerat Clavius; quod linea, cujus omnia puncta aeque distent a quadam supposita recta AB; qualis utique est (ex hypothesi praedictae descriptionis) linea DC; debet esse etiam ipsa linea recta; quia nempe ejusmodi erit, ut omnia ipsius puncta intermedia *ex aequo jaceant* (qualis est rectae lineae definitio) inter ejus extrema puncta D, & C; *ex aequo*, inquam, *jaceant*; cum omnia aeque distent ab ea supposita recta AB, nimirum quanta est longitudo ipsius BD, aut AC. Quo loco affert Clavius exemplum lineae circularis, de qua
[34] commodius infra disseremus; ubi ostendam clarissimam hac in parte disparitatem inter lineam rectam, & circularem. Nam interim dico non satis liquere, an linea descripta ab eo puncto D sit potius recta DC, quam curva quaedam DGC seu convexa, seu concava versus partes ipsius BA.

Si enim ex puncto F dividente bifariam ipsam BA intelligatur educta perpendicularis, quae occurrat rectae DC in E, & praedictis curvis in G, & G, constat sane (ex 2. hujus) rectos fore angulos hinc inde ad punctum E; qualiscunque tandem in eo motu intelligatur descripta linea DC a puncto D; ac praeterea (ex facili intellecta superpositione) aequales hinc inde fore angulos ad puncta G, prout alterutra curva DGC descripta fuerit.

Sed rursum; assumpto in AB quolibet puncto M; si educatur perpendicularis, quae occurrat rectae DC in N, & praedictis curvis in H, & H, paulo post demonstrabo rectos fore angulos hinc inde ad punctum N, quatenus quidem recta ipsa DC genita supponatur in suo illo motu a puncto D, seu quatenus recta MN aequalis censeatur ipsi BD. Sin vero alterutra curva DHC genita putetur; ex facili itidem praescripta superpositione demonstrabitur aequales rursum hinc inde fore angulos MHD, MHC, ubivis in ea alterutra descripta curva sumptum fuerit punctum H, ex quo ad subjectam rectam lineam AB demissa intelligatur perpendicularis HM. Verum hac de re fusius, ac diligentius in altera parte hujus libri, ubi locum proprium habet.

Quorsum igitur, inquies, praecox ista anticipatio? In eum, inquam, finem; ut ne ex ista lineae eo modo genitae verissima, & a me exactissime in praecitato loco demonstranda proprietate; & quidem citra omnem defectum quomodolibet infinite parvum; praecipitanter censeremus non nisi rectam lineam esse posse. Scilicet hic inquiritur penitior rectae lineae
[35] natura, sine qua vix infantiam praetergressa Geometria subsistere ibi deberet. Non igitur hac in re vituperari potest major quaedam exactissimae veritatis inquisitio.

I acknowledge the fitness of this Assertion, that thence a transit may be made to demonstrating that other Euclidean Assertion, upon which certainty at length must be supported all remaining geometry. For Clavius had previously declared; that a line, of which all other points are equally distant from a certain assumed straight AB; as assuredly is (from the hypothesis of the aforesaid construction) the line DC; this line also must be straight; because certainly it will be of such sort, that all its intermediate points lie evenly (such is the definition of a straight line) between its extreme points D, and C; lie evenly, say I, since all are equally distant from this assumed straight AB, truly as much as the length is of this BD, or AC.[8] In this place Clavius introduces the example of the circular line, of which we shall speak more conveniently below; where I shall show the clearest disparity in this regard between the straight line and circle. But meanwhile I say it is not sufficiently evident, whether the line described by this point D is rather the straight DC than a certain curve DGC either convex or concave toward the side of this BA.

For if from the point F bisecting this BA a perpendicular is supposed erected, which meets the straight DC in E, and the aforesaid curves in G, and G, it follows surely (Proposition 2) that the angles at the point E will be right, whatever line DC is understood at the length as described in this motion by the point D; and moreover (from an easily understood superposition)[9] the angles at the points G will be equal according as the one or the other curve DGC may be described.

But again; any point M in AB being assumed; if a perpendicular is erected, which meets the straight DC in N, and the aforesaid curves in H and H, I shall prove a little later that the angles on both sides at the point N will be right, in so far indeed as this straight DC is supposed generated by the point D in that motion of its, or in as far as the straight MN is decided equal to this BD. But if one or the other curve DHC is supposed generated; from the like aforesaid easy superposition will be demonstrated that again the angles MHD, MHC on both sides will be equal, wherever in the one or the other described curve the point H may be assumed, from which to the underlying straight line AB the perpendicular HM is understood as let fall.[10] But of this more fully and more scrupulously in the second part of this Book, where it has its proper place.

To what end therefore, will you say, this untimely anticipation? To this end, say I; lest from this indubitable property of the line generated in this manner, proved by me most rigorously in the aforesaid place; and indeed beyond any defect of any sort infinitely small; we may decide precipitately that the line can be only the straight. Obviously the nature of the straight line must here be investigated more profoundly, without which geometry scarcely grown beyond infancy must there remain. Therefore in this affair cannot be blamed a certain greater investigation of a most exact verity.

Neque tamen hic renuo, quin diligentissima aliqua experientia physica deprehendi possit, quod linea DC eo motu genita non nisi recta linea censenda sit. Sed quatenus ad experientiam physicam provocare hic liceat; tres statim afferam demonstrationes Physico-Geometricas ad comprobandum Pronunciatum Euclidaeum. Ubi non loquor de experientia physica tendente in infinitum, ac propterea nobis impossibili; qualis nempe requireretur ad cognoscendum, quod puncta omnia junctæ rectae DC aequidistent a recta AB, quae supponitur in eodem cum ipsa DC plano consistens. Nam mihi satis erit unicus individuus casus; ut puta, si juncta recta DC, assumptoque uno aliquo ejus puncto N, perpendicularis NM demissa ad subjectam AB comperiatur esse aequalis ipsi BD, sive AC. Tunc enim anguli hinc inde ad punctum N aequales forent (ex 1. hujus) angulis sibi correspondentibus ad puncta C, & D, qui rursum (ex eadem 1. hujus) aequales inter se forent. Quare anguli hinc inde ad punctum N, atque ideo etiam reliqui duo recti erunt. Igitur unum habebimus casum pro hypothesi anguli recti; ac propterea (juxta quintam, & decimamtertiam hujus) demonstratum habebimus Pronunciatum Euclidaeum. Atque haec esse potest prima demonstratio Physico-Geometrica.

Transeo ad secundam. Esto semicirculus, cujus centrum D, & diameter AC. Si ergo (Fig. 17) in ejus circumferentia assumatur punctum aliquod B, ad quod junctæ AB, CB comperiantur continere angulum rectum, satis erit hic unicus casus (prout demonstravi in 18. hujus) ad stabiliendam hypothesim anguli recti, ac propterea (ex praedicta 13. hujus) ad demonstrandum notum illud Pronunciatum.

[36] Superest tertia demonstratio Physico-Geometrica, quam puto omnium efficacissimam, ac simplicissimam, utpote quae subest communi, facillimae, paratissimaeque experientiae. Si enim in circulo, cujus centrum D, tres coaptentur (Fig. 22) rectae lineae CB, BL, LA, aequales singulae radio DC, comperiaturque juncta AC transire per centrum D, satis id erit ad demonstrandum intentum. Nam junctis DB, DL, tria habebimus triangula, quae (ex 8. & 5. primi) tum inter se invicem, tum etiam in se ipsis singula erunt aequiangula. Quoniam igitur tres simul anguli ad punctum D, nimirum ADL, LDB, BDC aequales sunt (ex 13. primi) duobus rectis; duobus etiam rectis aequales erunt tres simul anguli cujusvis illorum triangulorum, ut puta trianguli BDC. Quare (ex 15. hujus) stabilita hinc erit hypothesis anguli recti; ac propterea (ex jam nota 13. hujus) demonstratum manebit illud Pronunciatum.

Sin vero, ante omnem attentatam seu demonstrationem, seu figuralem exhibitionem, conferre inter se placeat duo illa Pronunciata, fateor sane Euclidaeum videri posse obscurius, aut etiam falsitati obnoxium. At post figuralem exhibitionem, quam Scholio IV. consequenti reservo, constabit viceversa Pronunciatum quidem Euclidaeum retinere posse dignitatem, ac nomen Pronunciati, alterum vero inter Theoremata computari tutius debere.

Nor yet do I here deny, but that by some most accurate physical experimentation may be discovered, that the line DC generated by this motion can only be adjudged a straight line. But in so far as may be here permissible to cite physical experimentation, I forthwith bring forward three demonstrations physico-geometric to sanction the Euclidean Assertion. Therewith I do not speak of physical experimentation extending into the infinite, and therefore impossible for us; such as of course would be requisite to the cognizing, that all points of the straight join DC are equidistant from the straight AB, which is supposed be in the same plane with this DC. For a single individual case will be sufficient for me; as suppose, if, the straight DC being joined, and any one point of it N being assumed, the perpendicular NM let fall to the underlying AB is ascertained to be equal to BD or AC.[11] For then the angles on both sides at the point N would be equal (Proposition 1) to the angles corresponding to them at the points C and D, which again (from the same Proposition 1) would be equal among them. Wherefore the angles on both sides at the point N, and therefore also the remaining two (C and D) will be right. Therefore we shall have a case for the hypothesis of right angle; and consequently (by Propositions 5 and 13) we shall have demonstrated the Euclidean Assertion. And this may be the first demonstration physico-geometric.

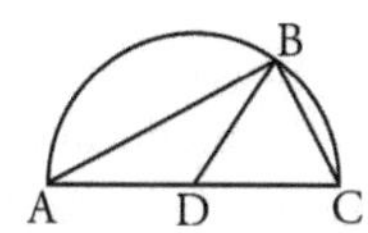

Fig. 17

I pass over to the second. Let there be a semi-circle, of which the center is D, and diameter AC. If then (Fig. 17) any point B is assumed in its circumference, to which AB, CB joined are ascertained to contain a right angle, this single case will be sufficient (as I have demonstrated in Proposition 18) for establishing the hypothesis of right angle, and consequently (from the aforesaid Proposition 13) for demonstrating that famous Assertion.

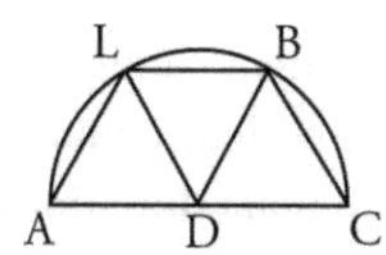

Fig. 22

There remains the third demonstration physic-geometric, which I think is the most efficacious and the most simple of all, inasmuch as it rests upon an accessible, most easy, and most convenient experiment. For if in a circle, whose center is D, are fitted (Fig. 22) three straight lines CB, BL, LA, each equal to the radius DC, and it is ascertained that the join AC goes through the center D, this will be sufficient for demonstrating the assertion. For, DB, DL being joined, we will have three triangles, which (from *Elements* I, 8 and 5) not only will be equiangular to one another, but also singly for themselves. Therefore since the three angles together at the point D, indeed ADL, LDB, BDC are equal (by *Elements* I, 13) to two right angles; also the three angles together of each of these triangles will be equal to two right angles, as suppose of the triangle BDC. Wherefore (Proposition 15) will be established hence the hypothesis of right angle; and consequently (from the already admitted Proposition 13) that Assertion will be demonstrated.

But if, before all attempt whether at demonstration or at graphic exhibition, one wishes to compare among them those two Assertions, I grant indeed the Euclidean may appear more obscure or even liable to objection. But after the graphic exhibition which I reserve for Scholium 4 following, it will appear *vice versa* that the Euclidean Assertion indeed can retain the dignity and name of Assertion, but the other ought rather to be reckoned among the Theorems.[12]

Sed hic explicare debeo (prout paulo ante me facturum spopondi) manifestam isto in genere disparitatem inter lineam circularem, & lineam rectam. Disparitas autem ex eo oritur; quod recta quidem linea dicitur ad se ipsam; circularis vero, ut puta (Fig. 23) MDHNM, non ad se ipsam, sed ad alterum dicitur, nimirum ad quoddam alterum in eodem cum ipsa plano existens punctum A, quod est ejusdem centrum. Consequens igitur est, prout
[37] optime demonstratur a Clavio, quod linea FBCL in eodem cum illa plano consistens, & cujus omnia puncta aequidistent a praedicta MDHNM, sit & ipsa circularis, nimirum omnibus suis punctis aequidistans a communi centro A. Quod enim BD, quae sit continuatio in rectum ipsius AB, sit mensura distantiae illius puncti B ab ea circulari MDHNM, ex eo constat; quia (ex 7. tertii, quae est independens a Pronunciato hic controverso) minima omnium ipsa est, quae ab eo puncto in eam circumferentiam cadere possint. Idem valet de reliquis CH, LN, FM. Quoniam igitur & totae AM, AD, AH aequales sunt, utpote radii ex centro A ad suppositam lineam circularem MDHNM; atque item aequales sunt abscissae FM, BD, CH, LN, quae nempe mensura sunt aequalis distantiae omnium punctorum illius lineae FBCLF ab ea supposita linea circulari MDHNM; consequens plane est, ut aequales pariter sint residuae AF, AB, AC, AL, ac propterea ipsa etiam linea FBCLF sub eodem centro A circularis sit.

Numquid autem uniformiter, ad demonstrandum, quod linea DC (Fig. 21) eo tali motu genita a puncto D sit linea recta, satis erit aequidistantia omnium ipsius punctorum a subjecta recta AB? Nullo modo. Nam linea recta dicitur absolute ad se ipsam, sive in se ipsa, nimirum ita *ex aequo jacens inter sua puncta*, ac praesertim extrema, ut manentibus istis immotis nequeat ipsa revolvi ad occupandum novum locum. Nisi haec passio aliquo pacto demonstretur de ea DC, nunquam constabit eam esse lineam rectam, qualiscunque tandem supponatur, aut demonstretur omnium ipsius punctorum relatio ad subjectam in eodem plano rectam AB; praesertim vero, ne uniformiter dicamus nullam aliam in eo plano fore lineam rectam, quae omnibus suis punctis non aequidistet ab ea supposita recta linea AB.

[38] Neque tamen dictum hoc meum ita accipi volo, quasi putem demonstrari non posse, quod linea sic genita ipsa sit linea recta, nisi post demonstratam veritatem controversi Pronunciati; cum magis ego ipse prope finem hujus Libri demonstraturus id sim, ad confirmandum ipsum tale Pronunciatum.

Scholion III.

In quo expenditur conatus Nassaradini Arabis, & simul idea super eodem negotio Clariss. Viri Joannis Vallisii.

Conatum istum Nassaradini Arabis latino idiomate typis vulgavit praelaudatus Vir Joannes Vallisius, cum animadversionibus opportuno loco adjectis. Duo autem in rem suam postulat sibi concedi Nassaradinus.

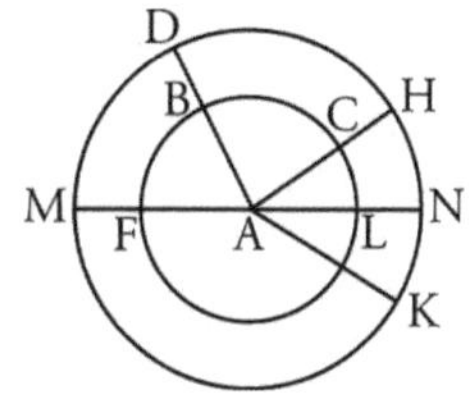

Fig. 23

But here I must explain (as a little above I have promised I was about to do) the manifest disparity in this relation between the circular line and the straight line. Now the disparity arises from this; that a line is called straight in reference to itself; but is called circular, as suppose (Fig. 23) MDHNM, not in reference to itself, but to something else, forsooth to a certain other point A existing in the same plane with it, which is its center. The consequence therefore is, as is most excellently demonstrated by Clavius, that the line FBCL existing in the same plane with it, and whose points area all equidistant from the aforesaid MDHNM, is also itself circular, a truly equidistant in all its points from the common center A. That in fact BD, which is the continuation in a straight of AB, is the measure of the distance of that point B from this line MDHNM follows from this; because (from *Elements* III, 7, which is independent of the Assertion here in controversy)[13] this is the smallest of all, which can fall from this point upon this circumference. The same holds of the remaining CH, LN, FM. Since therefore also the wholes AM, AD, AH, are equal as radii from the center A to the line assumed circular MDHNM; and also the sections FM, BD, CH, LN are equal, which obviously are the measure of the equal distance of all points of that line FBCLF from this line presumed circular MDHNM; the consequence plainly is, that equal likewise are the remainders AF, AB, AC, AL, and therefore also this line FBCLF is a circle with the same center A.

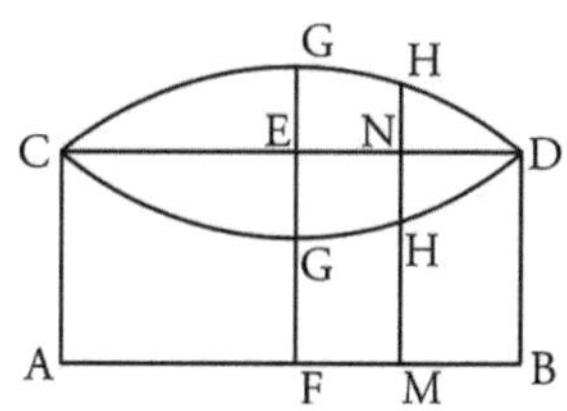

Fig. 21

But now likewise, for demonstrating that the line DC (Fig. 21) generated through such a motion by the point D is a straight line will the equidistance of all its points from the underlying straight AB be sufficient? In no way. For a line is called straight absolutely in reference to itself, or in itself, doubtless *lying evenly between its points*, and especially end points, so that these remaining unmoved it cannot be revolved into occupying a new place.[14] Unless this state in some way be demonstrated of this DC it will never be certain that this is a straight line, whatever relation is finally supposed or demonstrated of all its points to the underlying straight AB in the same plane; but especially we must not say analogically that no other line in this plane will be straight which in all its points is not equidistant from this line AB supposed straight.

Nor finally do I wish this dictum of mine so taken, as if I think it cannot be demonstrated, that the line thus generated is itself a straight line, except after truth demonstrated of the controverted Assertion; since rather I myself will demonstrate it toward the end of this Book, for confirming such Assertion itself.

Scholium 3.

In which is weighed the endeavor of the Arab Nasīr ad-Dīn, and likewise the idea of the illustrious John Wallis upon the same affair.

This endeavor of the Arab Nasīr ad-Dīn the above eulogized John Wallis has translated and published in Latin with remarks added in opportune place.[1] Nasīr ad-Dīn requires two things to be conceded to him for his purpose.

Primum est; ut duae quaelibet rectae lineae in eodem plano positae, in quas aliae quotlibet rectae lineae ita incidant, ut uni quidem earum perpendiculares semper sint, alteram vero ad angulos inaequales semper secent, nimirum versus unam partium sub angulo semper acuto, & versus alteram sub angulo semper obtuso; ut, inquam, priore loco dictae lineae censeantur semper magis (quandiu se mutuo non secent) ad se invicem accedere versus partes illorum angulorum acutorum; & vicissim semper magis a se invicem recedere versus partes angulorum obtusorum.

At ego quidem, si nihil aliud moratur Nassaradinum, libens permitto, quod postulat; cum istud ipsum, quod ab eo indemonstratum relinquitur, intelligi possit exactissime a me demonstratum in Cor. II. post 3. hujus.

Alterum Nassaradini Postulatum est reciprocum primi; ut nempe acutus semper sit
[39] angulus versus eas partes, ad quas jam dictae perpendiculares supponantur fieri semper breviores; obtusus autem versus alias partes, ad quas eaedem perpendiculares supponantur evadere semper longiores.

Verum hic latet aequivocatio. Cur enim (dum ab una aliqua statuta tanquam prima perpendiculari procedatur ad alias) consequentium perpendicularium anguli, ad eandem partem acuti, non fiant semper majores, quo usque incidatur in angulum rectum, nimirum in talem perpendicularem, quae ipsa sit utriusque praedictarum rectarum commune perpendiculum? Et istud quidem si accidat, evanescit latebrosa ista Nassaradini praeparatio, postquam ingeniose quidem, sed magno cum labore Euclidaeum Pronunciatum demonstrat.

Quod si Nassaradinus jure quodam suo praesumere velit tanquam per se notam consistentiam illam ad eandem partem angulorum acutorum: Cur non etiam (dicam cum Vallisio) concipi potest tanquam per se clarum: *Duas rectas in eodem plano convergentes* (in quas nempe alia recta incidens duos ad easdem partes angulos efficiat minores duobus rectis, ut puta unum rectum, & alterum quomodolibet acutum) *tandem occursuras, si producantur?* Neque enim opponi potest, quod major ista ad unas partes convergentia subsistere semper possit intra quendam determinatum limitem, adeo ut nempe tanta quaedam distantia inter eas lineas ad eam partem semper intersit, etiamsi caeteroquin una ad alteram semper propius accedat. Non, inquam, opponi id potest; quoniam ex hoc ipso demonstrabo, post XXV. hujus, omnium talium rectarum ad finitam distantiam occursum, juxta Pronunciatum Euclidaeum.

The first is; that any two straight lines lying in the same plane, upon which ever so many other straight lines so strike, that they are always perpendicular to one indeed of these, but always cut the other at unequal angles, truly toward one part always under acute angles, and toward the other under obtuse angles; that, I say, the above-mentioned lines be supposed always more (as long as they do not mutually cut) to approach each other toward the side of those acute angles; and on the other hand always more to recede from one another toward the parts of the obtuse angles.[2]

But I indeed, if nothing else impedes Nasīr ad-Dīn, willingly permit what he postulates; since just that, which with him remains undemonstrated, can be recognized as most rigorously demonstrated by me in Corollary 2 to Proposition 3.

The other postulate of Nasīr ad-Dīn is the reciprocal of the first; that indeed the angle may always be acute toward those parts where the just mentioned perpendiculars are supposed to become shorter; but obtuse toward the other parts where these perpendiculars are supposed to go out always longer.

But here lurks an ambiguity. For why (while we from any one perpendicular prescribed as the first we proceed to the others) may not the angles of the consequent perpendiculars, on the same side acute, not become even greater, even to where one strikes upon a right angle, consequently upon such a perpendicular as is itself the common perpendicular to each of the aforesaid straights? And if indeed that happens, evanishes this subtle preparation of Nasīr ad-Dīn, by means of which ingeniously indeed, but with great labor he demonstrates the Euclidean Assertion.

And yet if Nasīr ad-Dīn with a certain justice may determine to presume as if known *per se* that persistence of acute angles on the same side: why cannot also (I speak with Wallis)[3] be assumed as if clear *per se*: *Two straights in the same plane converging* (upon which of course another straight striking makes towards the same parts two angles less than two right angles, as suppose one right, and the other in whatever way acute) *finally meet, if produced*? Nor in fact can it be objected, that this greater convergence toward one side can always subsist within a certain determinate limit, so that indeed a certain so much of the distance always intervenes between these lines on this side, even if still one approaches always more nearly to the other. That cannot, I say, be objected; since from this itself I shall demonstrate, after Proposition 25, the meeting at a finite distance of all such straights, in accordance with the Euclidean Assertion.[4]

Jam transeo ad praelaudatum Joannem Vallisium, qui nempe, ut morem gereret tot Mag-
[40] nis Viris, Veteribus pariter, ac Recentioribus, & rursum ex onere Cathedrae suae Oxoniensi imposito, hoc idem pensum aggredi voluit demonstrandi saepe dictum Pronunciatum. Unice autem assumit tanquam certum, quod sequitur: nimirum *Datae cuicunque figurae similem aliam cujuscunque magnitudinis possibilem esse.* Et id quidem praesumi posse de qualibet figura (etiam si in rem suam unice assumat triangularem rectilineam) bene argumentatur ex circulo, quem scilicet sub quantolibet radio describi posse omnes agnoscunt. Deinde acutus Vir cautissime observat praesumptioni huic suae non obstare, quod praeter correspondentium angulorum aequalitatem requiratur etiam correspondentium omnium laterum proportionalitas, ut habeatur una figura rectilinea, v. g. triangularis, alteri rectilineae triangulari similis; cum tamen Proportionalium, ac subinde similium Figurarum definitio ex Quinto, ac Sexto Euclidis Libro desumendae sint: *Poterat enim Euclides* (inquit ipse) *utramque Libro Primo praemisisse.* Porro autem, hoc stante (quod tamen negari a quopiam posset, nisi demonstretur) intentum suum pulchro sane, atque ingenioso molimine exequitur laudatus Vir.

Sed nolo oneri a me suscepto in quoquam deesse. Itaque assumo duo triangula, unum ABC, & alterum DEF (Fig. 24) invicem aequiangula: Non dico plane similia; quia non indigeo proportionalitate laterum circa angulos aequales, immo neque ulla ipsorum laterum determinata mensura. Solum igitur nolo triangula invicem aequilatera, quia tunc sufficeret sola octava primi, sine ulla praesumptione. Itaque anguli ad puncta A, B, C, aequales sint angulis ad puncta D, E, F; sitque latus DE minus latere AB; assumaturque in AB portio AG aequalis ipsi DE, atque item in AC portio AH aequalis ipsi DF. Debere autem DF minorem esse ipsa AC infra declarabo. Tum (juncta GH) constat (ex 4. primi)
[41] aequales fore angulos ad puncta E, & F, ipsis AGH, AHG. Quapropter; cum modo dicti anguli una cum aliis BGH, CHG, aequales sint (ex 13. primi) quatuor rectis; quatuor itidem rectis aequales erunt anguli ad puncta B, & C, una cum eisdem angulis BGH, CHG. Igitur quatuor simul anguli quadrilateri BGHC aequales erunt quatuor rectis; ac propterea (ex 16. hujus) stabilietur hypothesis anguli recti; & simul (ex 13. hujus) Pronunciatum Euclidaeum.

Porro supposui latus DF, sive AH sumptum ipsi aequale, minus fore latere AC. Si enim aequale foret, & sic punctum H caderet in punctum C; tunc angulus BCA aequalis foret (ex hypothesi) angulo EFD, sive GCA (qui tunc fieret) totum parti; quod est absurdum. Sin vero majus foret, & sic juncta GH secaret in aliquo puncto ipsam BC; jam angulus ACB externus aequalis foret ex hypothesi (contra 16. primi) angulo interno, & opposito (qui tunc fieret) AHG, sive GHA. Itaque bene supposui latus DF unius trianguli minus fore latere AC alterius trianguli, juxta hypothesim jam stabilitam.

Quare ex duobus quibusvis invicem aequiangulis triangulis, sed non etiam invicem aequilateris, stabilitur Pronunciatum Euclidaeum. Quod intendebatur.

Now I go over to the aforesaid John Wallis, who, to fulfill the desire of so many great men, ancient as well as recent, and the obligation imposed on his Oxford professional chair, determined to undertake this same duty of demonstrating the oft mentioned Assertion. Now solely he assumes as if certain, what follows: namely that *to any given figure another similar of any magnitude is possible.* And that this indeed may be presumed of any figure (although in his affair he assumes solely a rectilineal triangle) is well argued from the circle, which of course all admit can be described with any-sized radius. Further the acute man observes most cautiously it does not thwart this his presumption, that besides the equality of corresponding angles also the proportionality of all corresponding sides is required, in order that a rectilineal figure, for example a triangle, may be similar to another rectilinear figure; though still the definition of proportion, and forthwith of similar figures are to be taken from the Fifth, and the Sixth Books of the *Elements*: *For* (says he himself) *Euclid could have put each in front of the First Book.*[5] Hereafter, this standing (which nevertheless can be denied by anyone, unless it is demonstrated)[6] the famous man carries out his intent with really beautiful and ingenious effort.

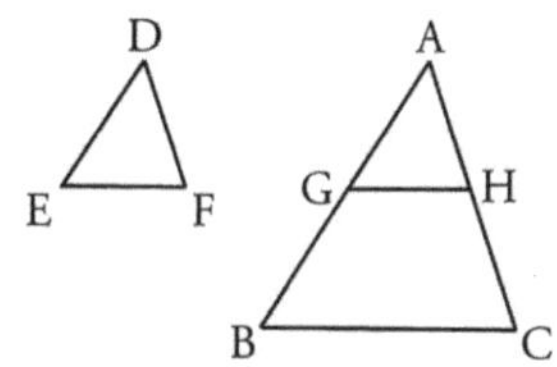

Fig. 24

But I am unwilling to fail in anything to the charge undertaken by me. Therefore I assume two triangles, one ABC, and the other DEF (Fig. 24) mutually equiangular. I do not say wholly similar; because I do not need the proportionality of the sides about the equal angles, nay nor any determinate measures of the sides themselves.[7] Merely therefore I do not wish triangles mutually equilateral, since then *Elements* I, 8 would alone suffice, without any assumption. So let the angles at the points A, B, C be equal to the angles at the points D, E, F; and let the side DE be less than the side AB; and in AB is assumed the portion AG equal to this DE, and likewise in AC the portion AH equal to DF. But that DF must be less than AC I will make clear below. Then (GH joined) follows (from *Elements* I, 4) the angles at the points E, and F will be equal to AGH, AHG. However since the just mentioned angles, together with the others BGH, CHG, are equal (*Elements* I, 13) to four right angles; likewise will be equal to four right angles the angles at the points B, and C, together with these same angles BGH, CHG. Therefore the four angles of the quadrilateral BGHC will be together equal to four right angles; and consequently (Proposition 16) is established the hypothesis of right angle; and at the same time (Proposition 13) the Euclidean Assertion.[8]

Moreover I have supposed the side DF, or AH assumed equal to it, to be less than the side AC. For if it were equal, and so the point H should fall upon the point C, then the angle BCA would be equal (by hypothesis) to the angle EFD, or GCA (which then it would become) a part to the whole; which is absurd. But if it were greater, and so the join GH should cut BC itself in some point,[9] now the external angle ACB would be from the hypothesis equal (against *Elements* I, 16) to the internal and opposite angle (which then would become) AHG, or GHA. Therefore I have rightly supposed the side DF of one triangle to be less than the side AC of the other triangle, in accordance with the hypothesis now established.

Wherefore from any two triangles mutually equiangular, but not also mutually equilateral, the Euclidean Assertion is established. This is what we wanted.

Scholion IV.

In quo exponitur figuralis quaedam exhibitio, ad quam fortasse respexit Euclides, ut suum illud Pronunciatum tanquam per se notum stabiliret.

Praemitto primo: sub quolibet angulo acuto BAX (recole ex hac Tab. Fig. 12) educi posse [42] ex aliquo puncto X ipsius AX quandam XB, quae sub quovis designato etiamsi obtuso angulo R, qui nimirum cum eo acuto BAX deficiat a duobus rectis; quandam, inquam, educi posse XB, quae ad finitam distantiam occurrat ipsi AB in quodam puncto B. Nam id ipsum jam demonstravi in Scholio post XIII. hujus.

Praemitto secundo: eas AB, AX (Fig. 25) intelligi posse in infinitum protractas usque in quaedam puncta Y, & Z; atque item praedictam XB (in infinitum & ipsam protractam usque in quoddam punctum Y) intelligi posse ita moveri super ea AZ versus partes puncti Z, ut angulus ad punctum X versus partes puncti A aequalis semper sit dato cuivis obtuso angulo R.

Praemitto tertio: nulli jam dubitationi obnoxium fore illud Pronunciatum Euclidaeum, si antedicta XY in eo quantocunque motu super recta AZ secet semper illam AY in quibusdam punctis B, D, H, P, atque ita consequenter in aliis punctis remotioribus ab eo puncto A. Ratio evidens est; quia sic duae quaelibet in eodem plano existentes rectae AB, XH, in quas recta quaelibet incidens AX duos ad easdem partes angulos BAX, HXA, duobus rectis minores efficiat, convenire tandem ad eas partes deberent in uno eodemque puncto H.

Praemitto quarto: nulli item dubitationi locum fore super veritate praecedentis hypothetici assumpti; si posteriores illi externi anguli YDH, YHP, & sic alii quilibet consequentes, aut aequales semper sint priori externo angulo YBD, aut saltem non ita minores semper sint, quin corum unusquisque major semper sit parvulo quopiam designato acuto angulo K: Hoc enim stante manifestum fiet, quod ea XY, in suo illo quantocunque motu versus partes puncti Z, nunquam cessabit secare praedictam AY; quod utique (ex praecedente notato) satis [43] est ad stabiliendum Pronunciatum controversum.

Unice igitur superest, ut quidam Adversarius dicat angulos illos externos in majore, ac majore distantia ab illo puncto A fieri semper minores sine ullo determinato limite. Inde autem fiet, ut illa XY in suo illo motu super recta AZ occurrere tandem debeat ipsi AY in quodam puncto P sine ullo angulo cum segmento PY, adeo ut nempe segmentum ejusmodi commune sit duarum rectarum APY, & XPY. At hoc evidenter repugnat naturae lineae rectae.

Scholium 4.

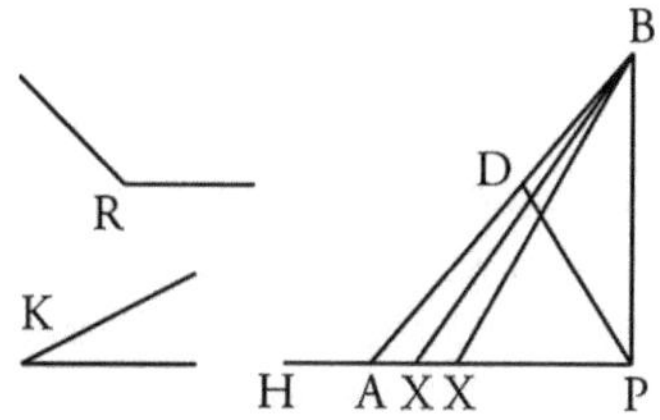

Fig. 12

In which is expounded on a figure a certain consideration, of which Euclid probably thought, in order to establish that Assertion of his as per se evident.

I premise first: within any acute angle BAX (Fig. 12) can be drawn from any point X of AX a certain straight XB, under any designated (even obtuse) angle R (provided only that R with the acute BAX falls short of two right angles); I say, a certain XB can be drawn, which at a finite remove meets AB in a certain point B. For just that I have already demonstrated in the Scholium after Proposition 13.

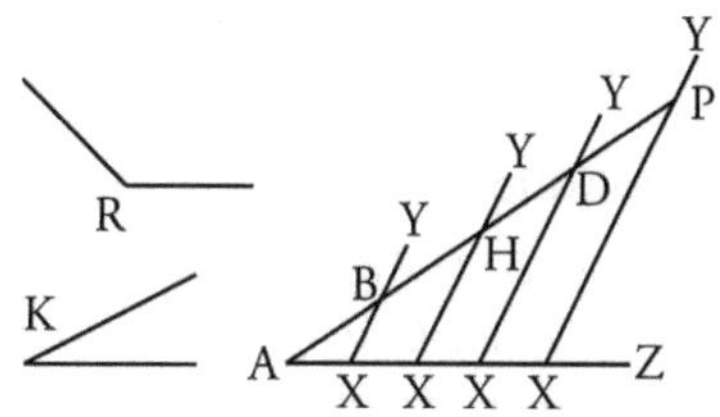

Fig. 25

I premise secondly: these AB, AX (Fig. 25) can be understood as produced *in infinitum* even to certain points Y, and Z, and likewise the aforesaid XB (produced *in infinitum* even to a point Y) can be understood to be so moved above this AZ toward the parts of the point Z, that the angle at the point X toward the parts of the point A is always equal to the certain given obtuse angle R.

I premise thirdly: that Euclidean Assertion would be liable now to no doubt, if the aforesaid XY in this however great motion above the straight AZ cuts always that AY in certain points B, H, D, P, and so successively in other points more remote from this point A. The reason is evident; since thus any two straights AB, XH lying in the same plane, upon which any straight AX cutting makes two angles toward the same parts BAX, HXA, less than two right angles, must at length meet toward those parts in one and the same H.

I premise fourthly: likewise will no doubt about the truth of the preceding hypothetical assumption, if the later external angles YHD, YDP[1] and so any other succeeding ones, either always are equal to the preceding eternal angle YBD, or at least always will be not so much less but that any one of them always will be greater than any little designated acute angle K. For, this holding, it is manifest that this XY in that however great motion of its toward the parts of the point Z, never will cease to cut the aforesaid AY; which assuredly (from the preceding remark) is sufficient for establishing the controverted Assertion.

Solely therefore remains, that some Adversary may say those external angles at greater and greater distance from the point A may become always less without any determinate limit. But thence would follow, that XY in its motion above the straight AZ would at length meet AY in a certain point P without any angle with the segment PY, so that indeed a segment of the two straights APY, and XPY would be in this way common. But this is evidently repugnant to the nature of the straight line.

Sin vero cuiquam minus opportunus videatur angulus obtusus ad illud punctum X versus partes puncti A, nullo negotio supponi poterit rectus; adeo ut nempe (in motu praedictae XY ad angulos semper rectos super recta AZ) manifestius appareat singula illius XY puncta aequabiliter semper moveri relate ad subjectam AZ; ac propterea nequire jam dictam XY transire de secante in non secantem alterius indefinitae AY, nisi eam aut aliquando in aliquo puncto praecise contingat, aut ipsi occurrat in aliquo puncto P, ubi cum eadem AY commune obtineat segmentum PY; quorum utrunque adversari naturae lineae rectae ostendam ad XXXIII. hujus. Igitur juxta veram ideam lineae rectae, debebit illa XY, in quantacunque distantia puncti X a puncto A, occurrere semper in aliquo puncto ipsi AY. Atque id quidem (quantumlibet parvus supponatur acutus angulus ad punctum A) satis esse ad demonstrandum, contra hypothesim anguli acuti, Pronunciatum Euclidaeum, constabit ex XXVII. hujus.

Propositio XXII.

Si duae rectae AB, CD in eodem plano existentes perpendiculariter insistant cuidam rectae
[44] *BD; ipsa autem AC jungens ea perpendicula internos (in hypothesi anguli acuti) acutos angulos cum eisdem efficiat: Dico* (Fig. 26) *rectas terminatas AC, BD commune aliquod habere perpendiculum, & quidem intra limites designatis punctis A, & C praefinitos.*

Demonstratur. Si enim aequales sint ipsae AB, CD; constat (ex 2. hujus) rectam LK, a qua bifariam dividantur illae duae AC, & BD, commune fore eisdem perpendiculum. Sin vero alterutra sit major, ut puta AB: demittatur ad BD (juxta 12. primi) ex quovis puncto L ipsius AC perpendicularis LK, occurrens alteri BD in K. Occurret autem in aliquo puncto K, consistente inter puncta B, & D; ne (contra 17. primi) perpendicularis LK secet alterutram AB, aut CD, perpendiculares eidem BD. Si ergo anguli ad punctum L recti non sunt, unus eorum acutus erit, & alter obtusus. Sit obtusus versus punctum C. Jam vero intelligatur LK ita procedere versus AB, ut semper ad rectos angulos insistat ipsi BD, atque item opportune aucta, aut imminuta, in aliquo sui puncto secet rectam AC. Constat angulos ad puncta intersectiva ipsius AC non posse omnes esse obtusos versus partes puncti C, ne tandem in ipso puncto A, dum recta LK congruet cum recta AB, angulus ad punctum A versus partes puncti C sit obtusus, cum ad eas partes positus sit acutus. Quoniam ergo angulus ad punctum L ipsius LK positus est obtusus versus partes puncti C, non transibit in eo motu recta LK ad faciendum in aliquo sui puncto cum recta AC angulum acutum versus partes praedicti puncti C, nisi prius transeat ad constituendum in aliquo sui puncto cum eadem AC angulum rectum versus partes ejusdem puncti C. Erit igitur inter puncta A, & L unum aliquod punctum intermedium H, in quo HK perpendicularis ipsi BD sit etiam perpendicularis alteri AC.

Simili modo ostendetur adesse aliquam XK inter ipsas LK, CD, quae sit perpendicularis
[45] & rectae BD, & rectae AC, dum scilicet angulus obtusus ad punctum L ponatur consistere versus partes puncti A.

But if to any one may seem less opportune the obtuse angle at the point X toward the parts of the point A, it may easily be supposed right; so that indeed (in the motion of the aforesaid XY at angles always right above the straight AZ) more manifestly may appear that the single points of that XY are always moved uniformly relatively to the basal AZ; and therefore the aforesaid XY cannot go over from a secant into a non-secant of the other indefinite AY, unless either once in some point it precisely touches it, or meets it some point P; where it has with this AY a common segment PY; each of which I shall show contrary to the nature of the straight line in Proposition 33. Therefore in accordance with the true idea of the straight line, must that XY, however great the distance of the point X from the point A, always meet in some point this AY. And that this indeed (however small is supposed the acute angle at point A) is sufficient for demonstrating, against the hypothesis of acute angle, the Euclidean Assertion, will follow from Proposition 27.

Proposition 22.

If two straights AB, DC existing in the same plane stand perpendicular to a certain straight BD; but AC joining these perpendiculars makes with them internal acute angles (in the hypothesis of acute angle): I say (Fig. 26) *the bounded straights AC, BD have a common perpendicular, and indeed within the limits fixed by the designated points A and C.*

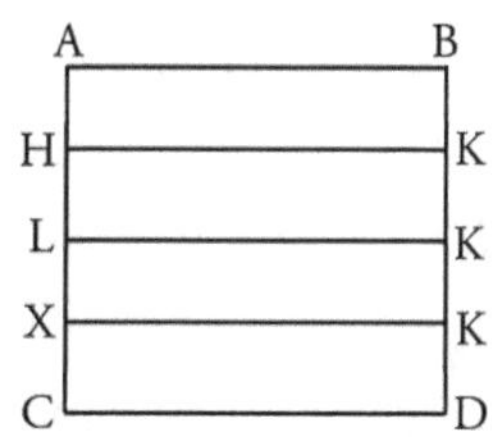

Fig. 26

Proof. For if AB, CD are equal, it follows (Proposition 2) that the straight LK, by which these two AC and BD are bisected, will be to them a common perpendicular. But if either be the greater, as suppose AB; let fall to BD (according to *Elements* I, 12) from any point L of AC the perpendicular LK, meeting the other BD in K. But it will meet in some point K existing between the points B and D; otherwise (contrary to *Elements* I, 17) the perpendicular LK would cut either AB, or CD, perpendicular to the same BD. If then the angles at the point L are not right, one of them will because and the other obtuse. Let the obtuse point be C. But now LK is understood as to proceed toward AB, that it always stands at right angles to BD, and likewise opportunely increased, or diminished, in some points of it cuts the straight AC. It follows that the angles at the intersection points with AC cannot all be obtuse toward the parts of the point C, lest at length in that point A, where the straight LK is congruent with the straight AB, the angle at the point A toward the parts of the point C should obtuse, when toward these parts it is by hypothesis acute. Since therefore the angle L of this LK is by hypothesis obtuse toward the parts of the point C, the straight LK will not change over in this motion so as to make in some point of it with the straight AC an angle acute toward the parts of the aforesaid point C, unless previously it changes over so as to make in some point of it with this AC an angle right towards the parts of this same point C. Therefore between the points A, and L will be some one intermediate point H, in which HK perpendicular to this BD is also perpendicular to the other AC.

In a similar manner is shown to be present a certain XK between LK, CD, which is perpendicular both to the straight BD, and to the straight AC, if namely an angle at the point L is assumed to be obtuse toward the parts of the point A.

Constat igitur rectas AC, BD commune aliquod habituras esse perpendiculum, & quidem intra limites designatis punctis A, & C praefinitos, quoties junctæ AB, CD in eodem plano existant, sintque perpendiculares ipsi BD. Quod erat &c.

Propositio XXIII.

Si duae quaelibet rectae AX, BX (Fig. 27) *in eodem plano existant; vel unum aliquod (etiam in hypothesi anguli acuti) commune obtinent perpendiculum; vel in alterutram eandem partem protractae, nisi aliquando ad finitam distantiam una in alteram incidat, semper magis ad se invicem accedunt.*

Demonstratur. Ex quolibet puncto A ipsius AX demittatur ad rectam BX perpendicularis AB. Si ipsa BA efficiat cum AX angulum rectum, habemus intentum communis perpendiculi. Caeterum vero ea recta efficiat ad alterutram partem, ut puta versus partes puncti X, angulum acutum. Itaque in praedicta recta AX designentur inter puncta A, & X quaelibet puncta D, H, L, ex quibus demittantur ad rectam BX perpendiculares DK, HK, LK. Si unus aliquis angulus ad puncta D, H, L acutus sit versus partes puncti A, constat (ex praecedente) unum aliquod adfuturum commune perpendiculum ipsarum AX, BX. Sin vero omnis hujusmodi angulus sit major acuto, vel unus aliquis erit rectus, & sic rursum habemus intentum communis perpendiculi, cum omnes anguli ad puncta K supponantur recti; vel omnes illi anguli ponuntur obtusi versus partes puncti A, ac propterea omnes itidem acuti versus partes puncti X, & sic rursum argumentor. Quoniam in quadrilatero KDHK recti
[46] sunt anguli ad puncta K, ponitur autem acutus angulus ad punctum D, erit (ex Cor. II. post 3. hujus) latus DK majus latere HK. Simili modo ostendetur latus HK majus esse latere LK; atque ita semper, conferendo inter se perpendiculares ex quolibet puncto altiore ipsius AX demissas ad alteram BX. Quapropter ipsae AX, BX semper magis versus partes puncti X ad se invicem accedent: Quae est altera pars propositi disjuncti.

Ex quibus omnibus constat duas quaslibet rectas AX, BX, quae in eodem plano existant, vel unum aliquod (etiam in hypothesi anguli acuti) commune habere perpendiculum, vel in alterutram eandem partem protractas, nisi aliquando ad finitam distantiam una in alteram incidat, semper magis ad se invicem accedere. Quod erat &c.

Corollarium I.

Hinc anguli versus basim AB erunt semper obtusi ad illud punctum ipsius AX, ex quo demittitur perpendicularis ad rectam BX: erunt, inquam, semper obtusi, quoties duae illae AX, & BX semper magis ad se invicem accedant versus partes punctorum X; quod quidem sano modo intelligi debet, nimirum de perpendicularibus demissis ante praedictum occursum, si forte ad finitam distantiam una in alteram incidere debeat.

It follows therefore that the straights AC, BD, will have a common perpendicular, and indeed within the limits fixed by the designated points A, and C, when the joins AB, CD exist in the same plane and are perpendicular to BD. This is what was to be demonstrated.

Proposition 23.

If any two straights AX, BX (Fig. 27) *are in the same plane; either they have (even in the hypothesis of acute angle) a common perpendicular; or prolonged toward the same part unless somewhere at a finite distance meets the other – they mutually approach ever more toward each other.*

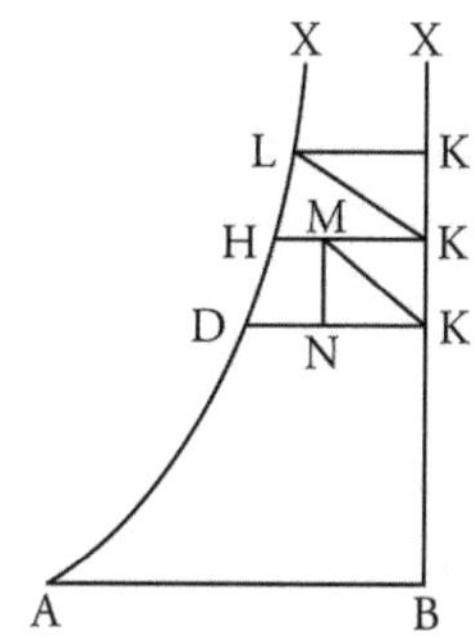

Fig. 27

Proof. From any point A of AX let fall to the straight BX the perpendicular AB. If BA makes with AX a right angle, we have the asserted case of a common perpendicular. But otherwise this straight makes toward one or the other, as suppose toward the parts of the point X, an acute angle. Accordingly in the aforesaid straight AX between the points A and X any points D, H, L are designated, from which are let fall to the straight BX the perpendiculars DK, HK, LK. If any one angle at the points D, H, L be acute toward the parts of the point A, it follows (Proposition 22) that AX, BX will have a common perpendicular. But if every angle of this sort be greater than acute; either some one will be right, and thus again we shall have the asserted case of a common perpendicular, since all angles at the points K are supposed right; or all those angles toward the parts of the point A are obtuse, and therefore all therewith acute toward the parts of the point X, and so again I argue: Since in the quadrilateral KDHK the angles at the points K are right, but the angle at the point D is acute, the side DK will be (Corollary 2 to Proposition 3) greater than the side HK. In a similar way the side HK is shown to be greater than the side LK; and so always, comparing to each other perpendiculars from any ever higher points of AX let fall upon the other BX. Wherefore AX, BX mutually approach each other ever more toward the parts of the point X: which is the second part of the disjunctive proposition. From which follows that any two straights AX, BX, which are in the same plane, either have (even in the hypothesis of acute angle) a common perpendicular, or produced toward either the same part, unless somewhere at a finite distance one meets the other, mutually approach each other ever more. This is what was to be demonstrated.

Corollary 1.

Hence the angles toward the base AB will be always obtuse at each point of AX, from which is let fall a perpendicular to the straight BX: will be, I say, always obtuse, as those two AX, and BX mutually approach each other ever more toward the parts of the points X; which of course should be understood in a sane way, of perpendiculars let fall before the mentioned meeting, if perchance one is to strike upon the other at a finite distance.

Scholion.

Video tamen inquiri hic posse, qua ratione ostendendum sit commune illud perpendiculum; quoties recta quaepiam PFHD (Fig. 28) occurrens duabus AX, BX in punctis F, & H, duos ad easdem partes efficiat internos angulos AHF, BFH, non eos quidem rectos, sed tamen
[47] aequales simul duobus rectis. Ecce autem commune illud perpendiculum geometrice demonstratum. Divisa FH bifariam in M demittantur ad AX, & BX perpendiculares MK, ML. Angulus MFL aequalis erit (ex 13. primi) angulo MHK, qui nempe supponitur duos rectos efficere cum angulo BFH. Praeterea recti sunt anguli ad puncta K, & L; ac rursum aequales sunt ipsae MF, MH. Igitur (ex 26. primi) aequales itidem erunt anguli FML, HMK. Quare angulus HMK duos efficiet rectos angulos cum angulo HML, prout cum eodem duos efficit rectos angulos (ex 13. primi) angulus FML. Igitur (ex 14. primi) unae erit recta linea continuata ipsa KML, commune idcirco perpendiculum praedictis rectis AX, BX. Quod erat &c.

Corollarium II.

Sed rursum docere hinc possum, quod illae duae AX, BX, in quas incidens recta PFHD, aut duos efficiat cum ipsis AX, BX internos ad easdem partes angulos aequales duobus rectis; aut consequenter (ex 13. & 15. primi) alternos sive externos, sive internos angulos inter se aequales; aut rursum, eodem titulo, externum (ut puta DHX) aequalem interno, & opposito HFX: quod, inquam, illae duae rectae neque ad infinitam earundem productionem coire inter se possint. Si enim ex quolibet puncto N ipsius AX demittatur ad BX perpendicularis NR, erit haec in ipsa hypothesi anguli acuti (quae utique sola obesse nobis posset) major (ex Cor. I. post 3. hujus) eo communi perpendiculo KL. Non igitur illae duae AX, BX convenire unquam inter se poterunt.

Porro autem demonstratas hinc habes Propos. 27. & 28. Libri primi Euclidis; & quidem
[48] citra immediatam dependentiam a praecedentibus 16. & 17. ejusdem primi, circa quas oriri posset difficultas, quoties sub basi finita infinitilaterum esset triangulum; ad quale nempe triangulum provocare non dubitaret, qui eas duas AX, BX ad infinitam saltem distantiam inter se coituras censeret, quamvis anguli ad incidentem PFHD tales forent, quales supposuimus.

Praeterea, propter demonstratum commune perpendiculum KL, nequirent sane illae duae KX, LX ad suam partem punctorum X simul concurrere, quin etiam (ex facili intellecta super positione) ad alteram etiam partem simul concurrerent reliquae & ipsae interminatae KA, LB. Quare duae rectae AX, BX clauderent spatium; quod est contra naturam lineae rectae.

Sed haec posteriora sunt. Nam in praecedentibus nusquam adhibui aut 16. aut 17. primi, nisi ubi clare ageretur de triangulo omni ex parte circumscripto, prout nempe in Proemio ad Lectorem ita me curaturum spoponderam.

Scholium.

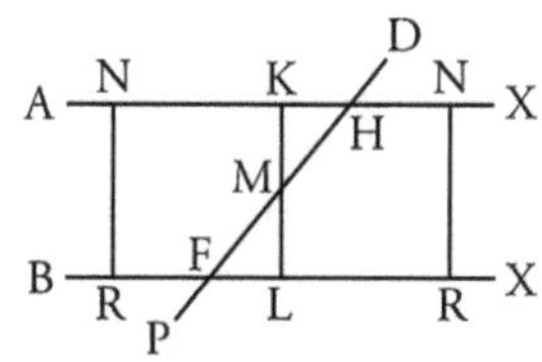

Fig. 28

I see indeed it may here be asked in what way that common perpendicular can be shown, when any straight PFHD (Fig. 28) meeting two AX, BX in points F, and H, makes toward the same parts two internal angles AHF, BFH, not themselves indeed right, but nevertheless together equal to two rights. But behold that common perpendicular geometrically demonstrated. FH being bisected in M, perpendiculars MK, ML are let fall to AX and BX. The angle MFL will be equal (*Elements* I, 13) to the angle MHK, which indeed is assumed to make up two right angles with the angle BFH. Moreover the angles at the points K, and L are right; and again MF, MH are equal. Therefore (*Elements* I, 26) so are the angles FML, HMK equal. Wherefore the angle HMK makes two right angles with the angle HML, since with this the angle FML (*Elements* I, 13) makes two right angles. Therefore (*Elements* I, 14) KML will be in one continuous straight line, consequently a common perpendicular to the aforesaid straights AX, BX. This is what was to be demonstrated.

Corollary 2.

But again I am able hence to show that those two straights AX, BX, meeting with which the straight PFHD makes with the said AX, BX either two internal angles toward the same parts equal to two right angles, or consequently (*Elements* I, 13 and 15) alternate external or internal angles equal to one another, or again, from the same cause, an external (as suppose DHX) equal to an internal and opposite (HFX); that, say I, those two straights not even in their infinite production can meet one another. For if from any point N of AX is let fall to BX the perpendicular NR, this will be in the hypothesis of acute angle (which alone in any case can hinder us) greater (Corollary 1 to Proposition 3) than the common perpendicular KL. Therefore those two straights AX, BX cannot ever meet one another.

But furthermore here you have *Elements* I, 27 and 28 demonstrated, and indeed without immediate dependence from the preceding *Elements* I, 16 and 17, about which difficulties could arise when the triangle should be of infinite sides on a finite base; to which sort of a triangle without doubt would refer one who believed that these two straights AX, BX met one another a least at an infinite distance, although the angles at the transversal PFHD were such as we have supposed.

Moreover, on account of the demonstrated common perpendicular KL, surely those two KX, LX cannot come together toward the part of the points X, since also (from a superposition easily understood) toward the other part also would meet at the same time the remaining and themselves unbounded KA, LB. Wherefore two straights AX, BX would enclose a space; which is contrary to the nature of the straight line.

But these things are later. For in the preceding I have never applied either *Elements* I, 16 or 17, except where clearly it treats of a triangle bounded on every side, as indeed I promised I would so take care to do in the Preface to the Reader.

Propositio XXIV.

Iisdem manentibus: Dico quatuor simul angulos (Fig. 27) *quadrilateri KDHK proximioris basi AB minores esse (in hypothesi anguli acuti) quatuor simul angulis quadrilateri KHLK remotioris ab eadem basi; atque ita quidem, sive illae duae AX, BX aliquando ad finitam distantiam incidant versus partes puncti* X; *sive nunquam inter se incidant; sed versus eas partes aut semper magis ad se invicem accedant, aut aliquando recipiant commune perpendiculum, post quod nempe (juxta Cor.* II. *praec. Propos.) ad easdem partes incipiant invicem dissilire.*

Demonstratur. Verum hic supponimus portiones KK sumptas esse invicem aequales.
[49] Quoniam igitur (ex praecedente) latus DK majus est latere HK, ac similiter HK majus latere LK; sumatur in HK portio MK aequalis ipsi LK, & in DK portio NK aequalis ipsi HK; junganturque MN, MK, LK; nimirum punctum K intermedium cum puncto L, & punctum K vicinius puncto B cum puncto M. Jam sic progredior. Quandoquidem latera trianguli KKL (initium semper ducam a puncto K viciniore puncto B) aequalia sunt lateribus trianguli KKM, & anguli comprehensi aequales, utpote recti; aequales etiam erunt (ex 4. primi) bases LK, MK; atque item aequales, qui correspondent invicem anguli, ad easdem bases, nimirum angulus KLK angulo KMK, & angulus LKK angulo MKK. Igitur aequales etiam sunt residui NKM, & HKL. Quare, cum latera NK, KM trianguli NKM aequalia itidem sint lateribus HK, KL trianguli HKL; aequales etiam erunt (ex eadem 4. primi) bases NM, HL; anguli KNM, KHL; ac tandem anguli KMN, KLH. Sunt autem in prioribus triangulis jam probati aequales anguli KLK, & KMK. Igitur totus angulus NMK aequalis est toti angulo HLK. Quare, cum omnes ad puncta K anguli sint recti, manifeste consequitur omnes simul quatuor angulos quadrilateri KNMK aequales esse omnibus simul quatuor angulis quadrilateri KHLK. Quoniam vero duo simul anguli ad puncta N, & M in quadrilatero KNMK majores sunt, in hypothesi anguli acuti, duobus simul angulis (ex Cor. post XVI. hujus) ad puncta D, & H in quadrilatero NDHM, seu quadrilatero KDHK; consequens inde est, ut (additis communibus rectis angulis ad puncta K) quatuor simul anguli quadrilateri KNMK, seu quadrilateri KHLK, majores sint (in hypothesi anguli acuti) quatuor simul angulis quadrilateri KDHK. Quod erat demonstrandum.

[50] **Corollarium.**

Sed opportune observari hic debet, nihil defuturum factae argumentationi, quamvis angulus ad punctum L poneretur rectus, juxta hypothesin anguli acuti. Nam adhuc illa communis perpendicularis LK minor foret (ex Cor. I. post III. hujus) altera perpendiculari HK, ex qua propterea sumi adhuc posset portio MK aequalis praedictae LK: Quo stante constat nullum posse obicem intercurrere.

Proposition 24.

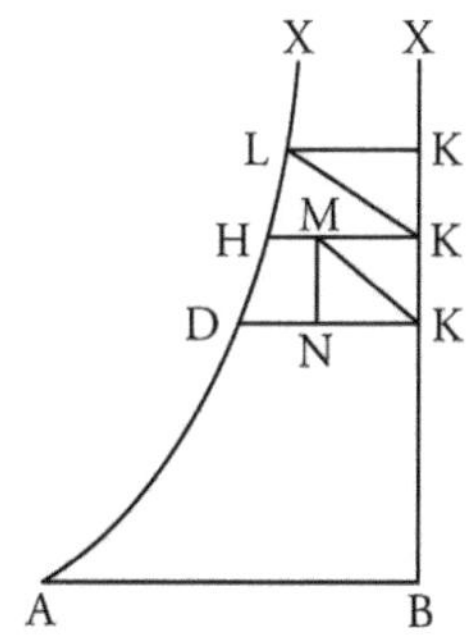

Fig. 27

The same remaining: I say the four angles together (Fig. 27) *of the quadrilateral KDHK nearer the base AB are less (in the hypothesis of acute angle) than the four angles together of the quadrilateral KHLK more remote from the same base; and indeed this is so, whether from those two AX, BC somewhere at a finite distance meet toward the parts of the point X; or never meet one another; but toward those parts either ever more mutually approach each other, or somewhere receive a common perpendicular, after which of course (in accordance with Corollary 2 of Proposition 23) toward the same parts they begin mutually to separate.*

Proof. Here however we suppose the portions KK assumed to be mutually equal. Since therefore (Proposition 23) the side DK is greater than the side HK, and similarly HK greater than the side LK, the portion MK in HK is assumed equal to LK, and in DK the portion NK equal to HK; and MN, MK, LK are joined, truly the intermediate point K with the point L, and the point K nearer to the point B with the point M. Now I proceed thus. Since indeed the sides of the triangle KKL (I make beginning always from the point K nearer to the point B) are equal to the sides of the triangle KKM, and the included angles equal, as being right, equal also will be (from *Elements* I, 4) the bases LK, MK, and likewise equal the angles which correspond mutually, at these bases, indeed the angle KLK to the angle KMK, and the angle LKK to the angle MKK. Therefore equal also are the remainders NKM and HKL. Wherefore, since the sides NK, KM of the triangle NKM are equal in the same way to the sides HK, KL of the triangle HKL, equal also will be (from the same *Elements* I, 4) the bases NM, HL, the angles KNM, KHL, and finally the angles KMN, KLH. But in the preceding triangles are already proved equal the angles KLK, KMK. Therefore the whole angle NMK is equal to the whole angle HLK. Wherefore, since all angles at the points K are right, it follows manifestly all four angles together of the quadrilateral KNMK are equal to all four angles together of the quadrilateral KHLK. But since the two angles together at the points N and M in the quadrilateral KNMK are greater, in hypothesis of acute angle, than the two angles together (Corollary to Proposition 16) at the points D and H in the quadrilateral NDHM, or the quadrilateral KDHK, the consequence thence is, that (the common right angles at the points K being added) the four angles together of the quadrilateral KNMK, or the quadrilateral KHLK are greater (in hypothesis of acute angle) than the four angles together of the quadrilateral KDHK. This is what was to be demonstrated.

Corollary.

But it ought here opportunely to be observed, nothing will fail in the argument made, although the angle at the point L is assumed right, together with the hypothesis of acute angle. For still that common perpendicular LK would be less (Corollary 1 to Proposition 3) than the other perpendicular HK, from which therefore still a portion MK could be assumed equal to the aforesaid LK. Which standing, it follows that no hindrance can intervene.

Scholion.

Dubitari nihilominus posset, an ex quolibet puncto K (assumpto nimirum in BX ante occursum ipsius BX in alteram AX) perpendicularis educta versus partes rectae AX occurrere huic debeat (Fig. 29) in aliquo puncto L; dum nempe illae duae, ante praedictum occursum, ponantur ad se invicem semper magis accedere. Ego autem dico ita omnino secuturum.

Demonstratur. Assignatum sit in BX quodvis punctum K. Sumatur in AX quaedam AM aequalis summae ex ipsa BK, & dupla AB. Tum ex puncto M ducatur ad BX (juxta 12. primi) perpendicularis MN. Erit MN (juxta praesentem suppositionem) minor ipsa AB. Quare AM (facta aequalis summae ex ipsa BK, & dupla AB) major erit summa ipsarum BK, AB, & NM. Jam ostendere oportet eandem AM minorem esse summa ipsarum BN, AB, & MN, ut inde constet eam BN majorem esse praedicta BK, ac propterea punctum K jacere inter puncta B, & N. Jungatur BM. Erit latus AM (ex 20. primi) minus duobus simul reliquis lateribus
[51] AB, & BM. Rursum latus BM (ex eadem 20. primi) minus erit duobus simul lateribus BN, & MN. Igitur latus AM multo minus erit tribus simul lateribus AB, BN, & NM. Hoc autem erat ostendendum, ut constaret punctum K jacere inter puncta B, & N. Inde autem consequens est, ut perpendicularis ex puncto K educta versus partes ipsius AX occurrere huic debeat in aliquo puncto L inter puncta A, & M constituto; ne scilicet (contra 17. primi) secare debeat alterutram AB, aut MN perpendiculares eidem BX. Quod &c.

Propositio XXV.

Si duae rectae (Fig. 30) *AX, BX in eodem plano existentes (una quidem sub angulo acuto in puncto A, & altera in puncto B perpendiculariter insistens ipsi AB) ita ad se invicem semper magis accedant versus partes punctorum* X, *ut nihilominus earundem distantia semper major sit assignata quadam longitudine, destruitur hypothesis anguli acuti.*

Demonstratur. Assignata sit longitudo R. Si ergo in ea BX sumatur quaedam BK quantumlibet multiplex propositae longitudinis R; constat (ex praecedente Scholio) perpendicularem ex puncto K eductam versus partes ipsius AX in aliquo puncto L eidem occursuram; ac rursum (ex praesente hypothesi) constat eam KL majorem fore praedicta longitudine R. Porro intelligatur BK divisa in portiones KK, aequales singulas ipsi R, usque dum KB aequalis sit ipsi longitudini R. Tandem vero ex punctis K erectae sint ad BX perpendiculares occurrentes ipsi AX in punctis L, H, D, M, usque ad punctum N proximius puncto A. Jam sic progredior.

Scholium.

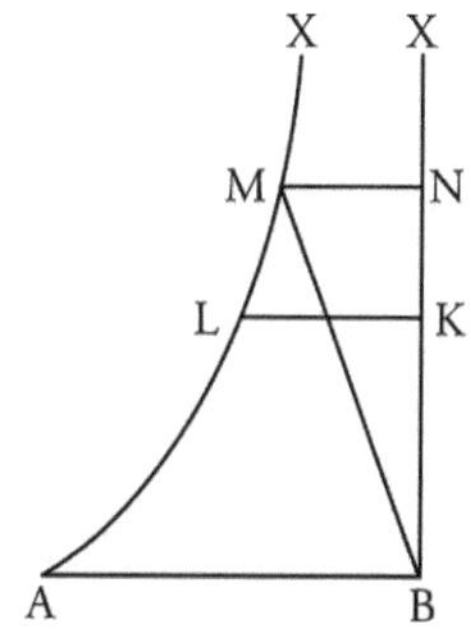

Fig. 29

Nevertheless it might be doubted, whether a perpendicular, from whatever point K (assumed indeed in BX before the meeting of this BX with the other AX) erected toward the parts of the straight AX, must meet this (Fig. 29) in some point L; provided of course those two, before the aforesaid meeting, are assumed ever more to approach each other mutually. But I say it will follow completely thus.

Proof. Let there be assigned in BX any point whatever K. In AX is taken a certain AM equal to the sum of this BK and of twice AB. Then from the point M is drawn to BX (according to *Elements* I, 12) the perpendicular MN. According to the present supposition, MN will be less than AB. Wherefore AM (made equal to the sum of BK, and of double AB) will be greater than the sum of BK, AB, and NM. Now it behooves to show this same AM to be less than the sum of BN, AB, and MN, that thence it may follow this BN is greater than the aforesaid BK, and therefore the point K lies between points B and N. Join BM. The side AM will be (from *Elements* I, 20) less than the two remaining sides together AB and BM. Again the side BM (from the same *Elements* I, 20) will be less than the two sides together BN and MN. Therefore the side AM will be by much less than the three sides together AB, BN, and NM. But this was to be shown, in order to deduce that the point K lies between the points B and N. Thence however it follows, that the perpendicular from the point K erected toward the parts of AX must meet this in some point L stationed between the points A and M; else obviously (against *Elements* I, 17) it must cut either AB or MN perpendiculars to BX. This is what was to be demonstrated.

Proposition 25.

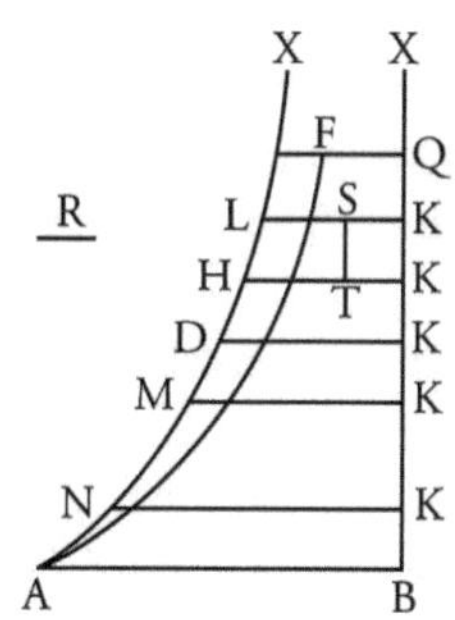

Fig. 30

If two straights (Fig. 30) *AX, BX existing in the same plane (standing upon AB, one indeed at an acute angle in the point A, and the other perpendicular at the point B) so always approach more to each other mutually, toward the parts of the point X, that nevertheless their distance is always greater than a certain assigned length, the hypothesis of acute angle is destroyed.*

Proof. Let R be the assigned length. If therefore in BX is assumed a certain BK any chosen multiple of the proposed length R; it follows (from the preceding Scholium) that the perpendicular erected from the point K toward the parts of AX will meet it at some point L; and again (from the present hypothesis) it follows that this KL will be greater than the aforesaid length R. Furthermore BK is understood divided into portions KK, each equal to R, even until KB is itself equal to the length R. Finally from the points K are erected to BX perpendiculars meeting AX in points L, H, D, M, even to the point N nearest the point A. Now I proceed thus.

Erunt (ex Prop. praecedente) quatuor simul anguli quadrilateri KHLK, remotioris ab ea
[52] basi AB, majores quatuor simul angulis quadrilateri KDHK, proximioris eidem basi; cujus itidem quadrilateri quatuor simul anguli majores erunt quatuor simul angulis subsequentis versus eandem basim quadrilateri KMDK. Atque ita semper usque ad ultimum quadrilaterum KNAB, cujus utique quatuor simul anguli minimi erunt, relate ad quatuor simul angulos singulorum ascendentium versus puncta X quadrilaterorum.

Quoniam vero tot aderunt praedicto modo recensita quadrilatera, quot sunt praeter basim AB demissae ex punctis ipsius AX ad rectam BX perpendiculares; expendenda est summa omnium simul angulorum, qui comprehenduntur in illis quadrilateris. Ponamus esse novem ejusmodi perpendiculares demissas, ac propterea novem itidem quadrilatera. Constat (ex 13. primi) aequales esse quatuor rectis angulos hinc inde comprehensos ad bina puncta illarum octo perpendicularium, quae mediae jaceant inter basim AB, & remotiorem perpendicularem LK. Itaque summa horum omnium angulorum erit 32. rectorum. Restant duo anguli ad perpendiculum LK, & duo ad basim AB. At anguli, unus quidem ad punctum K, & alter ad punctum B, supponuntur recti; angulus autem ad punctum L (ex Cor. post XXIII. hujus) est obtusus. Quapropter (etiam neglecto angulo acuto ad punctum A) summa omnium angulorum, qui comprehenduntur ab illis novem quadrilateris, excedet 35. rectos. Inde autem fit, ut quatuor simul anguli quadrilateri KHLK, remotioris a basi, minus deficiant a quatuor rectis, quam sit nona pars unius recti; & id quidem etiam si aequalis portio praedictae omnium angulorum summae contingeret singulis illis quadrilateris. Ergo minor adhuc erit insinuatus defectus, cum summa quatuor simul angulorum illius quad-
[53] rilateri KHLK ostensa sit omnium maxima, relate ad quatuor simul angulos reliquorum quadrilaterorum.

Sed rursum; juxta suppositionem, in qua procedit haec Propositio; assumi potest tanta longitudo ipsius BK, ut confici semper possint non tot quin plura quadrilatera sub basibus KK, aequalibus singulis illi assignatae longitudini R. Quare defectus quatuor simul angulorum illius remotioris quadrilateri KHLK a quatuor rectis ostendetur semper minor & una centesima, & una millesima, & sic sub quolibet assignabili numero una portiuncula unius recti.

Porro autem erunt semper (juxta praedictam suppositionem) ipsae LK, & HK majores designata longitudine R. Si ergo in KL, & KH sumantur KS, & KT aequales ipsi KK, seu longitudini R; erunt, juncta ST, duo simul anguli KST, KTS majores, in hypothesi anguli acuti, duobus simul angulis (ex Cor. post XVI. hujus) ad puncta H, & L in quadrilatero THLS, seu quadrilatero KHLK; ac propterea (additis communibus rectis angulis ad puncta K, K) erunt quatuor simul anguli quadrilateri KTSK majores quatuor simul angulis illius quadrilateri KHLK.

The four angles together of the quadrilateral KHLK, more remote from the base AB, will be (Proposition 24) greater than the four angles together of the quadrilateral KDHK, nearer to this base; of which quadrilateral in the same way the four angles together will be greater than the four angles together of the quadrilateral KMDK subsequent toward this base. And so always even to the last quadrilateral KNAB, whose four angles together assuredly will be the least, in reference to the four angles together of each of the quadrilaterals ascending toward the points X.

But since are present as many quadrilaterals described in the aforesaid manner, as are, except the base AB, perpendiculars let fall from points of AX to the straight BX; the sum of all the angles together, which are comprehended in these quadrilaterals can be reckoned. We assume that there are nine such perpendiculars let fall, and therefore so nine quadrilaterals. We get (from *Elements* I, 13) as equal to four rights the angles comprehended hither and yon at the two points of those eight perpendiculars, which lie in the middle between the base AB and the more remote perpendicular LK. So the sum of all these angles will be 32 rights. There remain two angles at the perpendicular LK, and two at the base AB. But the angles one indeed at the point K and the other at the point B are supposed right; but the angle at the point L (from the Corollary to Proposition 23) is obtuse. Wherefore (even neglecting the acute angle at the point A) the sum of all the angles which are comprehended by these nine quadrilaterals exceeds 35 rights. But hence follows, that the four angles together of the quadrilateral KHLK, more remote from the base lack less from four rights than the ninth part of one right; and that indeed even if an equal portion of the aforesaid sum of all the angles pertained to each of those quadrilaterals. Therefore less yet will be the occurring defect, since the sum of the four angles together of this quadrilateral KHLK was shown the greatest of all, in relation to the four angles together of the remaining quadrilaterals.

But again; in consequence of the supposition upon which this Proposition proceeds, so great a length of BK can be assumed, that as many quadrilaterals as we choose may be made on bases KK, each equal to the assigned length R. Wherefore the defect of the four angles together of this more remote quadrilateral KHLK from four rights is shown ever less both than a hundredth and than a thousandth, and thus under any assignable part of a right.

Further however, LK and HK will be always (in accordance with the aforesaid supposition) greater than the designated length R. Therefore if in KL and KH are assumed KS and KT equal to KK or the length R; ST being joined, the two angles together KST, KTS will be greater, in hypothesis of acute angle, than the two angles together (Corollary to Proposition 16) at the points H and L in the quadrilateral THLS, or the quadrilateral KHLK; and therefore (the common right angles at the points K, K being added) the four angles together of the quadrilateral KTSK will be greater than the four angles together of that quadrilateral KHLK.

Jam vero: cum ex una parte stabile sit, ac datum quadrilaterum KTSK, utpote constans data basi KK, quae nimirum aequalis ponitur assignatae longitudini R, ac rursum constans duobus perpendiculis TK, SK eidem basi aequalibus, ac tandem jungente TS, quae evadit omnino determinata; & ex altera quatuor simul anguli stabilis illius, ac dati quadrilateri, ostensi jam sint majores quatuor simul angulis quadrilateri KHLK quantumlibet distantis ab ea basi AB: consequens utique fit, ut quatuor simul anguli stabilis illius, ac dati quadrilateri KTSK majores sint qualibet angulorum summa, quae quomodolibet deficiat a quatuor rectis; quandoquidem ostensum jam est designari semper posse tale aliquod
[54] quadrilaterum KHLK, cujus quatuor simul anguli minus deficiant a quatuor rectis, quam sit quaevis designabilis unius recti portiuncula. Igitur quatuor simul anguli stabilis illius, ac dati quadrilateri, vel aequales sunt quatuor rectis, vel eisdem majores. Tunc autem (ex XVI. hujus) stabilitur hypothesis aut anguli recti, aut anguli obtusi; ac propterea (ex V. & VI. hujus) destruitur hypothesis anguli acuti.

Itaque constat destructum iri hypothesim anguli acuti, si duae rectae in eodem plano existentes ita ad se invicem semper magis accedant, ut nihilominus earundem distantia major semper sit assignata quadam longitudine. Hoc autem erat demonstrandum.

Corollarium I.

At (destructa hypothesi anguli acuti) manifestum fit, ex 13. hujus, controversum Pronunciatum Euclidaeum; prout a me hoc loco declaratum iri spopondi in Scholio III. post XXI. hujus, ubi de conatu Nassaradini Arabis locuti sumus.

Corollarium II.

Rursum ex hac Propositione, & ex praecedente XXIII. manifeste colligitur satis non esse ad stabiliendam Geometriam Euclidaeam duo puncta sequentia. Unum est: quod nomine parallelarum illas rectas censeamus, quae in eodem plano existentes commune aliquod obtinent perpendiculum. Alterum vero, quod omnes rectae in eodem plano existentes, quarum nullum commune sit perpendiculum, ac propterea quae juxta assumptam Definitionem parallelae non sint, debeant ipsae in alterutram partem semper magis protractae inter se
[55] aliquando incidere, si non ad finitam, saltem ad infinitam distantiam. Nam rursum demonstrare oporteret, quod duae quaelibet in eodem plano existentes, in quas recta quaepiam incidens duos ad easdem partes internos angulos efficiat minores duobus rectis, nusquam alibi possint ipsae recipere commune perpendiculum. Quod autem, hoc demonstrato, exactissime stabiliatur Geometria Euclidaea, infra constabit.

Propositio XXVI.

Si praedictae AX, BX (Fig. 31) *coire quidem inter se debeant, sed non nisi ad infinitam earundem productionem versus partes punctorum X: Dico nullum fore assignabile punctum T in ipsa AB, ex quo perpendicularis educta versus partes ipsius AX non occurrat ad finitam, seu terminatam distantiam eidem AX in aliquo puncto F.*

But now, since on one hand is stable and given the quadrilateral KTSK, inasmuch as constant in the given base KK, which indeed is taken equal to the assigned length R, and again constant in the two perpendiculars TK, SK equal to this base, and finally in the joining TS, which comes out completely determinate; and on the other hand the four angles together of this stable and given quadrilateral have now been shown greater than the four angles together of the quadrilateral KHLK distant as far as we choose from the base AB; assuredly it follows that the four angles together of this stable and given quadrilateral KTSK are greater than any sum of angles, which lacks however little you choose of being four right angles, since already it has been shown that a quadrilateral KHLK can always be designated such that its four angles together shall fall short of four rights by less than any assignable part of a right. Therefore four angles together of this stable and given quadrilateral either are equal to four rights or greater. But then (Proposition 16) is established the hypothesis either of right angle or of obtuse angle; and therefore (Propositions 5 and 6) the hypothesis of acute angle is destroyed.

So is established that the hypothesis of acute angle will be destroyed, if two straights existing in the same plane so approach each other mutually ever more, that nevertheless their distance is always greater than any assigned length. This is what was to be demonstrated.

Corollary 1.

But (the hypothesis of acute angle destroyed) the controverted Euclidean Assertion is manifest from Proposition 13; just as in Scholium 3 after Proposition 21, where we spoke of the attempt of the Arab Nasīr ad-Dīn, I promised would be disclosed by me in this place.

Corollary 2.

On the other hand from this Proposition 25, and from the preceding Proposition 23 is manifestly gathered that the two following points are not sufficient for establishing Euclidean geometry. One is: that we designate by the name of parallels those straights, which existing in the same plane possess a common perpendicular. The second indeed, that all straights existing in the same plane, of which there is no common perpendicular, and therefore which according to the assumed definition are not parallel, must, being produced toward either part ever more, somewhere meet each other, if not at a finite, at least at an infinite distance. For again it would be requisite to demonstrate, that any two straights existing in the same plane, upon which a certain straight cutting makes two internal angles toward the same parts less than two right angles, nowhere else can receive a common perpendicular. But that, this demonstrated, Euclidean geometry is most exactly established, will be shown below.

Proposition 26.

If the aforesaid AX, BX (Fig. 31) *must indeed meet each other, but only at their infinite production toward the parts of the point X: I say there will be no assignable point T in AB, from which a perpendicular erected toward the parts of AX does not at a finite, or bounded distance meet this AX in same point F.*

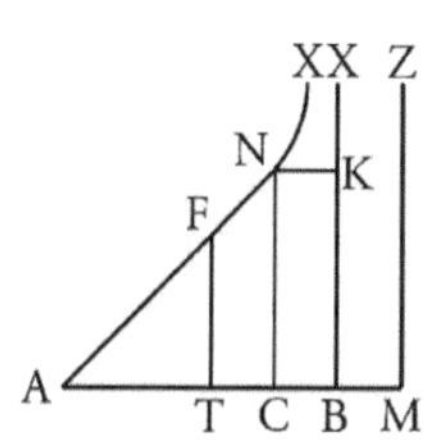

Fig. 31

Demonstratur. Nam (ex praecedente hypothesi) unum aliquod erit in ipsa AX punctum N, ex quo perpendicularis NK demissa ad BX minor sit qualibet assignata longitudine, ut puta ea TB. Tum vero sumatur in TB portio CB aequalis ipsi NK, jungaturque CN. Constat angulum NCB acutum fore, in hypothesi anguli acuti. Ergo (ex 13. primi) obtusus erit, qui deinceps est angulus NCT. Igitur recta, quae ex puncto T (inter puncta A, & C constituto) perpendiculariter educatur versus partes ipsius AX, non incidet (ex 17. primi) in ullum punctum ipsius CN; ac propterea (ne claudat spatium cum AT, aut cum TC) occurret ipsi terminatae AN in aliquo puncto F. Igitur in ipsa etiam hypothesi anguli acuti (quam scimus obesse unice hic posse) nullum erit assignabile punctum T in ea AB, ex quo perpendiculariter educta versus partes ipsius AX non occurrat ad finitam, seu terminatam distantiam eidem AX in quodam puncto F. Quod &c.

[56] **Corollarium I.**

Inde autem fit, ut assumpto in AB protracta quolibet puncto M, ex quo versus partes punctorum X educatur perpendicularis MZ, nequeat ipsa, etiamsi infinite producatur, occurrere praedictae AX; quia caeterum illa altera BX deberet (ex praemissa demonstratione) ad finitam distantiam occurrere eidem AX; quod est contra praesentem hypothesin.

Corollarium II.

Ex quo rursum consequitur omnem perpendiculariter eductam ex quolibet puncto illius quantumlibet continuatae AB, sed non tamen infinite dissito, debere ad finitam distantiam occurrere praedictae AX; quatenus nempe supponatur omnem talem perpendiculariter eductam semper magis, sine ullo certo limite accedere ad alteram semper continuatam AX.

Corollarium III.

Unde tandem fit, ut ab illa AX neque ad infinitam ejusdem productionem secari possit ipsa BX; quia caeterum ex quodam illius AX ultra praedictam sectionem puncto intelligi posset demissa ad AB productam quaedam perpendicularis ZM; unde rursum fieret, ut ipsa BX (contra praesentem hypothesim) non ad infinitam, sed omnino ad finitam distantiam occurreret praedictae AX. Sed hoc postremum dictum sit ultra necessitatem.

[57] **Propositio XXVII.**

Si recta AX (Fig. 32) *sub aliquo, ut libet, parvo angulo educta ex puncto A ipsius AB, occurrere tandem debeat (saltem ad infinitam distantiam) cuivis perpendiculari BX, quae ad quantamlibet ab eo puncto A distantiam excitari intelligatur super ea incidente AB: Dico nullum jam fore locum hypothesi anguli acuti.*

Proof. For (from the preceding hypothesis) there will be in AX some point N, from which the perpendicular NK let fall to BX is less than any assigned length as supposed than TB. But then is assumed in TB a portion CB equal to NK, and CN is joined. In the hypothesis of acute angle, it is known that the angle NCB will be acute. Therefore (from *Elements* I, 13) NCT, which is the adjacent angle, will be obtuse. Therefore the straight, which is erected toward the parts of AX perpendicularly from the point T (disposed between the points A and C), does not meet (*Elements* I, 17) CN at any point; and therefore (lest it should enclose a space[1] with AT, or with TC) it strikes the bounded AN in some point F. Therefore even in the hypothesis of acute angle (which we know can here alone hinder) there will be in this AB no assignable point T, from which the perpendicular erected toward the parts of AX does not, at a finite or bounded distance, meet this AX in a certain point F. This is what was to be demonstrated.

Corollary 1.

But theme follows, that, point M being assumed in AB produced, from which is erected toward the parts of the points X a perpendicular MZ, this cannot, even if infinitely produced, meet the aforesaid AX; because otherwise that other straight BX must (from the foregoing demonstration) at a finite distance meet this AX; which is against the present hypothesis.

Corollary 2.

From which again it follows, that every perpendicular, erected from any point (but not however infinitely removed) of this AB produced indefinitely, must at a finite distance meet the aforesaid AX, as soon as indeed it is assumed that every such perpendicular ever more, without any certain limit, approaches the other ever produced straight AX.

Corollary 3.

Whence finally follows, that BX is not cut by AX, not even at its infinite production; because otherwise from any point of that AX beyond the aforesaid intersection a certain perpendicular ZM could be supposed let fall to AB produced; whence again would follow, that BX (against the present hypothesis) met the aforesaid AX not at an infinite, but wholly at a finite distance. But this last remark is beyond necessity.[2]

Proposition 27.

If a straight AX (Fig. 32) *drawn at any however small angle from the point A of AB, must at length meet (anyhow at an infinite distance) any perpendicular BX, which is supposed erected at any distance from this point A upon the secant AB: I say there will then be no more place tor the hypothesis of acute angle.*

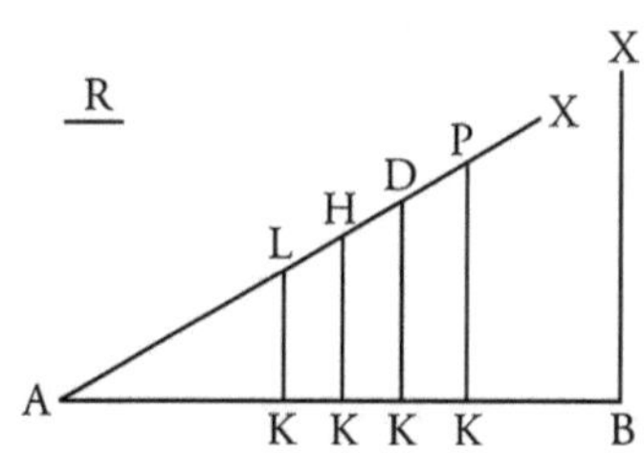

Fig. 32

Demonstratur. Ex quodam puncto K prope punctum A, ad libitum in ipsa AB designato, erigatur ad AB perpendicularis KL, quae utique (ex Cor. II. praecedentis Propositionis) occurret ipsi AX ad finitam, seu terminatam distantiam in aliquo puncto L. Jam vero constat sumi posse in KB portiones KK aequales singulas cuidam assignabili longitudini R, & eas plures quolibet assignabili numero finito; quandoquidem punctum B statui potest; juxta praesentem suppositionem; in quantalibet distantia ab eo puncto A. Itaque ex aliis punctis K erigantur ad AB perpendiculares KH, KD, KP, quae omnes (ex praecitato Corollario) occurrent rectae AX in quibusdam punctis H, D, P; atque ita circa reliqua puncta K uniformiter designata versus punctum B. Constat secundo (ex 16. primi) angulos ad puncta L, H, D, P, fore omnes obtusos versus partes punctorum X; atque item (ex 13. ejusdem primi) angulos ad praedicta puncta fore omnes acutos versus punctum A. Igitur (ex Cor. II. post 3. hujus) latus KH majus erit latere KL; latus KD majus latere KH; atque ita semper, procedendo versus puncta X. Constat tertio quatuor simul angulos quadrilateri KLHK majores fore quatuor simul angulis quadrilateri KHDK: nam id in simili demonstratum jam est in XXIV. hujus. Constat quarto idem similiter valere de quadrilatero KHDK relate ad quadrilaterum KDPK;
[58] atque ita semper, procedendo ad quadrilatera remotiora ab eo puncto A.

Quoniam igitur tot aderunt (ut in XXV. hujus) praedicto modo recensita quadrilatera, quot sunt, praeter primam LK, demissae ex punctis ipsius AX perpendiculares ad rectam AB; constabit uniformiter (si ponamus novem, praeter primam, demissas eiusmodi perpendiculares) summam omnium angulorum, qui comprehenduntur ab illis novem quadrilateris, excedere 35. rectos; ac propterea quatuor simul angulos primi quadrilateri KLHK, quod quidem in hac ratione ostensum est omnium maximum, minus deficere a quatuor rectis, quam sit nona pars unius recti. Quare; multiplicatis ultra quemlibet assignabilem finitum numerum eisdem quadrilateris, procedendo semper versus partes punctorum X; constabit similiter (ut in eadem praecitata) quatuor simul angulos stabilis illius quadrilateri KHLK minus deficere a quatuor rectis, quam sit quaelibet assignabilis unius recti portiuncula. Igitur quatuor simul illi anguli vel aequales erunt quatuor rectis, vel eisdem majores. Tunc autem (ex XVI. hujus) stabilitur hypothesis aut anguli recti, aut anguli obtusi; ac propterea (ex V. & VI. hujus) destruitur hypothesis anguli acuti.

Itaque constat nullum jam fore locum hypothesi anguli acuti, si recta AX sub aliquo, ut libet, parvo angulo, educta ex puncto A ipsius AB occurrere tandem debeat (saltem ad infinitam distantiam) cuivis perpendiculari BX, quae ad quantamlibet ab eo puncto A distantiam excitari intelligatur super ea incidente AB. Quod erat &c.

Proof. From any point K chosen at will in AB near the point A, the perpendicular KL is erected to AB, which certainly (Corollary 2 to Proposition 26) meets AX at a finite or bounded distance in some point L.

But now it holds, firstly, that there may be assumed in KB portions KK each equal to a certain assignable length R, and these more than any assignable finite number;[3] since indeed the point B can be situated, in accordance with the present supposition, at however great a distance from this point A. And accordingly from the other points K are erected to AB perpendiculars KH, KD, KP, which all (from the aforesaid Corollary 2 to Proposition 26) meet the straight AX in certain points H, D, P; and so about the remaining points K uniformly designated toward the point B.

It holds secondly (*Elements* I, 16) that the angles at the points L, H, D, P will all be obtuse toward the parts of the points X; and just so (*Elements* I, 13) the angles at the aforesaid points will all be acute toward the point A. Therefore (Corollary 2 to Proposition 3) the side KH will be greater than the side KL; the side KD greater than the side KH; and so always proceeding toward the points X.

It holds thirdly that the four angles together of the quadrilateral KLHK will be greater than the four angles together of the quadrilateral KHDK: for this in like case has already been demonstrated in Proposition 24.

It holds fourthly that the same is valid likewise of the quadrilateral KHDK in relation to the quadrilateral KDPK; and so on always, proceeding to quadrilaterals more remote from this point A.

Since therefore are present (as in Proposition 25) as many quadrilaterals described in the aforesaid mode, as there are, except the first LK, perpendiculars let fall from points of AX to the straight AB, it will hold uniformly (if we assume nine perpendiculars of the sort fall, beside the first) the sum of all the angles which are comprehended by these nine quadrilaterals will exceed 35 right angles; and therefore the four angles together of the first quadrilateral KLHK, which indeed in this regard has been shown the greatest of all, will fall short of four right angles by less than the ninth part of one right angle. Wherefore, these quadrilaterals being multiplied beyond assignable finite number, proceeding always toward the parts of the points X, it holds in the same way (as in the same already mentioned Proposition 25) that the four angles together of this stable quadrilateral KHLK will fall short of four right angles less than any assignable little portion of one right angle. Therefore these four angles together will be either equal to four right angles, or greater. But then (Proposition 16) is established the hypothesis of right angle or of obtuse angle; and therefore (Propositions 5 and 6) is destroyed the hypothesis of acute angle.

So then it holds, that there will be no place for the hypothesis of acute angle, if the straight AX drawn under however small angle from the point A of AB must at length meet (anyhow at an infinite distance) any perpendicular BX, which is supposed erected at any distance from this point A upon this secant AB. This is what was to be demonstrated.

Scholion I.

ET hoc est, quod praedixi in Cor. II. post XXV. hujus; nullum scilicet superfuturum locum
[59] hypothesi anguli acuti, seu stabilitum exactissime iri Geometriam Euclidaeam; si duae quaelibet in eodem plano existentes rectae, ut puta AX, BX, in quas incidens recta AB (sumpto puncto B in quantalibet distantia a puncto A) duos cum eisdem ad easdem partes punctorum X angulos efficiat minores duobus rectis; si (inquam) nusquam alibi (hoc stante) possint illae recipere commune perpendiculum. Tunc enim illae duae A X, B X semper magis ad se invicem accedent; nimirum vel intra quendam determinatum limitem, prout in XXV. hujus; vel sine ullo certo limite, ac propterea usque ad occursum saltem post infinitam productionem, prout in hac XXVII. Constat autem in utroque praedictorum casuum ostensam jam esse destructionem hypothesis anguli acuti. Quod intendebatur.

Scholion II.

Atque id rursum est, quod spopondi in fine Scholii IV. post XXI. hujus, prout ex ipsis terminis clare elucescit.

Scholion III.

Praeterea observari hic velim discrimen inter hanc Propos. & praecedentem XVII. Nam ibi (recole Fig. 15) ostensa est destructio hypothesis anguli acuti, si (existente, ut libet parva, recta AB) omnis BD sub quovis acuto angulo educta, occurrere tandem debeat in quodam puncto K ipsi perpendiculari AH productae. Hic autem (viceversa) permittitur quidem designatio cujusvis parvissimi acuti anguli ad punctum A, dum tamen interjecta AB, ad
[60] quam erigenda est perpendicularis indefinita BX, statui possit quantaelibet longitudinis.

Propositio XXVIII.

Si duae rectae AX, BX (quarum prior sub angulo acuto, & altera ad perpendiculum eductae sint versus easdem partes ex quantalibet recta AB) semper magis sine ullo certo limite ad se invicem accedant, praeterquam ad infinitam earundem productionem; Dico omnes angulos (Fig. 33) *ad quaelibet puncta L, H, D ipsius AX, ex quibus demittantur ad rectam BX perpendiculares LK, HK, DK; tum fore omnes obtusos versus partes puncti A; tum fore semper minores, qui magis distant ab eo puncto A; ac tandem angulos magis, ac magis distantes ab eodem puncto A, semper magis sine ullo certo limite accedere ad aequalitatem cum angulo recto.*

Scholium 1.

And this it is, that I said before in Corollary 2 to Proposition 25; obviously that no place would remain over for the hypothesis of acute angle, or Euclidean Geometry would be most exactly established, if any two straights existing in the same plane, as suppose AX, BX, which the straight AB meeting (the point B being assumed at a distance from the point A as great as you choose) makes with them toward the same parts of the points X two angles less than two right angles, if (I say) nowhere at another place (this standing) they can admit a common perpendicular. For then these two AX, BX mutually approach each other ever more, indeed either within a certain determinate limit, as in Proposition 25, or without any certain limit, and therefore even to meeting, anyhow after infinite production, as in Proposition 27. But it holds that in either of the aforesaid cases the destruction of the hypothesis of acute angle has now been shown. This is what we wanted.

Scholium 2.

And again this it is, that I promised at the end of Scholium 4 after Proposition 21, as from the very terms clearly appears.

Scholium 3.

Moreover I could wish here to be observed the difference between this Proposition 27 and the preceding Proposition 17. For there (recall Fig. 15) has been shown the destruction of the hypothesis of acute angle, if (the straight AB being as small as you choose) every BD erected at whatever acute angle, must at length meet in some point K the perpendicular AH produced. But here (*vice versa*) in fact is permitted the designation of however most small an acute angle at the point A while still the sect AB to which is to be erected the indefinite perpendicular BX, may be taken of any length whatever.

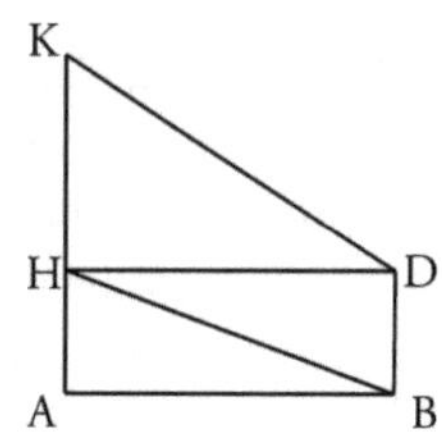

Fig. 15

Proposition 28.

If two straights AX, BX (produced from any-sized straight AB toward the same parts, the first under an acute angle, and the other perpendicularly) mutually approach each other ever more without any certain limit, save at their infinite production; I say all angles (Fig. 33) *at any points L, H, D of AX, from which are let fall to the straight BX perpendiculars LK, HK, DK, first* (1) *will all be obtuse toward the parts of the point A, secondly* (2) *will be ever less, the more distant from this point A, and finally* (3) *the angles more and more distant from this same point A ever more without any certain limit approach to equality with a right angle.*

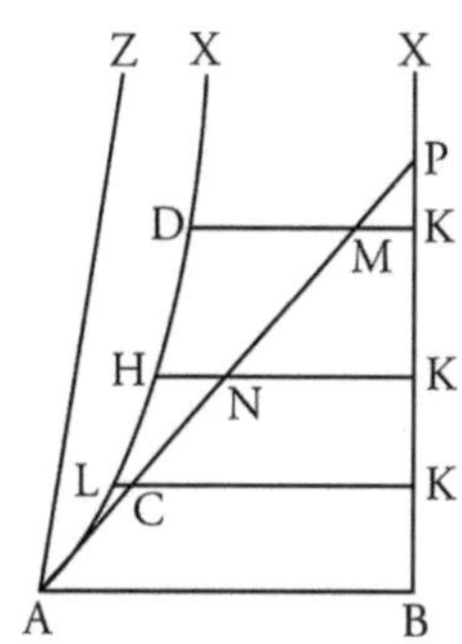

Fig. 33

Demonstratur. Et prima quidem pars constat ex Cor. I. post XXIII. hujus. Secunda vero pars ita evincitur. Nam duo simul anguli ad LK versus basim AB majores sunt (ex Cor. post XVI. hujus) duobus simul internis, & oppositis angulis ad HK versus eandem basim AB. Sunt autem inter se aequales, utpote recti, anguli ad utrumque punctum K versus basim AB. Ergo angulus obtusus ad L versus basim AB major est angulo obtuso ad H versus eandem basim AB. Simili modo ostendetur praedictum angulum obtusum ad H majorem esse angulo obtuso ad punctum D. Atque ita semper, procedendo versus puncta X.

Tertia tandem pars majore indiget disquisitione. Si ergo fieri potest, assignatus sit (Fig. 34) quidam angulus MNC, quo semper major sit, aut saltem non minor, excessus cujusvis ex praedictis angulis obtusis supra angulum rectum. Constat (ex XXI. hujus) latera NM, NC comprehendentia illum angulum MNC taliter produci posse, at perpendicularis
[61] MC, ex quodam puncto M ipsius MN demissa ad NC, major sit (in ipsa etiam hypothesi anguli acuti) qualibet finita assignata longitudine, ut puta praedicta basi AB. Hoc stante: assumatur in BX (Fig. 35) quaedam BT aequalis ipsi CN; educaturque ex puncto T versus AX perpendicularis TS, quae nempe (ex Scholio post XXIV. hujus) occurret ipsi AX in quodam puncto S. Deinde ex puncto S demittatur ad AB perpendicularis SQ. Cadet haec (propter 17. primi) ad partes anguli acuti SAB inter puncta A, & B. Porro acutus erit angulus QST in quadrilatero QSTB, cum reliqui tres anguli sint recti; ne (contra V. & VI. hujus) incidamus in hypothesin aut anguli recti, aut anguli obtusi. Hinc recta SQ major erit (ex Cor. I. post 3. hujus) recta BT, sive CN; ac rursum angulus ASQ major erit excessu, quo angulus obtusus AST excedit angulum rectum, & sic major angulo MNC. Ducatur igitur quaedam SF secans AQ in F, & efficiens cum SA angulum aequalem ipsi MNC. Deinde ex puncto A ducatur ad SF productam perpendicularis AO. Cadet punctum O (ex 17. primi) infra punctum F, cum angulus AFS (ex 16. ejusdem primi) sit obtusus. Tandem vero; cum FS major sit (ex 18. primi) ipsa QS, & sic multo major ipsa BT, sive CN; sumatur in FS portio IS aequalis ipsi CN, & ex puncto I erigatur ad FS perpendicularis IR occurrens in puncto R ipsi AS. Cadet autem punctum R inter puncta A, & S: si enim caderet in aliquod punctum ipsius AF, haberemus in eodem triangulo (contra 17. primi) duos angulos majores duobus rectis, cum angulus ad punctum F versus partes puncti A ostensus jam sit obtusus.

Proof. The first part follows indeed from Corollary 1 to Proposition 23.

The second part however is proved thus. For the two angles together at LK toward the base AB are greater (Corollary to Proposition 16) than the two internal and opposite angles together at HK toward the same base AB. But the angles at each point K toward the base AB are equal to each other, as being right. Therefore the obtuse angle at L toward the base AB is greater than the obtuse angle at H toward the same base AB. In like manner is shown that the aforesaid obtuse angle at H is greater than the obtuse angle at the point D. And thus ever, proceeding toward the points X.

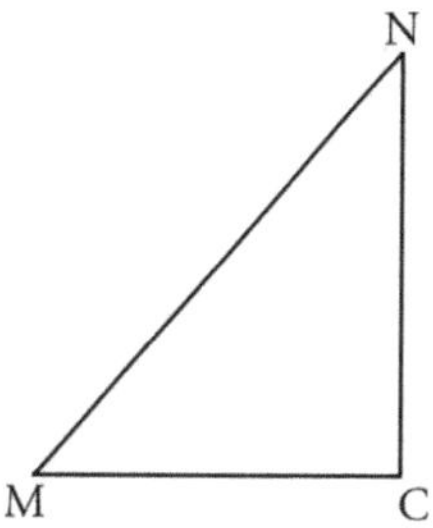

Fig. 34

Finally the third part requires a longer disquisition. If therefore it can be done, let there be assigned (Fig. 34) a certain angle MNC, than which is always greater, or anyhow not less, the excess of any of the aforesaid obtuse angles above a right angle. It follows (Proposition 21) that the sides NM, NC comprehending that angle MNC can be so produced that the perpendicular MC from a certain point M of MN let fall upon NC may be greater (even in the hypothesis of acute angle) than any assigned finite length, as for instance the aforesaid base AB. This standing; assume in BX (Fig. 35) a certain BT equal to CN, and erect from the point T toward AX the perpendicular TS, which obviously (from Scholium after Proposition 24) meets AX in a certain point S. Then from the point S let fall to AB the perpendicular SQ. This falls (because of *Elements* I, 17) toward the parts of the acute angle SAB between the points A and B. Again, acute will be the angle QST in the quadrilateral QSTB, since the remaining three angles are right; else (against Propositions 5 and 6) we come upon the hypothesis either of right angle or of obtuse angle. Hence the straight SQ will be greater (Corollary 1 to Proposition 3) than the straight BT, or CN; and again the angle ASQ will be greater than the excess by which the obtuse angle AST exceeds a right angle, and thus greater than the angle MNC. Draw therefore a certain SF cutting AQ in F and making with SA an angle equal to MNC. Then from the point A draw to SF produced the perpendicular AO. The point O falls (from *Elements* I, 17) below the point F, since the angle AFS (by *Elements* I, 16) is obtuse. Finally, however; since FS is greater (by *Elements* I, 18)[1] than QS and so much greater than BT or CN, assume in FS the piece IS equal to CN, and from the point erect to FS the perpendicular IR meeting AS in the point R. But the point R falls between the points A and S: for if it fell on any point of AF, we would have in the same triangle (against *Elements* I, 17) two angles greater than two right angles, since the angle at the point F toward the parts of the point A has already been shown obtuse.

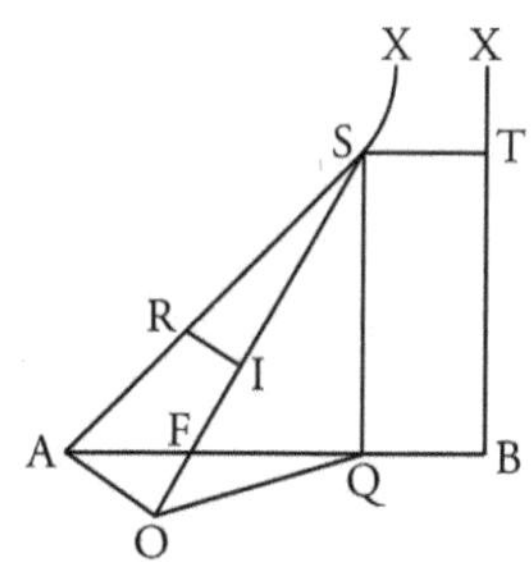

Fig. 35

Post tantum apparatum sic concludo. Quandoquidem in quadrilatero AOIR recti sunt anguli ad puncta O, & I; & est acutus angulus (ex 17. primi) ad punctum A, propter rectum
[62] angulum AOS; ac rursum est obtusus (ex 16. ejusdem primi) angulus IRA, cum rectus sit angulus RIS: consequens tandem est (ex Cor. II. post 3. hujus) ut latus AO majus sit latere IR. At (juncta OQ) latus AQ majus est (ex 18. primi) latere AO propter angulum obtusum in O, cum angulus AOS factus sit rectus. Igitur recta AQ multo major erit recta IR, sive (ex 26. primi) recta NC, & sic multo major recta AB, pars toto; quod est absurdum.

Non igitur ullus assignari potest angulus MNC, quo semper major sit, aut saltem non minor excessus cujusvis ex praedictis angulis obtusis supra angulum rectum. Quare anguli illi obtusi, magis ac magis distantes ab eo puncto A, semper magis sine ullo certo limite accedent ad aequalitatem cum angulo recto. Quod erat postremo loco demonstrandum.

Corollarium.

Hoc autem stante, quod postremo loco demonstratum est, manifeste consequitur, duas illas AX, BX, in infinitum protractas, commune tandem habituras, vel in duobus distinctis punctis, vel in uno, eodemque puncto X infinite dissito, perpendiculum. Rursum vero, quod non in duobus distinctis punctis haberi possit commune istud perpendiculum, ex eo manifeste liquet; quia caeterum (ex Cor. II. post XXIII. hujus) inciperent inde illae rectae invicem dissilire, & sic neque ad infinitam distantiam inter se concurrerent; quin etiam (contra expressam suppositionem) non ad se invicem, sine ullo certo limite, semper magis versus eas partes accederent. Itaque in uno, eodemque puncto X infinite dissito commune haberent perpendiculum.

[63] Propositio XXIX.

Resumpta Fig. 33 *praecedentis Propositionis: Dico omnem rectam AC, quae secet angulum BAX, aliquando ad finitam, seu terminatam distantiam (etiam in hypothesi anguli acuti) occursuram ipsi BX in quodam puncto P, dum nempe illa AC semper magis protrahatur versus partes punctorum* X.

After so much preparation thus I conclude. Since in the quadrilateral AOIR the angles at the points O and I are right, and the angle at the point A (by *Elements* I, 17) is acute because of the right angle AOS, and again the angle IRA (by *Elements* I, 16) is obtuse, since the angle RIS is right: the consequence finally is (by Corollary 2 to Proposition 3) that the side AO is greater than the side IR. But (OQ joined) the side AQ is greater (by *Elements* I, 18) than the side AO, because of the obtuse angle at O, since the angle AOS was made right. Therefore the straight AQ will be much greater than the straight IR, or (by *Elements* I, 26) than the straight MC,[2] and so much greater than the straight AB, the part than the whole; which is absurd.

Therefore it is not possible to assign any one angle MNC, than which always is greater, or anyhow not less, the excess of each of the aforesaid obtuse angles above a right angle. Wherefore those obtuse angles, more and more distant from this point A, ever more without any certain limit approach to equality with a right angle. This is what was to be demonstrated.

Corollary.

But this standing, which in the last case was demonstrated, it manifestly follows that those straights AX, BX, produced infinitely will finally have, either in two distinct points, or in one same point X infinitely distant, a common perpendicular. But again, that this common perpendicular cannot be had in two distinct points flows manifestly from this, because otherwise (by Corollary 2 to Proposition 23) those straights would thence begin mutually to separate, and so not meet each other at an infinite distance; so that also (against the express supposition) they would not mutually approach each other without any certain limit ever more toward those parts. So they must have the common perpendicular in one same point X infinitely distant.

Proposition 29.

Resuming Fig. 33 *of the preceding Proposition: I say every straight AC, which cuts angle BAX, finally at a finite, or bounded distance (even in the hypothesis of acute angle) will meet BX in a certain point P, if only AC be produced ever more toward the parts of the points X.*

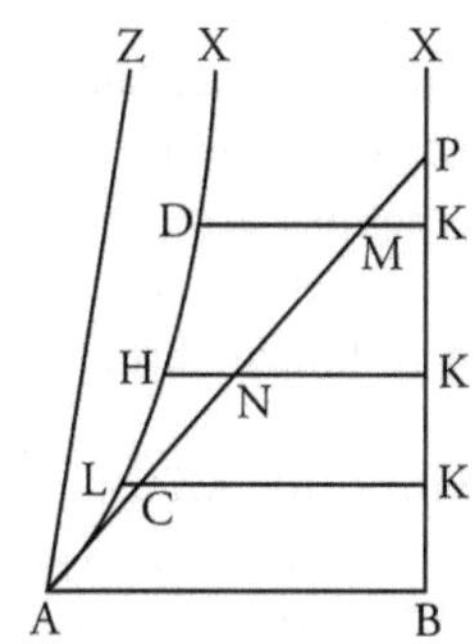

Fig. 33

Demonstratur. Et primo quidem (ne recta AC spatium claudat cum ea AX) occurret ipsa ad finitam distantiam rectis LK, HK, DK in quibusdam punctis C, N, M; occurret, inquam, nisi antea (ad finitam utique distantiam, prout intendimus) occurrat ipsi BX in aliquo puncto inter punctum B, & unum aliquod punctorum K constituto. Deinde (ex Cor. I. post XXIII. hujus) obtusi erunt anguli ACK, ANK, AMK. Praeterea anguli isti, semper obtusi, accedent (ex praecedente) sine ullo certo limite ad aequalitatem cum angulo recto, quoties nempe illa AC non nisi ad infinitam distantiam occursura putetur ipsi BX. Igitur deveniri posset ad talem ordinatam KMD, ad quam angulus AMK minus superaret angulum rectum, quam sit ille angulus DAC. Tunc autem angulus DAC, sive DAM, una cum angulo AMD major erit uno recto. Quare; addito obtuso angulo ADM; tres simul anguli trianguli ADM majores erunt duobus rectis, quod est contra hypothesin anguli acuti. Igitur omnis recta AC, quae secet illum angulum BAX, aliquando ad finitam, seu terminatam distantiam (etiam in hypothesi anguli acuti) occurret ipsi BX in quodam puncto P. Quod &c.

[64] **Corollarium I.**

Hinc nulla AZ, quae versus partes punctorum X angulum acutum efficiat majorem illo BAX, occurrere unquam poterit, sive ad finitam, sive ad infinitam distantiam ipsi BX. Quatenus enim ita contingeret, jam illa AX, dividens angulum BAZ, deberet (contra praemissam suppositionem) ad finitam distantiam occurrere ipsi BX; prout demonstratum id est de recta AC dividente angulum BAX.

Corollarium II.

Praeterea sequitur nullum fore determinatum acutum angulum omnium maximum, sub quo educta ex puncto A ad finitam distantiam occurrat illi BX. Si enim versus partes puncti X punctum quodvis assumas, quod sit altius puncto P, constat rectam jungentem punctum A cum illo puncto altiore majorem angulum effecturam cum ipsa AB, quam sit angulus BAP. Atque ita semper sine ullo termino intrinseco. Quare angulus BAX (dum scilicet ipsa AX, & semper accedat ad eam BX, & non nisi ad infinitam distantiam in eandem incidat) erit limes extrinsecus acutorum omnium angulorum, sub quibus rectae eductae ex illo puncto A ad finitam distantiam occurrunt praedictae BX.

Propositio XXX.

Cuivis terminatae AB insistat ad perpendiculum (Fig. 36) *quaedam indefinita BX. Dico primo rectam AY, perpendiculariter elevatam versus partes easdem super illa AB, fore limitem unum*
[65] *intrinsecum earum omnium, quae ex illo puncto A versus easdem partes eductae commune aliquod (juxta hypothesin anguli acuti) in duobus distinctis punctis obtinent perpendiculum cum altera indefinita BX. Dico secundo nullum fore acutum angulum omnium minimum, sub quo educta ex praedicto puncto A commune aliquod (juxta praedictam hypothesin) in duobus distinctis punctis obtineat perpendiculum cum eadem BX.*

Proof. And first indeed (lest straight AC include space with AX) it must meet at finite distance the straights LK, HK, DK in certain points C, N, M; must meet, I say, unless before (and that at a finite distance, just as we maintain) it meets BX in some point between the point B and one of the points K. Then (Corollary 1 to Proposition 23) the angles ACK, ANK, AMK will be obtuse. Moreover those angles, always obtuse, approach (Proposition 28) without any certain limit, to equality with a right angle, when indeed that AC is supposed to meet BX only at an infinite distance. Therefore such an ordinate KMD can be reached that at it the angle AMK exceeds a right angle by less than the angle DAC. But then angle DAC, or DAM, together with angle AMD will be greater than a right angle. Wherefore the obtuse angle ADM being added, the three angles together of the triangle ADM will be greater than two right angles, which is against the hypothesis of acute angle. Therefore every straight AC, which cuts that angle BAX, finally at a finite or bounded distance (even in the hypothesis of acute angle) must meet BX in a certain point P. This is what was to be demonstrated.

Corollary 1.

Hence no straight AZ, which toward the parts of the points X makes an acute angle greater than BAX can ever meet BX, either at a finite or at an infinite distance. For as far as so should happen, now AX, dividing angle BAZ, ought (against the premised supposition) to meet BX at a finite distance, as this is demonstrated of the straight AC dividing angle BAX.

Corollary 2.

Moreover it follows that no determinate acute angle will be the maximum of all under which a straight line produced from point A meets BX at finite distance. For if toward the parts of the point X you assume any point higher than point P, it follows that the straight joining point A with this higher point will make with AB a greater angle than angle BAP. And so ever without any intrinsic end. Wherefore angle BAX (since indeed AX both always approaches to BX, and meets it only at an infinite distance) will be the extrinsic limit of all acute angles under which straights produced from that point A meet the aforesaid BX at a finite distance.

Proposition 30.

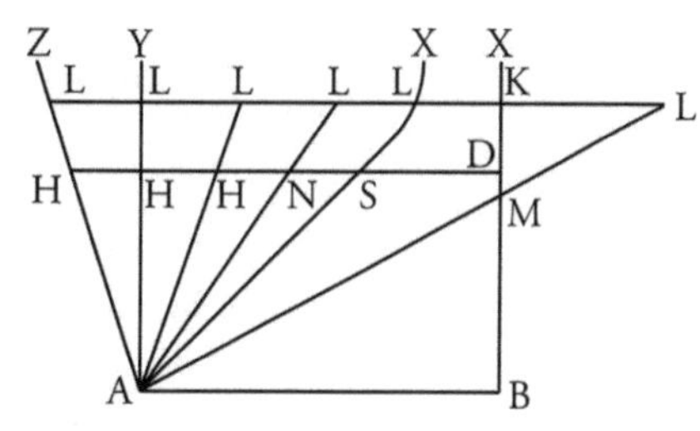

Fig. 36

To any bounded straight AB stands at right angles (Fig. 36) *a certain indefinitely long straight BX. I say firstly, that the straight AY, erected perpendicularly toward the same parts upon AB, will be one intrinsic limit of all those straights, which drawn from the point A out toward the same parts have (in the hypothesis of acute angle) a common perpendicular in two distinct points with the other indefinite straight BX. I say secondly, that no acute angle will be the minimum of all, produced under which a straight from the aforesaid point A (in the aforesaid hypothesis) has in two distinct points a common perpendicular with BX.*

Demonstratur prima pars. Quoniam enim illa AY commune obtinet cum altera BX perpendiculum AB in duobus distinctis punctis A, & B; si educatur versus easdem partes sub angulo obtuso quaepiam AZ, constat nullum ad eas partes esse posse in duobus distinctis punctis commune perpendiculum ipsarum AZ, BX; ne scilicet ex consecuturo quadrilatero continente quatuor angulos majores quatuor rectis incidamus (ex XVI. hujus) in hypothesin jam reprobatam anguli obtusi, contra suppositam hoc loco hypothesin anguli acuti. Igitur illa perpendicularis AY erit ex ista parte limes intrinsecus earum omnium, quae ex illo puncto A versus easdem partes eductae commune aliquod (juxta illam hypothesin anguli acuti) in duobus distinctis punctis obtineant perpendiculum cum altera indefinita BX. Quod erat primum.

Demonstratur secunda pars. Si enim fieri potest; esto quidam angulus acutus omnium minimus, sub quo educta AN commune habeat cum illa BX in duobus distinctis punctis perpendiculum ND. Tum assumpto in BX altiore puncto K, ex eo educatur ad BX perpendicularis KL, ad quam ex puncto A demittatur (juxta 12. primi) perpendicularis AL. Jam vero, si haec AL occurrat in quodam puncto S ipsi ND, constat sane angulum BAL minorem fore eo BAN, qui propterea non erit omnium minimus, sub quo educta AN com-
[66] mune habeat cum illa BX in duobus distinctis punctis perpendiculum ND. Porro autem ab ea perpendiculari AL secari praedictam ND in quodam ejus intermedio puncto S sic demonstratur.

Et primo quidem non posse ab ea AL secari ipsam BK in quodam puncto M constare absolute potest ex 17. primi, ne scilicet in eodem triangulo MKL duos habeamus angulos rectos in punctis K, & L; praeterquam quod in hoc ipso haberemus intentum contra illum angulum BAN, ne scilicet in hac tali ratione censeatur omnium minimus. Rursum vero nequit AL esse continuatio ipsius AN; quia caeterum in quadrilatero NDKL quatuor haberemus angulos rectos, contra hypothesim anguli acuti. Sed neque eam DN protractam secare potest in quovis ulteriore puncto H; quia angulus AHN (ex 16. primi) foret acutus, propter suppositum rectum angulum externum AND; ac propterea angulus DHL foret obtusus, & sic in quadrilatero DHLK quatuor haberemus angulos, qui simul sumpti majores forent quatuor rectis, contra praedictam hypothesin anguli acuti. Igitur constat ab ea AL secati debere angulum BAN, qui propterea nequit dici omnium minimus, sub quo educta AN commune habeat cum illa BX in duobus distinctis punctis perpendiculum ND. Quod erat secundo loco demonstrandum. Itaque constat &c.

Corollarium.

Inde autem observare licet, quod sub angulo minore BAL obtinetur (in hypothesi anguli acuti) commune LK perpendiculum, remotius quidem ab illa basi AB, prout constat ex ipsa constructione, sed rursum minus altero viciniore communi perpendiculo ND, quod
[67] obtinetur sub angulo majore BAN. Ratio hujus posterioris est, quia in quadrilatero LKDS angulus ad punctum S acutus est in praedicta hypothesi, cum reliqui tres supponantur recti. Quare (ex Cor. I. post 3. hujus) latus LK minus erit contraposito latere SD, & sic multo minus latere ND.

Proof of the first part. For since AY has in common at two distinct points A and B the perpendicular AB with BX; if any straight AZ is drawn toward the same parts under an obtuse angle, it follows there can be toward these parts in two distinct points no common perpendicular to AZ, BX. Otherwise from the resulting quadrilateral containing four angles greater than four right angles, we hit (Proposition 16) upon the already rejected hypothesis of obtuse angle, against the hypothesis of acute angle in this place assumed. Therefore that perpendicular A Y will be from that side an intrinsic limit of all the straights which drawn from the point A toward the same parts have (in the hypothesis of acute angle) at two distinct points a common perpendicular with the other unbounded straight BX. This is what was to be demonstrated in the first part.

Proof of the second part. For if it were possible, let a certain acute angle be the least of all, drawn under which AN has with BX in two distinct points the common perpendicular ND. Then in BX a higher point K being assumed, from this erect to BX the perpendicular KL, upon which from the point A let fall (by *Elements* I, 12) the perpendicular AL. But now, if this AL meets ND in any point S, it certainly follows that angle BAL will be less than BAN, which therefore will not be the least of all drawn under which AN has with BX in two distinct points a common perpendicular ND. But furthermore that the aforesaid perpendicular ND is cut by this perpendicular AL in same intermediate point of it S is thus demonstrated.

And first indeed, that BK cannot be cut by AL in any point M follows absolutely from *Elements* I, 17, since otherwise in the same triangle MKL we would have two right angles at the points K and L, apart from the fact that in this case we would have our assertion about that angle BAN, that it is not in such circumstances the least of all. But again AL cannot be the continuation of AN; because otherwise in the quadrilateral NDKL we would have four right angles, against the hypothesis of acute angle. But neither can it cut DN produced in any exterior point H; because angle AHN (from *Elements* I, 16) would be acute, on account of the external angle AND supposed right; and therefore angle DHL would be obtuse, and so in the quadrilateral DHLK we would have four angles, which taken together would be greater than four right angles, against the aforesaid hypothesis of acute angle. Therefore it follows that the angle BAN must be cut by this AL, and therefore cannot be declared the least of all, drawn under which AN has with BX in two distinct points a common perpendicular ND. This is what was to be demonstrated in the second part.

Corollary.

But hence is permitted to observe, that under a lesser angle BAL is obtained (in the hypothesis of acute angle) a common perpendicular LK, more remote indeed from the base AB, as follows from the construction, but moreover less than the other nearer common perpendicular ND, which is obtained under a greater angle BAN. The reason of this latter is because in the quadrilateral LKDS the angle at the point S is acute in the aforesaid hypothesis, since the three remaining angles are supposed right. Wherefore (Corollary 1 to Proposition 3) the side LK will be less than the opposite side SD, and so much less than the side ND.

Propositio XXXI.

Jam dico nullum fore praedictorum in duobus distinctis punctis communium perpendiculorum limitem determinatum, quo minus sub minore, ac minore acuto angulo, ad illud punctum A constituto, deveniri semper possit (juxta hypothesin anguli acuti) ad tale commune in duobus distinctis punctis perpendiculum, quod sit minus qualibet assignata longitudine R.

Demonstratur. Quatenus enim aliter res se habeat; si ex puncto K (recole Fig. 30) in quantalibet a puncto B distantia in ea BX assignato, educatur perpendicularis KL, ad quam ex puncto A (juxta 12. primi) demissa intelligatur perpendicularis AL, deberet ipsa KL major esse ea longitudine R. Ratio autem est; quia assumpto in eadem BX altiore puncto Q, ex quo educatur ad ipsam BX perpendicularis QF, ad quam (juxta eandem 12. primi) demittatur perpendicularis AF, deberet haec rursum saltem non esse minor ea longitudine R. Erit autem KL (ex Cor. praeced. Prop.) major ipsa QF. Igitur ea KL major foret praedicta longitudine R. Atque ita semper altius procedendo.

Jam vero: si illa quantacunque KB divisa intelligatur (prout in XXV. hujus) in portiones KK, aequales illi longitudini R, educanturque ex illis punctis K perpendiculares, quae occurrant ipsi AX in punctis H, D, M; non erunt anguli ad haec puncta, versus partes puncti
[68] L, aut recti, aut obtusi; ne in aliquo quadrilatero, ut puta KMLK quatuor simul anguli aequales sint, aut majores quatuor rectis, contra hypotesim anguli acuti, juxta quam procedimus. Omnes igitur hujusmodi anguli acuti erunt versus partes puncti L; ac propterea omnes itidem ad illa puncta obtusi versus partes puncti A. Quare (ex Cor. I. post 3. hujus) praedictarum perpendicularium minima quidem erit KL remotior a basi AB, maxima KM propinquior eidem basi; reliquarum vero propinquior remotiore semper major erit. Igitur (ex mea praeced. 24 ejusque Coroll.) quatuor simul anguli quadrilateri KHLK remotioris a basi AB majores erunt quatuor simul angulis reliquorum omnium quadrilaterorum eidem basi proximiorum. Quare (prout in XXV. hujus) destructa maneret hypothesis anguli acuti.

Itaque; constat nullum fore praedictorum in duobus distinctis punctis communium perpendiculorum limitem determinatum, quo minus sub minore, ac minore acuto angulo, ad illud punctum A constituto, deveniri semper possit (juxta hypothesin anguli acuti) ad tale commune in duobus distinctis punctis perpendiculum, quod sit minus qualibet assignata longitudine R. Quod erat &c.

Proposition 31.

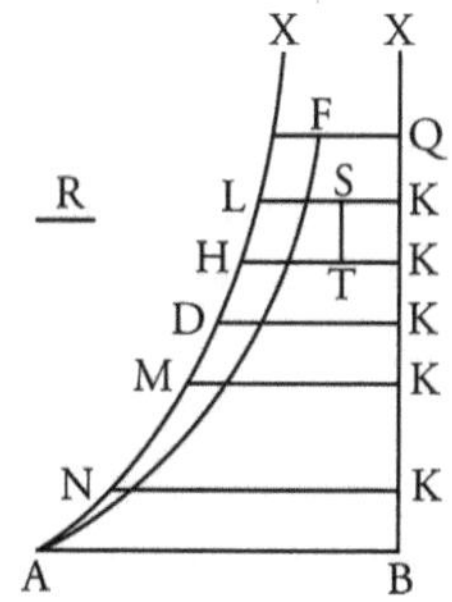

Fig. 30

Now I say there will be, of the aforesaid common perpendiculars in two distinct points, no determinate limit, such that under a smaller and smaller acute angle made at the point A, it would not always be possible to attain (in the hypothesis of acute angle) to such a common perpendicular in two distinct points as is less than any assignable length R.

Proof. For in so far as the thing were otherwise; if from the point K (resume Fig. 30) in BX assigned at any however great distance from the point B, a perpendicular KL is erected, to which from point A (by *Elements* I, 12) the perpendicular AL is supposed let fall, KL ought to be greater than the length R. The reason is; because a higher point Q being assumed in this BX, from which is erected to BX the perpendicular QF, to which (by the same *Elements* I, 12) a perpendicular AF is let fall, this again must anyhow not be less than the length R. But KL (Corollary to Proposition 30) will be greater than QF. Therefore KL would be greater than the aforesaid length R. And so ever proceeding higher.

But now, if this however great KB is supposed divided (as in Proposition 25) into portions KK, equal to the length R, and from these points K perpendiculars are erected, which meet AX in points H, D, M; the angles at these points, toward the parts of the point L, will neither be right nor obtuse; lest in same quadrilateral, as suppose KMLK, the four angles together should be equal to or greater than four rights, contrary to the hypothesis of acute angle, according to which we are proceeding. Therefore all such angles will be acute toward the parts of the point L; and therefore in like manner all at these points obtuse toward the parts of the point A. Wherefore (Corollary 1 to Proposition 3) of the aforesaid perpendiculars the least will indeed be KL more remote from the base AB, the greatest KM nearer this base. And of the remaining the nearer will be ever greater than the more remote. Therefore (from the preceding Proposition 25, and its Corollary) the four angles together of the quadrilateral KHLK more remote from base AB will be greater than the four angles together of all the remaining quadrilaterals nearer to this base. Wherefore (as in Proposition 25) the hypothesis of acute angle would be destroyed.

Therefore it holds, that of the aforesaid common perpendiculars in two distinct points there will be no determinate limit, such that under a smaller and smaller acute angle made at the point A, it would not always be possible to attain (in the hypothesis of acute angle) to such a common perpendicular in two distinct points as may be less than any assigned length R. This is what was to be demonstrated.

Propositio XXXII.

Jam dico unum aliquem fore (in hypothesi anguli acuti) determinatum acutum angulum BAX, sub quo educta AX (Fig. 33) *non nisi ad infinitam distantiam incidat in eam BX, ac propterea sit ipsa limes partim intrinsecus, partim extrinsecus; tum earum omnium, quae sub minoribus acutis angulis ad finitam distantiam incidunt in praedictam BX; tum etiam aliarum, quae sub majoribus angulis acutis, usque ad angulum rectum inclusive, commune obtinent in duobus distinctis punctis perpendiculum cum eadem BX.*

[69] Demonstratur. Nam primo constat (ex Cor. II. post XXIX. hujus) nullum fore determinatum acutum angulum, omnium maximum, sub quo educta ex illo puncto A ad finitam distantiam occurrat praedictae BX. Secundo constat nullum itidem esse (in hypothesi anguli acuti) acutum angulum omnium minimum, sub quo educta commune habeat in duobus distinctis punctis perpendiculum cum illa BX; quandoquidem (ex praecedente) nullus esse potest limes determinatus, quo minus sub minore acuto angulo ad illud punctum A constituto deveniri possit ad tale commune in duobus distinctis punctis perpendiculum, quod sit minus qualibet assignabili longitudine R.

Atque hinc tertio consequitur unum aliquem (in eae hypothesi) esse debere determinatum acutum angulum BAX, sub quo educta AX ita semper magis accedat ad eam BX, ut non nisi ad infinitam distantiam in eandem incidat.

Porro autem hanc ipsam AX fore limitem partim intrinsecum, partim extrinsecum utriusque praedictarum rectarum classis, sic demonstratur. Nam primo conveniet cum illis rectis, quae ad finitam distantiam occurrunt ipsi BX, cum ipsa etiam aliquando conveniat; discrepabit autem, quia ipsa non nisi ad infinitam distantiam. Secundo autem conveniet etiam, & simul discrepabit ab illis rectis, quae commune obtinent in duobus distinctis punctis perpendiculum cum illa BX; quia ipsa etiam commune obtinet perpendiculum cum eadem BX; sed in uno eodemque puncto X infinite dissito. Hoc autem postremum censeri debet demonstratum in XXVIII. hujus, prout moneo in ejusdem Corollario.

Itaque constat unum aliquem fore (in hypothesi anguli acuti) determinatum acutum
[70] angulum BAX, sub quo educta AX non nisi ad infinitam distantiam incidat in eam BX, ac propterea sit ipsa limes partim intrinsecus, partim extrinsecus; tum earum omnium, quae sub minoribus acutis angulis ad finitam distantiam incidunt in praedictam BX; tum etiam aliarum, quae sub majoribus angulis acutis, usque ad angulum rectum inclusive, commune obtinent in duobus distinctis punctis perpendiculum cum eadem BX. Quod erat &c.

Proposition 32.

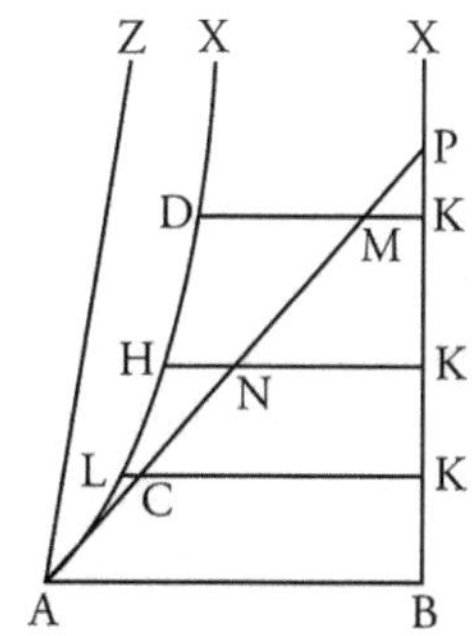

Fig. 33

Now I say there is (in the hypothesis of acute angle) a certain determinate acute angle BAX drawn under which AX (Fig. 33) *only at an infinite distance meets BX, and thus is a limit in part intrinsic, in part extrinsic; on the one hand of all those which under lesser acute angles meet the aforesaid BX at a finite distance; on the other hand also of the others which under greater acute angles, even to a right angle inclusive, have a common perpendicular in two distinct points with BX.*

Proof. First it holds (Corollary 2 to Proposition 29) that no determinate acute angle will be the greatest of all drawn under which a straight from the point A meets the aforesaid BX at a finite distance.

Secondly, it holds in like manner that (in the hypothesis of acute angle) no acute angle will be the least of all drawn under which a straight has a common perpendicular in two distinct points with BX; since indeed (Proposition 31) there can be no determinate limit, such that there cannot be found, under a lesser angle constituted at the point A, a common perpendicular in two distinct points, which is less than any assignable length R.

And hence follows thirdly, that (in this hypothesis) there must be a certain determinate acute angle BAX, drawn under which AX so approaches ever more to BX, that only at an infinite distance does it meet it.

But further that this AX is a limit in part intrinsic in part extrinsic of each of the aforesaid classes of straights is proved thus. First, it agrees with those straights which meet BX at a finite distance since it also finally meets;[1] but it differs, because it meets only at an infinite distance. But secondly it also agrees with, and at the same time differs from those straights which have a common perpendicular in two distinct points with BX; because it also has a common perpendicular with BX; but in one and the same point X infinitely distant. But this latter ought to be considered demonstrated in Proposition 28, as I point out in its corollary. Therefore it holds, that (in the hypothesis of acute angle) there will be a certain determinate acute angle BAX, drawn under which AX only at an infinite distance meets BX, and thus is a limit in part intrinsic, in part extrinsic; on the one hand of all those which under lesser acute angles meet the aforesaid BX at a finite distance; on the other hand also of the others which under greater acute angles, even to a right angle inclusive, have a common perpendicular in two distinct points with BX. This is what was to be demonstrated.

Propositio XXXIII.

Hypothesis anguli acuti est absolute falsa; quia repugnans naturae lineae rectae.

Demonstratur. Ex praemissis Theorematis constare potest eo tandem perducere Geometriae Euclideae inimicam hypothesin anguli acuti, ut agnoscere debeamus duas in eodem plano existentes rectas AX, BX, quae in infinitum protractae versus eas partes punctorum X in unam tandem eandemque rectam lineam coire debeant, nimirum recipiendo, in uno eodemque infinite dissito puncto X, commune in eodem cum ipsis plano perpendiculum. Quoniam vero de primis ipsis principiis agendum mihi hic est, diligenter curabo, ut nihil omittam quasi nimis scrupulose objectum, quod quidem exactissimae demonstrationi opportunum esse cognoscam.

Lemma I.

Duae rectae lineae spatium non comprehendunt.

Definit Euclides lineam rectam, quae *ex aequo sua interjacet puncta*. Esto igitur (Fig. 37) linea quaedam AX, quae ex puncto A per sua quaelibet intermedia puncta continuative
[71] excurrat usque ad punctum X. Non dicetur haec linea recta, si talis ipsa fuerit, ut circa duo illa immota extrema sua puncta possit ipsa in alteram partem converti, ut puta a laeva parte in dexteram: Non dicetur, inquam, linea recta; quia non jacebit ex aequo inter sua designata extrema puncta; quandoquidem vel in laevam partem declinabit, ubi ex puncto A excurrit ad punctum X per quaedam intermedia puncta B; vel declinabit in dexteram, ubi ex eodem immoto puncto A excurrit ad idem immotum punctum X per quaedam intermedia puncta C, quae alia plane sunt a praedictis punctis B. Scilicet illa sola linea AX dici poterit recta, quae excurrat ex puncto A ad punctum X per talia intermedia puncta D, quae ipsa, prout sic invicem continuata, revolvi nequeant, circa illa immota extrema puncta A, & X, ad novum & novum occupandum situm.

In hac autem rectae lineae idea manifeste continetur proposita veritas, duas nempe rectas lineas spatium non comprehendere. Si enim duae exhibeantur lineae claudentes spatium, quarum nempe communia sint extrema duo puncta A, & X, facile ostenditur vel neutram, vel unam tantum illarum linearum esse rectam. Neutra erit recta, ut puta ABBX, & ACCX, si circa duo extrema immota puncta A, & X, ita revolvi posse intelligantur ipsae ABBX, ACCX, ut reliqua ipsarum intermedia puncta ad novum, & novum occupandum locum pertranseant. Una tantum erit recta, ut puta ADDX, si circa illa immota extrema puncta ita revolvi intelligantur ipsae ABBX, ACCX, quae hinc inde cum illa ADDX spatium claudunt, ut ipsarum quidem ABBX, ACCX puncta intermedia ad novum, & novum occupandum locum pertranseant, ipsius vero ADDX puncta omnia etiam intermedia in eodem loco persistant. Non igitur fieri potest, ut duae juxta praemissam intelligentiam rectae lineae, spatium comprehendant. Quod erat propositum.

Proposition 33.

The hypothesis of acute angle is absolutely false; because repugnant to the nature of the straight line.

Proof. From the foregoing Theorems may be established, that at length the hypothesis of acute angle inimical to the Euclidean Geometry has as outcome that we must recognize two straights AX, BX, existing in the same plane, which produced *in infinitum* toward the parts of the points X must run together at length into one and the same straight line, truly receiving, at one and the same infinitely distant point a common perpendicular in the same plane with them.

But since I am here to go into the very first principles, I shall diligently take care, that I omit nothing objected almost too scrupulously, which indeed I recognize to be opportune to the most exact demonstration.

Lemma 1.

Two straight lines do not enclose a space.

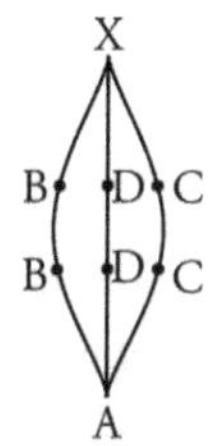

Fig. 37

Euclid defines a straight line as one which *lies evenly between its points.* Let there be therefore (Fig. 37) any line AX, which from the point A through any intermediate points of it runs continuously even to the point X. This line is not called straight if it be such, that it can be turned about its two end points into another region, as suppose from the left region into the right: I say it is not called a straight line; because it will not lie evenly between its designated extreme points; since either it will lean toward the left side, where from the point A it runs out to the point X through certain intermediate points B; or it bends to the right, where from the same fixed point A it runs out to the same fixed point X through certain intermediate points C which are wholly different from the aforesaid points B. Obviously only that line AX can be called straight, which runs out from the point A to the point X through such intermediate points D, as, in order one after another continued, cannot be revolved, about those fixed extreme points A, and X, to occupying new and new location.

But in this idea of the straight line is contained manifestly the announced truth, namely that two straight lines do not enclose a space. For if two lines are shown enclosing a space, which have in common the two extreme points A, and X, it is easily shown either that neither, or only one of them is straight. Neither will be straight, as for example ABBX, and ACCX, if it be supposed so that they can be revolved about two fixed extreme points A, and X, that their remaining intermediate points pass over to occupying new and new place. One only will be straight, as for example ADDX, if about those fixed end points we may suppose ABBX, ACCX, which on both sides with that ADDX enclose a space, so to be revolved, that indeed the intermediate points of ABBX, ACCX, pass over to the occupying of new and new position, but on the contrary all the intermediate points of ADDX remain in the same place. Therefore it cannot be, that two lines, straight in accordance with the previous concept, enclose a space. This is what was to be demonstrated.

[72] **Corollarium I.**

Hinc porro sequitur admitti oportere postulatum illud Euclidaeum: quod *a dato puncto ad quodlibet assignatum punctum rectam lineam ducere liceat.* Nam clare intelligitur, duas semper sine ullo certo limite duci posse lineas, praedictis punctis A, & X terminatas, quae propiores in vicem fiant, minusque idcirco spatium comprehendant, dum scilicet una quidem ducatur ad laevam partem, & altera uniformis ad dexteram, sive una sursum, & altera deorsum; duci, inquam, posse lineas ejusmodi semper invicem sine ullo certo limite propiores, quae utique omnino uniformes inter se sint, sibique invicem idcirco succedant, dum circa immota extrema puncta A, & X, revolvi ipsae intelligantur. Inde autem clare itidem intelligitur, sequi tandem debere (in semper majore harum uniformium linearum, unius ad alteram accessu) coitionem in unam, eandemque lineam ADX, quae circa immota extrema illa puncta revolvi nequeat ad occupandum novum locum. Et haec erit linea recta postulata.

Ubi rursum constat unicam esse, quae a dato puncto ad quodlibet alterum assignatum punctum potest duci linea recta.

Corollarium II.

Praeterea sequitur uniformem esse debere intelligentiam alterius Euclideae definitionis, in qua dicit planam superficiem esse, *quae ex aequo suas interjacet lineas.* Si enim superficies clausa praedictis lineis una ADX recta, & altera ABBX (sive haec sit unica, aut multiplex linea curva, sive sit composita ex duabus, aut pluribus lineis rectis, ut puta AB,
[73] BB, BX) si, inquam, superficies ejusmodi revolvi intelligatur circa immotam rectam ADX, usque dum ipsa linea ABX perveniat ad congruendum lineae ACX, in parte adversa locatae, quae utique ad omnimodam aequalitatem, & similis omnino sit ipsi ABX, & rursum cum eadem recta ADX claudat (versus eandem sive supernam, sive infernam partem) superficiem omnino aequalem, & similem antedictae: alterutrum sane continget; vel ita ut una superficies alteri adamussim congruat; vel ita ut intra duas illas superficies claudatur spatium trinae dimensionis. Et primum quidem si contingat, dicetur superficies plana; sin vero contingat secundum, non dicetur superficies plana; quia tunc aliae intermediae intelligi poterunt inter easdem extremas lineas interpositae superficies invicem aequales, ac similes, quae semper magis ad se invicem sine ullo certo limite accedant, ac propterea usque ad excludendum omne spatium intermedium. Tunc autem utraque illa superficies dicetur plana, quia vere jacebit ex aequo inter suas extremas lineas, sine ullo ascensu, aut descensu in partes adversas.

Corollary 1.

Hence moreover follows we should admit the Euclidean postulate: that *from a given point to any assigned point a straight line may be drawn.* For it is dearly understood, that always two lines without any certain limit can be drawn, bounded in the aforesaid points A, and X, which mutually approach, and therefore enclose less space, while indeed one is drawn toward the left side, and the other of the same shape toward the right, or one over, and the other under; I say, lines of this sort may be drawn always mutually approaching without any certain limit, which are completely of the same shape with each other, and therefore mutually succeed each other when supposed revolved about the fixed end points A, and X. Whence clearly in like manner is understood, at length (in ever greater approach of these like shaped lines, one to the other) should follow the coalescence into one, and the same line ADX, which cannot be revolved about those fixed extreme points so as to occupy a new position. And this will be the straight line postulated.

Where again is established to be unique the straight line, which can be drawn from a given point to any other assigned point.

Corollary 2.

Moreover it follows the interpretation should be the same of the other Euclidean definition, in which he says a surface is plane, *which lies evenly between its lines.* For if a surface enclosed by the aforesaid lines one ADX straight,[1] and another ABBX (whether this be a simple or broken curved line, i. e. composed of two, or several straight lines, as suppose AB, BB, BX) if, I say, a surface of this sort is supposed to be revolved about the fixed straight ADX, until the line ABX comes to congruence with the line ACX, located in the opposite part, which assuredly is in every way equal and wholly similar to ABX, and again with the same straight ADX encloses (toward the same part, whether upper or under) a surface wholly equal, and similar to the aforesaid: one of two things certainly happens; either one surface fits the other completely; or between those two surfaces is enclosed a three-dimensional space. And indeed if the first happens, the surface is called plane; but if the second happens the surface is not called plane; because then may be supposed other intermediate surfaces, mutually equal, and similar, interposed between the same extreme lines, which always mutually approach more to each other without any certain limit, and therefore even to the exclusion of every intermediate space. But then each surface is called plane, because truly it lies evenly between its extreme lines, without any ascent or descent into opposing parts.

Lemma II.

Duae lineae rectae non possunt habere unum & idem segmentum commune.

Demonstratur. Si enim fieri potest; unum & idem segmentum AX commune sit (Fig. 38) duabus rectis, per punctum X in eodem plano continuatis AXB, & AXC. Tum centro X, & intervallo XB, sive XC, describatur arcus BMC, ad cujus quodlibet punctum M jungatur ex puncto X recta XM.

[74] Dico primo, lineam AXM fore & ipsam, in facta hypothesi, lineam rectam, ex puncto A per punctum X continuatam. Si enim linea ejusmodi recta non sit, duci poterit (ex Cor. I. praecedentis Lemmatis) alia quaedam linea AM, quae ipsa sit recta. Haec autem vel secabit in aliquo puncto K alterutram ipsarum XB, XC; vel earundem alterutram, ut puta eam XB claudet intra spatium comprehensum ipsis AX, XM, & APLM. At horum prius manifeste repugnat praecedenti Lemmati; quia sic duae suppositae rectae lineae, una AXK, & altera ATK, spatium clauderent. Posterius autem uniformis absurdi statim convincitur.

Nam constat rectam XB, si per B ulterius protrahatur, occursuram tandem in aliquo puncto L ipsi APLM; unde rursum duae suppositae rectae, una AXBL, & altera APL, spatium claudent. Porro uniforme sequitur absurdum, si fingamus, quod recta XB, ulterius protracta per B, occurrat tandem in quovis alio puncto aut rectae XM, aut rectae XA.

Ex istis autem evidenter consequitur lineam AXM fore ipsam, in facta hypothesi, lineam rectam ex puncto A ad punctum M deductam. Quod erat propositum.

Dico secundo, eam suppositam rectam AXB (quatenus quidem intelligatur conservare suam illam qualemcunque continuationem ex puncto A per X versus B) non posse recipere duplicem aliam in eodem plano positionem, in quarum utraque portio quidem AX in eodem situ persistat, portio vero altera XB in una illarum duarum positionum congruat (exempli causa) ipsi XC, & in alia positione congruat ipsi XM.

Scilicet non hic renuo, quin portio XB, si intelligatur moveri in illo suo plano circa punctum X, adeo ut successive adamussim congruat (ex praecedente Lemmate) non modo
[75] ipsis XM, XC, verum etiam adamussim congruat infinitis aliis rectis, quae ex puncto X duci possunt ad reliqua intermedia puncta arcus BC: Non, inquam, hic renuo, quin illa XB in qualibet illarum positionum considerari debeat tanquam continuatio in rectum ipsius immotae AX; cum magis circa eam AXM jam demonstraverim id secuturum in facta hypothesi illius communis segmenti: Unice igitur hic assero, in una tantum novarum illarum positionum, ut puta dum congruit ipsi XC, retineri ab ea posse illam eandem qualemcunque continuationem, quam obtinet in prima positione, ubi ex puncto A per X procedit versus punctum B.

Lemma 2.

Two straight fines cannot have one and the same segment in common.

Proof. For if that is possible, let one and the same segment AX be common (Fig. 38) to the two straights AXB, and AXC produced through the point X in the same plane. Then with center X, and radius XB, or XC, describe the arc BMC, to any point of which M is drawn from the first point X the straight XM.

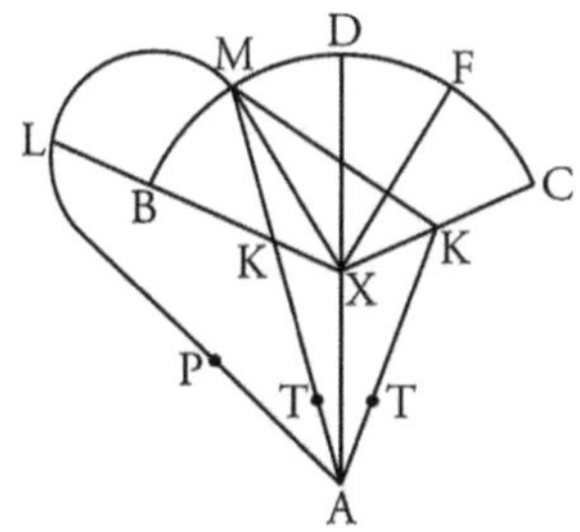

Fig. 38

I say first, under the assumed hypothesis also the line AXM will be a straight line, continued from the point A through the point X. For if a line of this sort be not straight, there can be drawn (Corollary 1 to Lemma 1) a certain other line AM, which itself is straight. But this either cuts in some point K one or the other of those straights XB, XC; or it closes one or the other of them, as suppose XB within the space bounded by AX, XM, and APLM. But the first of these is manifestly contrary to Lemma 1; because thus two lines supposed straight, one AXK, and the other ATK, would enclose a space. But the second is at once convicted of a like absurdity. For it is certain that the straight XB, if produced on through B, will at length meet this APLM in some point L; whence again two lines supposed straight, one AXBL, and the other APL, will enclose a space. But a like absurdity follows, if we assume, that the straight XB, produced on through B, at length meets in some other point either the straight XM, or the straight XA.

But from this evidently follows that the line AXM is itself, in the assumed hypothesis, the straight line drawn from the point A to the point M. This is what was to be demonstrated.

I say secondly, that the assumed straight AXB (inasmuch as it is understood to retain its arbitrary continuation from the point A through X toward B) cannot have two different positions in the same plane, in both of which the portion indeed AX persists in the same place, but the other portion XB in one of those two positions fits (for example) XC, and in the other position fits XM.

Of course I do not here deny, that the portion XB, if it is supposed to be moved in its plane about the point X, so that successively it fits exactly (Lemma 1) not merely XM, XC, but also exactly fits the other infinitely many straights, which from the point X may be drawn to the remaining intermediate points of the arc BC: I say, I do not here deny, that XB in any of its positions may be considered as the continuation in a straight of that fixed AX; when rather I have demonstrated already about AXM that this would happen in case of the hypothesis of a common segment: Solely therefore I here affirm, in one merely of those new positions, as suppose while it fits XC, may be retained by it the same arbitrary continuation, which it has in the first position, where from the point A it goes out through X toward the point B.

Et istud quidem sic demonstratur. Nam primo constat continuationem illam AXB nequire esse omnino similem, aut aequalem continuationi AXC, si utraque consideretur versus eandem seu laevam, seu dexteram partem; quia caeterum in ea tali positione deberent invicem congruere ipsae AXB, AXC; quod est contra hypothesim communis illius segmenti AX: Deberent, inquam, congruere; dum scilicet, relate ad eam immotam AX, aeque similiter in eandem seu laevam, seu dexteram partem convergerent in eo tali plano illae continuatae XB, & XC. Secundo constat nihil vetare, quin praedicta continuatio AXB, considerata versus unam partem, ut puta, ad laevam, similis plane sit, aut aequalis continuationi AXC, consideratae versus partem adversam, ut puta, ad dexteram, adeo ut propterea, sine ulla immutatione in ipsa AXB, locari haec possit ad congruendum in eodem plano alteri AXC. At manifeste repugnat, quod rursum, sine ulla immutatione illius suae continuationis, locari ea possit in eodem plano ad congruendum alteri AXM, quae nimirum dividat in X illum qualemcunque angulum BXC. Quod enim continuatio AXB alia plane sit a continu-
[76] atione AXM, si utraque consideretur versus eandem seu laevam, seu dexteram partem, ex eo manifestum esse debet; quia caeterum (ut in simili observatum jam est) in ea tali positione deberent invicem congruere ipsae AXB, AXM. Sed neque sustineri potest, quod continuatio AXB versus unam partem, ut puta ad laevam, similis plane sit, aut aequalis continuationi AXM versus partem adversam, ut puta ad dexteram; quia caeterum continuatio AXM versus dexteram similis plane foret, aut aequalis continuationi AXC versus eandem dexteram partem, propter suppositam omnimodam similitudinem, aut aequalitatem inter modo dictam continuationem, & illam aliam AXB versus laevam. Tunc autem in ea tali positione (ut est praedictum) deberent invicem congruere ipsae AXM, AXC; quod est contra praesentem hypothesim.

Ex quibus omnibus infero: eam suppositam recta AXB (quatenus quidem intelligatur conservare suam illam qualemcunque continuationem ex puncto A versus B) recipere non posse duplicem aliam in eodem plano positionem, in quarum utraque portio quidem AX in eodem situ persistat, portio vero altera XB in una illarum duarum positionum congruat (exempli causa) ipsi XC, & in alia positione congruat ipsi XM. Quod erat propositum.

Dico tertio: eandem suppositam rectam AXB non alia ratione conservare posse suam illam qualemcunque continuationem, dum ejusdem portio XB intelligitur transferri per nova, & nova loca usque ad congruendum in illo quodam plano ipsi XC, persistente interim in eodem suo loco portione AX; non posse, inquam, conservare suam illam qualemcunque continuationem, nisi quatenus portio ipsa XB intelligatur ascendere, aut descendere ad existendum cum illa immota AX in novis, & novis planis, usque dum redeat ad antiquum planum, congruens ibi praedictae XC.

[77] Id enim censeri potest jam demonstratum; quia scilicet nulla alia in eodem illo plano reperiri potest positio, juxta quam ipsa AXB (persistente portione AX in suo eodem loco) conservet suam illam qualemcunque continuationem, praeterquam ubi deveniat ad congruendum praedictae AXC.

And this indeed is demonstrated thus. For first it is evident that the continuation AXB cannot be wholly similar, or equal to the continuation AXC, if each is considered toward the same part whether left or right; because otherwise in such position AXB, AXC must mutually coincide; which is against the hypothesis of that common segment AX: I say, must coincide; provided that of course, in relation to the same fixed AX, the continuations XB, and XC in the plane concerned extend just similarly toward the same part whether left or right. Secondly is evident that nothing prevents the aforesaid continuation AXB, considered toward one part, as suppose, toward the left, being precisely similar, or equal to the continuation AXC, considered toward the opposite part, as suppose, toward the right, so that consequently, without any change in AXB, this may be brought to congruence with the other AXC in the same plane. But it is manifestly contradictory, that on the other hand, without any change of its prolongation, this can be brought in the same plane into congruence with the other AXM, which indeed at X divides that arbitrary angle BXC. For that the prolongation AXB is plainly other than the prolongation AXM, if each is considered toward the same part, whether left or right, must be manifest from this; because otherwise (as already observed in like case) in such a situation AXB, AXM must mutually fit. But neither can it be maintained, that the prolongation AXB toward one part, as suppose toward the left, is wholly similar, or equal to the prolongation AXM toward the opposite part, as suppose toward the right; because otherwise the prolongation AXM toward the right would plainly be similar, or equal to the prolongation AXC toward the same right side, because of the assumed complete similitude, or equality between the just cited prolongation, and that other AXB toward the left. But then in such a situation (as previously remarked) AXM, AXC should mutually fit; which is against the present hypothesis.

From all which, I infer: the assumed straight AXB (in so far as it is understood to retain its arbitrary prolongation from the point A toward B) cannot have two different positions in the same plane, in both of which the portion indeed AX remains in the same location, but the other portion XB in one of those two positions fits (for example) XC, and in the other position fits XM. This is what was to be demonstrated.

I say thirdly: the assumed straight AXB can in no other way retain its arbitrary prolongation, while its part XB is supposed to be transferred through new and new positions even to fitting XC in that one plane, the portion AX remaining meanwhile in the same place; I say it cannot retain its chosen continuation, except in so far as the portion XB is understood to ascend, or descend to be with the fixed AX in new, and new planes, until it returns to the old plane, fitting there the aforesaid XC.

For this may be adjudged already demonstrated; because obviously no position in that same plane can be found, at which AXB (the portion AX remaining in its place) retains its chosen prolongation, except where it comes to congruence with the aforesaid AXC.

Dico quarto: designari posse in eo arcu BC tale punctum D, ad quod si jungatur XD, jam ipsa AXD non modo recta linea sit, sed rursum ita se habeat, ut continuatio AXD, considerata versus laevam, aequalis plane sit, aut similis eidem continuationi consideratae versus dexteram.

Demonstratur. Et prior quidem pars (qualecunque sit illud punctum D in arcu BC designatum) eo modo ostenditur, quo supra usi sumus circa continuatam AXM. Posterior vero pars ita evincitur. Nam hic supponimus duas rectas AXB, AXC, sub eodem communi segmento AX. Praeterea supponimus continuationem AXB versus laevam non esse omnino similem, aut aequalem eidemmet continuationi versus dexteram; quia stante omnimoda ejusmodi similitudine, aut aequalitate, facile ostenditur nulli alteri rectae lineae commune esse posse illud segmentum AX, prout nempe sic demonstrabimus de illa continuata AXD. Tandem consequenter supponimus continuatam illam AXB ita locari posse in eodem plano, ut sub eodem immoto segmento AX congruat cuidam alteri AXC, in qua nimirum continuatio ipsa AXC versus dexteram similis plane sit, aut aequalis continuationi AXB versus laevam, ac rursum continuatio AXC versus laevam similis plane sit, aut aequalis continuationi AXB versus dexteram.

His stantibus: si ad quodvis punctum M sumptum in eo arcu BC jungatur XM; vel
[78] continuatio AXM erit sibi ipsi plane uniformis relate ad laevam; ac dexteram partem ipsius AX; vel non. Si primum; demonstrabo de ista AXM, quod statim demonstraturus sum de illa continuata AXD. Si secundum, ergo praedicta AXM ita rursum locari poterit in eodem plano, ut sub eodem immoto segmento AX congruat cuidam alteri AXF, in qua nimirum continuatio ipsa AXF versus dexteram similis plane sit, aut aequalis continuationi AXM versus laevam, ac rursum continuatio AXF versus laevam similis plane sit, aut aequalis continuationi AXM versus dexteram. Porro, cum punctum M supponi possit vicinius puncto B, quam punctum C, non cadet punctum F in ipsum punctum C; quia sic continuatio AXM versus laevam similis plane foret, aut aequalis continuationi AXF, sive AXC versus dexteram, ac propterea similis plane, aut aequalis continuationi AXB versus laevam, quod est absurdum, cum illae duae XM, XB non sibi invicem congruant in sua tali positione. Sed neque etiam existet punctum F ultra punctum C in eo arcu BC ulterius producto; quia sic uniformi ratiocinio ostendetur, contra hypothesim, quod etiam punctum M deberet existere in eo arcu CB ulterius producto, adeo ut nimirum ipsa XM divideret versus laevam eum qualemcunque angulum AXB, prout XF poneretur dividere versus dexteram eum qualemcunque angulum AXC: Deberet, inquam, sic existere, ad eum utique finem, ut ea AXM sub eodem immoto segmento AX locari rursum possit in eodem plano ad congruendum illi alteri AXF, in qua nimirum continuatio ipsa AXF versus dexteram similis plane sit, aut aequalis continuationi AXM versus laevam, ac rursum continuatio AXF versus laevam similis plane sit, aut aequalis continuationi AXM versus dexteram.

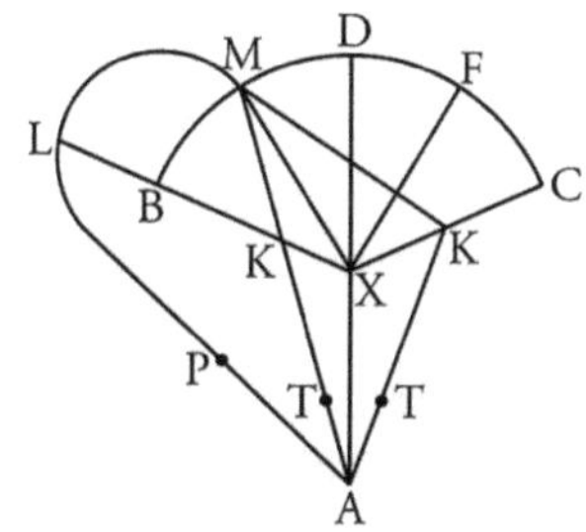

Fig. 38

I say fourthly: in the arc BC such a point D can be designated that, if XD be joined, then this AXD not only is a straight line, but moreover it lies so, that the prolongation AXD, considered toward the left, is wholly equal, or similar to the same prolongation considered toward the right.

Proof. The first part (whatever be the point D designated in the arc BC) is shown by the method used above in regard to the prolongation AXM. But the second part is proved thus. We suppose here two straights AXB, AXC with the same common segment AX. Further we suppose the prolongation AXB toward the left not to be wholly similar, or equal to the same prolongation toward the right; because, such a complete similitude or equality holding good, it is easily shown that segment AX can be common to no other straight line, just as we shall demonstrate of the prolongation AXD. Finally in consequence we suppose the prolongation AXB may so be located in that plane, that with its fixed segment AX it fits a certain other, AXC, in so far as truly the prolongation AXC toward the right is exactly similar, or equal to the prolongation AXB toward the left, and moreover the prolongation AXC toward the left is precisely similar, or equal to the prolongation AXB toward the right.

This remaining: if, assuming any point M in the are BC, we join XM; either the prolongation AXM will be precisely uniform in relation to the left, and the right side of AX; or not. If the first; I shall demonstrate of AXM, what immediately I shall have demonstrated of the prolongation AXD. If the second, therefore the aforesaid AXM can in turn be so located in the same plane, that with the same fixed segment AX it fits a certain other AXF, in which truly the prolongation AXF toward the right is precisely similar, or equal to the prolongation AXM toward the left, and moreover the continuation AXF toward the left is precisely similar, or equal to the prolongation AXM toward the right. Furthermore, since the point M may be supposed nearer to the point B than is the point C, the point F does not fall upon the point C; because thus the prolongation AXM toward the left would be precisely similar, or equal to the prolongation AXF, or AXC toward the right, and therefore precisely similar, or equal to the prolongation AXB toward the left, which is absurd, since the two XM, XB do not mutually fit each other in such position of theirs. But neither also is the point F beyond the point C in the arc CB produced farther on; because thus by like reasoning is shown, against the hypothesis, that also the point M must be in the arc CB produced farther on, so that XM would divide toward the left the arbitrary angle AXB, just as XF would be posited to divide toward the right the arbitrary angle AXC: I say must so lie, to the end, that AXM with its fixed segment AX can again be so placed in that plane as to fit the other, AXF in so far as truly the prolongation AXF toward the right is precisely similar, or equal to the prolongation AXM toward the left, and moreover the prolongation AXF toward the left is precisely similar, or equal to the prolongation AXM toward the right.

[79] Quoniam vero arcus BC major est ejusdem portione MF, designarique uniformiter possunt in ea portione MF alia duo puncta cum minore, sine ullo certo termino, intercapedine; alterutrum sane in hac praedictorum punctorum approximatione contingere debet. Unum est, si tandem incidatur in unum idemque intermedium punctum D, ad quod si jungatur XD, talis habeatur continuatio AXD, cui soli conveniat (facta comparatione inter laevam, ac dexteram partem) esse sibi ipsi omnino similem, aut aequalem. Alterum est, si duo talia inveniantur distincta puncta M, & F, ad quae junctæ XM, & XF, duas exhibeant continuationes, unam AXM, & alteram AXF, quarum utraque sit sibi ipsi, modo jam explicato, omnino similis, aut aequalis. Hoc autem secundum impossibile esse sic demonstro. Nam ex ipsis terminis constare potest, quod recta linea, ex puncto A per X ulterius producta, unicam tantum sortiri potest in eo tali plano positionem, dum scilicet quaedam superaddita XF aeque omnino se habeat in laevam, & in dexteram partem praesuppositae AX, seu non magis in laevam, quam in dexteram ejusdem partem convergat. Non ergo alia erit continuatio AXM, quae rursum aeque omnino se habeat in laevam, & in dexteram partem ejusdem AX. Scilicet constat subsistere simul non posse; & quod continuatio AXF versus dexteram similis plane sit, aut aequalis sibi ipsi consideratae versus laevam; & quod alia quaedam continuatio AXM versus laevam (quae, ex ipsa positione, minor sit continuatione AXF versus eandem laevam) aequalis iterum sit eidem continuationi versus dexteram, quae certe, ex ipsa rursum positione, major est praedicta continuatione AXF versus eandem dexteram.

Non ergo in eo arcu BC duo talia inveniri possunt puncta M, & F, ad quae junctæ XM,
[80] & XF, duas exhibeant continuationes, unam AXM, & alteram AXF, quarum utraque sit sibi ipsi, modo jam explicato, omnino similis, aut aequalis. Unde tandem consequitur incidi aliquando debere in unum, idemque punctum D, ad quod juncta XD talem exhibeat continuationem AXD, cui soli conveniat (facta comparatione inter laevam, ac dexteram partem) esse sibi ipsi omnino similem, aut aequalem. Quod erat hoc loco demonstrandum.

Dico tandem quinto: eam solam AXD fore lineam *rectam*, nimirum ex A per X *directe* continuatam in D. Quamvis enim ly *ex aequo*, in definitione lineae rectae, applicari primitus debeat punctis intermediis relate ad puncta ipsius extrema; unde utique jam elicuimus, *duas lineas rectas non claudere spatium*; intelligi tamen etiam debet de ejusdem rectae lineae continuatione in directum. Itaque ea sola XD (in eodem cum AX plano existens) dicetur esse continuatio *recta*, sive *in rectum* praedictae AX, quando ipsa neque in laevam, neque in dexteram illius partem convergat, sed utrinque *ex aequo* procedat; adeo ut nempe continuatio illa AXD versus laevam similis plane sit, aut aequalis eidem continuationi consideratae versus dexteram. Inde enim fiet, ut illi soli AXD conveniat non posse ab ea suscipi in eo tali plano aliam positionem sub illa immota AX; cum certe (ex jam demonstratis) illae aliae AXB, & AXM, citra omnem suarum talium continuationum immutationem, suscipere possint sub eadem immota AX alias in eodem plano positiones, quales sunt ipsarum AXC, & AXF. Igitur illa sola AXD, cujus nempe continuatio XD tum in eodem cum ipsa AX plano existat, tum etiam aeque omnino se habeat in laevam, ac dexteram partem praedictae AX, est linea *recta* juxta explicatam definitionem, seu continuatio *in rectum* ejusdem praesuppositae rectae AX.

Ex quibus omnibus tandem constat evenire non posse, ut unum quodpiam sit commune
[81] segmentum duarum rectarum. Quod erat demonstrandum.

But since the arc BC is greater than its part MF, and in this portion MF in like way may be designated two other points with an interval less, without any certain limit; truly one of two things must happen in this approximation of the aforesaid points. One is, if at length is attained one and the same intermediate point D, to which if XD is joined, such a prolongation AXD is obtained, as alone is such as to be wholly similar, or equal to itself (comparison made between the left and the right side). The other is, if two such distinct points M, and F are found, to which XM, and XF being joined, two prolongations arise, one AXM, and the other AXF, of which each is, in the way just explained, wholly similar, or equal. But this second I prove to be impossible thus. For from the very terms can be established, that a straight line produced from the point A on through X, can take in the plane only a single position, whilst obviously the superadded XF lies altogether equally toward the left, and toward the right side of the assumed AX, or deviates not more toward the left, than toward the right side of it. Therefore there will not be another prolongation AXM, which also lies altogether equally toward the left and toward the right of this AX. Obviously it holds that it cannot happen at the same time, both that the prolongation AXF toward the right is wholly similar, or equal to itself considered toward the left, and that another prolongation AXM toward the left (which, from its very position, is less than the prolongation AXF toward the same left) again is equal to the same continuation toward the right, which truly, again from its very position, is greater than the aforesaid prolongation AXF toward the same right.

Therefore in the are BC cannot be found two such points M, and F, that the joins XM, and XF, present two prolongations, one AXM, and the other AXF, of which each is to itself, in the way just explained, wholly similar, or equal. Whence at length follows, that somewhere must be attained one and the same point D, to which the join XD presents such a prolongation AXD, that to it alone belongs to be wholly similar, or equal to itself (comparison made between left, and right side). This is what was to be demonstrated.

At length I say fifthly: this AXD alone is a *straight* line, namely from A through X continued *in a straight* to D. For though the phrase "evenly", in the definition of the straight line,[1] should primarily be applied to points intermediate in relation to its extreme points; whence in particular we have just deduced, *two straight lines do not enclose a space;* nevertheless it should also be understood of the *direct* prolongation of this straight line. Therefore alone this AD (lying in the same plane with AX) is said to be the *straight* prolongation (or *in a straight*) of the aforesaid AX, when that deviates neither toward the left, nor toward the right side of it, but from each side proceeds evenly; so that the prolongation AXD is toward the left clearly similar, or equal to the same prolongation considered toward the right. For thence it will follow, that alone to AXD pertains that another position cannot be taken by it in the plane, while AX is fixed; when truly (from what has just now been proved) those others, AXB, and AXM, without any change of their prolongations, can, with the same fixed AX, take other positions in the same plane, such as AXC and AXF. Therefore alone AXD, whose prolongation XD not only is in the same plane with AX, but also lies altogether in like manner toward the left, and the right side of the aforesaid AX, is a *straight* line in accordance with the discussed definition, or the prolongation *in a straight* of the assumed straight AX.

From all which finally is established as impossible, that one segment can be common to two straight lines. This is what was to be demonstrated.

Corollarium.

Ex duobus praemissis Lemmatis tria opportune subnotare licet. Unum est: duas rectas, neque sub infinite parva inter ipsas distantia, claudere spatium posse. Ratio est, quia (prout in primo Lemmate) vel utraque illarum sub duobus illis communibus extremis punctis immotis revolvi posset ad novum situm occupandum, & sic (ex jam tradita lineae rectae definitione) neutra foret linea recta: vel una tantum in suo eodem situ persisteret, & sic illa sola recta linea foret. Quod autem nequeat utraque in eodem ipso situ persistere, dum aliquod concludant spatium, etiamsi infinite parvum, manifestum fiet consideranti posse faciem illius plani, in quo illae duae consistunt, converti de superna in infernam, manentibus caeteroquin in suo eodem loco duobus illis extremis punctis.

Alterum est: neque item ullam lineam rectam, in quantalibet ejusdem productione in directum, diffindi posse in duas, quamvis sub infinite parva intercapedine. Ratio est; quia (prout in praecedente Lemmate) continuatio in directum praesuppositae cujusdam simplicis rectae AX non alia esse intelligitur praeter unam XD, quae *ex aequo* utrinque procedat relate ad laevam, ac dexteram partem praedictae AX; ex quo utique fiat, ut sub ea immota AX non aliam ipsa immutata habere possit in eo plano positionem. Quod autem in eodem plano alia quaedam ad laevam decerni possit XM, infinite parum dissiliens ab ipsa XD, nihil suffragatur. Nam rursum alia item ad dexteram designari poterit XF, quae uniformiter
[82] infinite parum dissiliat ab eadem XD. Quare (prout in praecitato Lemmate) illa sola AXD erit linea recta a nobis definita.

Tertium tandem est: in hoc ipso secundo Lemmate censeri posse immediate demonstratam 4. undecimi; quod nempe ejusdem rectae nequeat pars una quidem in subjecto plano existere, & altera in sublimi.

Lemma III.

Si duae rectae AB, CXD sibi invicem occurrant (Fig. 39) *in aliquo ipsarum intermedio puncto* X, *non ibi se invicem contingent, sed una alteram ibidem secabit.*

Demonstratur. Si enim fieri potest, tota CXD ad unam eandemque partem ipsius AB consistat. Jungatur AC. Non erit porro AC eadem cum ipsa veluti continuata AXC; quia caeterum (contra praecedens Lemma) duarum rectarum, unius AXC, & alterius praesuppositae DXC, unum idemque foret commune segmentum XC. Itaque jungatur BC. Non erit rursum haec BC continuatio ipsius BA usque in punctum C; ne duae rectae, una XAC, portio ipsius BAC, & altera XC spatium claudant, contra praemissum Lemma primum. Igitur ea BC vel secabit in aliquo puncto L ipsam XD, sive praesuppositam rectam DXC; & tunc rursum duae rectae lineae, una LC portio ipsius BC, & altera LXC portio praedictae DXC, spatium claudent; vel alterutrum extremum punctum sive A ipsius BA, sive D ipsius CXD, claudetur intra spatium comprehensum ipsis CX, XB, & alterutra vel BFC, vel BHC. At in utroque casu idem absurdum consequitur: Sive enim BA protracta per A occurrat ipsi BFC in aliquo puncto F; sive CXD protracta per D occurrat ipsi BHC in aliquo puncto H: in idem semper absurdum incidimus, quod duae rectae spatium claudant; nimirum aut
[83] recta BF portio ipsius BFC una cum altera BAF; aut recta HC, portio ipsius BHC, una cum altera praesupposita recta continuata CXDH.

Corollary.

From the two preceding Lemmata three things may opportunely be noted. One is: not even with an infinitely small distance between them can two straights enclose a space. The reason is, because (just as in Lemma 1) either each of them with the two common extreme points fixed can be revolved into occupying a new position, and so (from the definition of the straight line already given) neither will be a straight line: or only one remains in the same place, and so it alone is a straight line. But that both cannot remain in the same place, while they enclose any space, even if infinitely little, will be manifest from considering that a face of the plane, in which the two are, can be converted from upper to lower, the two extreme points withal remaining in the same place.

Another is: nor moreover can any straight line, in any production of it in straight, split into two, although with an interval infinitely small. The reason is, because (just as in Lemma 2) the prolongation in straight of any assumed simple straight AX cannot be understood to be other than the one XD, which proceeds evenly on both sides in relation to the left, and right side of the aforesaid AX; from which assuredly follows, that with AX fixed it cannot, itself unchanged, have another position in this plane. But that in the same plane a certain other XM can be designated to the left, splitting infinitely little from XD, nothing avails. For again another, XF, likewise to the right could be designated, which just so splits infinitely little from the same XD. Wherefore (as in Lemma 2) alone AXD will be the straight line defined by us.

The third finally is: in this Lemma 2 may be judged immediately demonstrated *Elements* XI, 1; that of the same straight one part cannot be in a lower plane,[2] and another in an upper.

Lemma 3.

If two straights AB, CXD meet each other (Fig. 39) *in any intermediate point* X *of theirs, they do not there touch each other, but one cuts the other there.*

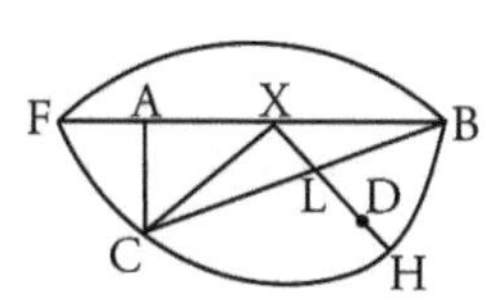

Fig. 39

Proof. For if that were possible, the whole CXD lies on one and the same side of AB. Join AC. Then AC will not be the same with AXC as if prolonged; because otherwise (against Lemma 2) of two straights, one AXC, and the other the assumed DXC, there would be one and the same common segment XC. And so join BC. Again this BC will not be a prolongation of BA to the point C; lest two straights, one XAC, portion of this BAC, and the other XC enclose a space, against the preceding Lemma 1. Therefore this BC either will cut XD (or the assumed straight DXC) in some point L; and then again two straight lines, one LC, portion of this BC, and the other LXC, portion of the aforesaid DXC, enclose a space; or one of the extreme points whether A of BA, or D of CXD, is enclosed within the space bounded by CX, XB, and either BFC, or BHC. But in either case the same absurdity follows: For whether BA produced through A strikes BFC in a point F; or CXD produced through D strikes BHC in a point H; always we came upon the same absurdity, that two straights enclose a space; forsooth either the straight BF portion of BFC together with the other BAF; or the straight HC, portion of BHC, together with the other assumed straight prolonged CXDH.

Porro idem, aut majus absurdum consequitur, si illa BA protracta per A occurrat in aliquo puncto vel ipsi CX, vel sibi ipsi in aliquo puncto suae portionis XB. Atque id similiter valet, si altera CXD protracta per D occurrat in aliquo puncto vel ipsi XB, vel sibi ipsi in aliquo puncto suae portionis CX.

Itaque constat, quod duae rectae AB, CXD sibi invicem occurrentes in aliquo ipsarum intermedio puncto X, non ibi se invicem contingent, sed una alteram ibidem secabit. Quod erat &c.

Lemma IV.

Omnis diameter dividit bifariam suum circulum, ejusque circumferentiam.

Demonstratur. Esto circulus (recole Fig. 23) MDHNKM, cujus centrum A, & diameter MN. Intelligatur illius circuli portio MNKM ita revolvi circa immota puncta M, & N, ut tandem accommodetur, seu coaptetur reliquae portioni MNHDM. Constat primo totam diametrum MAN quoad omnia ipsius puncta in eodem situ esse mansuram: ne duae rectae lineae (contra praecedens Lemma primum) spatium claudant. Constat secundo nullum punctum K circumferentiae NKM casurum vel intra, vel extra superficiem clausam diametro MAN, & altera circumferentia NHDM; ne scilicet contra naturam circuli, unus radius v. g. AK minor sit, aut major altero ejusdem circuli radio v. g. AH. Constat tertio
[84] quemlibet radium MA continuari unice posse in rectum per alterum quendam radium AN, ne (contra praecedens Lemma secundum) duae suppositae rectae lineae, ut puta MAN, MAH, unum idemque commune habeant segmentum MA. Constat quarto (ex proxime antecedente Lemmate) omnes cujusvis circuli diametros se invicem in centro secare, & ex nota natura circuli bifariam.

Ex quibus omnibus constare potest, quod diameter MAN tum dividit exactissime suum circulum, ejusque circumferentiam in duas aequales partes, tum etiam assumi universim potest pro qualibet ejusdem circuli diametro. Quod erat &c.

Scholion.

Hanc eandem veritatem demonstratam leges apud Clavium a Thalete Milesio, sed fortasse non exhausta omni qualibet objectione.

Furthermore the same, or a greater absurdity follows, if BA produced through A meets in any point either CX, or its own self in any point of its portion XB. And this likewise holds, if the other CXD produced through D meets in any point either XB, or its own self in any point of its portion CX.

Therefore is established, that two straights AB, CXD meeting each other in any intermediate point X of theirs, do not there touch each other, but one will cut the other there. This is what was to be demonstrated.

Lemma 4.

Every diameter bisects its circle, and the circumference of it.

Proof. Let there be a circle (recall Fig. 23) MDHNKM, A its center, and MN a diameter. Of this circle the portion MNKM is thought so to revolve about the fixed points M, and N, that at length it is superimposed upon, or applied to the remaining portion MNHDM. It is certain first that as to all its points the whole diameter MAN will remain in the same place; lest two straight lines (against the preceding Lemma 1) enclose a space. It is certain secondly that no point K of the circumference NKM will fall either within, or without the surface enclosed by the diameter MAN, and the other circumference NHDM; lest obviously against the nature of the circle, one radius, for example AK, be less, or greater than another radius of the same circle, for example AH. It is certain thirdly that any radius MA can alone be prolonged in a straight line by a certain other radius AN, lest (against the preceding Lemma 2) two lines assumed straight, as suppose MAN, MAH, should have one and the same common segment MA. It is certain fourthly (from the immediately preceding Lemma 3) that all the diameters of the circle cut one another in the center, and from the known nature of the circle bisect.

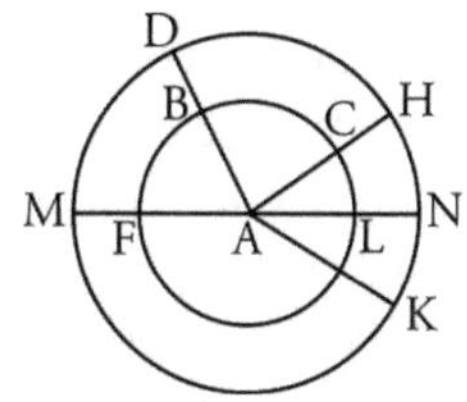

Fig. 23

From all which can be established, that not only the diameter MAN most exactly divides its circle, and the circumference of it into two equal parts, but also that this may be assumed universally for any diameter of this circle. This is what was to be demonstrated.

Scholium.

We read in Clavius that this truth was demonstrated by Thales of Miletus, but perhaps not to the exhaustion of every objection.

Lemma V.

Inter angulos rectilineos omnes anguli recti sunt invicem exactissime aequales, sine ullo defectu etiam infinite parvo.

Demonstratur. Angulum inter rectilineos rectum definit Euclides: *qui est aequalis suo deinceps*. Non hunc postulat ipse sibi concedi, sed problematice demonstrat in sua Prop. XI. Libri primi. Ibi enim ex dato in recta BC quolibet puncto A (Fig. 40) docet excitare perpendicularem AD, ad quam anguli DAB, DAC sint invicem aequales. Porro illos duos angulos esse invicem exactissime aequales, sine ullo defectu etiam infinite parvo, constare
[85] potest ex Corollario post duo priora praemissa Lemmata si nempe ipsae AB, AC designatae sint exactissime aequales.

Sed aliqua oriri potest dubitatio, si duo alii ad quandam alteram FM recti anguli LHF, LHM (Fig. 41) conferantur cum praedictis rectis angulis DAB, DAC. Itaque HL aequalis sit ipsi AD, ac rursum posterior integra Figura ita intelligatur superponi priori, ut punctum H cadat super punctum A, & punctum L super punctum D. Jam sic progredior. Et primo quidem (ex praecedente Lemmate) ipsa FHM non praecise continget alteram BC in eo puncto A. Ergo vel adamussim procurret super illa BC, vel eandem ita secabit, ut unum ejus punctum extremum v. g. F cadat supra, & alterum M deorsum. Si primum: jam clare habemus exactissimam inter omnes rectilineos angulos rectos aequalitatem intentam. At non secundum; quia sic angulus LHF, hoc est DAF, minor foret angulo DAB, ejusque supposito exactissime aequali DAC, & sic multo minor angulo DAM, sive LHM; contra hypothesin. Deinde vero nihil suffragatur, quod angulus DAF infinite parum deficiat ab angulo DAB, sive ejus exactissime aequali DAC, qui rursum solum infinite parum superetur ab angulo DAM. Nam semper angulus DAF, sive LHF, non erit exactissime aequalis angulo DAM, sive LHM, contra hypothesin.

Itaque constat omnes rectilineos angulos rectos esse invicem exactissime aequales, sine ullo defectu etiam infinite parvo. Quod &c.

Corollarium.

Inde autem fit, ut quae ex uno dato cujusvis rectae lineae puncto perpendiculariter in aliquo plano ad eandem educitur, ipsa sit in eo tali plano unica exactissime linea recta, nec potens diffindi in duas.

[86] *Post quinque praemissa Lemmata, eorumque Corollaria, progredi jam debeo ad demonstrandum principale assumptum contra hypothesin anguli acuti.*

Lemma 5.

Among rectilinear angles, all right angles are exactly equal to one another, without any deviation even infinitely small.

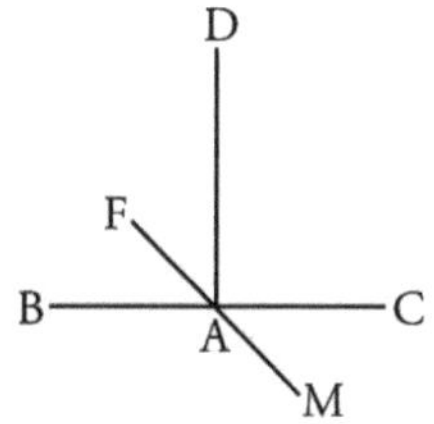

Fig. 40

Proof. Euclid defines a rectilinear angle as right: *which is equal to its adjacent.*[1] This he does not postulate as conceded to him, but demonstrates through a problem in *Elements* I, 11. For there he teaches from any given point A (Fig. 40) in the straight BC to erect a perpendicular AD at which the angles DAB, DAC are equal to each other. Moreover that those two angles are precisely equal to each other, without any difference even infinitely small, follows from the Corollary to the preceding Lemmata 1 and 2, if AB, AC are taken exactly equal.

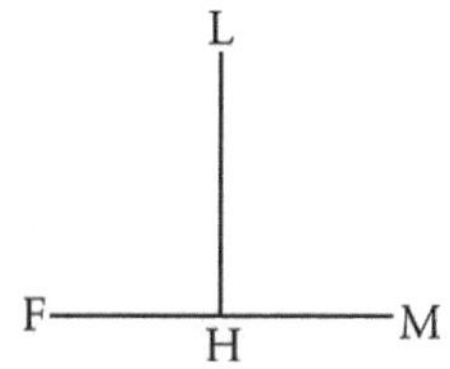

Fig. 41

But some doubt may arise, if two other right angles LHF, LHM (Fig. 41) at any other straight FM are compared with the aforesaid right angles DAB, DAC. Therefore let HL be equal to AD, and then the whole latter figure is thought to be superposed upon the former so, that point H falls upon point A, and point L upon point D. Now I proceed thus. And first indeed (Lemma 3) this FHM does not exactly touch the other BC in the point A. Therefore either it runs forward precisely upon BC, or will cut it so that one of its end points for example F falls above, and the other M below. If the first case holds: now clearly we have the exact equality asserted between all rectilinear right angles. But the second does not hold: because thus the angle LHF, here it is DAF, will be less than the angle DAB, and its supposed exact equal DAC, and thus much less than the angle DAM, or LHM; contrary to the hypothesis. Then it helps nothing that angle DAF differ infinitely little from angle DAB, or its exact equal DAC, which again would be exceeded only infinitesimally by the angle DAM. For always angle DAF, or LHF, will not be exactly equal to angle DAM, or LHM, against the hypothesis.

Therefore is established that all rectilinear right angles are exactly equal to one another, without any difference even infinitely small. This is what was to be demonstrated.

Corollary.

Thence follows that the straight line erected from a given point of any straight perpendicularly to it in a plane, is, in such plane, wholly unique; nor can it split in two.

After the five premised Lemmata, and their Corollaries, I must now go on 1to proof of the principal objection against the hypothesis of acute angle.

Ubi statuere possum, tanquam per se notum, non minus repugnare, quod duae rectae lineae (sive ad finitam, sive ad infinitam earundem productionem) in unam tandem, eandemque rectam lineam coeant; quam quod una eademque linea recta (sive ad finitam, sive ad infinitam ejusdem continuationem) in duas rectas lineas diffindatur, contra praecedens Lemma secundum, ejusque Corollarium. Quoniam ergo naturae lineae rectae (ex praecedente Corol. proximi Lemmatis) oppositum itidem est, quod duae rectae lineae ad unum, idemque punctum cujusdam tertiae rectae, perpendiculares ipsi sint in eodem communi plano; agnoscere oportet tanquam absolute falsam, quia repugnantem naturae praedictae, hypothesin anguli acuti, juxta quam duae illae AX, BX (Fig. 33) in uno eodemque communi puncto X perpendiculares esse deberent cuidam tertiae rectae, quae in eodem cum ipsis plano existeret. Hoc autem erat principale demonstrandum.

Scholion.

Atque hic subsistere tutus possem. Sed nullum non movere lapidem volo, ut inimicam anguli acuti hypothesim, a primis usque radicibus revulsam, sibi ipsi repugnantem ostendam. Iste autem erit consequentium hujus Libri Theorematum unicus scopus.

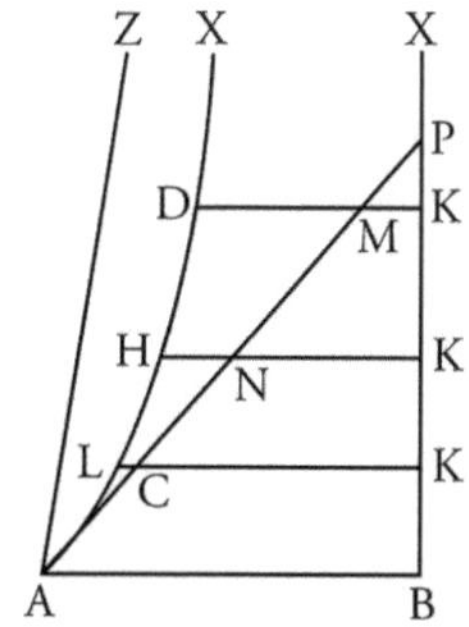

Fig. 33

Here I may set up, as known *per se*, it is not less contradictory, that two straight lines (whether at a finite, or at an infinite prolongation of them) at length run together into one and the same straight line, than that one and the same straight line (whether at a finite, or at an infinite prolongation of it) splits into two straight lines, against the preceding Lemma 2, and its Corollary. Since therefore it is in like manner opposed to the nature of the straight line (from the preceding Corollary to Lemma 5) that two straight lines at one and the same point of a third straight be perpendicular to this in the same common plane; it is proper to recognize as absolutely false, because repugnant to the aforesaid nature, the hypothesis of acute angle, according to which those two AX, BX (Fig. 33) in one and the same common point X must be perpendicular to a third straight, which is in the same plane with them. This is what was to be principally demonstrated.

Scholium.

And here I might safely stop. But I do not wish to leave any stone unturned, that I may show the hostile hypothesis of acute angle, torn out by the very roots, repugnant to itself. However this will be the single aim of the subsequent Theorems of this Book.

Euclidis ab omni naevo vindicati Libri Primi

Pars Altera

In qua idem Pronunciatum Euclidaeum contra hypothesin anguli acuti redargutive demonstratur.

[87] **Propositio XXXIV.**

In qua expenditur curva quaedam enascens ex hypothesi anguli acuti.

Recta CD jungat aequalia perpendicula AC, BD cuidam rectae AB insistentia. Tum divisis bifariam in punctis M, & H (Fig. 42) ipsis AB, CD, jungatur MH (ex 2. hujus) utrique perpendicularis. Rursum in hac hypothesi supponuntur acuti anguli ad junctam CD. Quare in quadrilatero AMHC erit MH (ex Cor. I post 3. hujus) minor ipsa AC. Hinc autem; si in MH protracta sumas MK aequalem ipsi AC; puncta C, K, D spectabunt ad curvam hic expensam. Deinde anguli ad junctam CK erunt & ipsi (ex 7. hujus) acuti. Igitur juncta LX, quae bifariam, atque ideo (ex 2. hujus) ad angulos rectos dividat ipsas AM, CK, erit similiter (ex Cor. I. post 3. hujus) minor eadem AC. Quapropter; si in LX protracta sumas LF aequalem ipsi AC, aut MK; etiam punctum F spectabit ad eam curvam. Praeterea jungens CF, & FK invenies similiter duo alia puncta ad eandem curvam spectantia. Atque ita semper. Quod autem dico pro inveniendis punctis inter puncta C, & K, idem etiam uniformiter valet
[88] pro inveniendis punctis inter puncta K, & D, scilicet curva CKD, enascens ex hypothesi anguli acuti, est linea jungens extremitates omnium aequalium perpendiculorum super eadem basi versus eandem partem erectorum, quae utique venire possunt sub nomine rectarum ordinatim applicatarum; est, inquam, linea ejusmodi, quae propter ipsam, ex qua nascitur, hypothesim anguli acuti, semper est cava versus partes contrapositae basis AB. Quod quidem hoc loco declarandum, ac demonstrandum a nobis erat.

Euclid Vindicated from every Blemish
Book One

Second Part

In which the same Euclidean Assertion is demonstrated by redargution against the hypothesis of acute angle.

Proposition 34.

In which is investigated a certain curve arising from the hypothesis of acute angle.

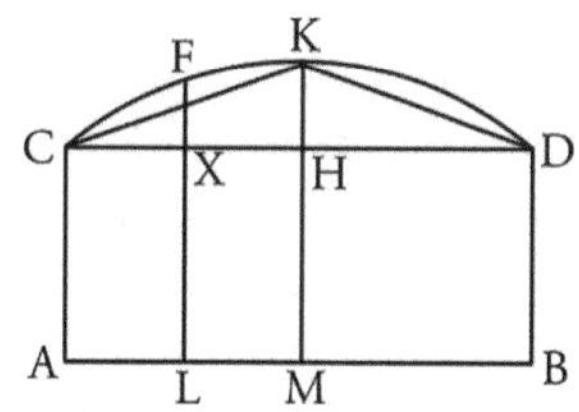

Fig. 42

Let the straight CD join equal perpendiculars AC, BD standing upon a certain straight AB. Then AB, CD being bisected in the points M and H (Fig. 42), MH is joined perpendicular (by Proposition 2) to each. Again in this hypothesis the angles at the join CD are supposed acute. Therefore in the quadrilateral AMHC (by Corollary 1 to Proposition 3) MH will be less than AC. Hence now, if in MH produced MK be taken equal to AC, the points C, K, D pertain to the curve here investigated. Then, the angles at the join CK will be themselves acute (by Proposition 7). Therefore the join LX, which bisects, and therefore (by Proposition 2), is at right angles to AM, CK, will be likewise (by Corollary 1 to Proposition 3) less than AC. Wherefore, if in LX produced we assume LF equal to AC or MK, the point F also will pertain to this curve. Further, joining CF, and FK we find likewise two other points pertaining to the same curve. And so on forever. But what I say for finding points between the points C and K, the same also holds good uniformly for finding points between the points K and D; of course the curve CKD, arising from the hypothesis of acute angle, is the line joining the extremities of all equal perpendiculars erected upon the same base toward the same part, which assuredly can come under the name ordinates; it is, I may say, a line of such sort, that on account of the hypothesis of acute angle, from which it arises, it always is concave toward the parts of the opposite base AB. This is what was to be demonstrated and clarified.

Propositio XXXV.

Si ex quolibet puncto L basis AB ordinatim applicetur ad eam curvam CKD recta LF: Dico rectam NFX perpendicularem ipsi LF cadere totam ex utraque parte debere versus partes convexas ejusdem curvae, atque ideo eam fore ejusdem curvae tangentem.

Demonstratur. Si enim fieri potest, cadat quoddam punctum X (Fig. 43) ipsius NFX intra cavitatem ejusdem curvae. Demittatur ex puncto X ad basim AB perpendicularis XP, quae protracta per X occurrat curvae in quodam puncto R. Jam sic. In quadrilatero LFXP non erit angulus in puncto X aut rectus, aut obtusus: Caeterum (ex 5. & 6. hujus) destrueretur praesens hypothesis anguli acuti. Ergo praedictus angulus erit acutus. Quare erit PX (ex Cor. I. post 3. hujus) & sic multo magis PR major ipsa LF. Hoc autem absurdum est (ex praecedente) contra naturam istius curvae. Itaque illa NF protracta cadere tota debet versus partes convexas, atque ideo ipsa erit ejusdem curvae tangens. Quod erat demonstrandum.

[89] **Propositio XXXVI.**

Si recta quaepiam XF (Fig. 44) *acutum angulum efficiat cum quavis ordinata LF, non cadet punctum* X *extra cavitatem curvae, nisi prius ipsa XF in aliquo puncto O curvam secuerit.*

Demonstratur. Constat sumi posse in ipsa XF punctum quoddam X adeo vicinum ipsi puncto F, ut juncta LX prius curvam secet in aliquo puncto S: caeterum ipsa XF vel non cadet tota extra cavitatem curvae, & sic habemus intentum; vel adeo non efficiet cum FL angulum acutum, ut magis censenda jam sit in unicam rectam cum altera LF coire. Itaque ex puncto S demittatur ad basim AB perpendicularis SP. Erit haec (ex 34. hujus) aequalis ipsi LF. Est autem SP (ex 18. primi) minor ipsa LS. Ergo etiam LF minor est eadem LS, ac propterea multo minor ipsa LX. Hinc in triangulo LXF acutus erit angulus in puncto X, quia minor (ex 18. primi) angulo LFX supposito acuto. Jam demittatur ad FX perpendicularis LT. Cadet haec (propter 17. primi) ad partes utriusque anguli acuti. Quare punctum T jacebit inter puncta X, & F. Deinde ex puncto T demittatur ad basim AB perpendicularis TQ. Erit LF (propter angulum rectum in T) major ipsa LT, & haec (propter angulum rectum in Q) major altera QT. Igitur LF multo major erit ipsa QT. Hinc autem; si in QT protracta sumatur QK aequalis ipsi LF; punctum K (ex 34. hujus) ad praesentem curvam spectabit, cadetque idcirco punctum T intra cavitatem ejusdem curvae. Non ergo recta FT, quae secat duas rectas QK, & LT in T, promoveri potest ad secandam protractam LS in puncto X, constituto extra cavitatem praesentis curvae, nisi prius ipsa protracta FT secet in aliquo puncto O por-
[90] tionem ejusdem curvae inter puncta S, & K constitutam. Hoc autem erat demonstrandum.

Proposition 35.

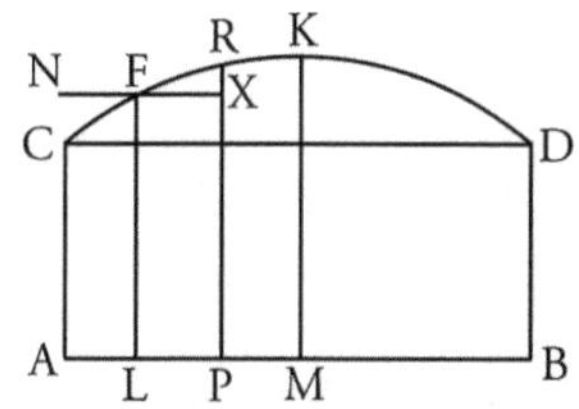

Fig. 43

If from any point L of the base AB the ordinate LF is drawn to this curve CKD: I say the straight NFX perpendicular to LF must on both sides fall wholly toward the convex parts of this curve, and therefore it will be tangent to this curve.

Proof. For if possible, let a certain point X (Fig. 43) of NFX fall within the cavity of this curve. Let fall from the point X to the base AB the perpendicular XP, which prolonged through X meets the curve in a certain point R. Now thus. In the quadrilateral LFXP the angle at the point X will be neither right nor obtuse: else (Proposition 5 and Proposition 6) would be destroyed the present hypothesis of acute angle. Therefore the aforesaid angle will be acute. Wherefore (Corollary 1 to Proposition 3) PX and so much more PR will be greater than LF. But this is absurd (Proposition 35) against the nature of this curve. So NF produced must fall wholly toward the convex parts, and so it will be tangent to this curve. This is what was to be demonstrated.

Proposition 36.

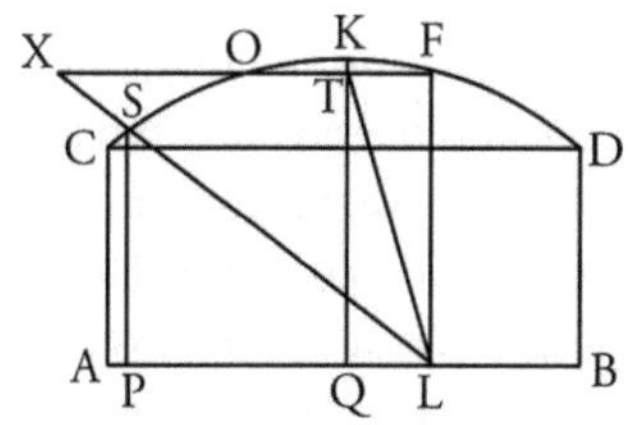

Fig. 44

If any straight XF (Fig. 44) *makes an acute angle with an ordinate LF whatsoever, the point X does not fall without the cavity of the curve, unless previously XF has cut the curve in some point O.*

Proof.[1] It is sure that some point X may be assumed in XF so near to the point F, that the join LX previously cuts the curve in some point S: otherwise XF either does not fall wholly without the cavity of the curve, and so we have our assertion; or so far is it from making with FL an acute angle, that now rather it must be supposed to combine with LF in one straight.[2] Accordingly from the point S let fall to the base AB the perpendicular SP. This will be (Proposition 34) equal to LF. But SP is (from *Elements* I, 18)[3] less than LS. Therefore also LF is less than LS, and consequently much less than LX. Hence in triangle LXF the angle at point X will be acute, because less (*Elements* I, 18) than the angle LFX supposed acute. Now let fall to FX the perpendicular LT. This will fall (because of *Elements* I, 17) toward the parts of each acute angle. Wherefore point T will lie between points X, and F. Then from the point T let fall to the base AB the perpendicular TQ. LF will be (because of the right angle at T) greater than LT, and this (because of the right angle at Q) will be greater than QT. Therefore LF will be much greater than QT. But hence; if in QT produced QK is taken equal to LF; the point K (Proposition 34) will pertain to the present curve, and therefore point T falls within the cavity of this curve. Therefore the straight FT, which cuts the two straights QK, and LT, in T, cannot be extended to cut LS prolonged in the point X, situated without the cavity of the present curve, unless previously the prolonged FT cuts in same point O the portion of this curve situated between the points S, and K. This is what was to be demonstrated.

Corollarium.

Atque hinc manifeste liquet, inter tangentem hujus curvae, & ipsam curvam locari non posse quandam rectam, quae tota ad hanc, vel illam tangentis partem extra curvae cavitatem cadat; quandoquidem recta sic locata efficere debet (ex praecedente) angulum acutum cum demissa ex puncto contactus ad contrapositam basim perpendiculari.

Propositio XXXVII.

Curva CKD, ex hypothesi anguli acuti enascens, aequalis esse deberet contrapositae basi AB.

Ante demonstrationem praemitto sequens axioma.

Si duae lineae bifariam dividantur, tum earum medietates, ac rursum quadrantes bifariam, atque ita in infinitum uniformiter procedatur; certo argumento erit, duas istas lineas esse inter se aequales, quoties in ista uniformi in infinitum divisione comperiatur, seu demonstretur, deveniri tandem debere ad duas illarum sibi invicem respondentes partes, quas constet esse inter se aequales.

Jam demonstratur propositum. Intelligantur erecta ex basi AB ad eam curvam CKD (Fig. 45) quotvis perpendicula NF, LF, PF, MK, TF, VF, IF; sintque aequales in ipsa basi AB portiones AN, NL, LP, PM, MT, TV, VI, IB.

Constat primo angulum ipsius AC cum ea curva aequalem fore singulis hinc inde ad puncta F, sive ad punctum K, aut punctum D, praedictarum perpendicularium angulis cum
[91] eadem curva. Si enim mistum quadrilaterum ANFC superponi intelligatur misto quadrilatero NLFF, constituta basi AN super aequali basi NL, cadet AC super NF, & NF super LF, propter aequales angulos rectos ad puncta A, N, L. Deinde propter aequalitatem rectarum (ex 34. hujus) AC, NF, LF, cadet punctum C super punctum F ipsius NF, & hoc super alterum punctum F ipsius LF. Praeterea curva CF congruet adamussim ipsi curvae FF: si enim una illarum, ut CF introrsum, aut extrorsum cadat; sumpto quolibet puncto Q inter puncta N, & L, ductaque perpendiculari secante unam curvam in X, & alteram in S, aequales forent (ex nota hujus curvae natura) ipsae QX, QS, quod est absurdum. Idem valebit, si in dicta superpositione maneat in suo situ recta NF, & recta AC cadat super LF. Rursum idem valebit, si idem quadrilaterum mistum ANFC utrovis modo superponi intelligatur cuivis reliquorum quadrilaterorum usque ad ipsum inclusive postremum quadrilaterum BDFI. Itaque angulus ipsius AC cum ea curva aequalis est singulis hinc inde ad puncta F, sive ad punctum K, aut punctum D, praedictarum perpendicularium angulis cum eadem curva.

Constat hinc secundo aequales adamussim inter se esse portiones ipsius curvae ab istis perpendicularibus hinc inde abscissas.

Corollary.

And hence it is clear that between the tangent of this curve, and the curve itself cannot be placed any straight, which, on one or the other side of the tangent wholly falls without the cavity of the curve; since a straight so located must (Proposition 35) make an acute angle with the perpendicular let fall from the point of contact to the opposite base.

Proposition 37.

The curve CKD, arising from the hypothesis of acute angle, must be equal to the opposite base AB.

Before the demonstration I premise the following Axiom.

If two lines be bisected, then their halves, and again their quarters bisected, and so the process be continued uniformly *in infinitum;* it will be safe to argue, those two lines are equal to each other, as often as is ascertained, or demonstrated in that uniform division *in infinitum,* that at length must be attained two of their mutually corresponding parts, of which it is certain they are equal to each other.[1]

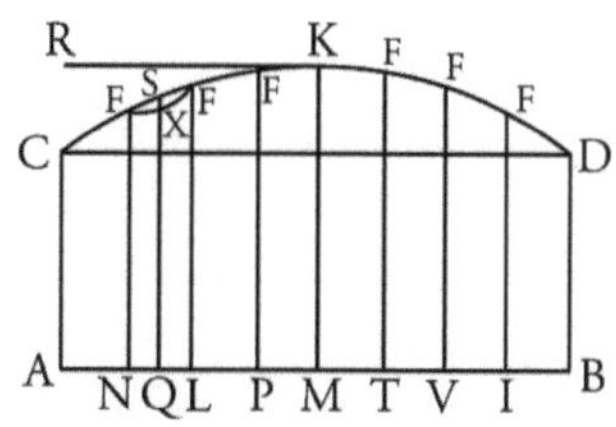

Fig. 45

Proof of the Proposition. Suppose erected from the base AB to the curve CKD (Fig. 45) any number of perpendiculars NF, LF, PF, MK, TF, VF, IF; and on the base AB take as equal the portions AN, NL, LP, PM, MT, TV, VI, IB.

First is certain the angle of AC with the curve will be equal to each of the angles of the aforesaid perpendiculars with the curve on either side at the points F, or at the point K, or at the point D. For if the mixed quadrilateral ANFC is supposed to be superposed upon the mixed quadrilateral NLFF, the base AN lying upon the equal base NL, AC falls upon NF, and NF upon LF, because of the equal right angles at the points A, N, L. Then because of the equality (Proposition 34) of the straights AC, NF, LF, the point C falls upon point F of NF, and this upon the other point F of LF. Moreover the curve CF exactly fits the curve FF; for if one of these, as CF, fell within or without; any point Q being assumed between points N, and L, and the perpendicular being drawn cutting one curve in X, and the other in S, QX, QS would be equal (from the known nature of this curve), which is absurd. The same will hold,[2] if in the said superposition the straight NF remains in its place, and the straight AC falls upon LF. Again the same will hold, if the same mixed quadrilateral ANFC in either mode is supposed to be superposed to any of the remaining quadrilaterals even to the last quadrilateral BDFI inclusive. Therefore the angle of AC with the curve is equal to either of the angles with this curve of the aforesaid perpendiculars on either side at the points F, or at the point K, or point D.

Hence follows secondly that the portions of the curve cut off on each side by these perpendiculars are exactly equal to one another.

Si ergo basis AB divisa sit bifariam in M, & medietas AM bifariam in L; tum quadrans LM bifariam in P; atque ita in infinitum, procedendo semper versus partes puncti M; constabit tertio, etiam curvam CKD bifariam dividi in K a perpendiculari MK, medietatem CK bifariam itidem dividi in F a perpendiculari LF, quadrantem FK bifariam in F a perpendiculari PF; atque ita in infinitum, procedendo semper uniformiter versus partes ipsius puncti K.

[92] Quoniam vero in ista basis AB in infinitum divisione considerare possumus rem devenisse ad portionem ipsius AB infinite parvam, quae nempe exhibeatur per latitudinem infinite parvam perpendicularis MK, constabit quarto (ex praemisso axiomate) aequalitas intenta totius basis AB cum tota curva CKD, dum alias ostendam portionem infinite parvam abscissam ex basi AB a perpendiculari MK aequalem esse adamussim portioni infinite parvae, quam eadem perpendicularis abscindit ex curva CKD. Et hoc quidem postremum sic demonstro.

Nam RK perpendicularis ipsi KM tanget (ex 35. hujus) curvam in K, atque ita eandem tanget in K, ut inter ipsam tangentem (ex Cor. post 36, hujus) & curvam, ex neutra parte locari possit recta, quae ipsam curvam non secet. Igitur infinitesima K, spectans ad curvam, aequalis omnino erit infinitesimae K spectanti ad tangentem. Constat autem infinitesimam K spectantem ad tangentem, nec majorem, nec minorem, sed omnino aequalem esse infinitesimae M spectanti ad basim AB; quia nempe recta illa MK intelligi potest descripta ex fluxu semper ex aequo ejusdem puncti M usque ad eam summitatem K.

Quare (juxta praemissum axioma) curva CKD, ex hypothesi anguli acuti enascens, aequalis esse deberet contrapositae basi AB. Quod erat demonstrandum.

Scholion I.

Sed forte minus evidens cuipiam videbitur enunciata exactissima aequalitas inter illas infinitesimas M, & K. Quare ad avertendum hunc scrupulum sic rursum procedo. Cuidam rectae AB insistant ad rectos angulos in eodem plano (Fig. 48) duae rectae aequales AC, BD. Rursum in eodem plano intelligatur existere circulus BLDH, cujus diameter BD; sitque
[93] semicircumferentia BLD aequalis praedictae AB. Praeterea idem circulus ita in eo plano revolvi concipiatur super ea recta AB, ut motu semper continuo, & aequabili perficiat, seu describat suae ipsius semicircumferentiae punctis praedictam BA; quousque nempe punctum D, ad illam semicircumferentiam spectans, perveniat ad congruendum ipsi puncto A, ita ut propterea punctum B, ejusdem semicircumferentiae alterum extremum punctum, deveniat ad congruendum illi puncto C.

If therefore the base AB be bisected in M, and the half AM bisected in L; then the quarter LM bisected in P; and so *in infinitum*, proceeding always toward the parts of the point M; it will follow thirdly, also the curve CKD is bisected in K by the perpendicular MK, the half CK in like manner bisected in F by the perpendicular LF, the quarter FK bisected in F by the perpendicular PF; and so *in infinitum*, proceeding always uniformly toward the parts of the point K.

But since in this division of the base AB *in infinitum* we may consider the thing to have arrived at a portion of AB infinitely small, which obviously may be exhibited by the infinitely small breadth[3] of the perpendicular MK, fourthly (from the premised Axiom) will follow the asserted equality of the whole base AB with the whole curve CKD, if only I now can show the infinitely small portion cut off from the base AB by the perpendicular MK to be exactly equal to the infinitely small portion, which the same perpendicular cuts off from the curve CKD.

And this last I thus demonstrate. For RK perpendicular to KM touches (Proposition 35) the curve at K, and touches this in K so, that between the tangent (Corollary to Proposition 36) and the curve from neither side can be placed a straight, which does not cut the curve. Therefore infinitesimal K, regarding the curve, will be wholly equal to infinitesimal K regarding the tangent. But it is certain the infinitesimal K regarding the tangent is neither greater nor less than, but exactly equal[4] to the infinitesimal M regarding the base AB; because obviously the straight MK may be supposed described by the flow always uniform of the point M up to the summit K.

Wherefore (according to the premised Axiom) the curve CKD, born of the hypothesis of acute angle should be equal to the opposite base AB. This is what was to be demonstrated.

Scholium 1.

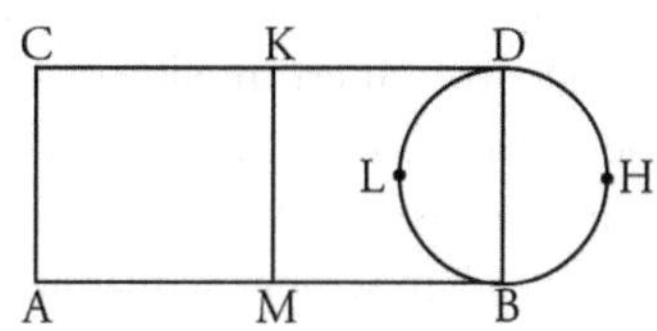

Fig. 48

But perchance to some one will seem by no means evident the enunciated exact equality between the infinitesimals M, and K. Wherefore to remove this scruple I again proceed thus. To a certain straight AB let two equal straights AC, BD (Fig. 48) stand at right angles in the same plane. Again in the same plane suppose there is a circle BLDH, whose diameter is BD; and let the semicircumference BLD be equal to the aforesaid AB. Further let the same circle be conceived so to be revolved in that plane upon the straight AB that with motion always continuous and uniform it achieves or describes with the points of its semicircumference the aforesaid BA, until indeed point D pertaining to that semicircumference comes to congruence with point A, so that moreover point B, the other extreme point of the same semicircumference comes to congruence with point C.

His stantibus; si in semicircumferentia BLD designetur quodvis punctum L, cui in descripta recta linea BA correspondeat punctum M, ex quo in eo tali plano educatur perpendicularis MK, aequalis ipsi BD: Dico illud punctum K fore ipsum punctum H diametraliter oppositum illi puncto L. Nam ibi in puncto M, sive L recta AB continget praedictum circulum. Igitur MK eidem AB perpendicularis transibit (ex 19. tertii, quae utique independens est ab Axiomate controverso) per centrum ejusdem circuli. Quare; ubi punctum L in ea tali circuli BLDH revolutione perveniat ad congruendum cum puncto M ipsius AB, etiam punctum H, diametraliter oppositum praedicto puncto L, incidet in punctum K illius MK.

Porro constat idem similiter valere de reliquis punctis semicircumferentiae BLD, & horum diametraliter correlativis in altera semicircumferentia BHD. Quare linea, eo tali modo successive descripta a punctis semicircumferentiae BHD, erit illa eadem jam expensa DKC, quae nempe suis omnibus punctis aequidistet ab illa recta BA; sitque idcirco (juxta hypothesin anguli acuti) semper cava versus partes ejusdem AB.

Inde autem fit, ut punctum M in ea BA censendum sit exactissime aequale puncto K in
[94] altera DKC, propter omnimodam istorum aequalitatem cum punctis L, & H diametraliter oppositis in eo circulo BLDH. Quare; cum idem valeat de omnibus punctis descriptae rectae BA, si conferantur cum aliis uniformiter contrapositis in praedicta supposita curva DKC; consequens plane est, ut ipsa talis curva, ex hypothesi anguli acuti enascens, censenda sit aequalis contrapositae basi AB. Atque id est, quod nova hac methodo iterum demonstrandum susceperam.

Scholion II.

Rursum vero: quoniam recta BA intelligitur successive descripta a punctis semicircumferentiae BLD motu illo semper aequabili, & continuo; cui nempe descriptioni correspondet descriptio illius lineae DKC a punctis diametraliter correlativis alterius semicircumferentiae BHD: obvium est intelligere, quod ipsa recta BA motu illo semper aequabili, & continuo describatur ab eo unico puncto B, quod nempe (veluti replicatum) intelligatur cum ipsa tali semicircumferentia semper excurrere super ea BA; dum interim eodem ipso tempore, motu eodem semper aequabili, & continuo, describitur illa altera DKC ab altero diametraliter correlativo unico puncto D, quod ipsum rursum (veluti replicatum) intelligatur cum sua altera semicircumferentia BHD semper excurrere super praedicta DKC. Tunc autem facilius intelligitur intenta aequalitas inteream DKC, & eidem contrapositam rectam BA; quippe quae duae aequali ipso tempore, & aequali motu intelliguntur descriptae a duobus exactissime inter se aequalibus punctis, seu mavis infinitesimis. Ubi constat hanc ipsam exactissimam praedictorum punctorum aequalitatem non esse mihi in ista nova contemplatione necessariam.

This abiding; if in the semicircumference BLD is designated any point L, to which in the described straight line BA corresponds point M, from which in that plane is erected the perpendicular MK, equal to BD: I say that point K will be the point H diametrically opposite the point L. For there in the point M, or L the straight AB touches the aforesaid circle. Therefore MK perpendicular to AB will go (from *Elements* III, 19, which is assuredly independent of the controverted Axiom)[5] through the center of the same circle. Wherefore; where point L in that revolution of the circle BLDH comes to congruence with the point M of AB, also point H, diametrically opposite the aforesaid point L, falls upon point K of MK.

Furthermore it is certain the same holds in like manner of the remaining points of the semicircumference BLD, and of those diametrically correlative in the other semicircumference BHD. Wherefore the line, in that way successively described[6] by the points of the semicircumference BHD, will be the already considered DKC, which in all its points is equidistant from the straight BA; and which therefore (in accordance with the hypothesis of acute angle) is always concave toward the side of AB.

But thence follows, that the point M in BA may be considered exactly equal to point K in DKC, because of their equality in every way with the points L, and H diametrically opposite in the circle BLDH. Wherefore; since the same holds of all points of the described straight BA, if they be compared with the other uniformly opposite in the aforesaid assumed curve DKC; the consequence evidently is, that this curve, born of the hypothesis of acute angle, is to be thought equal to the opposite base AB. And that is what I had undertaken again to demonstrate by this new method.

Scholium 2.

But again: since the straight BA is discerned as successively described by the points of the semicircumference BLD by that motion always uniform and continuous; to which description corresponds the description of that line DKC by the diametrically correlative points of the other semicircumference BHD: it is easy to understand, that this straight BA by that motion always uniform, and continuous is described by the one point B, which of course (as if unrolled) is thought always to run out with that semicircumference upon BA; whilst meanwhile in exactly the same time, by the same motion always uniform, and continuous, is described that other DKC by the other one diametrically correlative point D, which again itself (as if unrolled) is thought with its other semicircumference BHD always to run out upon the aforesaid DKC. But then is more easily understood the asserted equality between DKC, and the straight BA opposite it; since the two are imagined to be described in equal time, and equal motion by two exactly equal points, or, if you prefer, infinitesimals. Where it holds that this exact equality of the aforesaid points is not necessary for me in that new consideration.[7]

Propositio XXXVIII.

Hypothesis anguli acuti est absolute falsa, quia se ipsam destruit.

Demonstratur. Nam supra ex ipsa hypothesi anguli acuti evidenter elicuimus, curvam CKD (Fig. 46) ex ea prognatam aequalem esse debere contrapositae basi AB. Nunc autem contradictorium ex eadem hypothesi elicimus, quod curva CKD nequeat esse aequalis illi basi, cum certe sit eadem major. Quod enim curva CKD major sit recta CD ejus extremitates jungente, notio est omnibus communis, quam etiam demonstrare possumus ex vigesima primi, quod duo trianguli latera reliquo semper sunt majora; junctis nimirum CK, & KD; ac rursum junctis similiter apicibus, primo quidem duorum, tum quatuor, & sic in infinitum, duplicato numero enascentium segmentorum, quousque intelligatur hoc pacto absumi, seu desinere in ipsas infinite parvas seu chordas, seu tangentes, tota curva CKD. Sed hic procedere possumus ex sola communi notione. Quod autem juncta CD major sit basi AB, demonstratum a nobis est in 3. hujus ex ipsis visceribus hypothesis anguli acuti. Igitur curva CKD, ex hypothesi anguli acuti enascens, est certe major basi AB, quia est major, saltem ex communi notione, recta CD, quae ex hac ipsa hypothesi anguli acuti demonstratur major basi AB. Non igitur potest simul consistere, quod curva ista CKD aequalis sit basi AB. Itaque constat hypothesim anguli acuti esse absolute falsam, quia se ipsam destruit.

Scholion.

Observare tamen debeo, quod etiam ex hypothesi anguli obtusi enascitur curva quaedam
[96] CKD, sed convexa versus partes basis AB. Nam MH (Fig. 47) bifariam dividens ipsas AB, CD erit (ex 2. hujus) eisdem perpendicularis; & major (ex Cor. I. post 3. hujus) ipsis AC, BD, in hypothesi anguli obtusi. Quare ipsius MH portio quaedam MK aequalis erit ipsi AC, aut BD. Tum junctis CK, & KD, divisisque bifariam in punctis X, P, Q, N rectis CK, AM, MB, KD, constat (ex eadem 2. hujus) junctas PX, QM, perpendiculares fore ipsis rectis divisis. At rursum erunt illae (ex eodem Cor. I. post 3. hujus) majores ipsis AC, MK, BD. Hinc; assumptis earundem portionibus PL, QS, quae praedictis aequales sint; habebimus curvam, ex hypothesi anguli obtusi enascentem, quae transibit per puncta C, L, K, S, D. Atque ita semper, si decernere velimus reliqua puncta ejusdem curvae. Inde autem constat eam fore convexam versus partes basis AB. Jam fateor demonstrari uniformi plane methodo potuisse aequalitatem hujus curvae cum ipsa basi AB. At quis fructus? Nullus sane. Quemadmodum enim curva ista CKD censeri debet, ex communi saltem notione, major recta CD; ita etiam (in 3. hujus) basis AB demonstratur major eadem CD, dum stet hypothesis anguli obtusi. Nullum ergo ex hac parte absurdum, si basis AB aequalis sit praedictae curvae. Aliter vero rem procedere in hypothesi anguli acuti, constat ex dictis supra.

Proposition 38.

The hypothesis of acute angle is absolutely false, because it destroys itself.

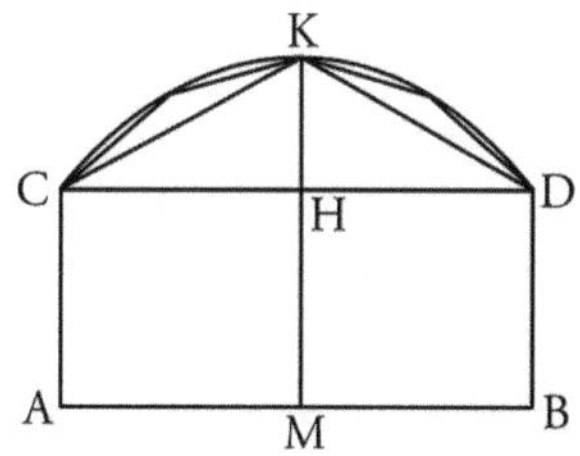

Fig. 46

Proof. Assuredly we have above clearly deduced from the hypothesis of acute angle, that the curve CKD (Fig. 46) born of it must be equal to the opposite base AB. But now we deduce the contradictory from the same hypothesis, that the curve CKD cannot be equal to that base, since surely it is greater than it. For that the curve CKD is greater than the straight CD joining its extremities, the notion is common to all, which also we may demonstrate from *Elements* I, 20, that two sides of a triangle are always greater than the third; join CK, and KD; and again join likewise the apices, first of two, then of four, and so on *in infinitum,* the number of the produced segments doubling, until the whole curve CKD is understood in this way to be exhausted, or to end in those infinitely small chords, or tangents.[1] However here we may proceed from the common notion alone.[2] But that the join CD is greater than the base AB, has been demonstrated by us in Proposition 3 from the very viscera of the hypothesis of acute angle. Therefore the curve CKD, born of the hypothesis of acute angle, is certainly greater than the base AB, because it is greater, anyhow from the common notion, than the straight CD, which from the hypothesis of acute angle is demonstrated greater than the base AB. Therefore cannot at the same time stand, that the curve CKD is equal to the base AB. Consequently is established that the hypothesis of acute angle is absolutely false, because it destroys itself.

Scholium.

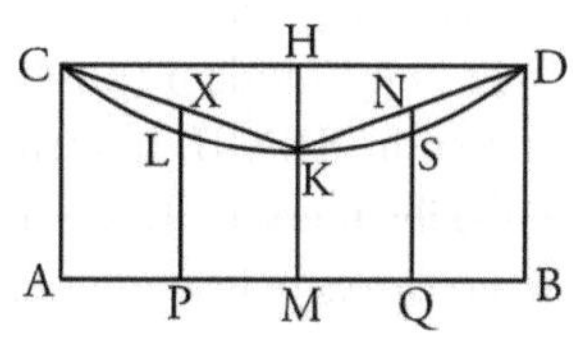

Fig. 47

I should still observe, that also from the hypothesis of obtuse angle is born a certain curve CKD, but convex toward the side of the base AB. For MH (Fig. 47) bisecting AB, CD will be (Proposition 2) perpendicular to them; and greater (Corollary 1 to Proposition 3) than AC, BD, in the hypothesis of obtuse angle. Wherefore a certain portion MK of MH will be equal to AC, or BD. Then CK and KD being joined, and the straights CK, AM, MB, KD bisected in the points X, P, Q, N, it follows (again from Proposition 2) that the joins PX, QN will be perpendicular to the divided straights. But again they will be (from the same Corollary 1 to Proposition 3) greater than AC, MK, BD. Hence; taking of them the portions PL, QS, which are equal to the aforesaid; we shall have a curve, born of the hypothesis of obtuse angle, which will go through the points C, L, K, S, D. And so on always, if we wish to determine remaining points of the same curve. But thence follows it will be convex toward the side of the base AB. Now I grant in just the same way could have been demonstrated the equality of this curve with its base AB. But what good? None at all. For just as the curve CKD must be thought, anyhow from the common notion, greater than the straight CD; so also (in Proposition 3) the base AB is proved greater than CD, when the hypothesis of obtuse angle holds. Therefore from this side is nothing absurd,[3] if the base AB be equal to the aforesaid curve. But that the thing goes otherwise in the hypothesis of acute angle, follows from what is said above.

Ex hoc igitur Scholio, & ex altero post 13. hujus intelligi potest, diversam plane viam iniri debuisse ad refellendam utranque falsam hypothesim, unam anguli obtusi, & alteram anguli acuti.

Praeterea facile itidem est ex istis dignoscere, non nisi rectam lineam CD esse posse, quae omnibus suis punctis aequidistet ab ea supposita recta linea AB.

[97] **Propositio XXXIX.**

Si in duas rectas lineas altera recta incidens, internos ad easdemque partes angulos duobus rectis minores faciat, duae illae rectae lineae in infinitum productae sibi mutuo incident ad as partes, ubi sunt anguli duobus rectis minores.

Et hoc est notum illud Axioma Euclidaeum, quod tandem absolute demonstrandum suscipio. Ad hunc autem finem satis erit recolere nonnullas praecedentium Demonstrationum. Itaque in meis Propositionibus, usque ad VII. hujus inclusive, tres secrevi hypotheses circa rectam jungentem extrema puncta duorum aequalium perpendiculorum, quae uni cuidam rectae, quam basim appello, in eodem plano insistant. Porro circa has hypotheses (quas invicem secerno ex specie angulorum, qui ad eam jungentem fieri censeantur) demonstro unam quamlibet earum, nimirum aut anguli recti, aut anguli obtusi, aut anguli acuti, si vel in uno casu sit vera, semper & in omni casu illam solam esse veram. Tum in XIII. ostendo universalem veritatem Axiomatis controversi, dum consistat alterutra hypothesis aut anguli recti, aut anguli obtusi. Hinc in XIV. declaro absolutam falsitatem hypothesis anguli obtusi, quia se ipsam destruentis, utpote quae praedicti Axiomatis veritatem infert, ex quo contra reliquas duas hypotheses soli hypothesi anguli recti locus relinquitur. Igitur sola restat hypothesis anguli acuti, contra quam diutius pugnandum fuit.

Et hujus quidem (post multa, ne dicam omnia, conditionate expensa) absolutam falsitatem in XXXIII. tandem ostendo, quia repugnantis naturae lineae rectae, circa quam multa ibi intersero necessaria Lemmata. Tandem vero in praecedente Propositione absolute
[98] demonstro sibi ipsi repugnantem hypothesin anguli acuti. Quoniam igitur unica restat hypothesis anguli recti, consequens plane est, ut ex praedicta XIII. hujus stabilitum absolute maneat praenunciatum Euclidaeum Axioma. Quod erat propositum.

Scholion.

Sed juvat expendere hoc loco notabile discrimen inter praemissas duarum hypothesium redargutiones. Nam circa hypothesin anguli obtusi res est meridiana luce clarior; quandoquidem ex ea assumpta ut vera demonstratur absoluta universalis veritas controversi Pronunciati Euclidaei, ex quo postea demonstratur absoluta falsitas ipsius talis hypothesis; prout constat ex XIII. & XIV. hujus. Contra vero non devenio ad probandam falsitatem alterius hypothesis, quae est anguli acuti, nisi prius demonstrando; quod linea, cujus omnia puncta aequidistent a quadam supposita recta linea in eodem cum ipsa plano existente, aequalis sit ipsi tali rectae; quod ipsum tamen non videor demonstrare ex visceribus ipsiusmet hypothesis, prout opus foret ad perfectam redargutionem.

From this Scholium therefore and from Scholium 1 after Proposition 13 may be realized, that a wholly different way was to be taken in refuting each false hypothesis, one of obtuse angle, and the other of acute angle.

Moreover it is easy in like manner to recognize from these, that it can only be a straight line CD, which in all its points is equidistant from the assumed straight line AB.

Proposition 39.

If upon two straight lines another straight striking makes toward the same parts angles less than two right angles, those two straight lines produced in infinitum meet each other toward those parts where are the angles less than two right angles.

This is the famous Euclidean Axiom,[4] which at length I undertake absolutely to demonstrate. For this end however it will be sufficient to recall some of the preceding demonstrations. Therefore in my Propositions, up to Proposition 7 inclusive, I have distinguished three hypotheses about the straight joining the extreme points of two equal perpendiculars, which stand upon a certain straight, that I call base, in the same plane. Furthermore in regard to these hypotheses (which in turn I distinguish from the species of the angles, which are supposed to be made at the join) I demonstrate that any one of them, forsooth either of right angle, or obtuse angle, or acute angle, alone is true always and in every case, if even in one case it be true. Then in Proposition 13 I show the universal truth of the controverted Axiom, when occurs either the hypothesis of right angle, or of obtuse angle. Hence in Proposition 14 I declare the absolute falsity of the hypothesis of obtuse angle, because it destroys itself, inasmuch as it occasions the truth of the aforesaid Axiom, from which against the present two hypotheses place is left for the hypothesis of right angle alone. Therefore remains only the hypothesis of acute angle, against which was longer to be fought.

And of this indeed (after many things, I do not say all, considered conditionally) at length in Proposition 33 I show the absolute falsity, because repugnant to the nature of the straight line, about which I there introduce many necessary Lemmata. Finally in Proposition 38 I absolutely prove the hypothesis of acute angle contradictory to itself. Since therefore the hypothesis of right angle alone remains, the consequence plainly is, that from the aforesaid Proposition 13 remains absolutely established the enunciated Euclidean Axiom.[5] This is what was to be demonstrated.

Scholium.

It is well to consider here a notable difference between the foregoing redargutions of the two hypotheses. For in regard to the hypothesis of obtuse angle the thing is clearer than midday light; since from it assumed as true is demonstrated the absolute universal truth of the controverted Euclidean Assertion, from which afterward is demonstrated the absolute falsity of this hypothesis; as is established from Proposition 13 and Proposition 14. But on the contrary I do not attain to proving the falsity of the other hypothesis, that of acute angle, without previously proving; that the line, all of whose points are equidistant from an assumed straight line lying in the same plane with it, is equal to this straight, which itself finally I do not appear to demonstrate from the viscera of the very hypothesis, as must be done for a perfect redargution.

Respondeo autem triplici medio usum me fuisse in XXXVII. hujus ad demonstrandam praedictam aequalitatem. Et primo quidem, in corpore illius Propositionis, demonstro eam curvam CKD, prout enascentem ex hypothesi anguli acuti (ac propterea semper cavam versus partes illius rectae AB) aequalem eidem esse debere, & quidem argumentum ducendo ex ipsis ejusdem curvae tangentibus. Deinde in duobus ejusdem Propositionis subsequentibus Scholiis, praecisive a qualibet speciali hypothesi, bis rursum demonstro aequalitatem illius genitae lineae CD cum subjecta recta linea AB, qualiscunque tandem censeatur esse ipsa linea CD eo modo genita.

[99] Jam vero quatenus illa curva CKD, prout enascens ex hypothesi anguli acuti, censeatur primo illo modo demonstrata aequalis subjectæ rectae lineae AB; manifesta evadit redargutio, cum ex eadem hypothesi evidenter demonstretur major. Sin autem alterutro ex duobus aliis modis ostensa censeatur aequalitas praedicta; neque tunc cessat redargutio contra hypothesin anguli acuti. Ratio est; quia nihil vetat, quin illa CD sit curva, & nihilominus aequalis sit illi rectae AB, dum tamen sit semper versus eas partes convexa, ac propterea recta jungens illa eadem puncta C, & D minor sit contraposita basi AB, prout in hypothesi anguli obtusi: At omnino repugnat, si versus easdem partes sit semper cava, ac propterea recta jungens praedicta illa puncta C, & D major sit eadem contraposita basi AB, prout in hypothesi anguli acuti. Atque ita declaratum jam est in Scholio praecedentis Propositionis. Scilicet contra hypothesin anguli obtusi manifestum est nullam hinc sequi redargutionem, quae propterea unice impetit hypothesin anguli acuti.

Hoc tamen loco aliquis fortasse inquiret, cur adeo sollicitus sim in demonstranda utriusque falsae hypothesis exacta redargutione. Ad eum, inquam, finem, ut inde magis constet non sine causa assumptum fuisse ab Euclide tanquam per se notum celebre illud Axioma. Nam hic maxime videtur esse cujusque primae veritatis veluti character, ut non nisi exquisita aliqua redargutione, ex suo ipsius contradictorio, assumpto ut vero, illa ipsa sibi tandem restitui possit. Atque ita a prima usque aetate mihi feliciter contigisse circa examen primarum quarundam veritatum profiteri possum, prout constat ex mea Logica demonstrativa.

Inde autem transire possum ad explicandum, cur in Proemio ad Lectorem dixerim: *non*
[100] *sine magno in rigidam Logicam peccato assumptas a quibusdam fuisse tanquam datas duas rectas lineas aequidistantes*. Ubi monere debeo nullum eorum a me hic carpi, quos in hoc meo Libro vel indirecte nominavi, quia vere magnos Geometras, hujusque peccati verissime immunes. Dico autem: *magnum in rigidam Logicam peccatum*: quid enim aliud est assumere tanquam datas *duas rectas lineas aequidistantes*: nisi aut velle; quod omnis linea in eodem plano aequidistans a quadam supposita linea recta sit ipsa etiam linea recta; aut saltem supponere, quod una aliqua sic aequidistans possit esse linea recta, quam idcirco seu per hypothesin, seu per postulatum praesumere liceat in tanta aliqua unius ab altera distantia? At constat neutrum horum venditari posse tanquam per se notum. Scilicet ratio objectiva lineae, quae omnibus suis punctis aequidistet a quadam supposita linea recta, non ita clare per se ipsam congruit cum definitione propria ipsius lineae rectae. Quare assumere duas rectas lineas sub ista *aequidistantiae* ratione inter se *parallelas* fallacia est, quam in praedicta mea Logica appello *Definitionis complexae*, juxta quam irritus est omnis progressus ad assequendam veritatem absolute talem.

But I reply I used a triple means in Proposition 37 for demonstrating the mentioned equality. And first, in the body of the Proposition, I prove the curve CKD, as born from the hypothesis of acute angle (and therefore always concave toward the side of the straight AB) must be equal to it, and indeed by drawing the argument from the tangents of the curve. Then in two subsequent Scholia of the Proposition, apart from any special hypothesis, twice again I demonstrate the equality of the generated line CD with the underlying straight line AB, of whatever kind the line CD so generated[6] is supposed to be.

But now; in so far as the curve CKD, as born from the hypothesis of acute angle, is judged to be proved by the first method equal to the underlying straight line AB, a manifest redargution arises, since from the same hypothesis it is evidently proved greater. But if the aforesaid equality is supposed shown in either of the two other modes; not even then does the redargution cease against the hypothesis of acute angle. The reason is; because nothing forbids, that CD may be curved, and nevertheless may be equal to the straight AB, while yet it may be always convex toward that side, and therefore the straight joining the points C, and D may be less than the opposite base AB, as in the hypothesis of obtuse angle. But, it is wholly contradictory, if toward that side it be always concave, and therefore the straight joining the points C, and D be greater than the opposite base AB, as in the hypothesis of acute angle. And so has just now been stated in the Scholium of Proposition 38. Of course against the hypothesis of obtuse angle it is manifest no redargution follows hence, which therefore only demolishes the hypothesis of acute angle.

In this place however some one perchance may inquire, why I am so solicitous about proving exact the redargution of each false hypothesis. To the end, say I, that thence may be more completely established that not without cause was that famous Axiom assumed by Euclid as known *per se*. For chiefly this seems to be as it were the character of every primal verity, that precisely by a certain perfect redargution based upon its very contradictory, assumed as true, it can be at length brought back to its own self. And I can avow that thus it has turned out happily for me right on from early youth in reference to the consideration of certain primal verities, as is known from my *Logica demonstrativa*.[7]

Thence now I may proceed to explain, why in the Preface to the Reader I have said: *not without a great sin against rigid logic two equidistant straight lines have been assumed by some as given*. Where I should point out that none of those is carped at, whom I have mentioned even indirectly in this book of mine, because they are truly great Geometers, and verily free from this sin. But I say: *great sin against rigid logic*: for what else is it to assume as given *two equidistant straight lines*: unless either to assume; that every line equidistant in the same plane from a certain supposed straight line is itself also a straight line; or at least to suppose, that some one thus equidistant may be a straight line, as if therefore it were allowable to make assumption, whether by hypothesis, or by postulate,[8] of any such distance of one from another? But it is certain neither of these can be made traffic of as if *per se* known. Forsooth the objective concept[9] of a line, which in all its points is equidistant from a certain supposed straight line, clearly is not thus *per se* congruent with the proper definition of the straight line. Wherefore to define two *parallel* straight lines under this relation of mutual *equidistance* is the fallacy, which in my aforesaid *Logica* I call *Complex Definition,* in connection with which every advance toward attaining truth absolutely such is ineffectual.[10]

Unam tamen superesse adhuc video necessariam observationem. Nam lineam jungentem extrema puncta omnium aequalium perpendiculorum, quae in eodem plano versus easdem partes erigantur a singulis punctis subjectæ rectae lineae AB, debere esse & aequalem praedictae AB, & rursum in seipsa rectam, fateri omnes volumus. Sed dico prius esse apud nos, quod aequalis sit; deinde autem, quod recta. Cum enim singula puncta illius rectae AB intelligi possint semper aequabiliter procedere per sua illa perpendicula ad formandam tandem illam qualemcunque CD; manifestum videri debet, quod illa qualiscunque genita CD aequalis sit eidem AB; praesertim vero, si respiciamus explicationem contentam in
[101] Scholio II. post XXXVII. hujus, ubi hoc punctum clarissime demonstratum est.

Sed postea magna adhuc restat difficultas in demonstrando, quod illa eadem sic genita CD non nisi recta linea sit. Atque hinc factum esse puto, ut ex communi quadam persuasione rectam lineam, pro faciliore progressu, maluerint praesumere, ut inde aequalem ostenderent illi basi AB, ac postea inferrent rectos angulos ad ipsam talem jungentem CD. Dico autem *magnam difficultatem*: Nam prius expendere oportebat tres hypotheses circa angulos ad illam junctam *rectam* CD, nimirum aut rectos, si ipsa aequalis sit basi AB; aut obtusos, si minor; aut acutos, si major. Tum vero ostendi debebat non nisi cavam esse posse versus basim AB lineam curvam, quae (in hypothesi anguli acuti) jungat extremitates illorum aequalium perpendiculorum, ac rursum non nisi convexam versus eandem basim aliam curvam, quae (in hypothesi anguli obtusi) jungat extremitates eorundem perpendiculorum. Deinde autem hypothesis quidem anguli acuti ex eo demonstranda erat falsa; quia linea jungens praedictorum perpendiculorum extremitates adeo non erit aequalis basi AB, ut immo (ex communi saltem notione) major sit illa juncta recta CD, quae ex natura ipsiusmet hypothesis major est praedicta basi AB. At hypothesis anguli obtusi aliunde ostendenda erat sibi ipsi repugnans, prout in XIV. hujus. Sed haec jam satis.

Finis Libri primi.

I see in addition there still remains one necessary observation. For we all are willing to grant the line joining the extreme points of all equal perpendiculars, which in the same plane are erected toward the same parts from the separate points of an underlying straight line AB, must be both equal to the aforesaid AB, and moreover in itself straight. But I say with us is first, that it is equal; then however, that it is straight. For since the single points of the straight AB may be thought always to proceed uniformly upon those perpendiculars of theirs to forming at length that certain CD; it should seem manifest, that the generated CD, of whatsoever kind, is equal to AB; but especially, if we consider the explication contained in Scholium 2 after Proposition 37, where this point is most clearly demonstrated.

But thereafter still remains a great difficulty in demonstrating, that this same generated CD cannot be anything but a straight line. And hence comes it I think, that from a certain common conviction, for more facile progress, they have preferred to presume the line straight, that thence they might show it equal to the base AB, and afterward infer right angles at the join CD. But I say *great difficulty:* For first it was necessary to consider three hypotheses about the angles at the *straight* join CD, forsooth either right, if it be equal to the base AB; or obtuse, if less; or acute, if greater. But then it had to be shown that the curved line, which (in the hypothesis of acute angle) joins the extremities of those equal perpendiculars, could only be concave toward the base AB, and again the other curve, which (in the hypothesis of obtuse angle) joins the extremities of the same perpendiculars, only convex toward the same base. But then the hypothesis indeed of acute angle from this was demonstrated false; because the line joining the extremities of the aforesaid perpendiculars was so far not equal to the base AB, as on the contrary (anyhow from the common notion) it is greater than the straight join CD, which from the nature of this hypothesis itself is greater than the aforesaid base AB. But the hypothesis of obtuse angle had to be shown from another source contradictory to itself, as in Proposition 14. But this now is enough.[11]

End of Book One

Euclidis ab omni naevo vindicati Liber Secundus

Solo plerumque ratiocinio opus erit in toto hujus Libri decursu. Nam hic Euclidem in eo unice accusant, quod in sexta Definitione Quinti magis obtenebret, quam explicet naturam magnitudinum aeque proportionalium, sibique idcirco onus imponat demonstrandi plures Propositiones, quae per se ipsas clarissimae videri possunt, vel certe allatis ab ipso demonstrationibus clariores: Tum vero, quod in quinta Definitione Sexti sub specie Definitionis Axioma quoddam assumat non facile permittendum, sine praevia demonstratione.

Pars Prima

In qua expenditur sexta Definitio Libri quinti Euclidaei.

[102] Definit ibi Euclides magnitudines aeque proportionales, prout sequitur: *In eadem ratione magnitudines dicuntur esse, prima ad secundam, & tertia ad quartam, cum primae, & tertiae aeque multiplicia, a secundae, & quartae, aeque multiplicibus, qualiscunque; sit haec*
[103] *multiplicatio, utrunque ab utroque vel una deficiunt, vel una aequalia sunt, vel una excedunt, si ea sumantur, quae inter se respondent.* Supersedeo ab omni exemplo, quia profiteor me scribere jam versatis in Geometria, non autem immaturis novitiis. Moneo tamen juste hoc loco reprehendi a Clavio Campanum, atque Orontium, quod ita hanc Euclidis definitionem interpretentur, quasi tunc solum illa *proportionalitas* subsistere debeat, quando praedicta aeque multiplicia vel una deficiant, vel una excedant *proportionaliter, sive in eadem proportione.* Nam id foret inducere Euclidem desipientem, qui nempe idem per idem definiret. Fateri igitur debemus, quod Euclides loquatur de quolibet defectu, aut excessu; dum tamen ad utramque partem juxta quamlibet multiplicationem alteruter idem semper consistat, & non etiam aliquando ex una quidem parte defectus, ex altera autem aut aequalitas, aut excessus.

G. Saccheri, *Euclid Vindicated from Every Blemish*, Classic Texts in the Sciences 1,
DOI 10.1007/978-3-319-05966-2_3,

Euclid Vindicated from every Blemish Book Two

For the most part, this book will require the use of reasoning only. For in this matter Euclid has been only accused for having obscured and not explained (in *Elements* V, def. 6) the nature of equiproportional magnitudes, and that in doing this, he took upon himself the burden of proving many Propositions which can be most clearly understood of themselves, or at least, when explanations have been brought to bear by him, understood more clearly. Moreover, he is accused of having assumed, in *Elements* VI, def. 5, under the cloak of definition, a certain Axiom not easily granted without proof.

First Part

In which the Sixth Definition of the Fifth Book of the Elements is considered.

In that Book Euclid defines[1] equiproportional magnitudes as follows: *Magnitudes are said to be in the same ratio, the first to the second as the third to the fourth if, when equimultiples of the first and the third are compared with equimultiples of the second and fourth, whatsoever be the multiplier, the former equimultiples alike fall short, or alike equal, or alike exceed the second pair of equimultiples, taken in corresponding order.* I omit all examples, because I profess to be writing for those now versed in the study of Geometry, not, on the other hand, for immature novices. Nevertheless I take this opportunity to point out the reproof by Clavius[2] of Campanus as well of Oronce Fine for having interpreted this definition of Euclid to mean that the *proportion* exists only when the aforesaid products either fall short or exceed *proportionally, or in the same proportion.* Now one would have to judge Euclid to be quite ignorant to think he would define a word in terms of itself. We must therefore acknowledge that Euclid speaks of any defect or excess whatever; while nevertheless, according to any multiplication whatever, either one or the other holds for both sides alike, and not at times a defect on the one hand, and on the other, either an equality or an excess.

Audire jam oportet, quid reprehensione dignum invenerint in exposita Euclidaea Definitione. Itaque dicunt debuisse probari ab Euclide istud ipsum esse verum; quod nempe, quoties praedicta qualiacunque aeque multiplicia vel una excedunt, vel una deficiunt, vel una aequalia sunt, toties quatuor illae magnitudines in eadem sint ratione. Ego autem non satis miror, quomodo Viri apprime docti in hunc lapidem impegerint. Quid enim, quaeso, probari debuit ab Euclide? An forte, quod nusquam asserat, seu praesumat quatuor magnitudines in eadem esse ratione, nisi quando aliunde constet praedicta qualiacunque earundem aeque multiplicia vel una excedere, vel una deficere, vel una aequalia esse? At id ipsum tam religiose exequitur, ut nimius hac in parte videri quibusdam potuerit in demonstrandis primis undecim Propositionibus Libri quinti, quae nimirum, sub alio quodam magis vulgari aeque Proportionalium conceptu, immediatam sibi ipsis facerent fidem.

Rursum vero: Quid per se ipsum clarius, si praecise attendamus ad vulgarem quandam
[104] acceptionem, quam quod Partes cum pariter Multiplicibus in eadem sint ratione; ut puta, quod 4. eandem habeat rationem ad 6. quam habet 8. duplus ipsius 4. ad 12. duplum illius 6. ? Nihilominus magna ista claritas satis non fuit Euclidi, qui in Propos. 15. praedicti Libri istud ipsum demonstrat ex Definitione ab eo assumpta aeque Proportionalium; ne scilicet accusari posset de petitione Principii, hoc est de duplici assumpta Definitione. Nihil ergo probare debuit Euclides, quod facto ipso non probet.

Sed in hoc ipso (inquiet aliquis) peccare ipse videtur, quod rem per se claram sua illa Definitione obscurare voluerit. Quasi vero (inquam ego) non potuerit acutus Euclides magnitudines invicem commensurabiles ab aliis non talibus separare, illasque idcirco ita definire, prout in Libro VII. Defin. XX. numeros proportionales apud Clavium definit: tunc nimirum quatuor magnitudines proportionales fore: *Cum prima secundae, & tertia quartae aeque multiplex est; vel eadem pars, vel eaedem partes; Vel certe, cum prima secundam, & tertia quartam, aequaliter continet, eandemque insuper illius partem, vel easdem partes.* Ita sane: At quis fructus, cum tandem descendere debuerit ad magnitudines multis modis incommensurabiles? Decuit igitur, ut una, eademque Definitione ipsas etiam non invicem commensurabiles magnitudines complecteretur.

Nullus, inquiunt, refragatur; aliamque idcirco ipsis etiam quomodolibet incommensurabilibus magnitudinibus Definitionem assignant, quae sequitur: *Prima ad secundam eandem rationem habebit, ac tertia ad quartam, si prima secundae partes aliquotas quascunque contineat, quoties tertio quartae similes partes aliquotas continet*: Ut puta; si magnitudo A toties continet magnitudinis B partes centesimas, millesimas, centies millesimas, & quascunque alias aliquotas similes; ita ut nulla sit pars magnitudinis B, quae pluries contineatur in magnitudine A, quam similis pars aliquota ipsius D contineatur in C, licet in irrationalibus
[105] restet semper aliqua quantitas; tunc ita est A ad B, ut C ad D.

It is now necessary to give attention to an examination of what might be at fault in the Euclidean Definition that has been presented. In effect, it is claimed that the truth of that very relationship should have been proved by Euclid: namely, that as often as any equimultiples of the given magnitudes either exceed, or fall short, or are equal among themselves, then the four magnitudes are in the same ratio.[3] However, I cannot understand how men, especially learned ones, can have dashed their heads against this stone. What is it that his critics would have Euclid prove? What indeed, since he at no time makes the assertion, but rather takes as an assumption that four magnitudes are in the same ratio only when it can be ascertained from some other source that any whatever equimultiples, of the same type, alike exceed, or alike fall short, or alike are equal? As a matter of fact, he pursues this notion so scrupulously in proving the first eleven Propositions of *Elements* V, that it might seem to certain people[4] that he was belaboring the point especially since these same propositions under a more widely promulgated concept of Proportion, would undoubtedly have brought with them immediate conviction.

On the other hand would it be clearer if we were to go along with popular opinion and take for granted the statement that Parts are in the same ratio as Equimultiples, as for example, that 4 has the same ratio to 6, as 8 (the double of 4) has to 12 (the double of 6)? This great clarity, however, did not satisfy Euclid who in *Elements* V, 15 gave a proof that made use of the Definition of Equiproportionals that he had assumed. It is obvious, moreover, that he cannot be accused of *petitio principii*, i.e. of double definition. Euclid needed not prove anything more than he actually proved.

But in this very point (some will protest) he seems to be at fault, because he obscured a matter clear in itself by that Definition of his. As if really (I say) the clever Euclid could not distinguish commensurable magnitudes from those which are not of that type, and therefore he should have defined them as he defines proportional numbers in *Elements* VII, def. 20 (in Clavius' edition). In accordance with this definition four magnitudes are proportional *when the first is the same multiple, or the same part, or the same parts, of the second, that the third is of the fourth. Or, in fact, when the first contains the second, and its part or parts, as many times as the third contains the fourth and its parts.*[5] He could proceed in this way, of course. But to what end, since eventually he would have to consider incommensurable magnitudes[6] of many types? It was proper, furthermore, that one and the same Definition apply to incommensurable magnitudes as well.

No one, they say, contests this, and for that reason they would assign another Definition for incommensurable magnitudes, which is as follows: *The first has the same ratio to the second as the third has to the fourth, if the first contains any submultiple of the second as many times as the third contains a like submultiple of the fourth.*[7] As for example, if magnitude A contains the one hundredth part of magnitude B, the one thousandth, the hundred thousandth, and any other similar submultiple of B so that there is no part of magnitude B that is contained in magnitude A a greater number of times than a similar submultiple of D is contained in C, even if, in the case of irrationals, there should always be a remainder, then A is to B, as C is to D.

Agnosco eximium, meisque oculis familiarem Geometram sic loquentem. Sed pace ipsius dictum sit: nullum video hujus novae Definitionis commendabilem fructum. Et primo quidem in dignum puto homine Geometra aliud quidpiam intelligere in usu alicujus termini scientifici praeter id, quod in ejusdem Definitione exprimitur; ac praesertim, ubi Definitio tradita sit per voces minime ambiguas, quales sunt *multiplicatio*, *majus*, *minus*, *aequale*. Igitur Definitio Euclidaea repraehendi non debuit titulo obscuritatis. Deinde vero: Nemo ibit inficias, quin promptior sit ad usum, ac propterea opportunior ad rectam, expeditamque intelligentiam multiplicatio praescripta ab Euclide, quam divisio ab aliis substituta. Unice igitur expendendum superest, an nova illa aeque proportionalium Definitio opportunior sit ad demonstrandum, ubi applicari debeat in specialibus materiis. Ad hunc finem assumo Prop. primam Libri sexti, quae utique prima est omnium talium applicationum: ubi Euclides demonstrat: Triangula, & Parallelogramma, quorum eadem fuerit altitudo, ita se habere, ut bases.

Ecce autem ex Euclide nitidissimam demonstrationem. Sint duo triangula (Fig. 49) ABC, DEF, eandem habentia altitudinem, quorum bases BC, EF. Omitto parallelogramma, quia scribo jam versatis in Geometria. Jam dico ita esse triangulum ABC ad triangulum DEF, ut basis BC ad basim EF. Nimirum; si basis BC statuatur prima magnitudo, basis EF secunda, triangulum ABC tertia, & triangulum DEF quarta; Dico qualiacunque aeque multiplicia primae ac tertiae a quibusvis aeque multiplicibus secundae & quartae vel una deficere, vel una aequalia esse, vel una excedere, prout exigit Definitio sexta Libri quinti.

I know a distinguished and to my eyes familiar Geometer who spoke in that fashion.[8] With apologies to him, it may be said I see in this Definition no praiseworthy advantage. In the first place, I regard it as unfitting for a Geometer to knowingly make use of any scientific term besides the one which is described in the definition of the same, and furthermore, particularly when a definition is given in words that are not in the least ambiguous, such as *multiplication*, *greater*, *less*, and *equal*. Therefore Euclid's Definition ought not be criticized on the grounds of obscurity. In fact, no one will deny that the multiplication prescribed by Euclid is easier to apply and, on that account, more suitable for a correct and quickly gained understanding of the subject than the division substituted for it by others. Therefore it remains only to consider whether that new Definition of equiproportionals is more convenient in formulating proofs calling for its application to special topics. To his end, let us consider *Elements* VI, 1 which is, at any rate, the first of all such applications, and in which Euclid proves the that Triangles and Parallelograms which are under the same height are to each other as their bases.

Here is the very elegant proof given by Euclid.[9] Let ABC and DEF be two triangles (Fig. 49) with the same height, and bases BC, EF. I omit the case of parallelograms because I am writing for experienced Geometers. Now I say triangle ABC is to triangle DEF as base BC is to base EF. Certainly, if base BC is set up as the first magnitude, base EF the second, triangle ABC the third, and triangle DEF the fourth, then any whatever of the equimultiples of the first and third, and any whatever of the same equimultiples of the second and fourth either alike fall short, or alike are equal, or alike exceed, according to the requirements of *Elements* V, def. 6.

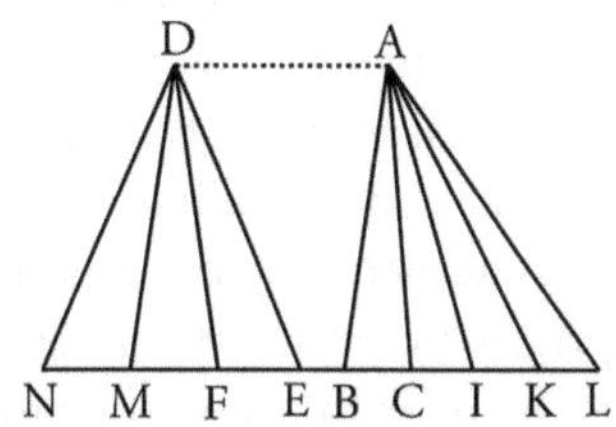

Fig. 49

[106] Intelligantur enim praedicta triangula tum in eodem plano existere, tum etiam constituta esse inter easdem parallelas AD, CBEF; adeo ut nempe (ex natura parallelarum) aequalem altitudinem habeant. Tum vero sumptis in BC indefinite protracta quotvis portionibus CI, IK, KL, aequalibus ipsi BC; atque item in EF similiter protracta quotvis FM, MN, aequalibus ipsi EF; jungantur AI, AK, AL; atque item DM, DN. Constat (ex 38. primi) aequalia inter se fore triangula ALK, AKI, AIC, ACB; atque item eadem ratione aequalia inter se fore triangula DEF, DFM, DMN. Igitur, quam multiplex est rectae CB quaecunque assumpta recta CL, tam multiplex quoque erit triangulum ACL trianguli ACB; & quam multiplex est recta EN rectae EF, tam quoque multiplex erit triangulum DEN trianguli DEF; quia in tot triangula aequalia sunt divisa tota triangula ACL, DEN, in quot rectas aequales sectae sunt totae rectae CL, EN. Quoniam vero, si basis CL aequalis fuerit basi EN, necessario (ex predicta 38. primi) etiam triangulum ACL aequale est triangulo DEN; ac proinde, si CL major fuerit quam EN, necessario ACL majus est quam DEN; & si minor, minus: consequens plane est, ut recta CL, & triangulum ACL, quae assumpta sunt pro quibusvis aeque multiplicibus primae magnitudinis BC, & tertiae ABC; si conferantur (prout inter se respondent) cum recta EN, & triangulo DEN, quae ipsa rursum assumpta sunt pro quibusvis aeque multiplicibus secundae magnitudinis EF, & quartae DEF; inveniantur semper vel una aequalia esse, vel una excedere, vel una deficere. Igitur (juxta sextam Definitionem quinti) quae proportio primae BC ad secundam EF, basis ad basim, ea est tertiae ABC ad quartam DEF, trianguli ad triangulum. Quod erat propositum.

[107] Conferre nunc debeo praemissam Euclidaeam Demonstrationem cum altera indicati eximii Geometrae, quam ipsis ejusdem vocibus statim exhibeo.

Sint duo triangula (Fig. 50) ABC, DEF, quorum altitudo sit eadem. Dico eandem esse rationem trianguli ABC ad triangulum DEF, quae basis BC ad basim EF. Demonstratur. Intelligatur basis EF divisa in quotcunque partes aequales, ut puta in quatuor, & ex puncto D ducantur lineae DG, DH, DI; & linea BC quadrantem lineae EF contineat ter, sintque lineae CM, ML, LK, quae sint singulae aequales lineis EG, GH, HI, IF, ducanturque lineae AK, AL, AM. Triangula DEG, DGH, DHI, DIF singula sunt aequalia (per 38. 1.) quia sunt supra bases aequales, & in eisdem parallelis. Igitur sunt quadrantes trianguli DEF: & quia triangula ABC, DEF habent aequales altitudines, etiam in eisdem parallelis intelligi possunt constituta, & sic triangula AKL, ALM, AMC, habentia bases aequales cum triangulis DEG, DGH, DHI, DIF, illis aequalia erunt. Ergo quot erunt quadrantes lineae EF in linea BC, tot erunt quadrantes trianguli DEF in triangulo ABC. Quod autem ostensum est de quadrantibus, potest similiter ostendi de quibuslibet partibus aliquotis. Igitur (juxta assumptam novam illam aeque proportionalium Definitionem) ut basis BC ad basim EF, ita triangulum ABC ad triangulum DEF. Quod erat &c.

It might be observed, to be sure, that the aforesaid triangles not only exist in the same plane, but also can be arranged between the same parallels AD, and CBEF, so that (from the nature of parallels)[10] they certainly have the same height. Furthermore, if in BC produced indefinitely, a number of segments are taken, CI, IK, KL, each equal to BC, and also in EF extended in the same manner, a number of segments FM, MN are taken equal to EF. Join AI, AK, AL, as well as DM, DN. It is evident (from *Elements* I, 38) that the triangles ALK, AKI, AIC, ACB are equal to each other; triangles DEF, DFM, DMN are also equal to each other, for the same reason. Therefore, whatever multiple line CL is of the line CB, triangle ACL will be that same multiple of triangle ACB; and whatever multiple line EN is of line EF, triangle DEN will be that same multiple of triangle DEF. For this reason, triangles ACL and DEN are divided into as many equal triangles as the lines CL and EN are divided into equal segments. Since actually, if base CL were equal to base EN, then of necessity (from the aforementioned *Elements* I, 38), triangle ACL would be equal to triangle DEN; and also, in like manner, if CL were greater than EN, of necessity, ACL would be greater than DEN; and if smaller, less. Consequently, it is clear that if line CL and triangle ACL, which are taken as any equimultiples of the first magnitude BC and of the third ABC, are compared (according to the correspondence indicated) with line EN and triangle DEN, which in turn are taken as any equimultiples of the second magnitude EF and of the fourth DEF, it would be found always that they either together are equal, or together are in excess, or together fall short. Therefore, according to *Elements* V, def. 6, the proportion of the first BC to the second EF, base to base, is the same as the third, ABC, is to the fourth, DEF, triangle to triangle. This is what was to be demonstrated.

I now compare the above Euclidean Proof with the one given by the previously mentioned great Geometer. This proof is herewith reproduced in his own words.[11]

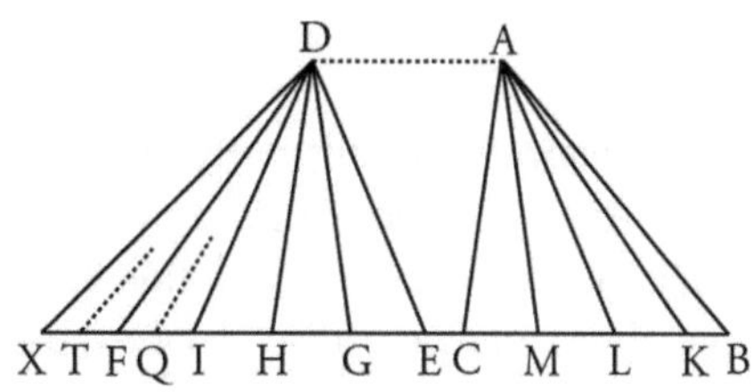

Fig. 50

Let ABC and DEF be two triangles having the same height (Fig. 50). I say that the ratio of triangle ABC to triangle DEF is the same as that of base BC to base EF.

Proof. Let EF be divided into any number of equal parts, as for example into four, and from point D lines are drawn DG, DH, DI. Suppose furthermore that BC contains the fourth part of EF three times, i. e. the lines CM, ML, LK each equal to lines EG, GH, HI, IF. Draw lines AK, AL, AM. Triangles DEG, DGH, DHI DIF are equal to each other (*Elements* I, 38) because they are on equal bases and between the same parallel lines. For this reason, they are each one fourth of triangle DEF. Since, moreover, triangles ABC, DEF have equal height it is evident that they can be placed between the same parallels, and so triangles AKL, ALM, AMC having equal bases as triangles DEG, DGH, DHI, DIF, are equal to them. Therefore, as many times as one fourth of line EF is contained in line BC, so many times is one fourth of triangle DEF contained in triangle ABC. Moreover, what can be shown for fourths can, in a similar manner, be proved for any number of submultiples whatsoever. Therefore (according to that newly assumed Definition of equiproportionals) as the base BC is to the base EF, so triangle ABC is to triangle DEF. This is what was to be demonstrated.

Jam fateor hanc etiam demonstrationem opportunam esse intento; sed puto non aeque claram esse, atque Euclidaeam. Ratio discriminis haec est; quia posterior Definitio procedit in *quid rei*; prior autem in *quid nominis*. Scilicet non eo inficias, quin probe demonstratum sit tot fore v. g. quadrantes (Fig. 50) trianguli DEF in triangulo ABC, quot erunt quadrantes basis EF in basi BC; atque ita uniformiter secundum quamlibet aliam partem aliquotam.

[108] At ubi venitur ad conclusionem, quod triangulum ABC ita sit ad triangulum DEF, ut basis BC ad basin EF, haeret statim mens; cum magis immediate debuerit inferri, ita esse basim KC ad basim EF, ut triangulum AKC ad triangulum DEF.

Neque juvat reponere, quod una veritas alteri officere non debeat. Nam dico ex hac ipsa innegabili veritate oriri dubium posse, an forte per nullam partem aliquotam ipsius EF quomodolibet multiplicatam exhauriri praecise possit ipsa BC; quod sane omnino continget, si ipsae EF, BC sint incommensurabiles; nisi forte provocare velimus ad partem aliquotam infinite parvam, circa quam rursum non levis oriri posset difficultas.

Cur autem (urgebit quispiam) licitum fuerit Euclidi tradere suam illam Definitionem aeque proportionalium, quae nulli communi notioni innititur; & non etiam aliis liceat quampiam aliam subrogare, quae ex quadam cum numeris aeque proportionalibus similitudine sibi ipsi faciat fidem? Per me (inquam) quidquid libeat, licitum etiam hac in parte puto; dum tamen observentur duo sequentia. Primo; ut ne (quatenus procedi velit per Definitionem *puri nominis*) accusetur ut mala quaepiam alia *puri nominis* assumpta Definitio; nisi aut ostendatur ex ipsis ejusdem vocibus nimis perplexa, aut aliunde constet longiore quadam indagine opus esse ad assequendam *rem ipsam*, cujus nempe veluti propria dignosci tandem debeat exposita Definitio. Secundo autem; ut (quatenus procedi velit per Definitionem *ipsius rei* aliunde cognitae) nunquam putemus *rem ipsam* assecutos nos esse, nisi clare deveniamus ad illam originem, unde sumpta est occasio, aut opportunitas talis Definitionis.

Post hanc non inutilem digressionem duo alia opportune subdo. Unum est; me non
[109] negare, quin posteriore loco exposita Demonstratio optima sit, & evidentissima; dum tamen ibi assumpta aeque proportionalium Definitio recipi unice debeat in *quid nominis*. Sed tunc observo nullam fuisse causam respuendi, ac multo minus damnandi Definitionem Euclidaeam: cum ex una parte & Demonstratio eidem innixa nulli tergiversationi obnoxia sit; & ex altera Definitio, certissime ibi assumpta in *quid nominis*, procedat per multiplicationem, quam unusquisque intelligit clariorem esse, ac faciliorem divisione, & quidem (suo quodam proprio jure) sine ullo periculo cuiusvis objectæ incommensurabilitatis; prout constare potest ex antea dictis.

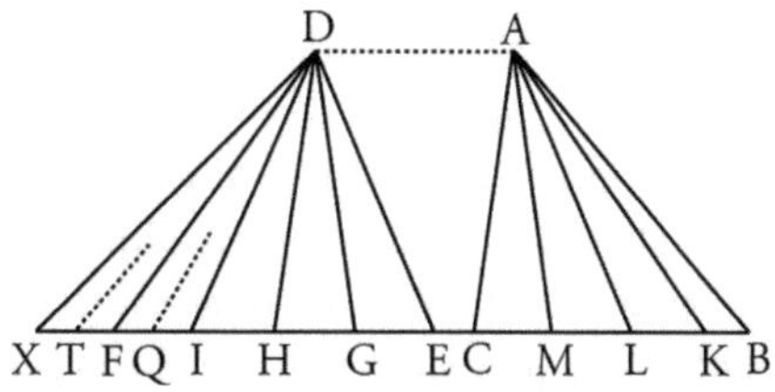

Fig. 50

Now I grant that this demonstration attains its aim, but in my opinion it is not as clear as the Euclidean. The ground of their difference is the following: the latter definition is presented as a *real* definition whereas the former is a *nominal* definition.[12] I do not deny that it has been properly proved that (Fig. 50) one fourth of triangle DEF is contained in triangle ABC as many times as one fourth of base EF is contained in base BC; and so on for any submultiple whatever. But when he arrives at the conclusion that triangle ABC is to triangle DEF as base BC is to base EF, then immediately we are at a loss, since we would more immediately infer that base KC is to base EF as triangle AKC is to triangle DEF.

Moreover, it would be useless to reply that one truth should not stand in the way of others. For I say from this undeniable truth, a doubt can arise as to whether, perhaps, no submultiple of this EF multiplied in any way whatever, can exactly exhaust this BC, an eventuality that would certainly occur if EF and BC were incommensurables, unless indeed, we would wish to consider infinitely small submultiples, which in turn would introduce difficulties and not small ones at that.[13]

Why, on the other hand (someone will insist) was Euclid permitted to expound his Definition of equiproportionals which depends upon no common notion, while others are not even permitted to make a substitution of something which from a certain analogy with proportional numbers would bring conviction? As for me (I say), I consider any definition acceptable, as long as the following two rules are followed. Firstly: to the extent that one may wish to assume a *purely nominal* definition, one should also accept any other purely nominal definition,[14] unless it has been shown to employ ambiguous terms, or that after a longer investigation it is clear that another Definition is needed to arrive at an understanding of the *thing itself*.[15] Secondly: to the extent that one may wish to employ a *real* definition of the thing (which is assumed as otherwise known), at no time should one consider that we understand the *thing itself* unless we were unmistakable to arrive at the source from which the occasion or opportunity for such a definition was taken.

After this digression, which was not without a purpose, I produce two other appropriate in another way.

One is the following: I do not deny that the proposed Demonstration, given last, is the best and most obvious one. Nevertheless, the Definition of equiproportionals assumed in the course of it ought to be considered as a *nominal* definition. But then I point out that there was no cause for disapproving and much less for censuring the Euclidean Definition, since on the one hand, the Demonstration is not guilty of resting on any subterfuge, and on the other, the Definition assumed there (as a *nominal* definition) is arrived at through multiplication, which everyone knows is clearer and easier than division,[16] and which (by its nature) carries with it no danger at all of any objection about incommensurability; as it is clear from what we have remarked.

Alterum est; me rursum non negare, quin aliquo tandem pacto demonstrari possit prima illa Propositio Libri sexti, etiamsi assumpta censeatur in *quid rei* praedicta aeque proportionalium nova Definitio. Verum arbitror hac sola ratione demonstrari eam posse, dum alias praesumatur unam aliquam esse rectam lineam determinatam, quae ad basim BC eam habeat rationem, quae est trianguli DEF ad triangulum ABC. Nam demonstrabo nullam EI, quae sit portio ipsius EF, ac similiter nullam EX, quae assumpta sit in EF producta, tales esse posse, ut eam habeant ad basim BC rationem, quae est trianguli DEF ad triangulum ABC.

Demonstratur prima pars. Nam constat primo (Fig. 50) talem ipsius BC assumi posse partem aliquotam BK, ut puta *octavam*, quae minor sit illa in EF residua portione IF. Constat hinc secundo; tot in EF designari posse portiones aequales ipsi BK, ita ut postrema designata portio desinat in punctum Q, quod jaceat inter puncta I, & F. Tunc autem (juncta QD) manifestum fiet (ex 38. 1.) tot contineri in triangulo DEQ octavas partes trianguli ABC,
[110] quot in basi EQ continentur octavae partes illius basis BC. Quare (ex nova illa assumpta in *quid rei* aeque proportionalium Definitione) ita erit triangulum DEQ ad triangulum ABC, ut basis EQ ad basim BC. Habet autem EQ ad BC majorem rationem (ex 8. 5.) quam illa EI ad eandem BC; Igitur (ex 13. 5.) etiam triangulum DEQ non aequalem, sed majorem rationem habebit ad triangulum ABC, quam sit ratio praedictae EI ad illam basim BC. Quod erat priore loco demonstrandum.

Demonstratur secunda pars. Nam rursum constat primo; talem illius BC assumi posse partem aliquotam BK, ut puta *decimam*, quae minor sit illo excessu FX, quo basis EF superatur ab ipsa EX. Atque hinc constat secundo; tot in EX designari posse portiones aequales ipsi BK, ita ut postrema designata portio desinat in punctum T, quod jaceat inter puncta F, & X. Tunc autem (juncta TD) manifestum fiet (ex eadem 38. 1.) tot in triangulo DET decimas partes trianguli ABC contineri, quot in basi ET continentur decimae partes ipsius basis BC. Quare (ex eadem assumpta in *quid rei* aeque proportionalium nova Definitione) ita erit triangulum DET ad triangulum ABC, ut basis ET ad basim BC; & invertendo ita erit triangulum ABC ad triangulum DET, ut basis BC ad basim ET. Habet autem basis BC ad basim ET rationem majorem (ex eadem 8. 5.) quam sit ratio ejusdem basis BC ad illam EX, quae supponitur major praedicta ET; ac propterea (ex eadem 13. 5.) majorem ratione trianguli ABC ad triangulum DEF. Non igitur ulla EX, quae sit major ipsa EF, talis esse potest, ad quam basis BC eam habeat rationem, quae est trianguli ABC ad triangulum DEF. Quod erat secundo loco demonstrandum.

Unde tandem fit, ut ratio trianguli ABC ad triangulum DEF, quod eandem cum ipso
[111] altitudinem habeat, eadem sit ac basis BC ad basim EF; dum scilicet alias praesumatur unam aliquam esse rectam lineam determinatam, ad quam basis BC eam habeat rationem, quae est trianguli ABC ad triangulum DEF. Quod erat principale intentum.

The second digression is as follows: I do not on the other hand deny that it is possible to demonstrate *Elements* VI, 1, even though the newly assumed Definition of equiproportionals is taken as a *real* definition. But I consider that it can be proved only if the following is assumed, namely, that there exists a determined straight line which would have the same ratio to base BC as triangle DEF has to triangle ABC. Now I shall show that no EI, which is a part of the very EF, and similarly, no EX, which is assumed to be in EF produced, can be such that they have the same ratio to base BC which triangle DEF has to triangle ABC.

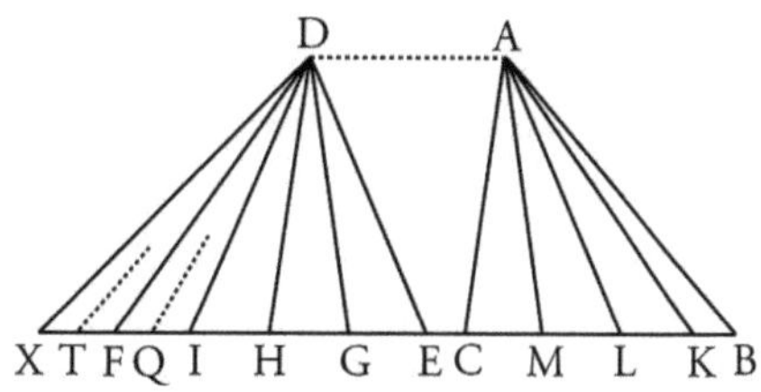

Fig. 50

Proof of the first part. In the first place, it is evidence (Fig. 50) that a submultiple BK of BC can be taken, as for example, one-eighth, which is less than IF, the residual part of EF. Hence it follows that segments equal to BK can be marked off on EF so that the last marked segment terminates in point Q, which lies between I and F. Therefore, joined QD, it is evident (*Elements* I, 38) that the eighth part of triangle ABC is contained in triangle DEQ as many times as the eighth part of base BC is contained in base EQ. For which reason (from the new Definition of equiproportionals assumed as a *real* definition) triangle DEQ is to triangle ABC as base EQ is to base BC. However, the ratio of EQ to BC is larger than that of EI to BC (*Elements* V, 8). Therefore the ratio of triangle DEQ to triangle ABC will not be equal but will be larger than the ratio of base EI to base BC (*Elements* V, 13). This is what was to be demonstrated in the first part.

Proof of the second part. Now it follows that such a submultiple BK of BC can be taken, as for example, *a tenth*, which is less than FX, the segment by which the base EF is exceeded by EX. And also it is evident that it is possible to mark off in EX segments equal to BK until the last segment marked off terminates in point T, which lies between points F and X. Then, joined TD, it is clear (from the same *Elements* I, 38) that the tenth part of triangle ABC is contained in triangle DET as many times as the tenth part of base BC is contained in ET. For which reason (again from the new Definition of equiproportionals assumed as a *real* definition) triangle DET is to triangle ABC as base ET is to base BC, and inverting, triangle ABC is to triangle DET as base BC is to the base ET. Moreover, the ratio of base BC to base ET is greater (from the same *Elements* V, 8) than the ratio of BC to EX, which was assumed to be greater than the aforementioned ET. For this reason (again from *Elements* V, 13) the ratio of BC to ET is greater than the ratio of triangle ABC to triangle DEF. Therefore, there is no segment EX greater than EF which is such that the base BC has the same ratio to EX that triangle ABC has to triangle DEF. This is what was to be demonstrated in the second part.

Under the assumption that there is a determined straight line to which the base BC would have the same ratio of triangle ABC to triangle DEF, it is obvious that the ratio of triangle ABC to triangle DEF, both triangles having the same height, is the same as the ratio of base BC to base EF. This is what was to be demonstrated.[17]

Post haec autem addere insuper debeo, non placere mihi necessitatem illius *Petitionis*, quod una aliqua sit recta linea determinata, ad quam praedicta basis BC eam habeat rationem, quae est trianguli ABC ad triangulum DEF. Nam certe verus naevus hic foret, maxime dolendus in primo ingressu novae hujus apud Geometras nobilissimae partis; quod scilicet reperiri demonstrative non possit recta linea, ad quam talis quaepiam data recta eam rationem habeat, quae est vel talis cujusdam dati trianguli rectilinei ad alterum datum triangulum rectilineum, vel talis cujuslibet datae rectae ad alteram quamlibet propositam rectam lineam; quod, inquam, talis recta linea reperiri non possit, nisi antea praesumendo, per modum primi Principii, quod una aliqua ejusmodi recta linea vere existat in rerum natura.

Sed hic audio quempiam reclamantem, quod ipse etiam Euclides, sine demonstratione, hoc veluti Axiomate utitur in Propos. 18. 5. ubi demonstrat compositionem rationis; hoc est, *duas magnitudines, quae divisae proportionales sint, has quoque compositas proportionales esse*. Quin etiam Clavius, acutissimus Euclidis Interpres, sub expresso *Axiomatis* nomine, ante omnes praedicti Libri Propositiones sic praemittit: *Quam proportionem habet magnitudo aliqua ad aliam, eandem habebit quaevis magnitudo proposita ad aliquam aliam; & eandem habebit quaepiam alia magnitudo ad quamvis magnitudinem propositam*. Hujus autem assumpti hanc reddit rationem. *Quamvis enim ignoremus interdum, quaenam sit quarta illa*
[112] *magnitudo, dubitandum tamen non est, eam esse posse in rerum natura, cum id contradictionem non implicet, ut Philosophi loquuntur, neque absurdi aliquid ex eo consequatur.*

Veruntamen, his ipsis stantibus, non desino agnoscere indecorum naevum, a quo, ex officio assumpto, vindicare debeam Euclidem; praesertim vero, cum ipse Clavius non Euclidi, sed interpretibus istud attribuat. Itaque dico illam 18. 5. demonstrari potuisse sine ullo recursu ad intrusum illud Axioma; nimirum provocando ad quasdam Libri sexti Propositiones, quae ab illo intruso Axiomate nullatenus pendent; & ex quibus illud ipsum Problematice demonstratur, saltem quoad lineas rectas. Cur autem oculatissimus Clavius non id animadverterit, in causa esse potuit, quod ibi fusior esse debuerit in explicanda vera doctrina proportionum, adversus quosdam non ignobiles Euclidis interpretes.

Jam pergo ad exequendum munus assumptum: Ubi, ut expeditior sim, resolvendo magis, quam componendo, rem ipsam paucioribus exhibere conabor. Praemitto autem, assumi hic posse tanquam a me juxta Euclidem jam demonstratam primam sexti, sine ulla dependentia a praefato Axiomate; cum ibi (si excipias doctrinam parallelarum) sola usui fuerit ex Libro primo Propositio 38. & ex quinto Euclidaea aeque proportionalium Definitio.

I must add, however, that I am not pleased with the necessity of this *Request*,[18] namely that there should exist a determined straight line, to which the aforesaid base BC would have the same ratio that triangle ABC has to triangle DEF. For certainly, it would be a real blemish, and one which would be deplorable in the introduction to a section considered by Geometers to be the most elegant of all, if it couldn't be demonstratively found[19] a straight line to which any given straight line would have the same ratio that either a certain given rectilinear triangle has to another given rectilinear triangle, or else any given straight line has to any other given straight line; or that such a straight line could not be found, unless it is assumed through a first Principle that a straight line of that type really exists in nature.

At this point, however, I hear certain strong objections to the effect that even Euclid used this Proposition without proof, as an Axiom, for example, in *Elements* V, 18, where he demonstrates the composition of ratios, that is: *two magnitudes which are in proportion when separated are also in proportion when composed.* Indeed, even Clavius, that extremely keen Commentator of Euclid, under the distinct heading of *Axiom*, and before the presentation of the Propositions of *Elements* V, writes as follows: *When some magnitude has a certain proportion to another, any given magnitude whatever will have that same proportion to a certain other magnitude, and some other magnitude will have that same proportion to any whatever given magnitude.* Moreover, he gives the following argument for this assumption. *For, if we ignore for the occasion, what the fourth magnitude is, nevertheless it is not to be doubted that it can exist in nature, since it does not imply a contradiction, as the Philosophers say, and neither does anything absurd follow from it.*[20]

Notwithstanding all this, I do not admit that I acknowledge the existence of an unseemly blemish of which I ought, out of a sense of duty, to vindicate Euclid, especially since Clavius himself attributes it not to Euclid but to commentators. I say therefore that it was possible for *Elements* V, 18 to have been proved without any recourse to that inserted Axiom, using, to be sure, certain propositions of *Elements* VI which do not depend on this Axiom, and out of which the statement itself is proved in the form of a Problem, at least for straight lines. The very discriminating Clavius, perhaps, did not call attention to the fact, as he had otherwise to expand on the true doctrine of proportion in opposition to certain not undistinguished commentators of Euclid.

I shall continue to pursue the obligation I assumed. In this task, in order to be more unencumbered, I shall try to explain the matter itself in a few passages, employing analysis more than synthesis.[21] I set forth in advance, moreover, *Elements* VI, 1, which can be assumed to have been proved as much by me as by Euclid, without any dependence on the stated Axiom, since in that matter (if we except the theory of parallels) we only used *Elements* I, 38 and the Euclidean Definition of equiproportionals in *Elements* V.

Deinde transeo ad exequendum Problema, quod continetur in 12. sexti; ubi datis tribus rectis lineis, ut puta AD, DB, AE, praecipitur quartae proportionalis inventio. Disponantur enim (Fig. 51) primae duae AD, DB secundum lineam rectam, quae sit AB: Tertia vero AE quemlibet angulum A efficiat cum prima AD. Deinde ex D ad E recta ducatur DE, cui per B parallela ducatur BC, occurrens rectae AE productae in C. Dico ipsam EC esse quartam
[113] proportionalem quaesitam. Cum enim in triangulo ABC acta sit parallela DE; erit (ex 2. 6.) ut AD ad DB, ita AE ad EC. Quare EC erit quarta proportionalis quaesita. Quod erat &c.

Verum obstat, quod nondum demonstravi independentiam secundae Sexti ab illo intruso Axiomate. Itaque (manente eadem Figura) jungantur EB, CD. Constat (ex 37. primi) aequalia inter se fore triangula DEB, DEC, utpote constituta inter easdem parallelas DE, BC, & super eadem basi DE. Quare (ex 7. 5.) ut triangulum ADE ad triangulum DEB, ita est idem triangulum ADE ad triangulum DEC. Atqui ut triangulum ADE ad triangulum DEB, ita est (ex 1. 6.) basis AD ad basim DB; propter aequalem eorundem altitudinem in puncto E: & eadem ratione, ut triangulum ADE ad triangulum CDE, ita est basis AE ad basim EC; propter aequalem eorundem altitudinem in puncto D. Igitur (ex 11. quinti) ut AD ad DB, ita est AE ad EC; cum hae duae proportiones eaedem sint proportionibus trianguli ADE ad triangulum DEB, & ejusdem trianguli ADE ad triangulum DEC. Quod erat hoc loco demonstrandum.

Hinc porro evidenter constat nihil vetuisse, quin statim (praeciso ordine materiae) post 17. 5. procederet Euclides ad demonstrandas primam, secundam, & duodecimam sexti, quae utique satis erant ad stabiliendam, saltem pro lineis rectis, illam 18. ejusdem quinti, pro qua sola induci ibi debuerat praefatum Axioma. Quod enim post illam 12. sexti progredi facile possimus (sine indigentia novae alicujus praesumptae veritatis) ad inveniendam rectam lineam, quae ita se habeat ad aliam quamlibet datam rectam, ut unum quodvis datum triangulum rectilineum ad alterum quodlibet datum triangulum rectilineum, etiamsi non aequalis altitudinis cum priore; solis Geometriae imperitis difficile videri poterit. Porro uni-
[114] formiter constabit, reperiri similiter posse rectam lineam, quae ita se habeat ad quamlibet aliam datam rectam, ut una quaelibet data figura rectilinea ad alteram quamlibet datam figuram rectilineam. Atque haec rursum utrovis modo vice versa.

Sed video opponi adhuc mihi posse ipsummet Euclidem, qui in 2. 12., ut ostendat eandem esse rationem cujusvis circuli ad alterum quemlibet exhibitum circulum, quae est quadratorum ab ipsorum diametris, tacite praesumit unam aliquam esse in rerum natura superficiem, vel aequalem, vel majorem, vel minorem exhibito circulo, ad quam alter circulus eam habeat rationem, quae est praedictorum ab illis diametris quadratorum.

Hence I pass over this to consider the Problem in *Elements* VI, 12, where the fourth proportional to three given straight lines AD, DB, AE is to be found. For let the first two AD, DB be set up along a straight line AB (Fig. 51). The third AE, in fact, makes with the first AD any angle whatever. Then line DE is drawn from D to E. Through B a line BC is constructed parallel to DE and meeting the line AE extended in C. I say that this EC is the required fourth proportional. Since certainly in triangle ABC, DE serves as a parallel, AD is to DB as AE is to EC (*Elements* VI, 2). Therefore EC is the required fourth proportional. This is what was to be demonstrated.

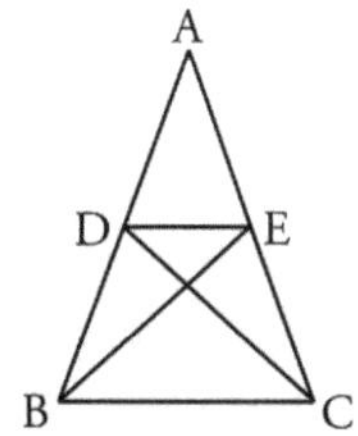

Fig. 51

Actually it still remains for me to prove the independence of *Elements* VI, 2 from that interpolated Axiom. Therefore (making use of the same figure), join EB, and CD. It is evident (*Elements* I, 37) that triangles DEB and DEC are equal to each other, inasmuch as they are situated between the same two parallels DE and BC, and on the same base DE. For which reason (*Elements* V, 7) as triangle ADE is to triangle DEB, so is the same triangle ADE to triangle DEC. But triangle ADE is to triangle DEB as base AD is to base DB (*Elements* VI, 1), since they have equal heights from point E; and, by the same reasoning, triangle ADE is to triangle CDE, as base AE is to base EC, because of their equal heights from point D. Therefore, (*Elements* V, 11) as AD is to DB, so is AE to EC, since these two proportions are the same as the proportion of triangle ADE to triangle DEB, and of triangle ADE to triangle DEC. This is what was to be demonstrated.[22]

Hence, it is established quite clearly that nothing prevented Euclid from proceeding immediately after *Elements* V, 17 (abridging the sequence of topics) to the proofs of *Elements* VI, 1 then *Elements* VI, 2 and finally *Elements* VI, 12, which were assuredly enough to establish, at least for straight lines, *Elements* V, 18, which was the only one for which he would have had to use the stated Axiom.[23] For indeed, after *Elements* VI, 12 we can easily advance (without need of any whatever new truth to be assumed beforehand) to finding a straight line, which has the same ratio to any other given line, as any whatsoever given rectilinear triangle is to any other given rectilinear triangle, even if they do not have equal heights as before. Only those unskilled in Geometry would have difficulty in seeing this. Next, it will be established, in one and the same manner, that it is possible to find a straight line, which is to any whatever other given line, as any given rectilinear figure is to any other given rectilinear figure; and *vice versa*.

But I see that Euclid himself might be opposed to me on this score in *Elements* XII, 2 in order to prove that the ration of any given circle whatever to any other given circle is the same as the squares of their diameters, he tacitly assumed the existence in nature of some one surface either equal to, or greater than, or less than the given circle, to which the other circle might have the same ratio as that of the squares of the aforesaid diameters.[24]

Et hic quidem duo respondenda censeo. Unum est; absonam utique videri potuisse, in ipso primo ingressu hujus materiae *de proportionibus*, praesumptionem ejusmodi; sed non etiam, postquam circa lineas rectas, ac figuras rectilineas demonstratum generaliter id sit (non obstante qualicunque earundem irrationalitate) aut incongruum, aut nimis remotum a veritate videri posse, quod simile quidpiam praesumatur circa duas figuras rectilineas ex una parte, & ex altera duas circulares. Alterum est; neque ibi necessariam esse Euclidi jam dictam praesumptionem. Nam antea in prima duodecimi sine ulla praesumptione demonstrat, duo quaelibet in circulis similia poligona esse inter se, ut a diametris quadrata. Deinde (in ipsa secunda) absolute rursum demonstrat, tale in exhibito circulo (duplicando, ac duplicando numerum laterum) inscribi posse poligonum, cujus defectus ab ipso integro circulo minor sit qualibet parvula assignata magnitudine; intacta nihilominus manente eadem semper ratione ad alterum simile poligonum, quod in alio circulo inscribatur. Quare
[115] ipsosmet circulos considerare jam potuit tanquam similia infinitilatera poligona, in quibus propterea firma adhuc consisteret jam demonstrata ratio, quae est quadratorum a diametris.

Agnosco tamen doctioribus Geometris nunquam plene satisfactum iri, nisi aliquo tandem pacto aut generaliter demonstrem praesumptum illud Axioma, aut saltem unum aliquod in ejus locum substituam, quod usui esse possit in decursu universae Geometriae. Ecce autem statim rem exequor.

Dico enim assumi posse hoc alterum Axioma; quod nempe *duae quaelibet, in quolibet eodem genere constitutae magnitudines, rationem inter se habebunt, quae vel aequalis sit, vel major, vel minor ratione duarum quarumlibet aliarum magnitudinum, quae vel in eodem cum prioribus genere, vel in alio quolibet ipsarum proprio constitutae sint*. Ubi constat duplex a me onus assumi; nimirum demonstrandi & veritatem propositi Axiomatis, & plenam ejusdem sufficientiam pro usu in decursu totius Geometriae.

Priori oneri sic satisfacio. Sint quatuor magnitudines (Fig. 52) A, B, C, D, quarum duae priores in suo proprio genere, ac similiter posteriores vel in eodem cum prioribus genere, vel in alio quodam suo proprio genere consistant. Dico rationem tertiae C ad quartam D vel aequalem fore, vel majorem, vel minorem ratione primae A ad secundam B.

And indeed, I believe that two answers can be given concerning this point. One is the following: certainly an assumption of this nature might seem unsuitable at the beginning of an essay *On proportions*; but actually, after it had been proved in all generality for straight lines as well as rectilinear figures (and also for those which are irrational in any way), it would not seem either incongruous or too remote from the truth to assume that something similar happens between two rectilinear figures and two circular ones. The other answer is that said assumption is not at all needed by Euclid, for earlier in *Elements* XII, 1 he proves without that assumption that any two similar polygons inscribed in circles are to each other as the squares of their diameters. From this, in *Elements* XII, 2 he again proves (without assumptions) that it is possible (by repeatedly doubling the number of sides) for a polygon to be inscribed in a circle in such a way that the difference between it and the whole circle is less than any whatsoever assigned small magnitude, nevertheless remaining always in the same ratio to the other similar polygon, which is inscribed in the other circle. Whereby he could consider the very circles as infinite-sided polygons,[25] so to speak, in which case the now proven ratio would hold, which is as the squares of the diameters.

I grant nevertheless that at no time would this be completely satisfactory to most experienced Geometers, unless finally I were to prove the assumed Axiom in full generality or under some condition, or at least were to substitute another in its place, one which could be used throughout the whole of Geometry.[26] The following shows how this object is accomplished with dispatch.

For I say this other Axiom can be assumed; that is: *any two magnitudes of the same kind will have a ratio which is either equal to, or greater than, or less than the ratio of any two other magnitudes whatever, which are of the same kind as the first, or anyhow of same kind among themselves*. Wherewith it is evident that I have assumed the double burden of proving both the truth of the proposed Axiom, and the complete sufficiency of the same for use in the whole course of Geometry.[27]

I discharge the first duty in the following manner. Let A, B, C, and D be four magnitudes (Fig. 52), the first two of which are of one kind, and the latter being either of the same kind as the first two, or of any other but same kind.[28] I say that the ratio of the third C, to the fourth D either is equal to, or greater than, or less than the ratio of the first A to the second B.

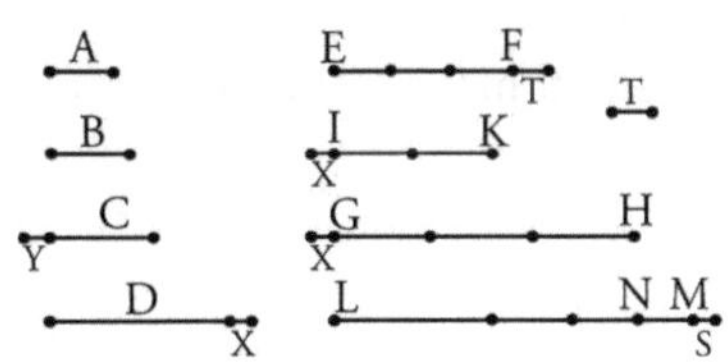

Fig. 52

Demonstratur. Sumantur enim ipsarum A primae, & tertiae C, duae quaelibet aeque multiplices EF, GH; atque item ipsarum B secundae, & quartae D, duae quaelibet aeque multiplices IK, LM. Constat primo, rationem ipsius A ad B aequalem fore rationi ipsius C ad D, si vel in uno casu talium assumptarum aeque multiplicium contingat, ut EF multiplex primae aequalis sit ipsi IK multiplici secundae, & GH multiplex tertiae aequalis sit ipsi LM multiplici quartae.

[116] Ratio evidens est; quia hic agitur de multiplici proprie tali secundum veram rationem numeri, juxta quam multiplicatum non transit in novam speciem entis, sed in eadem specie consistens ita se habet v. g. EF ad A, ut numerus ibi multiplicans ad unitatem. Quoniam igitur genitae magnitudines EF, & IK ponuntur aequales; consequens est (ex 19. 7.) ut numeri, per quos multiplicantur magnitudines A, & B, sint inter se reciproce, ut ipsae magnitudines multiplicatae. Simili modo ostendetur numeros, per quos multiplicantur magnitudines C, & D, esse inter se reciproce, ut sunt magnitudines multiplicatae; Porro (ex hypothesi) per eundem quempiam numerum multiplicatae intelliguntur magnitudines A, & C; ac similiter magnitudines B, & D. Igitur (ex 11. 5.) ita erit prima A ad secundam B, ut tertia C ad quartam D. Quod erat hic demonstrandum.

Constat secundo, rationem primae A ad secundam B majorem fore ratione tertiae C ad quartam D; si vel in uno casu talium assumptarum aeque multiplicium contingat, ut EF multiplex primae excedat ipsam IK multiplicem secundae, sed GH multiplex tertiae non excedat illam LM multiplicem quartae; aut illa EF aequalis sit praedictae IK (prout ego cum Clavio interpretor) dum altera GH minor est sibi correspondente LM.

Patet autem nulla mihi argumentatione opus hic esse. Nam omnes sciunt hanc ipsam esse rectam intelligentiam Definitionis octavae Libri quinti, juxta quam Euclides constantissime, ac rigidissime semper procedit. Ubi adverto (propter quosdam minus doctos) aliud esse accusare Euclidem de quadam veluti affectata in suis quibusdam definitionibus obscuritate;
[117] & aliud longe diversum, quod non rite juxta assumpta processerit; cum magis omnes docti consentiant accuratissimum hac in parte eundem fuisse. Unice igitur restat hoc loco, ut substitutum a me Axioma clarissime demonstrem.

Proof.[29] Take any two equimultiples EF and GH of the first A and of the third C; in the same way, take any two equimultiples IK and LM of the second B of the fourth D.

In the first place, it is established that the ratio of A to B is equal to the ratio of C to D, if indeed, in one case of equimultiples taken in that manner, it should happen that EF, the multiple of the first, is equal to IK, the multiple of the second, and GH, the multiple of the third, is equal to LM, the multiple of the fourth. The reasoning is evident, because in this case it is a question of multiples involving a true ratio of numbers, in which the multiplicand is not changed into a new species of being, but remains in the same species, in the same ratio (for example, EF is to A) as the multiplier number is to unity.[30] For which reason, therefore, if the magnitudes which are produced, EF and IK are considered as equal, then (from *Elements* VII, 19) the numbers, by which the magnitudes A and B are multiplied, are to each other as the very magnitudes that are produced by the multiplication. In a similar manner, it can be shown that the numbers by which the magnitudes C and D are multiplied, are to each other as the magnitudes that are produced. Furthermore (by hypothesis), the magnitudes A and C are understood to be multiplied by the same arbitrary number, and the same is the case with magnitudes B and D. Therefore (from *Elements* V, 11), the first A will be to the second B as the third C is to the fourth D. This is what was to be demonstrated.

In the second place it is established that the ratio of the first A to the second B is greater than the ratio of the third C to the fourth D if in one case of equimultiples taken in that manner, it should occur that EF, the multiple of the first exceeds IK, the multiple of the second, whereas GH, the multiple of the third does not exceed LM, the multiple of the fourth; or else EF is equal to the aforesaid IK (as I interpret, in the manner of Clavius)[31] whereas the other GH is less than the corresponding magnitude LM. Moreover, it seems to me that there is no occasion for disagreement on this point. For it is common knowledge that this is the correct interpretation of *Elements* V, def. 8, in accordance with which Euclid constantly, as well as rigidly, proceeded. For this reason I note (for the sake of certain less learned ones) that one matter is to accuse Euclid of a studied obscurity in a certain definition, and a very different matter is to claim that he did not proceed in accordance with the assumption; although those who are more learned agree that he was in this respect most careful. It only remains for me to clearly prove the substituted Axiom.

Sic autem procedo. Vel inter possibiles aeque multiplices primae A, & tertiae C, ac simul inter possibiles aeque multiplices secundae B, & quartae D, una quaepiam reperitur EF multiplex primae A, & IK multiplex secundae B invicem aequales; ac simul (in eodem casu) una quaedam GH multiplex tertiae C aequalis ipsi LM multiplici quartae D: vel nusquam talis aequalitas reperitur. Si primum; constat ex jam demonstratis ita fore A ad B, ut C ad D. Sin vero nusquam reperitur ejusmodi simul ex utraque parte aequalitas; vel saltem ad alterutram partem reperitur, ut puta ad partem primae A, vel nusquam. Si primum; ergo (ex praemissa Euclidaea majoris, ac minoris proportionis Definitione) habebit A ad B majorem, aut minorem proportionem, quam C ad D; prout GH multiplex tertiae C minor fuerit, aut major ipsa LM multiplici quartae D. Sin vero secundum; ergo ex una quidem parte v. g. ad ipsas A primam, & B secundam, contingere poterit, ut illa multiplex EF minor sit altera multiplici IK, dum vice versa ex altera parte illa multiplex GH major est altera multiplici LM. Tunc autem (sub eadem Euclidaea Definitione) ratio primae A ad secundam B erit minor ratione tertiae C ad quartam D; aut vice versa.

Igitur demonstratum manet substitutum illud Axioma; quod nempe duae quaelibet, in quolibet eodem genere constitutae magnitudines, rationem inter se habebunt, quae vel aequalis sit, vel major, vel minor ratione duarum quarumlibet aliarum magnitudinum, quae vel in eodem cum prioribus genere, vel in alio quolibet ipsarum proprio constitutae sint. Quod erat onus priore loco mihi impositum.

[118] Transeo ad secundum onus mihi adjectum. Ubi, ad exemplum aliarum in decursu Geometriae similium, illustrandam suscipio illam secundam duodecimi, prae missis (claritatis gratia) duobus sequentibus Lemmatis.

Lemma I.

Si quaepiam tertia magnitudo C habeat (Fig. 52) ad quartam D majorem rationem, quam sit ratio primae A ad secundam B; habebit etiam illa C ad eandem D, auctam quapiam magnitudine X, majorem rationem, quam sit praedicta ratio primae A ad secundam B.

Nam constat praedictam LM, multiplicem quartae D, tali magnitudine S augeri posse, ut adhuc minor sit illa GH multiplici tertiae C. Quare, si quarta D augeatur magnitudine X, cujus praedicta S ita sit multiplex, ut LM est multiplex quartae D, sive illa IK est multiplex secundae B; non ideo (ex defin. 8. 5.) ratio tertiae C ad quartam DX desinet esse major ratione primae A ad secundam B; quia adhuc GH multiplex tertiae C major erit illa LS multiplici quarte DX, dum interim EF multiplex primae A major non est illa IK multiplici secundae B. Quod erat &c.

I proceed therefore in this manner. Either among the possible equimultiples of the first A and of the third C, and similarly, among the possible equimultiples of the second B and of the fourth D, some EF, a multiple of the first A, and IK, a multiple of the second B, can be found that are equal to each other; and, at the same time, GH, a certain multiple of the third C, can be found equal to LM, the multiple of the fourth D; or on no occasion is such an equality to be found. If the first possibility holds, then it is evident from what has been proved, that A is to B as C is to D. If on the contrary, such an equality cannot be found for one as well as the other pair at the same time, either at the least it may be found for one (as for example for the first A) but not the other, or it cannot be found in any case. If the first of these possibilities holds, then (from the given Euclidean Definition of greater as well as lesser proportion) A will have a greater proportion to B, or lesser, than C to D according to whether GH, the multiple of the third C is less than, or greater than LM, the multiple of the fourth D. But if on the other hand possibility holds, then from one side, as for example to A the first and to B the second, it might happen that the multiple EF is less than the other multiple IK, while on the other hand, the multiple GH is larger than the multiple LM. In this case then (under the same Definition of Euclid), the ratio of the first A to the second B will be smaller than the ratio of the third C to the fourth D; or *vice versa*.

Therefore the substituted Axiom is to be regarded as having been proved: that any two magnitudes whatsoever of the same kind have a ratio to each other which is either equal to, greater than, or less than the ratio of any two other magnitudes which are of the same kind as the first, or anyhow of same kind among themselves. This was the task I assumed in the first place.

I now pass over the second task which I also assumed. In this I undertake to elucidate *Elements* XII, 2, as a model of all other similar ones in the course of Geometry, taking for granted (for the sake of clarity) the following two Lemmata.[32]

Lemma 1.

If the ratio of a third magnitude C to a fourth magnitude D (Fig. 52) is greater than the ratio of the first A to the second B, then the ratio of the same C to D augmented by some magnitude X, will be greater than the previously mentioned ratio of the first A, to the second B.

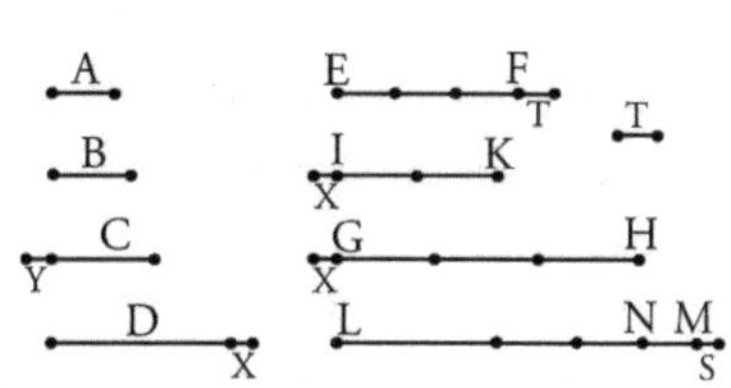

Fig. 52

For it follows that it is possible for the previously mentioned LM (the multiple of the fourth magnitude D), to be increased by such a magnitude S that the increased magnitude will still be less than GH (the multiple of the third magnitude C). For which reason, if the fourth D is increased by a magnitude X, of which the forementioned S is the same multiple that LM is of D, or, that IK is of B, then the ratio (from *Elements* V, def. 8) of the third C to the fourth DX would not cease to be greater than the ratio of the first A to the second B; because GH the multiple of the third C, would be greater than LS, the multiple of the fourth DX, while on the other hand EF, the multiple of the first A, is not greater than IK, the multiple of the second B. This is what was to be demonstrated.

Lemma II.

Porro autem (ne peritum Geometram in re facili moleste detineam) simili modo ostendam, quod tertia illa magnitudo C; quatenus ponatur habere ad quartam D minorem rationem, quam sit ratio primae A ad secundam B; habebit etiam ad eandem D, imminutam quapiam magnitudine T, minorem rationem, quam sit praedicta ratio primae A ad secundam B. Ratio est; quia (ex 26. 5.) habebit convertendo secunda B ad primam A minorem rationem,
[119] quam quarta D ad tertiam C. Igitur (ut in praecedente Lemmate) habebit adhuc quarta D ad tertiam C, auctam quapiam magnitudine Y, majorem rationem, quam sit secundae B ad primam A. Quapropter (ex eadem 26. 5.) habebit rursum convertendo prima A ad secundam B majorem rationem, quam tertia C, aucta illa magnitudine Y, ad quartam D.

Quoniam vero illa magnitudo Y sumi potest minor, ac minor, adeo ut sit pars quaedam ipsius C ab aliquo finito numero denominata; ac propterea sit pars ipsius YC a numero una unitate majore denominata: si rursum illius magnitudinis D sumatur pars quaepiam T, ab eodem numero denominata, a quo Y denominatur talis pars praedictae YC; erit (ex 15. 5.) Y ad T, ut YC ad D. Igitur (ex 13. 5.) ratio primae A ad secundam B major etiam erit ratione illius Y ad alteram T. Unde tandem fit (ex illa 13. & praedicta 15. 5.) ut illa tertia C, post sublatam ab YC additam illam portionem Y, minorem adhuc habeat rationem ad quartam D, imminutam nota illa magnitudine T, quam sit ratio primae A ad secundam B. Quod erat intentum.

Propositio Principalis:

In qua illustratur secunda Duodecimi; simulque exponitur substituti Axiomatis Major forsitan opportunitas pro similibus casibus in decursu totius Geometriae.

Demonstrat ibi Clavius ex Euclide, circulos esse inter se, quemadmodum a diametris quadrata. At supponit (ex communi ab Interpretibus intruso Axiomate) quod unus circulus
[120] habebit ad aliquam magnitudinem, quae vel aequalis sit, vel major, vel minor altero circulo, eam rationem, quae est praedicta quadratorum a diametris. Porro, hoc stante, clarissime exequitur intentum suum; quia optime demonstrat nullam esse posse magnitudinem aut majorem, aut minorem altero circulo, ad quam prior circulus eam habeat rationem, quam exposuimus.

Jam ego substituo alterum a me demonstratum Axioma, quod una cum duobus adjectis Lemmatis sic applicatur in nostra materia: Habebit unus circulus ad alterum circulum vel eandem rationem, quae est jam dicta quadratorum a diametris, vel habebit rationem ista majorem, non modo ad illum alterum circulum, sed etiam ad aliquam magnitudinem eodem altero circulo majorem, ut puta ad poligonum aliquod eidem circumscriptum: vel tandem habebit rationem minorem, non modo ad praedictum alterum circulum, sed etiam ad aliquam magnitudinem eodem minorem, ut puta ad poligonum aliquod eidem inscriptum.

Lemma 2.

Next, moreover (lest I annoyingly detain the skilled Geometer on a simple matter), I shall show in a similar manner that to whatever extent the third magnitude C is taken to be in lesser ratio to the fourth D than the first A is to the second B, it will be still in lesser ratio to the same D diminished by some magnitude T than the ratio of the first A to the second B. The reason is as follows: by conversion[33] (*Elements* V, 26), the second B will have to the first A a smaller ratio than the fourth D will have to the third C. Therefore (by the preceding Lemma 1), the fourth D will have to the third C increased by some magnitude Y a greater ratio than the second B will have to the first A. For which reason (from the same *Elements* V, 26) again by conversion, the first A will have to the second B, a greater ratio than the third C, increased by that magnitude Y, has to the fourth D.

Since, moreover, the magnitude Y can be taken smaller and smaller until it is a certain submultiple of C by a definite divisor,[34] an also, for this reason, it would be a submultiple of YC by the same divisor, plus one. If in turn, T is taken as a submultiple of D by the same divisor by which Y is a submultiple of YC, then (by *Elements* V, 15) Y will be to T as YC is to D. Therefore, the ratio of the first A to the second B will be greater than that of Y to T (*Elements* V, 13). Therefore, finally (*Elements* V, 13 and 15), the third C (after the magnitude Y has been subtracted from YC) will have a smaller ratio to the fourth D, decreased by the magnitude T, than the ratio of the first A to the second B. This is what was to be demonstrated.

Main Proposition

In which Elements XII, 2 is explained and, at the same time, the (perhaps) main advantage of the substituted Axiom for use in similar cases in the whole of Geometry is pointed out.

Clavius proves, from Euclid, that circles are to each other in the same manner as the diameters squared. He assumes (by the common Axiom inserted by commentators) that one circle will have to a certain magnitude which is either equal to, or greater than, or less than the other circle, the ratio which is that of the aforesaid diameters squared. Furthermore, this assumed, his intention is clearly executed, since he proves in the best possible manner that there can be no magnitude either greater or smaller than the other circle, to which the first circle has the ratio we have indicated.

Now I substitute the other Axiom which I have pointed out, and which, together with he two inserted Lemmata, has application to our subject. One circle will have to the other circle a ratio which is either that of the diameters squared, or it will have not only to the other circle, but to some magnitude greater than the other circle, as for example to any polygon circumscribed about it, a ratio which is greater than the ratio of the diameters squared; or finally, it might even have, not only to the other circle, but to some magnitude smaller than the other circle, as for example, to some polygon inscribed in it, a ratio that is smaller than the ratio of the diameters squared.

Tum unusquisque videt, quam facile (ex Prop. 8. 5.) demonstrari possit, non posse priorem circulum ad quoddam poligonum, posteriori circulo circumscriptum, habere majorem rationem, quam sit ratio illa quadratorum ex diametris; cum simile poligonum priori circulo circumscriptum habeat ad jam dictum poligonum (ex 1. 12.) eam solam rationem, quae est praedictorum quadratorum: ac similiter priorem jam dictum circulum habere non posse ad quoddam poligonum, eidem posteriori circulo inscriptum, minorem rationem, quam sit eadem ratio praedictorum quadratorum; cum simile poligonum priori circulo inscriptum habeat (ex eadem 1. 12.) ad modo dictum poligonum eam ipsam rationem, quae est illorum stabilium quadratorum: Atque hinc tandem fit, ut prior ille circulus (ex
[121] meo substituto, ac demonstrato Axiomate cum adnexis Lemmatis) eam habere debeat ad posteriorem circulum rationem, quae est jam dictorum quadratorum.

Constat autem pro similibus casibus procedi similiter posse in decursu totius Geometriae. Igitur substitutum, ac demonstratum a me Axioma, cum suis adnexis Lemmatis, non modo utile, verum etiam opportunius videri potest illo alio communi mere praesumpto, & indemonstrato. Quod &c.

Scholion I.

Quia tamen in meo secundo Lemmate bis assumpsi 26. 5. quae apud Clavium non demonstratur sine recursu ad illud commune Axioma, a me repudiatum in rigore Axiomatis; eam idcirco demonstro ex sola Definitione Euclidaea. Si enim prima A (Fig. 52) ponatur habere ad secundam B majorem rationem, quam tertia C ad quartam D; debebit quaedam EF multiplex primae A aequalis esse, aut major quadam IK multiplici secundae B; dum interim (ex aequo) quaedam GH aeque multiplex tertiae C, ut EF est multiplex primae A, vel minor est, vel non major quadam LM aeque multiplici quartae D, ut IK est multiplex secundae B. Tunc autem (considerando vice versa B ut primam, & A ut secundam; ac similiter D ut tertiam, & C ut quartam) debebit quaedam IK multiplex primae B aut aequalis esse, aut minor quadam EF multiplici secundae A; dum interim (ex aequo) quaedam LM aeque multiplex tertiae D, ut IK est multiplex primae B, vel major est, vel non minor quadam GH aeque multiplici quartae C, ut EF est multiplex secundae A. Quare (ex Def. 8. 5.) habebit ex opposito illa B
[122] statuta ut prima, ad alteram A statutam ut secundam, minorem rationem, quam sit ratio illius D statutae ut tertiae, ad alteram C statutam ut quartam. Quod erat &c.

Hence anyone can see how easily it can be proved (from *Elements* V, 8) that the first circle cannot have a greater ratio to the polygon which is circumscribed about the second circle than the ratio of the diameters squared, since a similar polygon circumscribed about the first circle is to the already mentioned polygon (from *Elements* XII, 1) as the diameters squared. Moreover, in a similar manner it can be shown that the said circle cannot have a ratio to that certain polygon inscribed in the second circle which is less than the ratio of the diameters squared, since a similar polygon inscribed in the first circle is to the already mentioned polygon (from the same *Elements* XII, 1) as the diameters squared. And hence finally it will be that the first circle (from the substituted Axiom which I proved, along with the two added Lemmata) has to the second the same ratio as the said squares.[35]

Moreover it holds that it is possible to proceed in a similar manner in similar cases throughout the whole of Geometry. Therefore the demonstrated Axiom which I substituted, together with the additional Lemmata, can be seen to be not only useful but also convenient; whereas that other common Axiom is entirely presupposed and never proved. This is what was to be demonstrated.

Scholium 1.

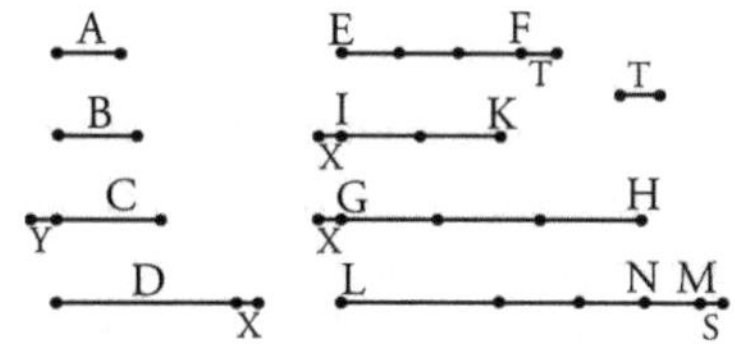

Fig. 52

Because *Elements* V, 26 was assumed twice in the proof of my Lemma 2, and because this proposition is not proved in Clavius without recourse to that commonly used Axiom which I reject as an axiom, for these reasons I will therefore prove this Proposition using only Euclid's Definition. Now, if a first A (Fig. 52) is assumed to have to the second B a greater ratio than the third C has to the fourth D, then a certain multiple EF of the first A will have to be equal to or greater than a certain multiple IK of the second B, while at the same time, a certain GH which is the same equimultiple of the third C that EF is of the first A, is either less than, or not greater than, LM, which is the same equimultiple of the fourth D as IK is of the second B. Then, moreover, considering B as first magnitude, A as second, and similarly D as third, and C as fourth; a certain multiple IK of the first B will either be equal to, or less than, EF, a multiple of the second A, while at the same time, a certain LM, the same equimultiple of the third D that IK is of the first B, is greater than or not less than GH, the same equimultiple of the fourth C that EF is of the second A. For which reason (*Elements* V, def. 8), if B is considered as first, will have to A, considered as second, a smaller ratio than D, considered as third, will have to C, considered as fourth. This is what was to be demonstrated.[36]

Scholion II.

Ne vero acutus quispiam Geometra accusare me possit, quasi de industria in rem meam immutaverim Definitionem Euclidaeam majoris, ac minoris proportionis; quoniam ego volo rationem primae A ad secundam B majorem esse ratione tertiae C ad quartam D; quoties (collatis inter se jam notis quibusvis illis aeque multiplicibus) vel semel inveniatur multiplex primae aequalis esse, aut major multiplici secundae, dum ex aequo multiplex tertiae vel minor est, vel non major multiplici quartae; cum Euclides (in illa Def. 8. 5.) unice id assumat, quoties multiplex primae major sit multiplici secundae, & non etiam multiplex tertiae major sit multiplici quartae: duas afferre hic debeo clarissimas responsiones, quae omnem dubitationem sustollant.

Prior responsio haec est; quod ipse Clavius (prout suo loco insinuavi) praedictam Definitionem sic interpretatur: *Quod si quando e contrario multiplex primae deficiat a multiplici secundae, non autem multiplex tertiae a multiplici quartae, dicetur prima magnitudo ad secundam minorem habere proportionem, quam tertia ad quartam.* Porro nemo omnium nescit *majus*, ac *minus* correlative invicem dici. Igitur, juxta hanc interpretationem, tertia magnitudo dicetur habere majorem rationem ad quartam, quam prima ad secundam; si quando contingat, ut multiplex tertiae vel aequalis sit, vel major multiplici quartae, dum multiplex primae minor est multiplici secundae; atque id propterea eodem jure vice versa. Cur vero (hoc stante) praefatus Clavius (cujus eximium in demonstrando nitorem magni a
[123] me fieri fateor) non demonstraverit eam 26. 5. ex sola Definitione; ex eo evenisse vaticinor, quia alias non dubitabat de veritate illius ab aliis Interpretibus jam intrusi Axiomatis; unde fieri potuit, ut nobilius, aut expeditius fortasse censuerit a prima ipsa Definitione aliquantulum recedere.

Scholium 2.

Actually, any clever Geometer might accuse me to having intentionally, and for my advantage, changed Euclid's Definition of greater as well as of lesser proportion. I say, in fact, that the ratio of the first A to the second B to be greater than the ratio of the third C to the fourth D, whenever (by comparison, as already noted, of any whatever equimultiples of them) either it is discovered at some time that a multiple of the first is equal to or greater than a multiple of the second, while a multiple of the third is either less than or not greater than a multiple of the fourth; although Euclid only considers (in *Elements* V, def. 8) the case in which the multiple of the first is greater than the multiple of the second, and at the same time the multiple of the third is not greater than the multiple of the fourth. I therefore have to bring to bear on this topic two arguments that will remove all doubt.[37]

The first argument is that Clavius himself (as I pointed out) interpreted the aforementioned Definition as follows: *Now if ever, on the contrary, the multiple of the first falls short of the multiple of the second, and the multiple of the third, however, does not fall short of the multiple of the fourth, we say that the first magnitude has to the second, a smaller proportion than the third to the fourth.* But no one at all is ignorant of the fact that the words *greater* and *less* are mutually related correlatively. Therefore according to this interpretation, the third magnitude is said to have a greater ratio to the fourth, than the first to the second, if it should happen that the multiple of the third is either equal to, or greater than, the multiple of the fourth, while the multiple of the first is less than the multiple of the second. Moreover by the same reasoning, it is true *vice versa*. Why, indeed (this being the case) did not Clavius (for whose distinguished elegance of proof, I must admit, I have great admiration) prove the same *Elements* V, 26 using only the Definition? On account of this, that since on other occasions he did not doubt the truth of that Axiom inserted by other Commentators, I judge that it is possible that perhaps he considered it superior or more rapid to abandon the first Definition somewhat.[38]

Posterior, & longe potior responsio est, quae sequitur. Licitum est unicuique definire, prout ipsi libuerit, terminos suae facultatis; dum tamen ex una parte eos nunquam usurpet, nisi juxta Definitiones jam stabilitas; & ex altera accusari istae non possint de confusione unius termini cum altero. Hoc autem loco nos esse in casu manifeste liquet. Nam constat, Definitiones aequalis, majoris, ac minoris proportionis traditas esse per membra opportune contradictoria, inter quae nec medium dari potest, nec inexpectata confusio. Quid enim clarius, minusque confusioni obnoxium; quam quod (sumptis quibusvis aeque multiplicibus primae, ac tertiae, ac rursum quibusvis secundum eandem, vel quamlibet aliam multiplicationem, aeque multiplicibus secundae, ac quartae) si fiat comparatio multiplicis primae cum multiplici secundae, ac similiter multiplicis tertiae cum multiplici quartae, vel una deficiant, vel una excedant, vel una aequalia sint? Similiter vero: quid clarius, minusque rursum confusioni obnoxium, quam quod (in defectu praedictae uniformitatis) aliquando contingat, ut multiplex primae vel major sit, vel saltem aequalis multiplici secundae, dum interim multiplex tertiae vel major non est, vel neque aequalis multiplici quartae; aut vice versa? Atque hinc (occlusa necessitate alterius subdivisionis) clarissime distinctas habemus Definitiones aequalis, majoris, ac minoris proportionis; inter primam, & secundam ex una parte; ex altera vero tertiam, & quartam. Plura alia in hanc rem hic omitto, quia spectantia ad sequens in hac materia Principale Scholium.

[124] **Scholion III.** ***omnino Principale.***

Nobilis quidam Nationis nostrae Italicae Scriptor, quem honoris caussa hic non nomino, quia vere eximium, ac magnum Geometram; nimius videri potest in accusando tam frequenter Euclide super Definitionibus terminorum. Assumo hic prae caeteris expendendam accusationem circa magnitudines proportionales. Praemittit laudatus Auctor neminem inficiari, quin sexta illa Libri quinti Definitio *perplexa sit, ignota, & ideo respuenda*, Rationem affert, quia *Definitio scientifica debet evidentissime exponere naturam rei definitae per passionem possibilem, veram, primam, & notissimam, per quam definitum distinguatur a quolibet alio subjecto.* At ego (cum bona venia tanti Viri) reponere possem, unum aliquem inter omnes esse Christophorum Clavium, ipsi optime notum, & alias in rem suam circa parallelas ab eodem citatum, qui tamen hanc ipsam circa proportionales magnitudines Euclidaeam Definitionem maxime commendat, tueturque ab adversis, aut extraneis interpretationibus. Sed praestat meritum causae diligentius inspicere.

Secondly, and by far the more important argument is the following. Anyone is permitted to define the terms of his own making in accordance with his wishes, as long as, however, he does not use incorrectly those Definitions already proposed, and these, moreover, cannot be accused of confusing one term with another. It is apparent that this is the case in this instance. For it is established that Definitions of equal, greater, as well as lesser proportions are set forth in clauses opportunely contradictory which can give rise to neither intermediate cases nor unexpected confusion. For what can be clearer, and present less danger of confusion than that (with any kind of equimultiples of the first as well as the third, and in turn, according to either the same or any other multiplication, with equimultiples of the second as well as of the fourth)[39] if a comparison were to be made of the multiple of the first with the multiple of the second, and similarly, of the multiple of the third with the multiple of the fourth, either together they should fall short, or together exceed, or together be equal? And similarly, what would be clearer and again present less danger of confusion, than that (lacking the aforementioned uniformity) it should happen at any time, that the multiple of the first either is greater than, or at least equal to the multiple of the second, while at the same time, the multiple of the third is either not greater or not even equal to the multiple of the fourth, or *vice versa*? And so, for this reason (with no need for further divisions of the terms) we evidently have separate Definitions for equal, greater, as well as lesser proportion, with the first and second magnitudes on the one side, and the other, indeed with the third and fourth. I omit here many other facts on this subject, because they are considered in the subject of the Principal Scholium which follows.

Scholium 3, *indeed the Principal one.*

A certain celebrated Writer of our Italian nation, whom I will not name here[40] out of respect for him, because, in truth, he is a well-known and great Geometer, seems to be intemperate in reproaching Euclid, on many occasions, in connection with Euclid's Definitions of terms. I take this up before the other complaints about proportional magnitudes that are to be considered. The esteemed author submits that there is no one who denies that *Elements* V, def. 6, is *involved, unknowledgeable*, and for that reason *to be rejected*. He gives this reason, that a *scientific Definition should most clearly explain the nature of the thing defined by a property, which is possible, true, first, and well known, by means of which the defined thing is distinguished from any other subject.*[41] But I (with no offence meant to so great a Man) might count as one person among many to be Christoph Clavius, himself highly esteemed and often quoted by this Geometer on the subject of parallel lines, who nevertheless had the highest praise for this Euclidean Definition of proportional magnitudes, and who guarded it against unfavorable and extraneous interpretations. The importance of this, however, warrants a more careful consideration of the reason.

Admitto *Definitionem scientificam* tradi debere, prout ibi describitur; omissa dumtaxat illa particula *primam*, de qua postea speciatim agemus. Nam multiplicatio propositarum quatuor magnitudinum, simulque praescripta collatio inter earundem multiplices, secundum rationem *majoris*, *minoris*, aut *aequalis*, quae semper, aut non semper consistat, passio est *possibilis*, *vera*, & *notissima*, immo etiam (propter membra contradictoria) cognita necessaria, *per quam definitum distinguitur a quolibet alio non tali subjecto*. Ubi indignum foret tanto Viro; si quis censeret eum respicere ad infinitas secundum quemlibet numerum
[125] propositarum magnitudinum praescriptas multiplicationes, quas certe mens humana assequi omnes distincte non potest, & sic neque inter se conferre secundum praescriptas rationes *majoris*, *minoris*, aut *aequalis*. Nam id necessarium non esse constare potest ex prima sexti, prout jam evidenter declaravi.

Venio ad ly *primam*. Atque hic (cum bona rursum venia) satis mirari non possum, quomodo laudatus Vir provocare velit ad passionem *primam* in sensu veluti metaphysico, ex qua nempe, tanquam primo fonte, reliquae proprietates ratione nostra promanare intelligantur. Si enim loquatur de passione *prima* quo ad nos, tam certum est assignatam ab Euclide esse verissime primam, quam certum est, ejusmodi eam esse, ut inde reliquae omnes huc usque notae proportionalium passiones cum summa certitudine eruantur; quod utique notum est omnibus Geometris. In hac proinde materia distinguendum sic puto. Vel agitur de subjecto quopiam per aliquam experientiam aliunde cognito, ut sunt elementa vulgaria, mista, planetae, stellae, aliaque hujuscemodi: & in istis dico Definitionem tradi debere per talem passionem, quae praecognita jam sit, eamque appello Definitionem *quid rei*. At ubi agitur de subjecto, quod nulli communi experientiae subjaceat, unde ab omni non ipso supponi debeat jam discretum, quales procul dubio sunt magnitudines inter se proportionales, & tunc dico opportunius omnino esse procedere per Definitionem *quid nominis*, dum scilicet observentur haec duo. Unum est; ut ne talis afferatur Definitio, quae repugnet, seu non compraehendat magnitudines invicem commensurabiles, quae nimirum sub datis quibusdam numeris exhiberi possint. Nam Definitio debet esse universalis, ac propterea non excludens magnitudines sub tali nomenclatura jam notas. Alterum est, ut nulla altera passio de ipso tali sic definito praesumatur, nisi quae ex ipsa
[126] tali statuta Definitione promanet. Atque ita certissime, nullo refragante, procedit in hac materia Euclides.

I admit that a *scientific Definition* ought to be given as it is described above, with the omission of the words "*first*", with which we will be concerned in particular at a later time. In fact, with the multiplication of the proposed four magnitudes, and at the same time, the prescribed comparison of their multiples according to the ratio of *greater, lesser,* or *equal*, which either always holds or does not always hold, the occurrence is *possible, true,* and *well known*, and even (because of the division in contradictory clauses) acknowledged necessary, and *by means of which the defined thing is distinguished from any other subject*. It would be unworthy of so great a Man, to think that he might have regarded the prescribed infinite multiplications of the proposed magnitudes, according to whatever number (all of which certainly the human mind cannot follow distinctly), as a hindrance to their comparison in accordance with the prescribed ratios of *greater, lesser*, or *equal*. In fact, it can be established from *Elements* VI, 1 that it is not necessary in practice, just as I have already declared.

I come to the words "*first*", and here (again without offence to him) I cannot cease to wonder how that esteemed Man would wish to invoke a *first* property in an almost metaphysical sense, from which (according to our reason) the remaining properties, as if from a first source, should flow. If indeed the subject of discussion is the *first* property for us,[42] we are as much certain that the one designated by Euclid is truly so, as from it all the remaining known properties of proportions derive with complete certainty – this is at any rate known to all Geometers. I believe therefore that the matter should be distinguished in the following manner. Either it concerns some subject which is known through another experience, as for example, common[43] and mixed elements, planets, stars, and other things of this kind. In this connection I say that the Definition should be presented through such an instance that it is known beforehand, and this I call a *real* Definition. But whenever it is a question of a subject which suggests no common experience, for which reason it is not already distinguished form other things, and magnitudes which are proportional to each other are without doubt in this category, then I say it is entirely more convenient to proceed by means of a *nominal* Definition, while certainly the following two rules should be observed. One is that a Definition should not be given if it contradicts or if, on the other hand, it isn't inclusive of commensurable magnitudes, which can be presented through some given numbers. For a Definition should be universal, and for that reason, should not exclude magnitudes under such nomenclatures as already noted. The other rule is that no other property of the very thing so defined should be assumed unless it comes out of the Definition already established.[44] And this is the manner, certainly, in which Euclid proceeded without exception.

Adeo igitur hac in parte repraehendi ipse non potest, ut immo a quadraginta circiter annis in fine meae Logicae Demonstrativae sic scripserim, ac demonstraverim. *Huc si respexissent doctissimi caeteroqui Geometrae, non tantum peccassent, ut in dubium revocarent Definitionem sextam Libri quinti Euclidis de aeque proportionalibus. Scilicet deponere noluerunt omnem praevium conceptum aeque proportionalium; unde factum est, ut quae recipienda erat tanquam Definitio* quid nominis, *respiceretur ut conclusio Theorematica aliunde confirmanda.* Paucis: In eodem meo Opusculo clarissime ostendo non Matrem, sed Filiam plurium praecedentium Demonstrationum esse debere illam *Definitionem*, quam nobis laudatus Vir tanquam origine primam commendare intendit.

Nondum tamen rem omnem a primis usque fibris discussi. Nam adhuc repraehendi ego possum, quod illam Def. 8. 5. ita assumpserim juxta Clavium; quasi una aliqua vice contingere non possit, ut multiplex primae aequalis sit multiplici secundae, dum multiplex tertiae minor est multiplici quartae; quin juxta quandam aliam assumptam ex aequo multiplicationem, multiplex primae excedat multiplicem secundae, dum multiplex tertiae non excedit multiplicem quartae; ac propterea unus quilibet priore loco dictus casus satis sit ad decernendum, quod ratio primae ad secundam major sit ratione tertiae ad quartam, cum aliis ab Euclide consequenter demonstratis. Quare hoc loco rem totam discutiendam assumo.

Sint igitur quatuor (Fig. 52) magnitudines, prima A, secunda B, tertia C, & quarta D; vel omnes quatuor in eodem genere; vel priores quidem in uno, & aliae duae posteriores
[127] in alio ipsarum communi genere constitutae. Assumptae etiam sint earundem, juxta praescriptum, aeque multiplices EF, IK, GH, LM; reperiaturque in uno tali casu EF quidem aequalis ipsi IK, at GH minor ipsa LM. Dico proportionem primae A ad secundam B majorem fore (juxta voces ipsas ab Euclide adhibitas in illa Def. 8. 5.) proportione tertiae C ad quartam D.

Demonstratur. Nam constat primo (ex Def. 6. 5.) hunc unicum casum sufficere, ut una proportio non dicatur esse alteri aequalis. Constat secundo (ex praedicta Def. 8. 5.) unicum item casum sufficere, ut prima A dicatur habere ad secundam B majorem proportionem, quam sit tertiae C ad quartam D; si probetur unam aliquam esse multiplicationem inter praescriptas, juxta quam multiplex primae A excedat multiplicem secundae B, dum interim multiplex tertiae C non excedit multiplicem quartae D. Hunc vero casum demonstrandum assumo ex illo ipso casu proposito, in quo EF multiplex primae A aequalis est ipsi IK multiplici secundae B, dum GH multiplex tertiae C minor est illa LM multiplici quartae D.

Indeed, therefore, in this sense Euclid cannot be blamed; as about forty years ago, at the end of my *Logica Demonstrativa*, I wrote as well as proved: *To this extent, the otherwise most learned geometers would not err so greatly if they did not cast doubt on the Sixth Definition of the Fifth Book of Euclid, that is, the definition of equiproportionals. Evidently they did not wish to set aside all previous concepts of equiproportionals so that what results is that a definition which was, as it were, to be regarded as a nominal definition, is to be considered as one whose conclusion is a type of theorem to be proved from something else.*[45] In a few words: In that same booklet of mine, I proved most conclusively that that *Definition* which the illustrious Man recommended to us as first in origin should be considered the Daughter and not the Mother of many of the preceding demonstrations.

However, I have not yet investigated the whole matter from the first part right on to the end. For until now I can be reproved for having taken *Elements* V, def. 8, in the form of Clavius, as if one other case could not occur, that we might have the multiple of the first equal to the multiple of the second, while the multiple of the third is less than the multiple of the fourth, as well as having, in consequence of a certain other equal multiplication taken in the same manner, the multiple of the first exceed the multiple of the second, while the multiple of the third does not exceed the multiple of the fourth; and therefore, either occurrence described above would be sufficient to judge that the ratio of the first to the second is greater than the ratio of the third to the fourth, with others proved in a suitable manner according to Euclid. For which reason, I take up the investigation of the whole matter at this point.[46]

Therefore let there be four magnitudes (Fig. 52), the first A, second B, third C, and the fourth D, with either all four of the same kind, or the first two of them of one kind, and the two latter of another, but same kind. Moreover, equimultiples EF, IK, GH, LM of these are taken according to the rule; and in one such case EF is taken equal to IK, while GH is less than LM. I say that the proportion of the first A to the second B will be greater (according to the words employed by Euclid in *Elements* V, def. 8) then the ratio of the third C to the fourth D.

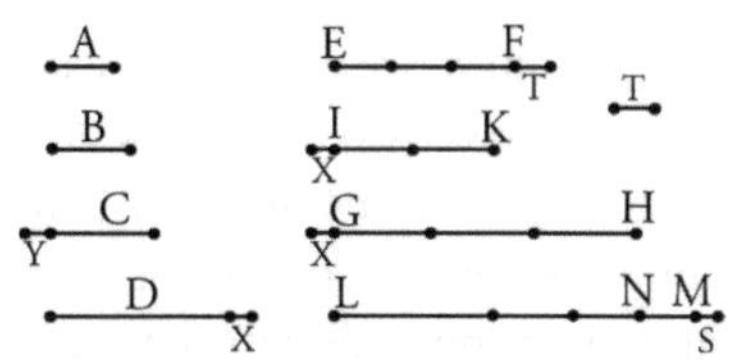

Fig. 52

Proof. First it has been established (from *Elements* V, def. 6) that this single case is sufficient for one proportion to be said to be not equal to another. Secondly it has been established (from the aforementioned *Elements* V, def. 8) likewise, that this single case is sufficient for it to be said that the first A has to the second B a greater proportion than the third C has to the fourth D, if it can be shown that some one multiplication among those prescribed is one for which the multiple of the first A will exceed the multiple of the second B, while, in the meantime, the multiple of the third C does not exceed the multiple of the fourth D. I consider this case proved, indeed, as a result of the case already proposed in which EF the multiple of the first A is equal to IK, the multiple of the second B, while GH, the multiple of the third C is less than LM the multiple of the fourth D.

Esto enim portio NM excessus quo LM superat illam GH. Tum praedicta ipsa GH tali magnitudine XG augeri intelligatur, quae & minor sit illo excessu NM, & sit pars quaedam tertiae C ab aliquo finito numero denominata. Deinde augeatur EF tali magnitudine FT, quae sit pars primae A, qualis XG pars est tertiae C. Quoniam igitur sermo est de multiplicibus, sive per numeros integros, sive per fractos; erit ET aeque multiplex primae A, ut XH est multiplex tertiae C. Praeterea manebunt, ut supra, IK, & LM aeque multiplices secundae B, & quartae D. Atqui ET multiplex primae A excedet tunc illam IK multiplicem secundae B, dum interim XH multiplex tertiae C adeo non excedit, ut immo deficiat a correspondente
[128] LM multiplici quartae D. Igitur a casu proposito transitur demonstrative ad alterum, qui exprimitur ab Euclide in illa Def. 8. 5. Quare constat de veritate, quam demonstrandam assumpseram.

Neque hic remorari quempiam debet, quod saepe assumam divisionem cujusvis datae rectae in quotlibet praescriptas aequales partes. Nam constat divisionem ejusmodi non indigere doctrina aeque proportionalium, ut videri potest apud Clavium, qui hanc divisionem demonstrat post Prop. 40. Libri primi. Neque etiam accusari hic possum, quod ordine quodam praepostero usus fuerim; quia nempe, resolvendo magis quam componendo demonstrare multa debui illustrando Euclidi necessaria. Nam facile est, si placeat, ipsum naturae ordinem sequi.

Et primo censeo Euclidaeam aeque proportionalium Definitionem pulcherrimam esse; & quia conceptam per voces communi intelligentiae facillimas, quales sunt *multiplicatio*, *majus*, *minus*, *aequale*; & quia compraehendentem, sine ulla necessaria discretione, omnes magnitudines, seu rationales, seu quomodolibet irrationales.

Secundo censeo excludendum esse indecorum illud Postulatum, sub nomine Axiomatis intrusum; cum sine illo, & ex solis aliunde notis, etiam problematice procedi possit circa omnes & lineas rectas, & figuras itidem rectilineas, opportune invicem comparatas; prout supra non modo declaravi, verum etiam ex parte demonstravi; unde utique opportunior postea, si opus sit, videri possit similis praesumptio circa reliquas omnes magnitudines.

Tertio censeo, sine ullo adjecto extraneo Postulato, ex sola Euclidaea aeque Proportionalium Definitione tale elici posse veluti Axioma, quod tutissime per omnem Geometriam versetur. Ejusmodi autem est: *Omnis magnitudo ad aliam quamlibet ejusdem generis magni-*
[129] *tudinem habet rationem vel aequalem, vel majorem, vel minorem illa, quae est cujusvis alterius magnitudinis ad aliam quamlibet in suo earundem proprio genere constitutam magnitudinem.* Tum vero, post duo alia a me demonstrata Lemmata, sic tandem stabilio integrum Axioma, quod sit immediate utile in qualibet materia: *Habebit omnis quaepiam tertia magnitudo C ad quamlibet aliam quartam magnitudinem D in eodem cum ipsa genere constitutam; vel rationem aequalem illi, quae est cujusdam primae magnitudinis A ad quamlibet secundam magnitudinem B, quae in eodem suo proprio cum magnitudine A genere consistat; vel tertia illa magnitudo C habebit majorem rationem non modo ad magnitudinem D, verum etiam ad aliquam magnitudinem majorem illa magnitudine D; vel tandem habebit rationem minorem non modo ad praedictam magnitudinem D, verum etiam ad aliquam magnitudinem eadem magnitudine D minorem.*

For let the segment NM be the excess by which LM is greater than GH. Then the aforesaid GH is known to be increased by such a magnitude XG, which is less than the excess NM, and is a certain submultiple of the third C under a finite divisor. Then let EF be increased by such a magnitude FT, that it is the same part of the first A that XG is of the third C. Since, therefore, it is a question of integer or fractional multiples,[47] ET will be the same multiple of the first A that XH is a multiple of the third C. Moreover, it will remain, as above, that IK and LM are equimultiples of the second B and of the fourth D. However ET, the multiple of the first A then exceeds that multiple IK of the second B, while in the meantime XH the multiple of the third C does not exceed, but on the contrary, falls short of the corresponding LM, the multiple of the fourth D. Therefore, it is proved that the proposed case may be transformed into the other one which is expressed by Euclid in *Elements* V, def. 8. All of which establishes the truth of the statement I undertook to prove.

Neither ought one linger on the fact that I often take for granted the division of a given straight line into any prescribed number whatsoever of equal parts. For it is established that a division of this type does not require the doctrine of equiproportionals, as can be seen in Clavius,[48] who proved this division after *Elements* I, 40. Neither can I be blamed for having made use of a somewhat reversed order, because it was necessary for me to prove many statements required for the clarification of Euclid through analysis rather then synthesis. Indeed, it is an easy matter, if it pleases one, to follow the natural order.

In the first place, I consider the Euclidean Definition of equiproportional to excel all others, because it is conceived in simple words of common knowledge, which are *multiplication, greater, less, equal,* and because it embraces all magnitudes without the necessity for distinguishing as to whether they are rational or irrational of any type whatsoever.

Secondly, I think that the improper Postulate, inserted under the name of Axiom, ought to be excluded. Because without it, and only referring to the other established Axioms, it is possible to proceed, as noted elsewhere, with problems concerning all straight lines and rectilinear figures, comparing them mutually and opportunely, in accordance with what I not only declared above but which I also proved in part. From this a similar assumption about all the remaining magnitudes can be made more convenient later, if the need should arise.[49]

In the third place, I believe that without the added extraneous Postulate, and only from the Euclidean Definition of equiproportionals, an Axiom of such a kind can be deduced that it can be used safely throughout the whole of geometry: *Every magnitude has to any other magnitude of the same kind a ratio that is either equal to, or greater than, or smaller than the ratio of any other magnitude to any magnitude whatsoever of the same type.* Then, indeed, with the other two Lemmata already established by me, I can express the whole Axiom, which is immediately useful in any subject: *Any third magnitude C will have to a fourth magnitude D constituted of the same kind as itself, either a ratio equal to the ratio of any first magnitude A to a second magnitude B, which is of the same kind as magnitude A, or the third magnitude C will have a greater ratio not only to the magnitude D but to some magnitude greater than D, or finally will have a smaller ratio not only to the aforementioned magnitude D but also to some magnitude smaller than D.*

Quod autem Axioma ejusmodi opportunissimum sit, jam supra ostendi, sumpta experientia a secunda duodecimi. Sed idem rursum experiri volo sub tota sua generalitate, circa illam 18. 5. ut magis constet nullam fuisse, in ipso ferme Geometriae initio, necessitatem illius intrusi Postulati.

Proponit ibi Euclides (Fig. 53) compositas magnitudines proportionales fore, si divisae proportionales sint; ut puta, ita fore AC ad BC, ut DF ad EF; si divisae proportionales sint; nimirum, si ita sit AB ad BC, ut DE ad EF.

Demonstratur. Et primo non erit AC ad quandam YC, minorem ea BC, ut DF ad EF; quia dividendo ita foret (ex praec. 17. 5.) AY ad YC, ut DE ad EF, sive (ex 11. 5.) ut AB ad BC; cum divisae istae magnitudines suppositae sint proportionales. Hoc autem absurdum est, contra 8. ejusdem 5. ex qua constat rationem illius AY ad YC majorem fore ratione praedictae AB ad BC.

[130] Simili modo ostendetur non esse AC ad quandam AX, minorem praedicta AB, ut DF ad EF; quia uniformiter (ex praedictis 17. & 11. 5.) deberet esse AX ad XC, ut AB ad BC; contra eandem 8. 5.

Tum ex meo illo Axiomate, sub adnexis Lemmatis perfecte constituto, demonstro principale intentum. Nam AC ad BC habebit vel aequalem, vel minorem, vel majorem rationem, quam DF ad EF. At non minorem, neque majorem; quia, in qualibet vicinitate punctorum Y, & X ad punctum B, erit semper (stante proportionalitate jam dictarum partium) ratio illius AY ad YC, major, & ratio illius AX ad XC, minor ratione praedictae AB ad BC, sive DE ad EF. Igitur; ne facta incidentia punctorum Y, & X in idem punctum B, ex libera permissa destinatione, debeat esse ratio AB ad BC & major, & minor ratione illius DE ad EF; consequens est ita fore AC ad BC, ut DF ad EF; dum scilicet divisae illae magnitudines AB, BC, & DE, EF supponantur proportionales. Quod &c.

Sed quia hic agitur de stabiliendo uno modo argumentandi circa magnitudines proportionales, qui frequentissimus est apud Geometras; nolo fidere (in hac summe abstracta rationum similitudine) illi soli communi notioni; quod, ubi consistitur in eodem infimo genere, non possit successive ordinatim transiri de majori in minus, nisi transeundo per aequale. Itaque sic rursum argumentor ex eodem meo, noviter illustrato Axiomate. Nequit AC ad BC habere v. g. majorem rationem, quam DF ad EF, quin majorem etiam habeat rationem ad quandam XC, majorem praedicta BC, dum nempe alias supponatur ita esse AC ad BC, ut DF ad EF. Haec autem simul stare non posse, ita demonstro.

That an axiom of this type is most convenient, I have already demonstrated above, having assumed it for the proof of *Elements* XII, 2. But I wish to prove the same once again in all its generality in connection with *Elements* V, 18, so that it might be established more completely that there is no necessity, in the very beginning of Geometry, for the Postulate which was inserted.

In this Proposition, Euclid proves that magnitudes are in proportion by composition if they are in proportion separately, as for example, (Fig. 53), AC would be to BC as DF is to EF, if the magnitudes are divided proportionally, that is, if AB is to BC, as DE is to EF.

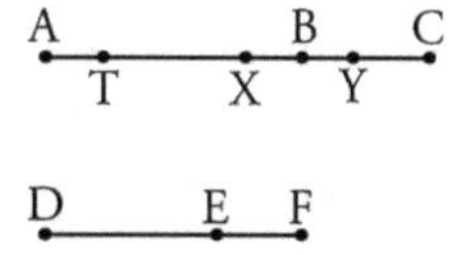

Fig. 53

Proof.[50] Firstly, AC cannot be to a certain YC, which is less than BC, as DF is to EF, because, by separation (from *Elements* V, 17) AY would be to YC, as DE is to EF, or (from *Elements* V, 11) as AB is to BC, since we assume that the separated magnitudes are in proportion. This, however, is absurd, contradicting *Elements* V, 8, from which it can be established that the ratio of AY to YC is greater than the ratio of the aforesaid AB to BC.

Similarly, it can be shown that AC cannot be to a certain AX, which is less than the previously named AB, as DF is to DE, because in one and the same manner (from the previously mentioned *Elements* V, 17 and 11), AX would be to XC, as AB is to BC, in contradiction to the same *Elements* V, 8.

Thereupon, from my Axiom, complemented with the added Lemmata, I prove the principle passage. For AC will have to BC a ratio either equal to, or less than, or greater than that DF to EF. But it can be neither less, nor greater, because however near to the point B are points Y and X taken, the ratio of AY to YC will always be greater, and the ratio of AX to XC smaller, than the ratio of the previously mentioned AB to BC, or DE to EF (the proportionality of the mentioned parts still holding). Therefore, making Y and X coincide with the point B, a determination that is freely permitted, the ratio of AB to BC would be both greater than, and less than, the ratio of DE to EF. As a consequence, AC would be to BC, as DF is to EF, while naturally the magnitudes AB, BC, and DE, EF are supposed to be divided proportionally.

But because at this point it is a question of establishing a method of arguing about proportional magnitudes which is most frequently used by Geometers, I am unwilling to trust (in this extremely abstract similarity of ratios)[51] to those common notions alone, namely that where the argument is about the *infimum* genus,[52] one cannot successively pass in an orderly fashion from greater to lesser without passing through equality. For this reason, therefore, I go on in the argumentation by means of my own newly illustrated Axiom. It is impossible, for example, for AC to have a ratio to BC which is greater than the ratio of DF to EF without also having a greater ratio to some XC greater than BC, when on the other hand it is assumed that AC is to BC as DF is to EF. That this cannot hold at one and the same time, I shall prove as follows.

[131] Quandoquidem rursum (ex illo eodem meo Axiomate) haberet AC ad quandam TC, majorem illa XC, rationem adhuc majorem, quam sit ejusdem DF ad EF; atque ita semper usque ad ipsum punctum A. Hoc autem ex ipso Euclide certissime absurdum est; quod nempe magnitudo aliqua ad aequalem habeat majorem rationem, quam tota quaepiam DF ad unam sui partem EF. Unice igitur restat, ut harum rationum majorum unus quispiam sit terminus, si non intrinsecus, at saltem mere extrinsecus, ut puta in eo puncto T; adeo ut nempe habeat quidem AC ad TC rationem minorem, quam DF ad EF; sed rursum ad quamlibet, minorem illa TC, majorem rationem obtineat. Verum hoc etiam repugnat meo illi Axiomati. Si enim habet AC ad TC minorem rationem, quam DF ad EF, habebit etiam AC ad aliquam, quae minor sit eadem TC, minorem adhuc rationem illius DF ad EF. Non igitur subsistere potest terminus ille extrinsecus constitutus in illo quolibet designato puncto T. Inde autem sit, ut ratio AC ad BC nequeat esse major ratione DF ad EF. At simili modo ostendetur rationem DF ad EF majorem non esse ratione AB ad BC. Quare (ex illo meo Axiomate) unice restat, ut ratio illius DF ad EF non nisi aequalis sit rationi ipsius AC ad BC, dum scilicet divisae magnitudines AB, BC proportionales supponantur divisis magnitudinibus DE, EF. Quod erat &c.

Atque haec satis jam sunt ad ostendendam illius Definitionis Euclidaeae non modo certitudinem, verum etiam opportunitatem ad repellendum intrusum illud, sub nomine Axiomatis, Postulatum.

Since indeed (from my Axiom) AC would still have to some TC greater than XC, a ratio that is greater than the ratio of DF to EF, and so on for a succession of points T right up to the very point A. This would certainly be absurd, according to Euclid himself, because evidently no magnitude can have to its equal a greater ratio than the ratio of the whole DF to one of its parts EF. Hence it follows that there is a limit of these greater ratios, if not an intrinsic limit, then at least an extrinsic one, as for example the point T, to the end that AC would have to TC a smaller ratio than DF to EF, but on the other hand would have to any magnitude less than TC a ratio that was greater than the ratio of DF to EF. This, however, is also inconsistent with my Axiom. For if AC has a smaller ratio to TC than the ratio of DF to EF, magnitude AC would also have to some magnitude less than TC, a ratio smaller than that of DF to EF. Therefore there is no external limit at any designated point T. Hence the ratio of AC to BC cannot be greater than the ratio of DF to EF. In a similar manner it can be proved that the ratio of DF to EF cannot be greater than the ratio of AC[53] to BC. For which reason (from my Axiom) it follows that the ratio of DF to EF is equal to the ratio of AC to BC when of course the separated magnitudes AB, BC are assumed to be in proportion to the divided magnitudes DE and EF. Which was etc.

Moreover these are already sufficient to illustrate not only the certainty of the Euclidean Definition, but also the opportunity offered for rejecting that Postulate inserted under the title of an Axiom.

Euclidis ab omni naevo vindicati Liber Secundus

Pars Secunda

In qua expenditur quinta Definitio Libri sexti Euclidaei.

[132] Definitio est, quae sequitur: *Ratio ex rationibus componi dicitur, cum rationum quantitates inter se multiplicatae aliquam effecerint rationem.*

Definitionem hanc egregie more suo elucidat Clavius, qui nempe ad Def. 10. Lib. 5. jam explicaverat, quo sensu una ratio dicatur penes Euclidem alterius cujusdam *duplicata*, *triplicata*, atque ita consequenter. Sed placet rem totam a primis usque initiis diligentius scrutari; nimirum hic addendo, quae in priore hujus Libri parte videri potuissent importuna.

Nam praelaudatus, Nationis nostrae Italicae, eximius Geometra ipsam etiam Libri quinti Def. 3. accusat, ubi legimus: *Ratio est duarum magnitudinum ejusdem generis mutua quaedam, secundum quantitatem, habitudo*; ac similiter Def 4. ubi habemus: *Proportionem esse harum rationum similitudinem.* Quasi vero (inquam ego) Definitiones istiusmodi quidquam amplius continere deberent, praeter abstractos quosdam terminos, grammatico more ibi explicatos, sed postea per voces communi usu notissimas philosophice explicandos, sine periculo ullius confusionis; prout sit in Definitionibus sexta, & octava ejusdem Libri.

Euclid Vindicated from every Blemish
Book Two

Second Part

In which the Fifth Definition of the Sixth Book of the Elements is explained.

The definition is the following: *A ratio is said to be compounded of ratios when the quantities of the ratios multiplied together, form some ratio.*[1]

Clavius explained this definition in his very elegant manner when he set forth in connection with *Elements* V, def. 10, the sense in which, for Euclid, a ratio is said to be the *duplicate*, or *triplicate*, and so on, of any other ratio. But it seems proper to me to examine the whole matter very carefully from first principles right up to the elements, with the addition here of whatever seemed unsuitable to the First Part of this Book.

Now the previously praised Geometer of distinction, one of our own Italians, finds fault with *Elements* V, def. 3, which reads as follows: *A ratio is a certain mutual relation of quantity between two magnitudes of the same kind*; and similarly in *Elements* V, def. 4, where we have: *A proportion is a similarity of these ratios.*[2] As if (I say) Definitions of this kind should contain something more than certain abstract terms explained in a grammatical manner, but later stated philosophically in known words of common usage, without danger of confusion, as it is in *Elements* V, def. 6 and 8.

Quid vero, si latinus interpres male posuerit *quantitatem*, cum magis debuerit scribere *quotitatem*, prout interpretatur, ac demonstrat ex Graeco Euclidaeo textu, omnimode lau-
[133] dandus Joannes Vallisius? Tunc enim multo magis, etiam ante subsequentes Definitiones, promptum foret intelligere non loqui ibi Euclidem de qualicunque habitudine, seu relatione unius magnitudinis ad alteram, sed de illa dumtaxat, juxta quam una vel est alteri aequalis, vel tali quodam modo altera major, aut minor. Ubi; ne quis erraret circa magnitudines invicem comparatas, notumque faceret se loqui de magnitudinibus ejusdem generis; subdit Definitionem quintam, in qua dicit: *Rationem inter se habere magnitudines, quae possunt multiplicatae sese mutuo superare*: Unde utique constat nullam esse v. g. cujusvis lineae rationem ad quamlibet superficiem, quia nulla linea quantumvis multiplicata superare potest vel minimam quampiam superficiem.

Quo loco fateor praeclarum Geometram Christianum Volfium bene observare in suis Elementis Arithmeticae Cap. III. Schol. I. quod tertia illa Euclidaea Definitio *videri potest incompleta*; quia nempe *dantur & aliae magnitudinum relationes, quae sunt constantes, nec tamen in rationum numero continentur*: Ubi ex Trigonometria in exemplum affert relationem *sinus recti ad sinum complementi*. Quamvis enim certissimum sit talem esse habitudinem, seu relationem cujusvis sinus recti ad sinum correspondentis complementi, ut quadrata utriusque sinus simul sumpta aequent quadratum sinus totius; aliunde tamen scimus non eandem esse *rationem* cujusvis sinus recti ad sinum correspondentis complementi, quae est alterius sinus recti ad sinum sibi correspondentis complementi. Unde infert non idem esse hac in re *habitudinem*, seu *relationem* ex una parte, & ex altera *rationem*, juxta communem Geometrarum intelligentiam.

Nihilominus dico, neque ex hoc capite accusari posse Euclidem. Nam in sua Definitione
[134] dicit; mutua *quaedam* secundum quantitatem *habitudo*. Qui autem dicit *quandam habitudinem*, certe non vult compraehendere omnes habitudines, seu relationes. Et hic rursum expendere debeo ly *secundum quantitatem*. Quis enim putet loqui ibi Euclidem de *quantitate* metaphysico more expensa, juxta quam unum corpus dicitur alteri naturaliter impenetrabile; & non magis loqui de *extensione* in suo tali quodam genere, juxta quam una magnitudo dicitur, relate ad alteram, vel aequalis, vel major, vel minor? Et sane ad interrogationem *quanta sit* quaepiam linea recta, respondebitur v. g. palmaris, bipalmaris, tripalmaris, aut alio quovis modo, cum relatione (seu per numeros integros, seu per fractos, sive etiam per minutiam) ad palmum, aut ad aliam longitudinem jam notam. Atque ita uniformiter circa alias cujusvis generis magnitudines.

What if, in truth, the Latin translator erroneously fixed upon he word *quantity*, when he should have written *multiplicity*, which is the word employed in the translation and explanation of the Greek text of Euclid made by John Wallis,[3] who is greatly deserving of praise in every respect? For indeed it would have been much easier, even before the subsequent Definitions, for Euclid to observe and not to speak about any condition or relation of one magnitude to another, except in so far as to describe whether or not one is equal to, or in some way greater than, or less than, the other. Moreover, so that no one will make a mistake about the magnitudes that are being compared with each other, and so that it is understood that we are speaking of magnitudes of the same kind, he added *Elements* V, def. 5, in which he says: *Magnitudes have a ratio to one another, which are capable, when multiplied, of exceeding one another*. From which it is established, for example, that a line of any kind whatsoever cannot have a ratio to any surface, because no matter by what number the line is multiplied, it cannot exceed even the smallest surface.[4]

On this matter, I admit that the celebrated Geometer Christian Wolff made a very apt observation in his *Elements of Arithmetic*, Chapter 3, Scholium 1, to the effect that *Elements* V, def. 3, *may seem to be incomplete*, because in fact *other relations are given which are constants, and are not included in the class of ratios*. In this connection he gives as an example from Trigonometry, the relation of the *sine* to the *cosine*. Although certainly it has been established that there is a relation between the sine and the cosine of an angle, i.e. that the square of one and of the other taken in a sum equals the square of the sine of a right angle; we know, however, that the *ratio* of the sine to the cosine of an angle is not the same as the ratio of the sine to the cosine of any other angle. From which Wolff gathers that, in the common use of Geometers, a *property* or *relation* is not the same thing as a *ratio*.[5]

Nevertheless, I say that Euclid cannot be reproached on this point. For in his Definition he says: *a certain mutual relation* according to quantity. Anyone, moreover, who says *a certain relation* assuredly does not wish to include all properties or relations. And again in this case I must explain the words "*according to quantity*". Who would think that Euclid speaks of *quantity* considered in this instance in a metaphysical sense,[6] just as one body is said to be naturally impenetrable to another, and not to speak rather of an extension in its own definite kind, in consequence of which one magnitude is said to be related to another as either equal to, or greater, or less than? And of course to the question *how great* is one line, the answer would have to be, for example, one hand, two hands, or three hands, or some other measure, with a relation (either in integers, or fractions, or fractions of the unity)[7] in hands or some other already known length. And so uniformly in connection with any type of magnitude whatsoever.

Unde infero tertiam illam Def. Euclidaeam nulli querelae obnoxiam esse, etiamsi inspiciatur seorsum a consequentibus. Quod enim praelaudatus Christianus Volfius *rationem* definiat *esse eam homogeneorum relationem, quae quantitatem unius determinat ex quantitate alterius sine tertio homogeneo assumpto*, verissime quidem dicit; quia hinc manifestum fit, non omnem *relationem*, etiam constantem, unius talis magnitudinis ad alteram, ut puta sinus recti ad sinum complementi, esse *rationem*; cum quantitas sinus complementi determinari non possit ex quantitate sinus recti, nisi assumatur tertium homogeneum, quale est sinus totus: sed non ideo repraehendi hinc potest Euclides, quasi *incomplete* definiverit; cum haec ipsa discretio manifesta sit in illis suis vocibus *secundum quantitatem*, accepta nimirum *quantitate* pro *quotitate*, prout certissime intelligi sic debere paulo ante declaravi, sumpto argumento ex interrogatione super *quantitate* talis cujusdam propositae magnitudinis. Atque ita a prima usque aetate intellectum a me fuisse Euclidem, fateri omnibus possum. Hoc
[135] autem stante vix intelligo, quomodo dubitari possit, an ex quantitate unius magnitudinis decerni possit immediate quantitas alterius, dum alias nota sit praedicta unius ad alteram *secundum quantitatem* habitudo: Nam qualiter una illarum nota erit, taliter rursum, sine alio extrinsecus assumpto, nota erit etiam altera ex illa sola praecognita habitudine.

Post haec gradum facio ad illam quintam Def. sexti: Ubi dico falsissimum esse, quod sub specie simplicis Definitionis Axioma quoddam intrudatur, non permittendum sine demonstratione.

Et primo: Si sermo sit de magnitudinibus in suo tali quodam ordine commensurabilibus, sive rationem inter se habentibus, quae est alicujus numeri (aut integri, aut fracti, aut cujusvis minutiae) ad alium quempiam hujusmodi numerum, ita ut nempe prima quaedam magnitudo A per talem quempiam seu numerum, seu minutiam multiplicata, aequalis fiat secundae magnitudini B; adeo clare, & immediate ostendetur veritas illius Definitionis, ut etiam Axiomatis loco aequissime censeri possit.

Sint enim quatuor quaelibet (Fig. 54) ejusdem generis magnitudines, prima A, secunda B, tertia C, & quartae D, praedicto modo invicem rationales. Dico rationem primae A ad quartam D componi ex rationibus magnitudinum intermediarum, hoc est primae A ad secundam B, secundae B ad tertiam C, & tertiae C ad quartam D. Scilicet dico magnitudinem A toties contineri in magnitudine D (sumpto ly *toties* pro quolibet numero, sive integro, sive fracto, sive unitate, aut qualibet ipsius unitatis minutia) quotus fuerit numerus ortus, aut quaelibet unitatis minutia, ex ductu inter se numerorum praedicto modo sumptorum, qui significent quoties magnitudo A continetur in magnitudine B, haec in magnitudine C, & ista
[136] in magnitudine D. Sint porro isti numeri T, X, Y, qui inter se ducti procreent numerum Z.

From which I infer that the Definition of Euclid in *Elements* V, def. 3, is not answerable to any complaint, even if it is examined apart form the others. Because certainly the previously praised Christian Wolff speaks with justification in defining *ratio* to be that *relation of homogenous magnitudes which determines the quantity of one by the quantity of the other, without assuming a third homogenous magnitude,*[8] because here it is evident that not every *relation*, even if determined, of one magnitude to another, as for example, the relation of the sine to the cosine, is a *ratio*; as the quantity of the cosine cannot be determined from the sine without a third homogenous magnitude being assumed, which is the sine of the right angle. Euclid cannot, however, be reproached in this instance for that reason, that is, for having given an *incomplete* definition, since this very distinction is evident in the words *according to quantity*, quantity meaning *multiplicity*, as I declared a short time back, about the question arising from the discussion on the *quantity* of any given magnitude.[9] I confess that this was my understanding of Euclid from the early years through the whole of my life. This being the case, I find it difficult to understand how it is possible to doubt that one can determine immediately the quantity of one magnitude from another, when the previously mentioned *relation according to quantity* of one to the other is known. For, in whatever manner one of them is known, in that same manner, in turn, without any outside assumption, will the other also be known, from that single relation already set forth.

After this I progress to *Elements* VI, def. 5. In this connection I say it is most erroneous, that, under the species of simple Definition, an Axiom is inserted which is inadmissible without proof.

In the first place.[10] If it generally concerns magnitudes which are commensurable in their own class so that they have a ratio which is the ratio of any kind of number (either integral, or fractional, or a fraction of the unity) to any other number, so that assuredly a first magnitude A multiplied by any number, fraction or fraction of the unity, is made equal to the second magnitude B, then clearly and immediately the truth of that Definition will be shown in order that it can be considered quite correctly as an Axiom.

Z
T · X · Y
A : B : C : D

Fig. 54

Let there be any four magnitudes, rational in relation with each other in the manner mentioned before, with A the first, B the second, C the third, D the fourth (Fig. 54). I say that the ratio of the first A to the fourth D is compounded out of the ratios of the intermediate magnitudes, that is, of the first A to the second B, of the second B to the third C, and of the third C to the fourth D. I say, of course, that the magnitude A is contained as many times in the magnitude D (taking the words "*as many times*" to mean any number whatever, integral, fractional, unity, or fraction of the unity) as the number (or the fraction of the unity), which is the result of the multiplication made upon the particular numbers taken in the aforementioned manner, that is, the numbers which indicate how many times the magnitude A is contained in the magnitude B, and this one in the magnitude C, and this last in the magnitude D. Let these be T, X, Y, and let an operation upon these generate the number Z.

Jam sic. Constat, quod magnitudo A multiplicata per numerum T facit magnitudinem B, & haec multiplicata per X facit magnitudinem C, quae rursum multiplicata per Y facit magnitudinem D. Igitur A in T, in X, in Y producit magnitudinem D. Ponitur autem, quod numeri T, X, Y inter se multiplicati faciant numerum Z. Igitur magnitudo A multiplicata per numerum Z facit eam magnitudinem D. Rursum constat, quod numerus T exprimit illam quantitatem, seu *quotitatem*, juxta quam prima magnitudo A taliter se habet ad secundam magnitudinem B, nimirum prout unitas se habet ad eum numerum T; atque ita uniformiter de secunda magnitudine B relate ad tertiam C, prout unitas se habet ad eum numerum X; ac tandem de hac tertia C ad quartam D, prout unitas se habet ad reliquum numerum Y.

Simili modo ostendetur exhiberi ab eo numero Z (qui nempe oritur ex ductu praedictorum inter se numerorum T, X, Y) illam quantitatem, seu *quotitatem*, juxta quam prima magnitudo A taliter se habet ad quartam magnitudinem D, nimirum prout unitas se habet ad eundem numerum Z. Cum ergo hic numerus Z compositus sit ex praedictis numeris T, X, Y, manifestum fit nos esse in casu illius Def. Euclidaeae. Si enim quantitates, seu quotitates rationum primae A ad secundam B, secundae B ad tertiam C, ac tandem tertiae C ad quartam D, invicem multiplices, gignaturque quantitas, seu quotitas Z, haec porro exhibebit rationem primae magnitudinis A ad quartam D, rationem idcirco compositam ex rationibus magnitudinum intermediarum. Quod utique erat demonstrandum.

Tum secundo: non diffiterer istud ipsum demonstrari a me posse, dum sermo sit de
[137] magnitudinibus quomodolibet irrationalibus. Sed non vacat tantum laborem impendere in re non necessaria. Nam dico non nisi inique hac in parte Euclidis nomen vexatum fuisse; quia nempe (ad ipsius usum) nullam ibi veritatem proponit praeter eam, quae in usu *puri nominis* consistit. Ad quod explicandum, seu mavis demonstrandum, capere licet exemplum ex Propos. 23. lib. 6. in qua Euclides demonstrat aequiangula parallelogramma eam inter se rationem habere, quae ex rationibus laterum componitur.

Sint enim duo talia parallelogramma (Fig. 55) unum ABCD, & alterum CEFG ita constituta, ut anguli ad punctum C sint aequales, ac propterea in unam rectam lineam coeant ipsae BCG, & DCE: Tum compleatur alterum parallelogrammum BCEH, fiatque, ut latus BC unius parallelogrammi ad latus CG alterius, ita recta quaepiam linea I ad K, & ut latus DC ad latus CE, ita illa K ad alteram L. Jam sic. Constat (ex 1. sexti) parallelogrammum AC ita fore ad parallelogrammum CH, ut basis DC ad basim CE, sive (ex 11. quinti) ut I ad K. Rursum, eodem jure, parallelogrammum CH ita erit ad parallelogrammum CF, ut basis BC ad basim CG, sive (ex eadem 11. quinti) ut K ad L. Igitur ex aequo (nimirum ex 22. quinti) ita erit parallelogrammum AC ad parallelogrammum CF, ut I ad L.

Now consider the following. It is evident that the magnitude A multiplied by the number T produces the magnitude B, and that this multiplied by X produces the magnitude C, which in turn multiplied by Y produces the magnitude D. Therefore A multiplied in turn by T, by X, by Y produces the magnitude D. Let it be assumed, moreover, that the numbers T, X, Y multiplied together give the number Z. Hence the magnitude A multiplied by the number Z produces the magnitude D. On the other hand, it is evident that the number T represents that *quantity* or better *multiplicity* of the first magnitude A with respect to the second B, which is the same as that of the unity with respect to the number T; and likewise the second magnitude B is to C as the unity is to X, and finally, the third C is to the fourth D as the unity is to Y.

In a similar manner it can be shown that the number Z (which is the product of the aforementioned numbers T, X, Y) is the quantity or *multiplicity* of the first magnitude A with respect to the fourth magnitude D, which is the same as that of the unity with respect to the number Z. Since, therefore, this number Z is composed of the aforementioned numbers T, X, Y, it is clear to us that this is a case where the Euclidean Definition applies. For if the quantities, or multiplicities, of the ratios of the first A to the second B, and of the second B to the third C, and finally of the third C to the fourth D, are multiplied together and the quantity or multiplicity Z is produced, this will give the ratio of the first magnitude A to the fourth D, a ratio, therefore, which is compounded from the ratios of the intermediate magnitudes. This is what was to be demonstrated.

Then secondly.[11] I would not deny the fact that if the discussion involved any kind of irrational magnitudes whatever, I could prove the same relation to hold. But I do not ask that so much labor be spent on a matter that is unnecessary. For I say that Euclid's reputation in this instance was unjustly attacked, because certainly (as was customary with him) he never proposed any truth that was not based on a *purely nominal* use of the Definition. To explain this, or better, to prove it, let me take as an example *Elements* VI, 23, in which Euclid proves that equiangular parallelograms have a ratio to each other which is compounded from the ratios of the sides.[12]

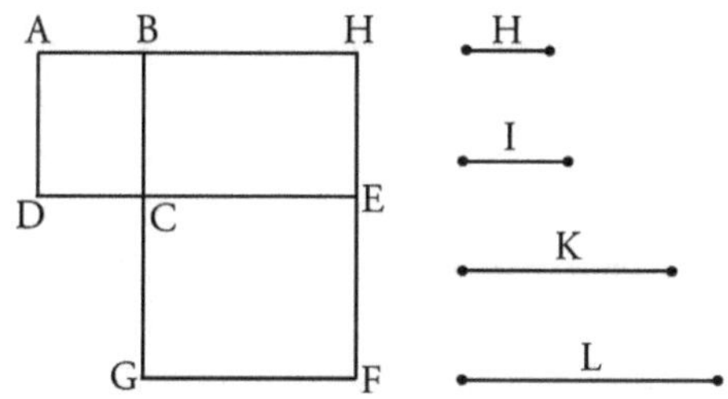

Fig. 55

Let there be two such parallelograms (Fig. 55) one, ABCD, and the other CEFG, so constituted that the angles at point C are equal, and in this way, the lines BCG and DCE are straight lines. Let the other parallelogram BCEH be completed.

Assume[13] that the side BC of one parallelogram is to the side CG of the other, as any straight line I is to a straight line K; and that the side DC is to the side CE, as the straight K is to another straight L. Now we proceed as follows. The parallelogram AC is to parallelogram CH as the base DC is to the base CE (*Elements* VI, 1), or rather (*Elements* V, 11), as I is to K. Again, by virtue of the same, parallelogram CH is to parallelogram CF as the base BC is to the base CG, or rather (from the same *Elements* V, 11) as K is to L. Therefore, *ex aequo* (more particularly from *Elements* V, 22),[14] parallelogram AC is to parallelogram CF as I is to L.

Atque id est, quod intelligit Euclides, dum dicit rationem unius parallelogrammi ad alterum aequiangulum parallelogrammum componi ex rationibus laterum: Id enim unice vult, ut ratio praedicta aequalis sit rationi cujusdam rectae lineae I ad alteram L, inter quas interponatur quaepiam K, per quam continuentur duae rationes aequales rationibus praedictorum laterum; dum scilicet ita sit I ad K, ut latus DC unius parallelogrammi ad latus CE alterius; & rursum ita sit K ad L, ut est prioris parallelogrammi alterum latus BC ad alterum posterioris parallelogrammi latus CG.

[138] Eodem plane modo interpretari debemus Propos. 19. & 20. ejusdem Sexti, in quibus legimus similia triangula, & quaelibet similia itidem poligona, duplicatam habere inter se eam rationem, quae est lateris homologi ad latus homologum. Ibi enim nihil aliud demonstrari debere intelligitur, nisi quod ratio unius trianguli, aut poligoni, ad alterum simile triangulum, aut poligonum, aequalis sit rationi cujusdam rectae lineae I ad alteram L, inter quas interposita sit quaepiam K, per quam continuentur duae rationes aequales illi, quae est cujusdam lateris unius trianguli, aut poligoni ad latus homologum alterius trianguli, aut poligoni.

Praeterea (ut nullus supersit dubitationi locus) simili itidem modo interpretari debemus Propos. 33. undecimi, ubi legimus: Similia solida parallelepipeda esse inter se in triplicata ratione laterum homologorum. Nam ibi nihil aliud demonstrandum assumitur, nisi quod ratio unius parallelepipedi ad alterum simile parallelepipedum aequalis sit rationi cujusdam rectae lineae H ad alteram L, inter quas duae quaedam I, & K interpositae sint, per quas continuentur tres rationes aequales illi, quae est cujusdam lateris unius parallelepipedi ad latus homologum alterius parallelepipedi.

Sed nolo dissimulare, quod jam inutilis fieret illa Definitio, super qua disputamus. Nam respondeo voluisse utique Euclidem rationem veluti reddere nominis ab ipso assumpti, ita ut nempe ad eum modum una aliqua ratio intelligatur ex pluribus rationibus componi, quo unus quispiam numerus ex pluribus numeris invicem multiplicatis exoriri intelligitur, & componi; sed ea tamen nusquam violata lege, ut nunquam ad demonstrandum eam definitionem adhibeat, nisi antea ita omnes terminos disponat, ut locum habere possit demonstra-
[139] tio *ex aequo* juxta praedictam 22. quinti. Atque ita semper faciunt omnes magni Geometrae tam veteres, quam recentiores; quod sane necessarium praesertim est, ubi componendae invicem sint rationes magnitudinum diversorum generum, ut puta linearum, planorum, solidorum, velocitatum, temporum, & ejusmodi. Tunc enim certum est has omnes rationes ex utraque parte reduci primum debere ad unam aliquam talium magnitudinum speciem, ut postea detur locus alicui argumentationi *ex aequo*; nimirum vel ad probandam (ex illa 22. quinti) aequalem rationem inter extremas; vel ad probandam inter easdem unam rationem altera majorem, ex una aliqua consequentium ejusdem Libri quinti Propositionum. Unde tandem constat, illam 5. Def. sexti nulli difficultati obnoxiam esse; utpote quae *solius nominis* impositionem decernit, nulli postea ad demonstrandum usui futuram.

This, therefore, is what Euclid meant when he said the ratio of one parallelogram to another mutually equiangular one is compounded from the ratios of the sides. For he intended this only, that the aforementioned ratio be equal to the ratio of some straight line I to another L, between which is inserted a certain K, through which the two ratios, equal to the ratios of the aforementioned sides, are formed; i. e. I is to K, as the side DC of one parallelogram is to CE, a side of the other, and in turn, K is to L, as the other side BC of the first parallelogram is to the side CG of the second parallelogram.

In the same way, evidently, we must interpret *Elements* VI, 19 and 20, where we read that similar triangles, and likewise, similar polygons are to each other in the duplicate ratio of their homologous sides.[15] For, in this instance, nothing else need be shown except that the ratio of one triangle, or polygon, to a similar triangle, or polygon, is equal to the ratio of some straight line I to another L, between the two of which has been inserted a certain K through which two ratios are constituted, that are equal to the ratio of a side of one triangle, or polygon, to the homologous side of the other triangle, or polygon.

For this reason (so that there will remain no room for doubt) we must interpret in a like manner *Elements* XI, 33, where we read that similar parallelepipeds are in triplicate ratio of their homologous sides. Now in this case nothing more has to be proved than that the ratio of one parallelepiped to a similar parallelepiped is equal to the ratio of some straight line H to another L, between the two of which two lines I and K have been inserted, through which three ratios are constituted, that are equal to the ratio of a side of one parallelepiped to the homologous side of the other.

However, I do not wish to conceal the fact that the Definition which we are discussing is useless.[16] But I answer that Euclid in this way gave a reason for his choice of a name, so that assuredly in that manner some ratio might be considered to be compounded out of many ratios, just as a certain number is known to arise out of the multiplication together of several numbers, and to be compounded. He never violated the rule by which one should not employ a definition in a proof, without having first so determined all terms; in a such way that he can have ground for a proof *ex aequo*, just as in the aforementioned *Elements* V, 22. This, moreover, is the way in which all great geometers, ancient and modern, have proceeded. Certainly it is especially necessary where ratios of different types of magnitudes are compounded together, as for examples, lines, surfaces, solids, velocities, times, and such.[17] As it is certain that all these ratios on both sides must first be reduced to one type of magnitude of such a kind that afterwards the subject will yield to the argument of *ex aequo*, especially in proving that either (from *Elements* V, 22) the ratios between extremes are equal, or in proving one ratio, among them, greater than the other, as a consequence of one of the following propositions of *Elements* V. From which, finally, it is evident that *Elements* VI, def. 5 is not subject to any difficulty, inasmuch as it is a *purely nominal* convention, with no future use for the purposes of demonstration.

Appendix

Atque hic opportunum est observare, nullius Analyticae ope decerni posse rationem datae cujusdam figurae, etiamsi rectilineae, ad alteram quamlibet datam figuram rectilineam, nisi prius stabilitum supponatur Euclidaeum illud Axioma, unde pendet doctrina parallelarum.

Demonstratur. Praemitto autem communes esse Analyticae, & Arithmeticae vulgari, regulas omnes additionis, subtractionis, divisionis, & extractionis radicum; quousque; nempe in eodem infimo jam stabilito entis genere consistitur. At ubi transire oporteat de genere in genus, ut puta (per multiplicationem, seu ductum cujusdam rectae lineae in alteram rectam lineam) de mera longitudine in superficiem planam; tum consimiliter de hac (per quandam
[140] rursum rectam lineam multiplicata) in solidum trinae dimensionis; atque ita ascendendo per novas multiplicationes ad altiores conceptibiles gradus plurium dimensionum; quod utique uniformiter valet de divisione, per quam ad inferiores gradus descenditur: Tunc enimvero censeo, nullum ab Analytica subministrari posse Principium, quo fulciantur praescriptae ab ipsa operationes ad assequendam veritatem.

Nam constat duas intelligi posse figuras rectilineas, unam (Fig. 55) DABC, & alteram CEFG; quarum & noti sint anguli ad puncta D, C, G, ut puta omnes quatuor recti; & rursum nota sint perpendicula DA, CB, CE, GF, nimirum aequalia singula uni palmo; ac tandem notae sint ipsae bases DC, CG; prior quidem v. g. unius palmi, & altera duorum. Ex his autem rursum constat, datas fore positione ipsas rectas AB, EF, nimirum jungentes ipsarum extrema puncta, quae supponuntur data in sua tali positione.

His positis: audire cupio ab Analytica Principium aliquod, ex quo decerni possit ratio prioris rectilineae figurae DABC ad alteram itidem rectilineam CEFG. Respondebit quispiam rationem esse, ut basis DC ad basim CG; addetque demonstrari id posse (jure quodam Analyticae proprio) ex 18. septimi; ubi habemus, numeros genitos ex duobus, unum quempiam multiplicantibus, eandem inter se habere rationem, quam multiplicantes.

Et ego quidem non renuo jus quoddam hac in parte proprium Analyticae prae Arithmetica vulgari. Itaque agnosco rectam DA, quae ad angulos rectos semper excurrat super recta DC, quoad usque congruat ipsi CB, toties multiplicari, quot sunt quomodolibet distinguibilia puncta in eadem DC; adeo ut propterea superficies quaedam DABC intelligi
[141] possit genita ex illa DA multiplicata per DC. Tum simili rursum modo agnosco, rectam CH, quae ad angulos rectos semper excurrat super recta CG, quoad usque congruat ipsi GF, toties multiplicari, quot sunt quomodolibet distinguibilia puncta in ea CG; adeo ut similiter superficies quaedam CEFG intelligi possit genita ex praedicta CE, sive ipsius aequali DA, multiplicata per CG.

Appendix

At this point it is proper to observe that the ratio of any given figure, even rectilinear to any other given rectilinear figure cannot be determined with the aid of Analysis,[1] unless the Euclidean Axiom which forms the basis of the doctrine of parallels is first assumed.

Proof. I submit now that all the rules of addition, subtraction, division, and of the extraction of roots are common to Analysis and to ordinary Arithmetic, as long as we remain in the same *infimum* genus of being. But if it is necessary to pass from genus to genus, as for example from a mere length to a plane surface (through multiplication or product of one straight line by another straight line), or similarly from this (multiplied in turn by a straight line) into a solid of three dimensions,[2] and so on, ascending through new multiplications to higher conceivable steps of greater dimensions (and this holds uniformly as well for division through which one can descend to lower steps); then I do believe that no Principle of Analysis can be taken by which the prescribed operations for pursuing the truth are grounded.

For it is evident that two rectilinear figures (Fig. 55) can be found, one DABC, and the other CEFG, for which the angles at points D, C, G are known, as for example all four of them right angles; and the perpendiculars DA, CB, CE, GF are also known, each one equal to one hand's breadth; and finally let the bases DC and CG be known, the first to be of one hand's breadth, and the second, of two hands' breadth. From this, moreover, it is evident that the given straight lines AB, EF have a definite position since certainly they join the end points of lines that are determined in a certain position.

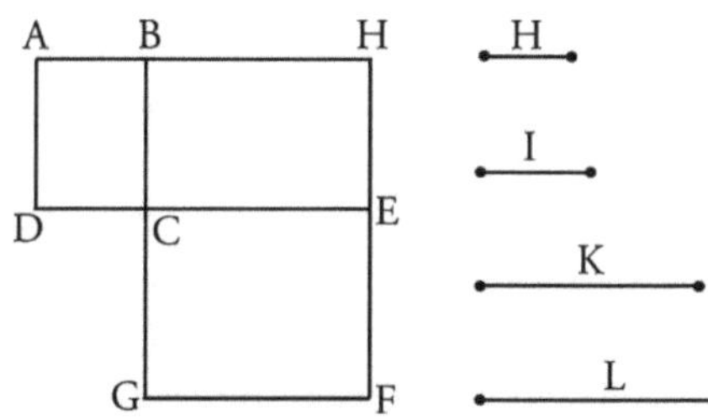

Fig. 55

With this hypothesis, I would like to ask a Principle of Analysis, from which the ratio of the first rectilinear figure DABC to the other rectilinear one of the same kind, CEFG can be determined. Some one will reply that the ratio is as the base DC is to be the base CG, and might add that it is possible to prove this (by means of Analysis) by *Elements* VII, 18, where we have the proposition that if two numbers by multiplying any number make certain numbers, the number so produced will have the same ratio as the multipliers.[3]

And I do not deny that in this respect some advantage may be gained by Analysis over common Arithmetic. I acknowledge, in fact, that the straight line DA, flowing upon DC at right angles up to be congruent to CB, is multiplied as many times as there are (anyhow) distinguishable points in that same DC, with the result that a certain surface DABC is known to be generated out of the multiplication of DA by DC. In a similar manner I acknowledge that the line CE, flowing upon CG at right angles up to be congruent to GF, is multiplied as many times a there are (anyhow) distinguishable points in CG so that in a manner similar to the above, a certain surface CEFG is known to be generated out of the aforesaid CE, or its equal DA, multiplied by CG.

At hoc opus, hic labor: Decernere enim oportet, quaenam sint istae superficies genitae, una DABC, & altera CEFG, circa quas demonstratum agnosco fore eas inter se, ut bases DC, CG.

Si enim praesumere quis velit non alias esse modo dictas superficies, praeter illas, quas jam supposuimus concludi a duabus illis rectis, una AB, & altera EF, jungentibus extremitates illorum quatuor aequalium perpendiculorum, quae supponuntur in eodem plano insistere rectis DC, CG: Is enimvero convinci posset de manifesta petitione principii; cum id ipsum maxime inquiratur, an scilicet utraque linea jungens extremitates etiam intermediorum perpendiculorum sit ipsa etiam linea recta, & non magis aut semper cava, aut semper convexa versus partes suae basis, juxta diversam hypothesin aut anguli acuti, aut anguli obtusi; quod quidem satis constat ex dictis in secunda Parte mei primi Libri.

Praeterea non renuo, quin demonstrari uniformiter possit ab Analytica, quod ratio unius superficiei genitae DABC ad alteram superficiem genitam CEFG (quamvis ipsae DA, & CE ponantur invicem inaequales) componatur ex rationibus perpendiculi DA ad perpendiculum CE, seu perpendiculi CB ad perpendiculum GF; ac rursum basis DC ad basim CG; dum scilicet ipsae AB, EF ponantur jungere extremitates omnium aequalium perpendiculorum a
[142] subjectis basibus erectorum: Atque id insuper multis aliis modis. At semper manebit quaestio circa jungentes extrema puncta illorum perpendiculorum. Quapropter tandem statuo recurri semper oportere ad Geometriam, quae nempe ex illo stabilito Euclidaeo Axiomate demonstret naturam talium linearum.

Ex quibus omnibus satis constat, nullius Analyticae ope decerni posse rationem datae cujusdam figurae, etiamsi rectilineae, ad alteram quamlibet datam figuram rectilineam, nisi prius stabilitum supponatur Euclidaeum illud Axioma, unde pendet doctrina parallelarum. Quod &c.

Atque haec sufficere jam possunt ad vindicandum Euclidem a naevis eidem objectis.

Finis totius operis.

But this the task, this the labor.[4] It is necessary to determine what type of surface has been generated in DABC and CEFG, which I acknowledge to be to each other as the base DC is to the base CG.

For if anyone should wish to assume that the named surfaces are not of any type other than those enclosed by the two straight lines AB and EF which join the extremities of the four equal straight lines drawn in the same plane perpendicular to DC and CG, then indeed this might be demonstrated by an obvious *petitio principii* when a particular inquiry is made as to whether each line joining the extremities of the intermediate perpendiculars is itself a straight line and not more apt to be either always concave or always convex towards the base, according to the different hypotheses of acute angle or of obtuse angle. This indeed was made sufficiently clear in the discussions of the Second Part of my First Book.

Besides I do not deny but that it is possible, in one and the same manner, to demonstrate from Analysis, that the ratio of one generated surface DABC to the other generated surface CEFG (insofar as DA and CE are taken as mutually unequal) is compounded of the ratio of the perpendicular DA to the perpendicular CE, or of CB to GF, and that of the base DC to the base CG while AB and EF are taken as joining the extremities of all the equal perpendiculars erected to the bases under discussion. We may arrive in several ways to this same result, but we always have the problem about the lines joining the ends of those perpendiculars. For this reason I finally conclude that it is always necessary to go back to Geometry, that establishes the nature of such lines by means of the Euclidean Axiom

From all of which it is sufficiently clear that it is not possible with the aid of Analysis alone to determine the ratio of any given figure, even if rectilinear, to any other given rectilinear figure unless that previously proved Euclidean Axiom upon which the whole doctrine of parallels depends is established. This is what was to be demonstrated.

And now all this is sufficient to clear Euclid of every blemish with which he has been reproached.

End of the whole work.

Notes to the text

Notes to the text

Saccheri's manuscript of *Euclid Vindicated* has not been found. The first and only edition of the book appeared in Milan in 1733. It is an *in quarto* volume published by Paolo Antonio Montano, with a dedication to the Senate of Milan, and an *imprimatur* by the Inquisition (July 13th, 1733) and the Society of Jesus (August 16th, 1733), that I don't translate. The diagrams are not inserted in the text, but collected in six tables at the end of the volume (here reproduced as Appendix 2). I have corrected the most obvious typos in the Latin text, as well as those signaled by Saccheri himself in the *Errata corrige* to the volume. I have only underlined in the *Notes* the changes that may have conceptual relevance. In the references, the page numbers always refer to the original edition of *Euclides vindicatus*, whose pagination is to be found in the margins of the Latin text. A reprint of the original work has been published in the 2011 Italian edition.

Notes to the Preface

[1] Saccheri employs the plural form *Mathematicas disciplinas,* which was more commonly used at the time, and betrays the effective difficulty of encompassing in one common category the collection of multiple subjects composing the body of Classical and Modern mathematics. To the Classical *quadrivium* disciplines (arithmetics, geometry, astronomy and music), many others were included throughout the centuries, such as mechanics, architecture, military engineering, optics, perspective or geodesy. Thus it did not mean much, in the eighteenth century, to be a mathematician, and professionals from many disparate sciences were classified under that broad name. The problem was also of a theoretical nature, as there was a need to pin down a reasonable definition of mathematics' sole, or main, object of study. It was related to the issue of precisely defining a general mathematical science not confined to a specific discipline. Identifying a common genus for the objects of the mathematical disciplines within the Aristotelian category of *quantity* was but the first step in this direction, and produced a variety of disputes and manifold interpretations throughout the Renaissance. Moreover, the need to establish but one mathematical discipline as the general science of quantity drove some mathematicians, logicians and philosophers from as early as the mid-sixteenth century to search for a *mathesis universalis* – be this understood as

G. Saccheri, *Euclid Vindicated from Every Blemish,* Classic Texts in the Sciences 1,
DOI 10.1007/978-3-319-05966-2_4, © Springer International Publishing Switzerland 2014

a science yet to be discovered, or a science that was to be found in common logic or Euclid's theory of proportions in Book V of the *Elements.* Although this search was already well under way by the time Saccheri wrote *Euclid Vindicated*, the Jesuit seems to position himself outside of the research tradition (no mention to the numerous projects of *mathesis universalis* is to be found in *Logica* or any of his other works). In fact, when in *Euclid* he discusses the theory of proportions of Book V of the *Elements*, he seems compelled by his own results to exclude the universality of its application (see the *Notes* to the Second Part of Book Two). It is therefore unsurprising that Saccheri reaffirms tradition and here refers to a plurality of mathematical disciplines.

A Classical reference regarding this question during the Renaissance is PROCLUS, *In primum Euclidis*, 38–39, who reports a list of mathematical sciences due to Geminus; on the Greek classification, see B. VITRAC, *Les classifications des sciences mathématiques en Grèce ancienne*, "Archives de philosophie", 68, 2005, pp. 269–301. On the Renaissance origins of universal mathematics and its relation to the theory of proportions, see CRAPULLI, *Mathesis Universalis.* A clear and enlightening perspective on the ambiguity of the term 'mathematics', and the multiplicity of disciplines that it encompassed in the mid-eighteenth century, is found in D'Alembert's article *Mathématique* in *Encyclopédie* (vol. 10, p. 188; published in 1765).

[2] As we mentioned in the *Introduction* (§ 1), this expression comes from Henry Savile's *Praelectiones tresdecim in principium elementorum Euclidis* (1621); the Classical reference is HOR. *Sat.* I, 6, 67. Saccheri probably read Savile's sentence in Wallis' 1663 work on parallel lines, for he nearly quotes him to the letter: cf. J. WALLIS, *Opera mathematica*, vol. 2, p. 665. Both Savile and Wallis identify *two* blemishes in Euclid – the Fifth Postulate and the definition of the composition of ratios. Saccheri here finds a third blemish, the well-known (and infamous) Euclidean definition of equiproportionals (topic of the First Part of Book Two of *Euclid Vindicated*) which many geometers of the time considered either incomprehensible or flawed. Although Saccheri explicitly mentions these three blemishes in the programmatic introduction to his work, there is no doubt that he believed himself to have 'vindicated' many others of lesser magnitude (as will later be shown).

[3] In his discussion of the *Elements,* Saccheri constantly refers to Clavius' work, which for a couple of centuries was considered the principal translation of and commentary on Euclid's work. Clavius' version of the Euclidean text had been extensively revised throughout his life, and many editions thereof were published. The first appeared in 1574. Saccheri reads a later edition to the (second) one published in 1589, for in the first edition Clavius does not offer any original discussion of the Fifth Postulate (therein indexed as Axiom 11 rather than Axiom 13 as stated by Saccheri). Here, however, Saccheri does not literally quote Clavius (whose adherence to Euclid's Greek is more faithful than Saccheri's). He prefers instead to provide his own version of the Postulate, and introduces further modifications in later passages of the volume (title of Book One; Proposition 39).

[4] I always translate the Latin word *pronunciatum* as 'assertion'. It might also be translated as 'axiom', as the Latin term was coined by Cicero as a means of contending with the Greek ἀξίωμα (cf. CIC. *Tusc. disp.* I, 14), and was later used by Renaissance mathematicians as an

exact synonym for *axioma* or *dignitas* or *notio communis*. Saccheri could have happened upon it in the works of Clavius and especially in those of Borelli who makes especially frequent use of it. In later passages of *Euclid Vindicated*, however, Saccheri also employs the Latin word *axioma*, as if to emphasize the existence of a subtle difference between the two terms: an axiom is a principle whose evidence is indubitable, while an assertion (always and only with reference to the Parallel Postulate) is one whose evidence (rather than truth) one sets out to unveil. Book One of *Euclid Vindicated* seems to consist of an attempt to turn an assertion into what it has always (implicitly) been, namely, an axiom.

[5] The definition of axiom as a proposition whose truth is evident as soon as its terms are understood is very classic. Saccheri already employed it in *Logica*; for instance: "Axioma seu dignitas est propositio universalis, prima, & immediata, v.g. *quodlibet est, vel non est, idem non potest simul esse & non esse*; quae sunt propositiones universales, primae, & immediatae, utpote quae ex ipsis terminis ritè intellectis clare constat" (*Logica demonstrativa*, ed. 1701, p. 118 [ed. 1697, p. 187]; cf. also p. 127 [p. 200]). The opinion was widespread in seventeenth-century mathematics. See for instance a similar opinion by Borelli: "Sunt axiomata, ut dictum est, propositiones speculativae, seu theoremata, quae ob sui maximam evidentiam non indigent demonstratione; sed statim, cognitis terminis, neceße est, ut intellectus illis aßentiatur; & propterea aßumuntur inter principia indemonstrabilia" (*Euclides restitutus*, p. 14).

[6] It is well known that Euclid first employs the Fifth Postulate when proving *Elements* I, 29, which is itself a proposition equivalent to it. Most interpreters have argued that the fact that Euclid does not employ the Postulate within the first 28 theorems of Book I, some of which could have been more promptly demonstrated by using it, implies that Euclid harbored scruples about the Postulate itself. More recently, however, I. Mueller, *Philosophy of Mathematics and Deductive Structure in Euclid's Elements*, Cambridge, MIT Press 1981, has argued that the arrangement of the theorems in the *Elements* is (obviously) deductive, but also organized in blocks of propositions regarding similar topics; in *Elements* I, 26–34, the theorems on parallels constitute a unity in itself, and it is thus natural that Euclid does not employ the Parallel Postulate earlier in the work. In any case, it is obvious that any attempt to prove the Fifth Postulate with reference to geometrical results obtained after *Elements* I, 28 would yield a vicious circle; here, however, Saccheri is probably referring to a passage in Clavius expressing an analogous statement ("ut illud ipsum Euclidis axioma demonstraremus ex ijs solum, quae ante propos. 29 primi libri demonstrata sunt", *Euclidis*, p. 50), and to the fact that he will, instead, attempt to prove the Postulate without relying on *Elements* I, 16, 17, 27, 28 (which Clavius freely employs).

[7] *Elements* I, def. 22, which appears in Clavius as Definition 34 (*Euclidis*, pp. 21–2; the wording differs from Saccheri's). Immediately after, Clavius himself states that parallel straight lines are also equidistant. In this respect he aligned himself with Commandino's 1572 edition of Euclid, which reads: "Parallelae, seu aequidistantes rectae lineae sunt, quae cum in eodem sint plano, et ex utraque parte in infinito producantur, in neutram partem inter se conveniunt" (*Euclidis*, p. 6; we have seen in § 2 of the *Introduction* that the terminology came from the Middle Ages). Such double definition, which adopts the term equidis-

tance to mean non-incidence was later used by many Renaissance interpreters to construct their own proofs of the Fifth Postulate on the basis of equidistance (see below, the *Notes* to Proposition 38 and 39). Commandino himself assumes in his proof of the Fifth Postulate the equidistance of the straight lines (*Euclidis*, p. 20).

[8] The theorems in *Elements* I, 27, 28 are the inverse of *Elements* I, 29 (itself equivalent to the Fifth Postulate). The fact that these former theorems can be proved from the first four postulates has always been seen as an argument in support of the demonstrability of the Fifth Postulate (cf. Proclus, *In primum Euclidis*, 183–4); during the Renaissance, several logicians, Ramus in particular, argued for the convertibility of all geometrical propositions, and some important geometers such as Jacques Peletier du Mans (1517–1582) went so far as to lose interest in proof of the Fifth Postulate, as proofs of *Elements* I, 27–28 were already available. I take this to be one of the reasons for Saccheri's conviction that these two propositions should never be used in his proofs; as if to say that the Fifth Postulate's dependence on its inverse would qualify it as less evident and immediate (and an axiom must be evident and immediate). Another reason, and perhaps a stronger one, is that *Elements* I, 27 and 28 resulted in two straight lines being parallel, and thus concluded in a property 'of the infinite' (i. e. straight lines do not meet when indefinitely extended). Both Ancient and Modern interpreters identified this as a possible source of difficulty. One of the Moderns, for instance, was Borelli, whose objections are directly discussed in *Euclid Vindicated* (cf. Scholium 2 to Proposition 21). Saccheri was probably susceptible to these kind of difficulties, given that his entire construction of hyperbolic geometry bears on the behaviour of straight lines meeting at infinity. He also returns to the matter a number of times, and employs his own theorems to re-demonstrate *Elements* I, 27 and 28, utilizing a different approach directly connected to the problem of an infinite limit (cf. Corollary 2 to Proposition 23).

However, Saccheri assumes the validity of *Elements* I, 27–28 in an implicit yet crucial way in Proposition 14, that is, when he refutes the obtuse angle hypothesis; see the *Notes* to that section for a more extensive discussion of the matter. It is worth noting, however, that Saccheri's self-imposed restriction on the use of *Elements* I, 27 and 28, by which he generally abides, is not relevant here, mainly because *Elements* I, 16 and 17 (which Saccheri does employ) carry most of the demonstrative power of those other two theorems, and also because it would be difficult to find a proposition of *Euclid Vindicated* that is more easily proved by directly resorting to *Elements* I, 27 and 28.

Saccheri's attempt to limit the use of the exterior angle theorem (*Elements* I, 16), and its immediate consequence that the sum of any two angles of a triangle is less that π (*Elements* 1, 17), is a more complex matter. These two theorems too are intricately related to the theory of parallelism (*Elements* I, 17 is itself, to a certain extent, the inverse of the Fifth Postulate), and are in fact the most important – not to say the only – tools employed by Euclid in the proof of *Elements* I, 27 and 28. Saccheri makes use of both theorems a countless number of times, and they allow him to easily refute the obtuse angle hypothesis. In the *Introduction*, we remarked that (with some qualifications) the validity of *Elements* I, 16, in fact, presupposes that a straight line has infinite length, and this is false in elliptic geometry: it thus offers the most adequate grounds for positing a contradiction between the obtuse angle hypothesis and the theorems preceding *Elements* I, 29. Let us see the qualifications: as GREENBERG, *Euclidean and Non-Euclidean Geometries*, pp. 164–66 and 189–90, rightly notes, the Euclidean proof of *Elements* I, 16 is based in fact on a diagrammatic inference about the position of a point, which is assumed to lay in the interior of an angle. This implicit assumption, of course, does not amount directly to the unboundedness of straight lines, and may be warranted through a couple of axioms of betweenness. This kind of axioms, however, are not explicitly stated in either Euclid or Saccheri, and in fact represented for many centuries the greatest foundational obstacle for a complete axiomatization of geometry; they were simply ignored by any foundational attempt before the nineteenth century (with the partial exception of Leibniz), and consistently supplied with diagrammatic inferences. It is true, however, that the principles of betweenness that are to be employed in the proof of *Elements* I, 16 are only true in hyperbolic and Euclidean geometry, while they fail in the elliptic plane. It is also true, however, that Euclid rules out elliptic geometry from the beginning, as he explicitly assumes that straight lines are unbounded and of in(de)finite length (the Second Postulate). But then, even if there is an actual gap in Euclid's proof of *Elements* I, 16, its truth is guaranteed by the Second Postulate. Moreover, this Postulate is in fact employed in the proof, since it requires one to double a given segment, and thus must assume that one can take straight lines of any length. A further complication arises from the fact that the axioms of betweenness implied in the proof of *Elements* I, 16 (and thus *Elements* I, 16 itself) are in fact compatible with the hypothesis of the obtuse angle if one denies the Archimedean Axiom (see below Dehn's non-Legendrian geometry in *Notes* to Propositions 11, 12 and 13). It is thus impossible to refute the obtuse angle hypothesis through *Elements* I, 16 in a non-Archimedean context.

In any case, Saccheri accepts the Axiom of Archimedes (and the almost equivalent Principle of Aristotle, as we will see in Proposition 21), and thus his obtuse angle hypothesis may only be modeled through spherical or elliptic geometry, in which *Elements* I, 16 fails and straight lines have finite length. Thus, Saccheri's claim that *Elements* I, 16 is to be employed only in the case of bounded triangles led interpreters to discuss the extent to which the geometer was aware that a space of constant positive curvature had to be compact, and, more generally, of the possibility of the existence of a geometry with closed geodesics. But any such expectation, I think, is doomed to be disappointed (cf. above, *Introduction*, § 6). The problem of *Elements* I, 16 with respect to elliptic geometry concerns the evaluation of the *finite* measure of the triangle at issue, not its boundedness or unboundedness; in a spherical space of unitary curvature the exterior angle theorem may be false for triangles with one side greater than $\pi/2$, not for unbounded triangles (which do not exist). In fact, by only focusing on the boundedness of triangles (not their measure), Saccheri obtains a refutation of elliptic geometry (in Proposition 14) by appealing to *Elements* I, 16. Restriction on *Elements* I, 16 in the case of bounded triangles is thus independent of the possibility of extending a straight line to infinity, and of the role played by such a theorem in the obtuse angle hypothesis; this is explained in Corollary 2 to Proposition 23 (see below, the *Notes* to the respective section).

Since the assumption of the infinite length of a straight line in *Elements* I, 16 is important for an evaluation of the deductive structure of *Euclid Vindicated*, specifically in the case of the obtuse angle hypothesis, I here offer a list of propositions where Saccheri employs the exterior angle theorem: Proposition 3, 6, 10, 13 Scholium 2, 21 Scholium 3, 23 Corollary 2, 27, 28. Saccheri employs *Elements* I, 17 in Propositions 6, 10, 11, 13, 14, 15, 17, 22, 24 Scholium, 26, 28, 30, 31. In a number of places, Saccheri also employs *Elements* I, 18 and 21, which in turn depend on *Elements* I, 16; but this dependence is not essential for our purpose, for both propositions are independent of the hypothesis of the infinite extensibility of the straight line and do hold in elliptic geometry. Saccheri uses *Elements* I, 26, which also depends on *Elements* I, 16 (this time in an essential way, in the sense that *Elements* I, 26 is false in elliptic geometry), in Propositions 11 and 17. As we have previously observed, he never employs *Elements* I, 27 and 28. Of all propositions of *Euclid Vindicated* directly dependent on *Elements* I, 16 and 17, those following Proposition 14 are most certainly correct, since at this point in the text Saccheri believes himself to have refuted the obtuse angle hypothesis and therefore always works in Euclidean or hyperbolic geometry (in which *Elements* I, 16 is unconditionally valid). As they stand, proofs of Propositions 3, 6 and 10 are wrong and require either corrections or additional hypotheses: see the *Notes* hereinafter. Proposition 11 is correct because the proof is carried out within the Euclidean hypothesis. Propositions 13, 13 Scholium 2, and 14 are correct because the proof is carried out locally by explicit hypothesis.

Some remarks concerning these problems (and an interpretation of the phrase "*de triangulo omni ex parte circumscripto*" which contradicts the one provided here) can be found in A.M. Dou, *Logical and Historical Remarks on Saccheri's Geometry*, "Notre Dame Journal of Formal Logic", 11, 1970, pp. 385–415.

[9] Again according to Clavius' version (cf. *Euclidis*, pp. 208–10 and 243–7). In Heiberg's critical edition of the Euclidean text, which is commonly used today, the first of these definitions appears as Definition 5 in Book V, and the second is missing, as it was most certainly interpolated at a later time. See hereinafter, the *Notes* to the Second Part of Book Two.

[10] Refers to the assumption of the existence of the fourth proportional, which Saccheri wants to replace with a different principle, for which he offers a proof. See below, pp. 111 and ff.

Saccheri hints here at a distinction between axiom and postulate which he had already dealt with in *Logica demonstrativa*. The geometrical tradition distinguished between these two principles in a variety of ways, especially starting from Proclus' commentary to Euclid (cf. Proclus, *In primum Euclidis*, 178–84): a postulate was either a principle with less evidentiary support than an axiom, or a constructive principle (while an axiom expresses a state of affairs), or a principle specific to a science (while an axiom is generic to all disciplines). In *Logica* (ed. 1701, pp. 118–9 and 127–34 [ed. 1697, pp. 187–8, 200–8]), Saccheri certainly accepts the idea that a postulate is a specific principle, though he also seems to admit that there can be specific axioms (with a restrictive *plerunque*, p. 188); he never speaks of construction, though he always provides examples of constructive postulates; and with regard to the question of evidence, which is the demarcation criterion that he probably has in mind here, he claims that an axiom is true in virtue of the terms' meanings, while postulates require more evidentiary support although they should not contradict given meanings. So, for instance, we could postulate (or better, hypothesize) that a straight line be one hand span long, as such a supposition does not contradict the notion of straight line (nor is it included in the definition thereof), or we could postulate (in the proper sense of the term) the existence of a straight line, for there is no contradiction between the definition of line and that of existence (though the one is not included in the other); cf. after *Note* 8 to Propositions 38 and 39. Saccheri seems to consider the existence of the fourth proportional a postulate in this last sense, and thus wants to do away with it via reference to a principle (axiom) that is equivalent to it and is found analytically in the very definition of proportion.

Notes to Propositions 1 and 2

Very similar results to these two theorems are to be found in the work on parallel lines by the great Persian scientist and scholar Umar Khayyām, whose *Explanations of the Difficulties in the Postulates of Euclid*, composed around 1077, was circulated during the Early Modern Age, mostly through Nasīr ad-Dīn's commentary on Euclid (see the *Notes* to Scholium 3 following Proposition 21). The original manuscript, however, was not discovered until 1936. It is therefore unlikely that Saccheri had any direct knowledge of it (cf. *Introduction*, § 4).

As for the content, Khayyām's Lemma 1 corresponds almost perfectly to Saccheri's Proposition 1, though it employs the hypothesis that both angles at the base of the quadrilateral (A and B in Saccheri, Fig. 1) are right angles. Saccheri's result is thus stronger. Furthermore, Khayyām's proof is slightly more involute and employs *Elements* I, 28 (which Saccheri assiduously avoids). In Lemma 2, Khayyām assumes AM and MB to be equal (without assuming the equality of CH, HD), and MH to be perpendicular to AB, and concludes that MH and CD are perpendicular.

Thābit ibn Qurra, another great Arab mathematician, also demonstrated Saccheri's Proposition 1 in his second work on parallel lines, the title of which is the Fifth Postulate's statement itself. In this case, the theorem makes the stronger statement also found in Saccheri's formulation. Thābit's text, however, does not seem to have circulated in the West any time before twentieth-century editions (see *Introduction*, § 2, for the references).

The properly historical precedent for Saccheri's theorems is certainly to be found in Clavius' attempt to demonstrate the Fifth Postulate. Clavius proves (with greater simplicity) Khayyām's Lemmata 1 and 2 in Lemma 4 of the Scholium to Proposition 28. Concerning Clavius' knowledge of Khayyām's work (through Nasīr ad-Dīn), see the *Introduction*, § 2.

Of greater importance is the fact that Saccheri's Proposition 1 is found (with the same hypotheses and the same proof) in Vitale Giordano; cf. Lemma 2 in the Remark after Proposition 26, in GIORDANO, *Euclide restituto*, pp. 46–7. Giordano, who is usually very scrupulous when it comes to mentioning the sources of his own work (he goes so far as to write a Scholium 3 to Proposition 31 in which he traces a short history of the attempts to prove the Fifth Postulate; cf. *Euclide restituto*, pp. 62–6), does not make any reference with regards to this Lemma. Since Giordano's hypotheses are different to Khayyām's, it seems reasonable to suppose that he was not aware of this Medieval attempt, and, accordingly, that the statement and proof of Lemma 2 are entirely a product of his own genius. It is difficult, on the other hand, to imagine that Saccheri arrived at an identical proof without having seen Giordano's book.

Formal proofs of Saccheri's first two Propositions are to be found as Theorem 36 in D. Hilbert, *Grundlagen der Geometrie*, Stuttgart, Teubner 1968 (1899[1]), pp. 42–3. Note that Hilbert proves such result in Khayyām's stronger hypotheses, and employs, just as the latter does, the exterior angle theorem (which is not actually necessary). Saccheri is never mentioned in any of Hilbert's work, though he appears in the lessons from which the 1899 volume was later derived; see M. Hallet, U. Majer, *David Hilbert's Lectures on the Foundations of Geometry 1891–1902*, Berlin, Springer 2004, p. 262; cf. M.M. Toeppell, *Über die Entstehung von David Hilberts "Grundlagen der Geometrie"*, Göttingen, Vanderhoeck 1986.

Finally note that in the absence of a hypothesis on the size of angles at A and B, nothing in spherical geometry guarantees the uniqueness of join CD (even when taken as a complete straight line). But, as we mentioned above, Saccheri always seems to assume that between any two points there is only one complete straight line (as in the single elliptic model). He also seems to assume, at least implicitly, the stronger principle that between any two points there is but one segment of a straight line – which is false in any model of space with constant positive curvature. This last situation requires an additional hypothesis: one must always consider that join (amongst the two existing ones) that forms with the sides angles C and D (on the same side as angles A and B) each of value less than π; and likewise for all other cases (at any rate, Classical geometry does not ever seem to consider angles greater than π; nor does Saccheri). Yet, independently from this last assumption, Propositions 1 and 2 remain valid if referred to *any* join CD.

Notes to Propositions 3 and 4

At this point in the text, Saccheri introduces quadrilaterals of equal opposite sides and right angles to the base (birectangular isosceles quadrilaterals), which allow him, immediately after Proposition 4, to differentiate between the three hypotheses and the three geometries. These figures are still known today as Saccheri Quadrilaterals, and are sometimes also referred to as Khayyām-Saccheri Quadrilaterals, as Umar Khayyām, in his Lemma 4 (see the *Note* above), demonstrated that in a quadrilateral with four right angles and two equal opposite sides, the two remaining sides are also mutually equal, and this corresponds to the first part of Saccheri's Proposition 3. Although Khayyām also discusses (in the proof of Lemma 3) the case of quadrilaterals with two obtuse or acute angles, he does not draw any consequences, and limits the discussion to an evaluation of the 'Euclidean' result of his Lemma 4. In other words, he does not see that which for Saccheri, and for us, is to be the fundamental point, i.e. an opening towards alternative geometrical hypotheses.

On the other hand, Clavius' and Giordano's attempts at proving the Fifth Postulate (see the preceding *Note*) are very interesting, for they hope to prove it by directly demonstrating that the remaining two angles of a Saccheri Quadrilateral are right (while Khayyām's four cited Lemmata had a completely different aim). Therefore, Saccheri's proof strategy must have drawn on the work of the former two authors. Both held that, in order to prove that a quadrilateral has four right angles, one must first show that bisector HM is equal to sides AC and BD. This last implication is certainly correct, and proof of the equality of side and bisector may be regarded as the front line of Saccheri's strategy (cf. Corollary 1, where Saccheri shows HM to be longer than AC and BD in the obtuse angle hypothesis, and shorter in the case of the acute angle hypothesis). Clavius believed that he had obtained this result – that is, that he had proved the equality of side and bisector in a quadrilateral – both by employing very intuitive reasoning about the rigid motion of a segment and by employing more complicated reasoning which he, perhaps, borrowed from Nasīr ad-Dīn; cf. Lemmata 3 and 4 in the above-mentioned Scholium to Proposition 28 (*Euclidis*, pp. 51–2). Giordano, who was obviously unsatisfied with Clavius' proofs, offers a third demonstration, this time based on the concept of convexity; cf. his Lemmata 5, 7 and 8 after Proposition 26 (*Euclide restituto*, pp. 48–53).

All three proofs are, of course, wrong. Saccheri discusses and criticizes Nasīr ad-Dīn's proof in Scholium 3 to Proposition 21, while those proofs that rely on the concept of rigid motion and convexity are the main topic of the Second Part of Book One of *Euclid Vindicated.*

Also noteworthy is Vitale Giordano's proof that if but one perpendicular to the base AB is equal to sides AC and DB, then all perpendiculars to the base are also equal to these sides (Lemma 6; *Euclide restituto*, pp. 51–2). Such a result, which is indeed correct, is important, as it allows geometers to move from considerations of the length of all lines perpendicular to AB (which Clavius considered with reference to the rigid and continuous motion of segment AC along base AB), to the consideration of just one intermediate segment. Indeed, this is how Giordano and Saccheri proceed. In other words, the Fifth Postulate is not only equivalent to the existence of a straight line everywhere equidistant from a given straight line (see below for Saccheri's discussion on this), but more simply to the existence of a straight line equidistant at three points to a given straight line. It is remarkable that in the entry *Parallèle* of the *Encyclopedie*, D'Alembert lamented the lack of a rigorous theory of parallels and hoped for precisely such a proof – namely, a proof showing that if a straight line is equidistant to a given straight line at two points then it is also equidistant at a third (vol. 11, p. 906; published in 1765, though there is no reference to Giordano nor to Saccheri). For more on this aspect of Giordano's work, see a note by R. Bonola, *Un teorema di Giordano Vitale da Bitonto sulle rette equidistanti*, "Bollettino di Bibliografia e Storia delle Scienze Matematiche", 8, 1905, pp. 33–6.

Finally, we should mention that, after Lobachevsky and Bolyai's trigonometric work, Saccheri's Propositions 3 and 4 can be refined through the quantitative determination of length c of the join CD in relation to length b of the base AB and length l of the equal sides AC and BD: in the acute angle hypothesis (in hyperbolic geometry) the following formula is valid: $\sinh(c/2) = \sinh(b/2)\cosh(l)$.

[1] At this point in the text, Saccheri employs the exterior angle theorem within a proof developed in the obtuse angle hypothesis, where the theorem is false. So, as it here stands,

Proposition 3 is wrong. In fact, if we consider double elliptic geometry, then straight lines AC and BD converge on both sides and meet at two different points (as opposed to *Elements* I, 27, which Saccheri, however, does not want to employ). At a distance $\pi/2$ (for unitary curvature) from any of these intersections, the join CD forms obtuse angles with AC and BD, and is nevertheless equal to AB – in contrast to what Saccheri here affirms. But Saccheri's error is formal rather than substantive, as it seems unlikely that Saccheri took non-convex quadrilaterals as a possibility to be taken into account. This is thus one situation in which the diagram seems to play a crucial role in the proof process. It is also important to note that, regardless of the diagram and of assumptions concerning the quadrilaterals' convexity, the theorem is correct if we take single elliptic space as a model for the obtuse angle hypothesis; in such space, AC and BD meet at only one point, and there is not enough space, if the lines are extended after the intersection, for join CD to become of the same length as AB. And there are good reasons to take the latter as the correct model for *Euclid Vindicated*'s spherical geometry. So the theorem is correct: (a) if we take figure ABCD to be convex; or (b) if it is considered only locally valid, i.e. in the case of sufficiently small quadrilaterals; or (c) if we include the principle that two straight lines do not enclose space (see Proposition 33, Lemma 1) and hence that, in the obtuse angle hypothesis, the topological structure of space is that of a projective plane.

Note that analogous concerns apply to the validity of later theorems, where Saccheri employs either *Elements* I, 16 or Proposition 3; in particular, the aforementioned changes apply with equal force to proof of Proposition 6 (which also employs *Elements* I, 16), Proposition 8 (which employs Proposition 3), Proposition 9 (which employs Proposition 8) and Proposition 10 (which again employs *Elements* I, 16).

[2] These quadrilaterals with 3 right angles and a fourth right, obtuse or acute angle (according to the three hypothesis) are today generally referred to as Lambert Quadrilaterals, because Lambert, who was more renowned than Saccheri for a long time, employed them in his attempts to prove the Fifth Postulate. Cf. Lambert, *Theorie der Parallellinien*, § 39, pp. 329–30. Lambert's indirect dependence on Saccheri's work is highly probable, even though he does't seem to have read *Euclid Vindicated* (see *Introduction*, § 7).

In this Corollary and elsewhere in *Euclid Vindicated*, Saccheri himself often employs Lambert Quadrilaterals – which are simply halves of Saccheri Quadrilaterals.

On the relevance of Saccheri-Lambert Quadrilaterals in the modern axiomatization of hyperbolic geometry, see V. Pambuccian, *Lambert or Saccheri quadrilaterals as single primitive notions for plane hyperbolic geometry*, "Journal of Mathematical Analysis and Applications", 346, 2008, pp. 531–2.

[3] This is the first, and certainly not the last, example of a remark on infinitesimals that has neither theoretical depth nor structural consequence. Corollary 3 will never be used explicitly in *Euclid Vindicated*, though Saccheri may have had it in mind when proving Proposition 25 and 27. Worse still, it appears to contradict an essential proof passage of Proposition 37, where Saccheri holds to have demonstrated the inconsistency of hyperbolic geometry (see *Note* 4 to Proposition 37).

Notes to Propositions 5, 6 and 7

Paul Mansion, who was among the first to discover *Euclid Vindicated*, gave the content of these three Propositions the name of 'Saccheri's Theorem', in recognition of its great importance within the work in which it appears, as well as for the development of non-Euclidean geometries in general. The theorem essentially shows that if the plane is assumed to be homogeneous and isotropic, then Saccheri's three hypotheses (Euclidean, hyperbolic and spherical geometry) are mutually exhaustive and exclusive. In other words, according to this homogeneity hypothesis only three (plane) geometries are possible.

The theorem is also logically significant because it is one of the first and most conscious examples of a quantified geometrical proof. In other words, Saccheri cannot just prove the theorem for a single figure and then proceed, meta-theoretically, to generalize this result for all equivalent figures (according to affinity, similarity or congruence), as was the standard procedure in Classical and Modern geometrical tradition. Instead, he has to prove a proposition that is itself (in certain respects) meta-theoretical, and in which universal quantification is part of the statement itself. Hence the novelty of the proof structure. Note, furthermore, that Saccheri appropriately uses double universal quantification procedure, for he must demonstrate the result's validity for every base and height of the quadrilateral, and this considerably complicates his proof, especially in Proposition 6. Through this double quantification, Saccheri arrives at the invariance of his result by affinity, whereas invariance by translations and rotations (i. e. by isometries) is taken for granted.

A similar universality proof was given, in the same context, by Legendre. The proof can be found in the well-known theorem which states that if the sum of the interior angles of *one* triangle is equal to, or less than (or greater than) two right angles, it will be such in *all* triangles. This result can be obtained by applying Saccheri's result to Proposition 9 and 15 of *Euclid Vindicated*. Concerning this proposition, generally referred to as Legendre's Second Theorem, see his memoirs on parallels published in 1833: LEGENDRE, *Réflexions*, pp. 375–8. Formal proofs of Saccheri's Theorem in the Euclidean case and of Legendre's Second Theorem are to be found as Theorems 28 and 29 in Hilbert's *Grundlagen* (pp. 43–5).

[1] From the perspective of the axiomatic strength employed, this passage represents the fundamental point of the entire theorem, because Saccheri uses the concept of rigid motion and hence implicitly assumes space to be homogeneous and isotropic. In the next paragraph, he will also offer a proof that does not directly reference the concept of rigid motion, but rather relies on an explicit and reiterated use of *Elements* I, 4, which is the first Euclidean proposition that uses movement (or at least superposition) as a demonstrative tool – for which reason it was heavily criticized by subsequent geometers (cf. especially J. PELETIER, *Demonstrationum in Euclidis elementa geometrica libri sex*, Lyon, Tornes 1557, pp. 15–7). Saccheri (just like his role model Clavius) does not seem to have any particular problem with accepting proofs that utilize the concept of rigid motion, and he himself employs them extensively in other parts of *Euclid Vindicated*: in Scholium 2 to Proposition 21, in Corollary 2 to Proposition 23, and (especially) in the Lemmata to Proposition 33, and in

Proposition 37. Foundational problems may arise in the latter two cases (for which see the *Notes* to Lemmata 4 and 5 in Proposition 33, and Proposition 37), but in all other instances Saccheri could no doubt easily avoid the superposition procedure. In any case, it is essential here. There are also several, more problematic, cases, where Saccheri employs the flow of points to draw lines and figures.

[2] The proof's structure thus demonstrates the theorem's validity for quadrilaterals whose height is an integer multiple of the initial height, reduced by a (continuous) magnitude smaller than the height itself; we thus obtain quadrilaterals of any height. Note that Saccheri is here implicitly adopting Archimedes' axiom, the very same axiom employed by Legendre in proving his Second Theorem. Such a principle, however, is not necessary to prove this result (see, in contrast, *Note* 3 to Propositions 11, 12 and 13), as has been shown by M. Dehn, *Die Legendre'schen Sätze über die Winkelsumme im Dreieck*, "Mathematische Annalen", 53, 1900, pp. 405–39. Alternative proofs that yield this same result, as well as proofs of the independence of the Saccheri-Legendre Theorem from Archimedes' axiom, are to be found, amongst others, in F. Schur, *Über die Grundlagen der Geometrie*, "Mathematische Annalen", 55, 1902, pp. 265–92; R. Bonola, *I teoremi del Padre Girolamo Saccheri sulla somma degli angoli di un triangolo e le ricerche di M. Dehn*, "Rendiconti dell'Istituto Lombardo", 38, 1905, pp. 651–62; J. Hjelmslev, *Neue Begrundung der ebenen Geometrie*, "Mathematische Annalen", 64, 1907, pp. 449–74. The above-mentioned Bonola, *Un teorema di Giordano Vitale*, calls attention to the fact that Vitale Giordano obtains *similar* results to Legendre's Theorem by employing a procedure that eliminates the need for Archimedes' axiom (though he nevertheless did not succeed in proving Legendre's Theorem).

[3] Saccheri employs the principle of continuity, which is certainly stronger than Archimedes' principle. He employs this kind of reasoning elsewhere too. Indeed, such reasoning was quite common at the time, and exemplifies the extent to which several approaches to geometry were not in the least 'constructivist' (for an even more obvious example, see the following discussion on the existence of the fourth proportional, pp. 111–2). Saccheri's theorem, which can be demonstrated without employing Archimedes' Axiom can, *a fortiori*, also be proved without resort to continuity. Lambert pointed this out in § 57 of his essay (*Theorie der Parallellinien*, p. 337).

[4] We have already seen that *Elements* I, 16 and 17, which are false in the hypothesis of the obtuse angle (as is here the case), pose problems to the soundness of the proof, and that auxiliary hypotheses have to be assumed. With reference to what has been said in *Note* 1 to Propositions 3 and 4, if, for instance, we accept that two opposite sides of a quadrilateral can intersect, then in (single) elliptic geometry we will also have quadrilaterals with two right angles and two acute angles, and quadrilaterals with four right angles; therefore, it is not true that if the right, obtuse or acute angle hypothesis hold in one case, then it holds in all cases.

Notes to Propositions 8, 9 and 10

As Saccheri specified in the Summary, these three technical Propositions are instrumental in proving Propositions 12 and 13.

Proposition 8 is a form of *Elements* I, 24 and 25 applied to triangles with one common side (AD); it is very interesting to see how Saccheri manages to re-interpret one of Euclid's classic theorems as a criterion for discerning between Euclidean and non-Euclidean geometry. To understand the connection between Proposition 8 and the theory of parallels, note how the Proposition states that although two straight lines AC and BD form with a transversal line AB internal angles whose sum on the same side is equal to π (in this case, they are both right angles), they nevertheless form equal alternate angles with another transversal AD only in the hypothesis of the right angle. So the hypotheses in which *Elements* I, 27 and *Elements* I, 28 prove parallelism are not generally equivalent. In hyperbolic geometry (as is also the case in Euclidean geometry), parallelism of AC and BD can be proven either by assuming right angles in A and B (*Elements* I, 28), or by assuming equal alternate angles in A and D (*Elements* I, 27), but the two conditions are not satisfied in the same point A (as is always the case in Euclidean geometry).

Proposition 9 can easily be generalized by removing the hypothesis that the triangle is right angled: it suffices to realize that every triangle can be decomposed in two right-angled triangles, and then to apply Saccheri's proof to each of these. In this more general form (which is not explicit in Saccheri), Proposition 9 states that, depending on whether the right, obtuse or acute angle hypothesis holds, the sum of the interior angles of a triangle (any, thanks to Propositions 5–7) is equal to, more than, or less than π respectively. We hence have the inverse of Proposition 15.

Also note that in the statement of this proposition (and in proof of Proposition 14) Saccheri assumes that, given a right angle in a triangle, the two remaining angles must be acute; in other words, the sum of two angles in a triangle must be smaller than π (for *Elements* I, 17). Saccheri is thus already in the peculiar situation of having to assume, within the obtuse angle hypothesis, that both the sum of any two angles in a triangle is less than π, and that the sum of three angles is greater than π. Thus the sum of the interior angles of an elliptic triangle would have an upper bound of $(3/2)\pi$ (instead of the correct value 3π). If we consider that the sum of the interior angles of an elliptic triangle is related to its area (Saccheri does not seem to be aware of this result; it will be proved in this context by Lambert, although it was already demonstrated in A. GIRARD, *Invention nouvelle en l'algebre*, Amsterdam, Blaeuw 1629), we find that Saccheri's theorems are unconditionally valid only in the case of triangles which are sufficiently small as to have the aforementioned sum of angles and thus, according to Lambert's formula, an area less than $\pi/2$. Thus, at this point, Saccheri is already considering an impossible geometry, for such a sub-system of elliptic geometry (which models a projective space with spherical metric of unitary curvature, though limited to triangles of area less than $\pi/2$) is not a complete system.

Proposition 10 does not rely on any prior result in *Euclid Vindicated*, and could easily be found in Euclid's work. It employs *Elements* I, 16 and 17, but is false in elliptic geometry only for non-convex triangles whose sides intersect before reaching the base.

[1] Saccheri means *Elements* I, 19.

Notes to Propositions 11, 12 and 13

This is one of the most important groups of propositions in the whole of *Euclid Vindicated*. Here, the Fifth Postulate is proved in the hypothesis of the right and obtuse angle. In the right angle hypothesis, Proposition 11 shows that two straight lines meet if they form with a transversal line two internal angles whose sum is less than π, in the limiting condition that one of the two incident lines is at right angles with the transversal line. Proposition 12 proves the same theorem in the obtuse angle hypothesis. Proposition 13 removes the restrictive condition for both cases, thereby arriving at a complete formulation of the Fifth Postulate.

We should note immediately that Saccheri's proof strategy in Propositions 12 and 13 is perfectly legitimate, for Euclid's formulation of the Postulate (from which Saccheri, displaying meticulous conservational concern, never strays) is trivially valid in elliptic geometry: since any two straight lines always have (in this hypothesis) a point in common, they will have such a point in common also when forming with a transversal line internal angles whose sum is less than π. Saccheri's reasoning is hence not flawed; it may sound strange to a contemporary ear that Euclid's Fifth Postulate holds good in the obtuse angle hypothesis, but this is because we have become used to Playfair's formulation of the Parallel Postulate ("given a line and a point not on the line, there is *one and only one* line through the point that is parallel to the given line"), which does not hold good in elliptic geometry.

Propositions 11 and 12 are both guided by the same underlying idea. We first consider projections of equal segments of the transversal line (AP) onto the oblique straight line (AD), and then prove that these projections cannot be longer than the projected segments. Since the length of the transversal segment (AP) between the two intersection points is arbitrarily chosen but finite (given), the projection of this entire segment AP onto the oblique straight line will also be of finite length; the oblique straight line will hence meet straight line PL at a finite distance. In modern terms, we could say: if we take curve AD to be the graph of a function whose x-axis is AP, Saccheri proves the function to be Lipschitzian and hence continuous over the whole domain; the resulting image of a compact space (the segment AP) is therefore itself compact, with finite value at P.

Although the details of Saccheri's proof belong to him alone, the original idea for such a proof originated with Nasīr ad-Dīn (Saccheri knew his work through Wallis; cf. *Opera mathematica*, vol. 2, pp. 671–3) and can also be found in Clavius (in the second part of Lemma 4 in the Scholium to Proposition 28; *Euclidis*, p. 52). Both authors believed that such demonstration proved the Fifth Postulate.

In this case, Saccheri's merit consists in having provided clear formulations of the three hypotheses (right, obtuse and acute angle hypothesis), and in understanding that such proof was not applicable to the case of the acute angle, for in this latter instance nothing guarantees that projections of segments of AP onto the oblique line AD could not grow indefinitely. Instead, in his proof Nasīr ad-Dīn clearly believed that projections onto AD could effectively be longer than the projected segments, though he always assumes a proportionality constant between the two (in other words, he assumes that the projection map is Lipschitz with a constant of value less than 1); on the basis of such assumption, he believes to have proved (as does Clavius) the Fifth Postulate for any possible case. In this respect, we can say that seventeenth-century developments of analysis, and studies on convergent series – which were available to neither the Persian astronomer nor Clavius – allowed Saccheri to formulate his own objection to the extrapolation of this procedure to the case of the acute angle. The objection goes as follows: if we take equal segments of the oblique curve AD, and consider projections of these elements on AP (an inverse though identical procedure to the one outlined above), it may be the case (in the acute angle hypothesis) that these projections are shorter than the projected segments, and that their length gradually decreases as they approach point P, a point in which the series converges; then curve AD does not meet, but rather is asymptotic to, ordinate PL, and this is exactly what happens in hyperbolic geometry. For a broader discussion on the matter, with explicit relation to Nasīr ad-Dīn, see Scholium 3 to Proposition 21.

The natural continuation of the discussion developed in this group of propositions is to be found in Proposition 17. In the intermediate propositions, Saccheri draws conclusions from his own results concerning the obtuse angle hypothesis, and then formulates, in the form of corollaries, results concerning triangles and quadrilaterals that are interesting in themselves.

[1] Due to a misprint, the Latin version states AP.

[2] This is one of the few explicit applications, in *Euclid Vindicated,* of Euclid's Fourth Postulate, which states that all right angles are equal to one another. This postulate was generally considered provable; Saccheri expounds his own proof in Lemma 5 to Proposition 33.

[3] Here, Saccheri employs Archimedes' Axiom, as did Nasīr ad-Dīn and Clavius before him. It is important to note that the use of this axiom does not depend on the chosen development of proof – it is actually crucial for the proposition to be correct. Max Dehn showed that it is possible to construct both a geometry in which the right angle hypothesis is valid, although Archimedes' Axiom is not, and in which the Fifth Postulate is false (he called this *semi-Euclidean* geometry); and a geometry in which the obtuse angle hypothesis is valid (as in the following Proposition 12), although Archimedes' Axiom is not, and in which, once again, the Fifth Postulate is false (*non-Legendrian* geometry). Cf. Dehn, *Die Legendre'schen Sätze*; and Hilbert's discussion on the matter in *Grundlagen*, p. 50. Note that the inverse propositions are valid independently of Archimedes' assumption. In other words, the right angle hypothesis can be proved from the Fifth Postulate (and a straight line's infinite extensibility) without resorting to Archimedes' Axiom; and the obtuse angle hypothesis can be proved from the Fifth Postulate (in addition to some other global assumption, for instance that the straight lines intersect at two points, or that they do not separate the plane) without resorting to Archimedes' Axiom.

These considerations are usually developed with reference to the so-called Legendre's First Theorem, which states (assuming the validity of the Exterior Angle Theorem, or that straight lines are infinite, or else that the obtuse angle hypothesis is false) that the sum of the interior angles of a triangle is less than or equal to π. Proof of this theorem requires Archimedes' Axiom and rests upon an elegant construction formulated by the great French mathematician that is nowhere to be found in *Euclid Vindicated*; it was first published in the third edition of Legendre's *Éléments*, and is also found in *Réflexions* (1833; pp. 369–71). The proposition's content, however, is very explicit in Saccheri's work: once Saccheri refutes the obtuse angle hypothesis (Proposition 14), it follows (Proposition 9) that the sum of angles in a triangle is less than or equal to π. Since Proposition 14 bears entirely on Propositions 11 and 12, both of which employ Archimedes' Axiom, we thus retrieve, in more devious a manner, the need for Archimedes' principle in proving Legendre's result. The statement will also be proved in Lambert's *Theorie der Parallellinien* (§§ 73–74, pp. 348–9), who however misses the all-important Theorem of Saccheri (our Propositions 5–7), or Second Theorem of Legendre, and thus may affirm that in hyperbolic geometry the interior angle sum of *every* triangle is less than π only relying on a naïve generalization – considering in his proof an unspecified triangle.

[4] Here Saccheri implicitly applies the so-called Pasch Axiom, which states that if one straight line intersects one side of a triangle, then it also intersects a second side. Ancient, Renaissance and seventeenth- and eighteenth-century geometrical axiomatizations had considerable difficulty in isolating Axioms of Order. Of these, Pasch's axiom is one of the most important. A detailed formulation was not put forward until the 1880's. Cf. M. Pasch, *Vorlesungen über die neuere Geometrie*, Berlin, Springer 1926 (Leipzig, Teubner 1882[1]), where the axiom is to be found as the Fourth Principle of plane geometry (p. 20); the 1926 edition of Pasch's *Vorlesungen* includes an appendix by Max Dehn on the history of geometry, in which Saccheri is mentioned a couple of times and is partially credited with the birth of non-Euclidean geometries, thus improving previous works both by Hilbert and Dehn himself. In Hilbert's *Grundlagen*, Pasch's principle appears as axiom II, 4. The principle does not hold in elliptic geometry.

[5] Here, as in the rest of the book, Saccheri writes *finita seu terminata* in hendiadys; he does not seem to differentiate between the two terms.

[6] The Latin text contains no reference to the diagram.

[7] There may be an important problem hiding beneath these "just noted perpendiculars", since it is essential for Saccheri's argument that the perpendicular lines described at the beginning of proof to Proposition 11 be unique. But the uniqueness of a perpendicular to a given straight line through a point not on the line is guaranteed in Euclidean geometry (hence in Proposition 11), though not in elliptic geometry, i. e. not in the obtuse angle hypothesis here at issue. Saccheri's proof is nevertheless correct, for the way in which the diagram has been constructed (i. e. the conditions on the angles at A and P) allows multiple perpendicular lines to be drawn at AP from points lying on AD only if these points lie beyond the intersection point L. There is thus is no need for Saccheri to take these into consideration.

There is another, more general, issue worth mentioning: Euclid never seems to concern himself with the uniqueness of his constructions. Specifically, in the First Postulate there seems to be no reference to the uniqueness of a straight line passing through two points, nor is there any reference to uniqueness when constructing a parallel to a given straight line passing through a point not on the line (*Elements* I, 31), nor when constructing a perpendicular to a given straight line through a point not on the line (*Elements* I, 12; which, in addition, is never employed in Book I of the *Elements*). Whereas the uniqueness of the straight line passing through two points attracted the attention of both Ancient and Modern commentators (who purposely included an additional axiom), and the uniqueness of the parallel line is focus of debate in works such as *Euclid Vindicated*, there seems to have been no particularly heated foundational debate concerning the uniqueness of the perpendicular line in either Antiquity or the Renaissance. Proclus gives (*In primum Euclidis*, 286–89) some additional constructions that may aim to this effect (he says that the uniqueness is provable in *In primum Euclidis*, 375), but Heath found Proclus' proof insufficient (see T. Heath, *The Thirteen Books of the Elements*, Cambridge, Cambridge University Press 1908, vol. 1, pp. 273–75).

[8] Should read *Elements* I, 5.

[9] Proof of Proposition 8 employs Proposition 3, which in turn (as we have already seen) requires the general validity *Elements* I, 16, which is false in the obtuse angle hypothesis. In fact, Saccheri's deduction, which concludes by stating that BM is greater than DF, is false in general: in elliptic geometry, it may be the case that for certain straight lines and angles, BM is equal to DF just as in Euclidean geometry (consider for instance a spherical model with straight line PL at the equator, point A as a pole, and AP and AL as *equal* meridians). On the other hand, the next part of the proof, where Saccheri shows that BM is not less than DF, is correct because it does not employ Proposition 8 in an essential way (it should also hold in the Euclidean case). Thus, Saccheri can only prove that BM is not less than DF. But this is sufficient for the purpose of this Proposition.

[10] Saccheri explains that the proof technique employed in Proposition 11 and 12 does not work in the acute angle hypothesis (as we have seen above), and, moreover, that it is useless to look for a proof of the Fifth Postulate in the acute angle hypothesis under certain limiting conditions (namely, that the incident straight line forms a right angle): because Proposition 13 (i. e. the Fifth Postulate in its full generality) is false in the acute angle hypothesis. To show this, Saccheri takes two hyperbolic straight lines at right angles to a transversal, which are thus parallel lines (by *Elements* I, 27; though Saccheri prefers to use *Elements* I, 16). These lines, however, form with another transversal line two interior angles whose sum is less than π, and should thus meet in virtue of the Fifth Postulate. Therefore the Postulate is proved false in the acute angle hypothesis.

In Proposition 17, Saccheri arrives at a stronger result in which the Fifth Postulate (hence Proposition 13) does not hold in hyperbolic geometry, nor in the limiting conditions of Propositions 11 and 12.

[11] These are the limiting hypotheses of Proposition 17, stating that if we assume the validity of the Fifth Postulate in the special case that one of the two angles of the incident straight line is a right angle, and that this line is of arbitrary length (thus possibly very small, if one

so chooses), then the Postulate is generally valid even when one does not assume these two additional hypotheses.

[12] The short proof of this Scholium is important, as it clearly demonstrates that the Fifth Postulate is trivially valid if interpreted as referring to an incident straight line of arbitrary fixed (not just any) length. Saccheri's proof is very similar to Wallis' (which Saccheri reproduces in pp. 40–1): clearly, if we allow for the possibility of constructing triangles of any dimension that are similar to a given triangle, then this simple proof can be immediately generalized to the case of incident straight lines of any length, thus proving the Fifth Postulate in its general form. But the hypothesis of the existence of triangles similar to a given triangle is in fact itself equivalent to the Fifth Postulate.

Notes to Proposition 14

Saccheri here draws conclusions from previous Propositions and refutes the obtuse angle hypothesis by means of an (alleged) *consequentia mirabilis*.

The general idea guiding the proof is clear: the hypothesis of the obtuse angle proves the Fifth Postulate, from which the truth of Euclidean geometry thus follows; the obtuse angle hypothesis thus refutes itself. What is not entirely clear, however, is how Saccheri intends to carry out the details of such a proof, at least when it comes to the first part of Proposition 14. In point of fact, he has not yet explicitly demonstrated that the truth of the right angle hypothesis follows from the truth of the Fifth Postulate. He states here that this latter passage "is manifest to all geometers", and though this may perhaps be true (or not), it nonetheless leaves us in doubt as to how he might have formalized his *consequentia*. I take it to be rather likely that he is here thinking of *Elements* I, 32 (or some similar proposition), where from the Fifth Postulate it is proved that the sum of the interior angles of a triangle is equal to π, which is incompatible with the obtuse angle hypothesis. To complete the proof, he would then need only Proposition 15, which is no way dependent on Proposition 14. Saccheri's reasoning is thus as follows (keep in mind that the three hypotheses are defined in reference to a single quadrilateral, call it a). 1: The obtuse angle hypothesis holds good for a given quadrilateral a. 2: The obtuse angle hypothesis holds good for all quadrilaterals (Proposition 6). 3: The Fifth Postulate holds good (Proposition 13). 4: The sum of the interior angles of a triangle is π (*Elements* I, 32). 5: The right angle hypothesis holds good for a given quadrilateral (Proposition 15). 6: The right angle hypothesis holds good for all quadrilaterals (Proposition 5). 7: The obtuse angle hypothesis does not hold for a given quadrilateral a. We therefore conclude, by *consequentia mirabilis* and 'anhypothetically', that the obtuse angle hypothesis is false.

The second demonstration is perfectly explicit, and requires neither conjectures nor additions. It does not seem, however, to employ *consequentia mirabilis*, but rather a common *reductio ad absurdum*: it proves that, in the obtuse angle hypothesis, straight lines AD and PL meet in virtue of Proposition 13, but then must also be parallel in virtue of *Elements* I, 27 (though Saccheri, as usual, employs the equivalent *Elements* I, 17).

In any case, it is absolutely crucial to realize that in both cases the refutation of the obtuse angle hypothesis, though resorting to proof of the Fifth Postulate, concludes by employing *Elements* I, 16 – which is in fact incompatible with elliptic geometry. In the second proof this is achieved by an explicit use of *Elements* I, 17; in the first, as I have reconstructed it, by relying on *Elements* I, 32, which proves that the sum of the interior angles of a triangle is equal to π by employing both the Fifth Postulate and *Elements* I, 31 (the latter relies on *Elements* I, 27, and hence *Elements* I, 16).

It is therefore useful to briefly discuss whether the proof of this Proposition (or at least the first of the two), is, as Saccheri claims, effectively a *consequentia mirabilis*. Most interpreters offer the following reconstruction of the proof passage: Saccheri proves: (a) that the Fifth Postulate is true in the obtuse angle hypothesis; and (b) that truth of the right angle hypothesis is deduced from the truth of the Fifth Postulate. This double step is, effectively, a *consequentia mirabilis*. However, these Modern interpreters all have Playfair's Axiom in mind, and hold that (b) is correct, while (a) holds good only because Saccheri employs *Elements* I, 16. In other words: since the obtuse angle hypothesis contradicts and the exterior angle theorem, Saccheri can derive Playfairs' Axiom (*ex falso quodlibet*) and hence conclude by *consequentia mirabilis*. But from a historical perspective such interpretation is surely flawed. Firstly: because Saccheri has Euclid's original formulation of the Fifth Postulate in mind, and this formulation does in fact hold in elliptic geometry and cannot therefore be used to prove the right angle hypothesis; which is to say, passage (b) will not work, no matter how its details are reconstructed. Secondly: because although Proposition 12 employs *Elements* I, 16, the way it is there used, as we have seen in previous *Notes,* contradicts neither the properties of elliptic space nor the obtuse angle hypothesis; thus in the proof of (a) one instance of the exterior angle theorem is used, that can be eliminated.

We thereby reconstruct the argument as is effectively found in *Euclid Vindicated.* Saccheri claims that his proof is conducted by *consequentia mirabilis* because from the obtuse angle hypothesis he (a') derives the Fifth Postulate (without, as we now know, any essential reference to *Elements* I, 16), and then (b') proves the right angle hypothesis from the Fifth Postulate (this is done in a somewhat obscure manner, which is left to the geometer to understand; but of course *Elements* I, 16 is essential here). Yet, from *Elements* I, 16, Saccheri (or his acute reader) can derive the right or acute angle hypothesis (as will later be done by Legendre), without recourse to the Fifth Postulate; in other words, deduction (b') employs the Fifth Postulate in an *irrelevant* way. So Saccheri's proof is: (a") the validity of the Fifth Postulate is proved via the obtuse angle hypothesis; (b") the right angle (or acute angle) hypothesis is proved via the validity of *Elements* I, 16. Step (b"), which is completely independent of (a"), is sufficient to refute the obtuse angle hypothesis: so the conclusions from Proposition 14 are correct (if we allow for Saccheri's premises), though its proof is not a *consequentia mirabilis*, but rather a *reductio ad absurdum*.

Were we to look at the issue from a reversed perspective, however, we would see that when refuting the obtuse angle hypothesis by calling upon the Fifth Postulate, Saccheri's proof suggests that the hypothesis here at stake is somehow related to the theory of parallels (in addition to the exterior angle, or the infinite extension of the straight line). This gestures towards what later resulted in the exploration of the three different isotropic geometric structures – which was not the case in Legendre's (or Lobachevsky's) formulation (see the *Introduction*, § 6).

In conclusion, we might also mention that Playfair's Axiom, which is a straightforward consequence of the Fifth Postulate and *Elements* I, 30, was already known and discussed at Saccheri's time. It was, perhaps, formulated for the first time by Al-Haytham, though Saccheri could also have read it in the Corollary to Proposition 31 of Vitale Giordano's *Euclide restituto* (p. 60).

Notes to Propositions 15 and 16

These are two very standard theorems of absolute geometry, which Saccheri seems to prove not only for later use, but also for their own individual significance. Note that Saccheri also proves the theorems in the hypothesis of the obtuse angle, despite having refuted such hypothesis already in Proposition 14 (he must not have been troubled by *ex falso quodlibet*).

As we have already seen, Proposition 15 is the inverse of Proposition 9 (or rather, of its generalization). Therefore, the right, obtuse, or acute angle hypothesis is valid if and only if the sum of the interior angles of a triangle (and hence, by Proposition 5–7, of all triangles), is equal to, greater than or less than π. By combining this result with Proposition 14 we obtain Legendre's First Theorem that the sum of the interior angles of any triangle is less than or equal to π. From the previous *Notes* discussion, we see that in a way Saccheri proves, but does not understand, that Legendre's Theorem follows from *Elements* I, 16.

1 Saccheri's formulation may at first glance seem to employ the inverse of Proposition 9 (which he sets out to prove here) in an unjustified way. His reasoning, however, is legitimate: if the acute angle hypothesis were valid, it would not be possible (by Proposition 9) for the sum of the angles of a triangle to be greater than π, which is in fact the case in one of the two triangles here at issue; while if the obtuse angle hypothesis were to hold good, then it would not be possible (again, by Proposition 9) for the sum of the angles of a triangle to be less than π, as is the case in the other triangle; it follows, by exclusion, that the right angle hypothesis must hold good.

2 This is a generalization for the three geometries of the first part of Euclid's statement in *Elements* I, 32.

Notes to Proposition 17

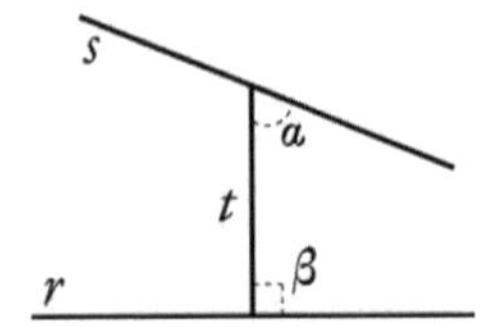

This Proposition is the obvious continuation of Proposition 13, and (as Saccheri himself says) will find its natural conclusion in Proposition 27. It is important because it clearly demonstrates that in hyperbolic geometry two straight lines r and s forming with an incident straight line t two internal angles α and β whose sum is less than π (according to Saccheri's example α is acute and β is right) sometimes meet and sometimes do not. But above all, the theorem reveals Saccheri's awareness of the existence of a functional relation between the length of segment t and the size of angle α.

In fact, both his statement and his proof clearly state that, *given* any incident straight line t ('however small'), it is always possible to find a sufficiently large acute angle α (depending on the length of t) such that in hyperbolic geometry the straight lines r and s do not meet. Conversely, in Proposition 27 he shows that, *given* any angle α ('however small'), one can always find a length of segment t (depending on the size of angle α) such that in hyperbolic geometry the straight lines r and s do not meet. But Saccheri's Classical and synthetic (as opposed to analytic and algebraic) geometrical style precludes him from attaining the more important result (at least according to our modern conception) later obtained by Lobachevsky and Bolyai, namely, an exact determination of this function, which in hyperbolic geometry characterizes the so-called *angle of parallelism* (the smallest angle α for which r and s are parallel): $\tanh(t) = \cos(\alpha)$. Still, we cannot deny (and Proposition 17 bears witness to this), that Saccheri did in fact grasp a functional relation of this kind.

There is a typical uncertainty in the Latin use of 'quantifiers', in the sense that (both here and in the related Proposition 27), Saccheri employs the formula *quaevis recta* ('any straight line') to signify the universal quantification of 'every straight line', while he elsewhere and more often (for instance, in Proposition 36) uses it to refer to modern existential quantification ('any straight line' meaning 'a straight line whatsoever'). In Proposition 29, which presents a similar context, Saccheri more appropriately employs the term *omnis recta* for the same purpose.

Proposition 17 is never employed in later parts of *Euclid Vindicated.*

[1] In Scholium 3 to Proposition 27.

Notes to Propositions 18 and 19

As in the case of Propositions 15 and 16, Saccheri proves Propositions 18 and 19 because they yield simple and interesting results of elementary geometry which prove useful for a comparison of the three hypotheses. Note, once again, that Saccheri discusses these three theorems also in the hypothesis of the obtuse angle, which he had already refuted, and that neither Proposition 18 nor 19 will be applied in any other section of *Euclid Vindicated.*

Proposition 18 is a generalization (to the three geometries) of Euclid's *Elements* III, 31; Proposition 19 is a generalization of *Elements* VI, 2. Both of these very famous Euclidean theorems were attributed to Thales (although – at least in the case of the latter – with scarce philological evidence).

We should note that Proposition 19 will be the starting point for non-Euclidean statics, as the classical treatment of the matter is grounded in the application of *Elements* VI, 2 (or equivalent) to find the barycenter of a triangle (in Archimedes' *De planorum aequilibriis*). For a Modern discussion, see the very beginning of Lagrange's *Mécanique analytique* (1788), where he proves a theorem of statics on an isosceles triangle, implicitly employing

Saccheri's Proposition 19 in the right angle hypothesis; a complete discussion of the matter is in BONOLA, *La geometria non-euclidea*, pp. 174–6 (English ed., pp. 182–4). As we have remarked (*Introduction*, § 7), some mathematicians attempted to ground the Fifth Postulate on the laws of (Euclidean) statics; but Saccheri probably remained unaware of this connection.

Notes to Propositions 20 and 21

Proposition 20 is a technical lemma which Saccheri only employs when proving the subsequent Proposition. In Proposition 21 he employs it both in the hypothesis of the acute angle (as is here proved), and in the hypothesis of the right angle (where BD is exactly half of MC), which is a simple Euclidean result, and an immediate consequence of Thales' Theorem (the mentioned *Elements* VI, 2).

Proposition 21, on the other hand, is of great historical importance. It was generally referred to as Aristotle's Principle because of a passage in *De Caelo* (A5, 271b30–32), which states that two infinite incident straight lines (two radii of an infinite circle) bound an infinite space. This principle, interpreted as stating that the distance between two incident lines infinitely increases, was applied by Proclus to the theory of parallels in order to prove the Fifth Postulate (see the following Scholium 1), and was employed in the same fashion many times thereafter. For instance, Umar Khayyam's proof, which we mentioned in the *Notes* to Propositions 1 and 2, concluded by adopting this principle. Such a move can be found in many other proofs that assumed parallel straight lines to be equidistant (see the following Scholium 2). In 1572, Commandino had explicitly assumed Aristotle's Principle among the axioms, in order to prove the Fifth Postulate: "Axioma. Si ab uno puncto duae rectae lineae angulum facientes in infinitum producantur, ipsarum distantia omnem finitam magnitudinem excedit" (*Euclidis*, p. 19v). Two years later, Clavius attempted to prove Aristotle's Principle (*Euclidis*, ed. 1574, p. 49), but already in the 1589 edition of his *Euclidis* he realized not to have succeeded (see below, *Notes* to Scholium 2). Saccheri was the first to provide a correct proof of this Principle.

Proof of Proposition 21 relies in an important way on Archimedes' Axiom. Recently, however, it has been shown that (assuming axioms I, II and III of Hilbert's *Grundlagen*, which model a so-called *Hilbert plane*) Archimedes's Axiom is in fact stronger than Aristotle's Principle, and one can conceive non-Archimedean Hilbert planes in which Aristotle's Principle holds (but not *vice versa*). See M.J. GREENBERG, *Aristotle's Axiom in the Foundations of Geometry*, "Journal of Geometry", 33, 1988, pp. 53–57.

Also note that in the Corollary Saccheri correctly identifies Proposition 21 as false in the obtuse angle hypothesis (a *non-Legendrian* plane, in fact, is not consistent with Aristotle's Principle). Note that although according to such a hypothesis incident straight lines diverge, reach a maximum distance, and then converge, Saccheri never claims that these lines end up meeting: because he has in mind the axiom according to which two straight lines share no more than one common point (as in single elliptic geometry).

In conclusion, we note that from a contemporary axiomatic perspective, if we allow for the validity of Aristotle's Principle and assume (as does Saccheri; cf. my *Notes* to Lemma 4 in Proposition 33) that a segment must intersect a circumference if one end of it lies inside and the other outside the circumference ('Line-Circle continuity'), then the hyperbolic axiom on parallel lines is equivalent to saying that rectangles (quadrilaterals with four right angles) do not exist. Thus, Saccheri's acute angle hypothesis, which indeed denies the existence of rectangles, is equivalent to the hyperbolic postulate also in its more rigorous and formal sense. For an overview of contemporary axiomatizations, see V. PAMBUCCIAN, *Axiomatizations of Hyperbolic and Absolute Geometries*, in *Non-Euclidean Geometries*, edited by András Prékopa and Emil Molnár, New York, Springer 2006, pp. 119–53. Stressing the elegant role of Aristotle's Principle in various axiomatizations, see also V. PAMBUCCIAN, *Zum Stufenaufbau des Parallelenaxioms*, "Journal of Geometry", 51, 1994, pp. 79–88; and M.J. GREENBERG, *Old and New Results in the Foundations of Elementary Plane Euclidean and Non-Euclidean Geometries*, "The American Mathematical Monthly", 117, 2010, pp. 198–219.

Notes to Proposition 21, Scholium 1

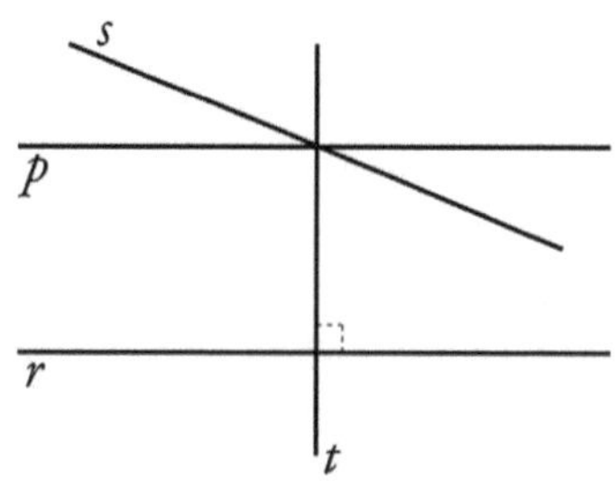

In this Scholium, Saccheri voices some objections to Proclus' proof of the Fifth Postulate (and to those of several mathematicians that followed him). Apart from some minor variations, these proofs follow the same simple schema: (1) take two straight lines r and s forming with a transversal line t internal angles whose sum is less than π (for instance, t forms a right angle with r and an acute angle with s); (2) draw straight line p perpendicular to t passing through the intersection point of s and t; (3) by *Elements* I, 27, p is parallel to r; (4) assume line p, parallel to r, to be equidistant to r; (5) according to Aristotle's Principle, the distance between p and s increases; (6) such distance, in fact, tends to infinity; (7) we conclude that r and s will meet at some point.

It is important to note that, in the first edition (1574) of his commentary on Euclid, Clavius did in fact accept Proclus' proof (as it is found in Commandino's *Euclidis*). In the second edition (1589), however, he changes his mind, probably as a consequence of having discovered the work of Nasīr ad-Dīn; here Clavius strives to provide an alternative proof and advances critiques of Proclus – critiques of which Saccheri will later make use.

First of all, Aristotle's Principle must be proved: otherwise we would just be replacing one axiom with another axiom – an approach that would yield no real advantage. Clavius was well aware of this, and consequently provides his own proof of Aristotle's Principle. He offers a very naïve proof in his first edition, and complements it in the subsequent editions (at the beginning of his Scholium after *Elements* I, 28; cf. *Euclidis*, p. 49). Saccheri proves this Principle in Proposition 21 of *Euclid Vindicated.*

Secondly, we must be very careful to understand the Principle correctly. It is in fact easy to assume – and prove – that the distance between two straight incident lines increases as the lines are extended (step 5 in the above-mentioned proof sketch). But in order to prove the Fifth Postulate, something more is needed: such distance must tend to infinity (step 6). It is in fact possible to imagine a line *s* whose distance from a straight line *p* always increases although it never reaches the straight line *r* (for instance, if *s* is asymptotic to *r* or to another parallel line between *r* and *p*). This is where the example of the Conchoid – an asymptotic line – comes in useful. Clavius' proof of Aristotle's Principle (which improves Proclus' exposition, as the latter takes the Principle to be an axiom) only demonstrates that the distance between *p* and *s* increases; he was not able to show this distance to tend to infinity. Clavius was well aware of his proof's shortcomings, and it is for this reason that he ultimately rejected Proclus' demonstration and proposed an entirely different one. Saccheri, on the other hand, overcomes the difficulties faced by Clavius and successfully proves that the distance between *p* and *s* tends to infinity.

Thirdly, and finally, Saccheri notes he has not provided an adequate proof of the Fifth Postulate, because it is still possible to raise objections to step 4 (see above), which assumes parallel straight lines (namely, non-incident straight lines) to be equidistant. In this respect, there is a long tradition of attempts to prove the Fifth Postulate (including Clavius') on the basis of the equidistance of parallel lines. Saccheri here proves himself well aware of the fact that parallel lines are not in fact equidistant in hyperbolic geometry, and hence that such a proof (with which he deals in the Second Part of Book One of *Euclid Vindicated*) is in fact equivalent to proof of the Euclidean postulate.

A historical discussion of equidistant straight lines can be found in the following Scholium 2.

[1] Clavius, *Euclidis*, p. 49, which Saccheri here quotes to the letter; Clavius himself was quoting Commandino. Proclus' passage can be found in *In primum Euclidis*, 371.

[2] Proclus does not provide any such proof. Saccheri refers here to a proof by Clavius whose text is sufficiently ambiguous, as to imply a connection back to Proclus. This shows that Saccheri probably did not have any first-hand knowledge of Proclus' commentary.

[3] Again: Clavius, *Euclidis*, pp. 49–50. The Conchoid of Nicomedes is a fourth-degree curve with Cartesian equation $(y-a)^2(x^2+y^2)=k^2y^2$ with fixed $a,k>0$, used by Nicomedes of Alexandria (third century BC) to trisect angles; it was the subject of extensive algebraic and analytic study in the seventeenth century. It is interesting to note that, in elementary geometry, using a Conchoid is equivalent to using a marked ruler; for a demonstration see Hartshorne, *Geometry: Euclid and Beyond*, p. 264.

[4] Saccheri once again uses the technical term *redargutio* to refer to *consequentia mirabilis* (see Scholium after Proposition 33): if, from the truth of the Fifth Postulate, it is proved that a straight line parallel to another straight line can behave like a conchoid, it follows that the Fifth Postulate itself is *absolutely false*.

Notes to Proposition 21, Scholium 2

In this rather long and somewhat disorderly Scholium, Saccheri addresses a variety of different topics connected by a common reference to the notion of equidistant lines. His main references here are Borelli and Clavius. We start by noting that neither of these authors employs the notion of equidistance in his definition of parallel lines, whereas such approach was very common in Classical Antiquity and the Modern Age. Instead of surreptitiously assuming the notion of equidistance within a definition, both authors state that such property has to be proved or postulated – and, as we have seen in the Preface to *Euclid Vindicated* (see also the Scholium to Proposition 39), Saccheri indeed agrees with their approach.

Clavius accepts the Euclidean definition of parallel lines as non-incident straight lines, and then attempts to prove the Fifth Postulate by showing that these lines are in fact equidistant. Borelli, on the other hand, rejects the Euclidean definition, for he sees in it a reference to the notion infinity (parallel lines are straight lines that never meet, even when extended to infinity), which makes it obscure and useless; he thus replaces this definition with an equivalent one which he regards as clearer. Borelli then postulates parallel lines to be equidistant (he does not believe the Fifth Postulate can be proved).

Saccheri starts with Clavius' (quite naïve) demonstration of equidistance, which he has no difficulty refuting (although he maintains great respect for the Jesuit mathematician all the while). He then proceeds to refute (both here and elsewhere) Borelli's methodological proposal that a definition should not be adopted if (through it) the existence of its object remains uncertain (for further discussions of Borelli, see *Euclid Vindicated*, pp. 124–7), and reiterates that, throughout the *Elements*, Euclid always constructively demonstrates the existence of the object to be defined. For a historical contextualization of this type of discussion, see the *Introduction*, § 5. He also rejects Borelli's equidistance postulate, firstly because he believes that the Fifth Postulate can be proved without appealing to additional (and equivalent) principles, and secondly, and more importantly, because he holds that a postulate must in some way be justified, which is to say, the possibility (consistency) of its application must be demonstrated: see *Logica demonstrativa*, ed. 1701, pp. 129–30 [ed. 1697, pp. 203–6], and *Note* 10 to the Preface. Whereas there are no such attempts by Borelli, Saccheri sets out to do so in the Second Part of Book One of *Euclid Vindicated*. Some references to the notion of parallelism as equidistance can be found in the *Notes* to Propositions 38 and 39.

The correspondence between Tommaso Ceva and Guido Grandi contains a comment by Saccheri (dated 1713) on Borelli's 'real' definition of parallels. He states that the main point is to prove that if a straight line is perpendicular to two straights, also any other straight line perpendicular to one of them is perpendicular to the other: "Ma il punto è dimostrare senza petizione di principio che se una retta è perpendicolare a due rette, qualunque altra che sia perpendicolare ad una di esse, debba essere perpendicolare alla compagna. Io non mi arrendo che il Borelli abbia ciò dimostrato, quando non abbia fatto capo al concetto formale obiettivo di linea retta. Fin qui il Saccheri ..." (Ceva to Grandi, 9 August 1713, in Tenca, *Relazioni fra Gerolamo Saccheri e il suo allievo Guido Grandi*, p. 38). Ceva often wrote letters on behalf of Saccheri, who "is a hero of laziness" (7 March 1702, here in p. 24) and "cannot move a hand, having the gout of sloth" (3 August 1712, p. 33). Saccheri's remark is completely correct: according to Borelli's definition of parallel lines, the Fifth Postulate can be reformulated to state that any line perpendicular to a given straight line must also be perpendicular to any line parallel to the given line (in Borelli's sense). Borelli himself, however, preferred a formulation of the Postulate that employed the notion of equidistance.

In the second part of the Scholium, Saccheri digresses, to discuss physical experiments that can demonstrate the truth of the Fifth Postulate. Concerning this topic, see the *Introduction,* § 7. On this Scholium in general, see M.E. Di Stefano, M. Frasca Spada, *Logica, metodo, geometria : Saccheri, Borelli e l'equidistante da una retta*, "Epistemologia", 8, 1985, pp. 33–76.

[1] *Euclides restitutus*, 1658, by Giovanni Alfonso Borelli. We know that Saccheri read the third edition of this book (Roma, Mascardi 1679); all quotes will be taken from this edition: see L. Maierù, *Il Quinto Postulato Euclideo da C. Clavio [1589] a G. Saccheri [1733]*, "Archive for History of Exact Sciences", 27, 1982, pp. 297–334 (esp. p. 325 n. 74). The first Italian translation of the title, as 'Euclid Renewed' (*Euclide rinnovato*, 1663), is perhaps more adequate than the usual and correct 'Euclid Restored', as it better reflects the book's content, which does not 'restore' Euclid's *Elements* so much as it sharply criticizes and improves them. The Galilean writer Cosimo Noferi, in *Disceptatio pro Euclide* (written in 1658–63 ca., and never published), advances the following comments concerning Borelli's title: "… quare potius mihi videtur *Euclides mendax, permutatus, dirutus, pervolutus, inversus*, et simili appellatione inscribendus, ipsius volumini magis propria, quam *Euclides restitutus*" (manuscript transcription in Giusti, *Euclides Reformatus*, p. 154). A less controversial but similar observation is put forward by Dechales: "Quamvis autem opus sit bonum, & utile, malè tamen ei titulus est praefixus, cum vix Euclidem in Euclide restituto agnosces" (*Cursus*, p. 25).

[2] Saccheri presents here the Euclidean definition as formulated by Clavius in his Latin translation. Clavius in fact refers to two straight lines that *in infinitum producantur* (Definition 34; *Euclidis*, p. 21). The Greek text, on the other hand, simply reads ἐκβαλλόμεναι εἰς ἄπειρον which may not refer to actual infinity, but rather to 'indefinitely extended' lines. Note that Saccheri himself later admits that adopting insufficiently clear terms is a definitional error (see *Euclid Vindicated*, p. 108), even in the case of a nominal definition. It follows that Saccheri – to answer Borelli' complains – should show that Euclid's definition is both nominal (i. e. that Euclid does not assume, but actually demonstrates, the existence of parallel lines in his definition), and clear enough even though it appeals to the concept of infinity. Saccheri does not, however, put forward any argument about the latter issue.

[3] Borelli's quote is to be found in the second page of his Preface *ad lectorem geometram* in *Euclides restitutus*. A similar formulation (in both content and wording) is found in Giordano, *Euclide restituto*, pp. 12–3.

[4] The Euclidean definition of a square is found in *Elements* I, def. 22 and in Clavius' *Euclidis*, def. 29, p. 20. In Antiquity there had been some debate over proof of *Elements* I, 46 (cf. Proclus, *In primum Euclidis,* 423) in relation to the concept of construction (here ἀναγράφειν). Such proof, of course, bears upon the Fifth Postulate.

[5] *Elements* I, 31 explains how to construct a parallel to a given straight line passing through a point that is not on the line. It does not depend on the Fifth Postulate because it is always possible to perform such a construction in hyperbolic geometry (where the parallel line in question is not unique). Instead, it draws on *Elements* I, 27, which relies on the exterior angle theorem and does not hold in elliptic geometry (in which no parallel line can be drawn). In the *Elements*, Euclid never explicitly discusses the uniqueness of the parallel line. Such uniqueness, however, is guaranteed by the preceding *Elements* I, 30 (which requires the Fifth Postulate). In *Data* 28, however, a straight line parallel to a given straight line and passing through a given point is said to be *given in position*; this may be an (almost) explicit hint to the uniqueness of the parallel.

[6] In Borelli, *Euclides restitutus*, third page of the mentioned Preface. Borelli there states that his own definition of parallel lines is better than the Euclidean "cum exponatur per passionem possibilem & evidentissimam". He then mentions *Elements* I, 10 and 11, which allow him to draw a parallel (according to his definition) to a given straight line.

Note that this definition of parallel lines is equivalent to Euclid's definition only if we assume the validity of the Fifth Postulate (which Borelli, of course, accepted without objection). The equivalence, in this case, can easily be demonstrated by combining *Elements* I, 28 and 29, which show that if two straight lines are perpendicular to a third line, then they are parallel to each other (in the Euclidean sense of non-incidence), and, if they are parallel to each other, they are also perpendicular to a third line.

Borelli's definition is not, however, generally equivalent to Euclid's definition, without assuming the Fifth Postulate. In hyperbolic geometry, while Borelli's definition is verified by a proper subset of the non-incident straight lines, those normally termed *ultraparallel lines*, we also have non-incident straight lines (parallels, in Euclid's sense) with no perpendicular line in common (namely non-incident asymptotic lines, normally called hyperbolic parallel lines in the proper sense). Saccheri was well aware of this problem. In later parts of *Euclid Vindicated* (see, in particular, Corollary 2 to Proposition 25) he deals extensively with non-incident asymptotic straight lines. Here, however, he does not see the need to put forward any critique of Borelli, as he takes his definition to be *nominal* and thus arbitrary (neither true, nor false, but nonetheless legitimate, as, in virtue of *Elements* I, 11, it is not inconsistent).

[7] Saccheri quotes directly (indeed, even the diagram and letters are identical), Clavius' Lemma 2 after the Scholium to Proposition 28 (Clavius, *Euclidis*, p. 51). The whole of Clavius' subsequent reasoning depends on this lemma (and his reasoning is correct once we allow for this premise). Borelli's formulation of the Fifth Postulate (in his work: Axiom 13) is slightly different, but nonetheless equivalent to Clavius': "Si recta linea, in suo extremo semper perpendiculariter constituta super aliam rectam lineam, moveatur in transversum in eodem plano: alterum punctum extremum translatae rectae lineae in eius fluxu rectam lineam describet" (*Euclides restitutus*, p. 23; cf. a similar exposition in the book's Preface). Borelli's Axiom 13 is in fact equivalent to the Fifth Postulate only if Archimedes' Axiom is assumed.

[8] The difficulty of this passage lies in the infamous obscurity of Euclid's definition of straight line: a line that *ex aequo sua interiacet puncta* (according to Clavius' translation, *Euc-*

lidis, p. 14; but one of the main problems is precisely that we do not know how to translate the original formula ἐξ ἴσου ἐφ' ἑαυτῆς σημείοις κεῖται). Indeed, it is in virtue of the definition's vagueness that Clavius was able to prove his Lemma 2 (mentioned in the previous *Notes*), from which he easily deduced the Fifth Postulate. Saccheri clearly shows that Clavius' demonstration proves nothing at all; it resolves in some sort of *petitio principii* in which *ex aequo sua interiacet puncta* is interpreted so as to conclude with the Lemma's truth. Saccheri attempts to lend more substance to Euclid's definition of a straight line, so as to transform it, towards the end of this Scholium and in the Lemmata to Proposition 33, into a proper definition.

[9] Saccheri is referring to the superposition of quadrilateral CGFA and GDBF, which (if we assume that line CGD, though not straight, is equidistant to AB) are equal, since AF = FB, AC = FG = BD, and angles at A, F and B are right-angles. It follows that if we overlap the bases AF and FB, then CG will overlap with GD, and angles at G should be the same on either side (that is, towards C and towards D). Details of this proof by superposition are to be found in Proposition 37.

[10] For details of the proof, see, once again, Proposition 37. The point of this brief proof is to show that an equidistant curve (or *hypercycle*) does in fact have all properties that Clavius ascribes to it (namely, the property of forming right angles with all lines perpendicular to AB which intersect it), though this does not imply that it is a straight line.

[11] As already mentioned in the *Introduction*, the important point here is that Saccheri's Theorem (Propositions 5, 6, and 7) allows for a single, and local, measurement that determines the curvature of the entire space (assumed to be constant). In this respect, *Euclid Vindicated* is the first work to propose a possible empirical verification of the geometrical hypothesis.

[12] This axiom is, of course, Borelli's formulation of the Parallel Postulate, which states that a line equidistant to a straight line is itself a straight line.

[13] *Elements* III, 7 is in fact a theorem of absolute geometry, the proof of which only requires propositions previous to *Elements* I, 24. The theorem produced certain debate amongst Ancient and nineteenth-century interpreters; see Heath's commentary to Euclid (vol. 2, pp. 15–7). Clavius offers a rather long commentary in *Euclidis*, pp. 110–1.

[14] This 'mechanical' explanation of the Euclidean definition of a straight line as an axis of rotation was quite common in the seventeenth century; it suggested a certain co-origination of the straight line and the circle (that is, the two most simple and perfect curves), since the motion of a body with two fixed points is described by a straight line axis, and every point not on the axis draws a circumference as its orbit. Saccheri also employs this interpretation of the Euclidean definition in Lemma 1 to Proposition 33 (see the *Notes* to the respective Proposition).

Notes to Proposition 21, Scholium 3

As we have seen in the *Introduction*, the first print version of the demonstration of the Fifth Postulate attributed to Nasīr ad-Dīn (found in the *Longer Version* of his commentary to the *Elements* and presumably written in 1298 by one of his disciples) was published in Rome in 1594, and was later translated into Latin by the orientalist Edward Pocock on the occasion of Wallis' 1651 Oxonian lectures. Wallis himself later published Nasīr ad-Dīn's proof, alongside his own demonstration (dated 1663), in *De Postulato Quinto* (in the Appendix to the 1693 edition of *Algebra*). Clavius, too, frames it as an alternative to his own proof (see the *Introduction*, § 2) which relies on equidistant straight lines (the proof discussed by Saccheri in the previous Scholium 2).

Nasīr ad-Dīn's *Longer Version* of the proof does in fact bear upon a Lemma 1, which was unproved and held to be self-evident, which Saccheri decomposes here into two 'postulates'. Wallis (in *Opera Mathematica*, vol. 2, p. 673) believes these principles to be no more evident than the Euclidean Fifth Postulate itself. It is worth noting, however, that Nasīr ad-Dīn's original demonstration, found in *A Treatise to Cure Doubts Regarding Parallel Lines* and in the *Shorter Version* of his commentary to the *Elements*, does not assume these two Lemmata to be self-evident but instead sets about attempting to prove them, as Clavius himself does in turn (probably drawing on manuscript sources).

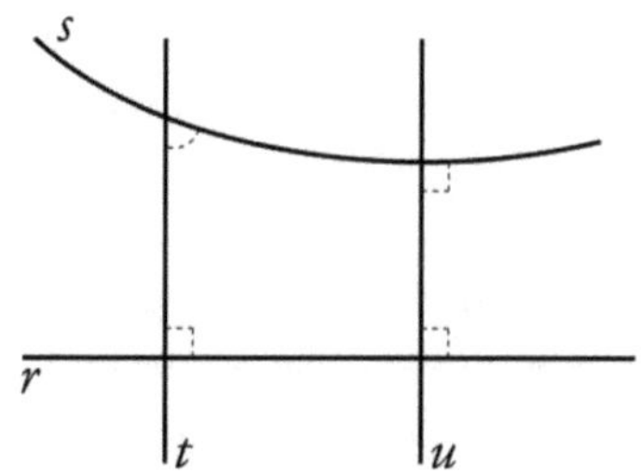

In this Scholium, Saccheri's reasoning is significant in that he demonstrates an awareness of the fact that one of Nasīr ad-Dīn's principles does not hold in hyperbolic geometry. In his two Lemmata, the Persian mathematician states that if two straight lines form with a third incident line two angles whose sum is less than π (in the restricting hypothesis that one be a right-angle), then these two straight lines will approach each other; and, inversely, that converging straight lines form with a transversal line angles whose sum is less than π. From here he goes on to prove the Fifth Postulate, i. e. that those straight lines (in addition to approaching each other) will actually meet. But in his proof Nasīr ad-Dīn also assumes the equality of the acute angles mentioned in Lemma 1 (namely, his 'postulate'). This, of course, holds in Euclidean geometry. Saccheri, however, goes much further, and proves Nasīr ad-Dīn's Lemma in hyperbolic geometry (as a consequence of Corollary 2 to Proposition 3; for the inverse case, see Corollary 1 to Proposition 23, and Proposition 28). The problem, however, is that in hyperbolic geometry those acute angles are not all the same – they can approach and even exceed the size of a right angle. Nasīr ad-Dīn's Lemma (and Saccheri's demonstration, as he himself is aware) is thus only locally, not globally, valid: it may be the case that hyperbolic straight lines r and s initially approach each other (forming with a transversal line t internal angles whose sum is less than π), reach a point of minimum distance (forming right angles with the transversal line u) and finally diverge; it may also be the case that as the hyperbolic lines extend, they tend to form with a transversal line two internal right-angles, though they never reach this configuration at any finite distance, and the hyperbolic lines are thus asymptotic to each other.

The demonstrative power of Nasīr ad-Dīn's attempt of proof does not depend in fact on the Lemmata that he postulates, which are true and can be demonstrated; and that he had himself proved, although defectively, in his original writings (the ones unknown to Saccheri). His proof rather depends on the implicit assumption on the equality of those acute angles, which is equivalent to the Fifth Postulate. This is the same mistake we find in Clavius' second part of the proof in Lemma 4 in the Scholium to Proposition 28 (*Euclidis*, p. 50). Consequently, he does not make any real progress with respect to Nasīr ad-Dīn. As we observed in the *Notes* to Propositions 11, 12 and 13, Saccheri is probably more able than Nasīr ad-Dīn or Clavius to grasp the problem of the extremal point for the distance between the straight lines because eighteenth-century mathematics was familiar with limiting procedures. On the other hand, Wallis, who was certainly capable of grasping this difficulty, makes no mention thereof.

It is worth noting that, if Nasīr ad-Dīn's Lemma were globally valid, it would establish the impossibility of ultraparallels, which form with a transversal line interior angles whose sum is less than π without ever (globally) approaching each other. While the principle according to which two straight lines will eventually meet if they approach each other is necessary for demonstrating the impossibility of the existence of asymptotic parallel lines (Nasīr ad-Dīn's Lemma does not talk about these latter lines). What allows the Persian mathematician and his successors to charge a single theorem with the two different results outlined above is again the above-mentioned error about acute angles. *Note* 3, below, shows how Arnauld (who nonetheless commits the same mistake with respect to the angles) had already distinguished between the two principles. In any case, Nasīr ad-Dīn's error of proof remained for a long time difficult to identify, and the posterity of *Euclid Vindicated* was so meager that, at the end of the eighteenth century or beginning of the nineteenth century, many geometers were still attempting to prove the Fifth Postulate by employing a similar strategy (cf. in particular, Pagnini's Third Method of proof in *Theoria rectarum parallelarum*, pp. 24–7).

In conclusion, we mention yet another proof strategy which is somehow linked to this concept of mutually approaching straight lines: Mercator's approach in *Euclidis Elementa Geometrica novo Ordine ac Methodo fere demonstrata* (1678). Although this book enjoyed widespread success at the time at which Saccheri was studying and teaching, the Jesuit never makes any reference to it. Mercator began by defining parallel lines as straight lines that are not 'inclined' towards one another, though without specifying the meaning of 'inclination': "Parallelae lineae sunt, quae non inclinantur ad se mutuo" (Definition 11; p. 2). He proceeded to take on an axiom according to which two lines are not inclined to one another if they are not inclined to a third: "Duo lineae non inclinantur ad se mutuo: quando earum una non magis quàm altera versus eandem partem inclinatur ad aliquam tertiam", and explains: "Sicuti duo homines, numquam assequentur sese mutuo, quando eorum unus, non magis celeriter quam alter, per eandem viam currit ad eundem terminum" (Axiom 3; p. 2). Next, he easily deduces (Theorem 7; p. 7) the transitivity of parallelism (*Elements* I, 30), then *Elements* I, 29 and the Fifth Postulate. Although Saccheri was quite inclined to abstract logic, he does not seem to have ever considered proving the Fifth Postulate by demonstrating the transitivity of the parallelism relation, an approach that will be attempted many times throughout the following decades; see especially Hindenburg, *Ueber die Schwürigkeit bey der Lehre von den Parallellinien*, "Leipziger Magazin zur Naturkunde, Mathematik und Oekonomie", 1781, pp. 145–68.

The second part of the Scholium is dedicated to Wallis' demonstration of the Fifth Postulate, which relies on the possibility of constructing a triangle similar (but not congruent) to a given triangle. It is hard to underestimate the importance of Wallis' proof, which kept many eighteenth- and nineteenth-century mathematicians and philosophers (from Leibniz to Lambert, to Delboeuf, to Bertrand Russell) quite busy. The proof had significant metaphysical, in addition to mathematical, implications. Saccheri, however, only concerns himself with its geometrical and technical aspects, and does not seem to think of it as bearing much importance. He may have believed that the key part of Wallis' proof was already to be found in Clavius (see L. Maierù, *Il quinto postulato euclideo in Cristoforo Clavio*, "Physis", 20, 1978, pp. 191–212; cf. p. 197 n. 12), or even in Euclid, who, in *Elements* VI, 2, seems to explicitly reconnect the theory of parallelism to the theory of similarity (though not, of course, by means of such a strong characterizing theorem); cf. also Euclid's *Data*, 32–38. Saccheri also provides his own proof of Wallis' characterization of the Fifth Postulate through the notion of similarity. As his proof rests upon previous propositions of *Euclid Vindicated*, it results much shorter and more elegant than the original 1663 version. I think this very fine demonstration can be considered, in itself, a very important result of Saccheri's work. In any case, the major difference between Wallis and Saccheri is that the former did not take his proof to be a mere reduction of the Fifth Postulate to another principle (namely, that of the possibility of transformation by similarity). Rather, he took it as proof of the Postulate itself, as he believed those transformations to be possible. For Saccheri, on the other hand, the possibility of similarity, though obvious (as is the Fifth Postulate, after all), remains a mathematical hypothesis in need of proof. Finally, note that neither Wallis nor Saccheri explicitly provide a proof of the inverse proposition, according to which the Fifth Postulate implies the possibility of transformations by similarity; yet this is in fact a very elementary proof.

[1] Wallis, *Opera Mathematica*, vol. 2, pp. 669–73.

[2] Here, Saccheri quotes Pocock's translation found in Wallis, *Opera Mathematica*, vol. 2, p. 669 almost word-for-word.

[3] Wallis' comment on Nasīr ad-Dīn's exposition of the principle: "Esto. At, inquam, ecquis non facilius conceperit ut clarum, *Duas rectas (in eodem plano) convergentes, tandem (si producantur) occursuras*, quam hunc totum apparatum" (Wallis, *Opera Mathematica*, vol. 2, p. 670). It may be the case that, given his greater familiarity with limiting procedures, Wallis recognized what Nasīr ad-Dīn and Clavius failed to grasp, namely, that acute angles can increase in size until they become right angles. This brings the argument back to the Ancient problem (already identified by Geminus) of asymptotic straight lines. Wallis, however, does not provide any further specific comment on Nasīr ad-Dīn's proof (or on Clavius', which he does not even discuss).

In this context, it is worth mentioning Arnauld's theory of parallel lines, to which Saccheri never explicitly refers but which may have in part provoked discussion in *Euclid Vindicated.* In his *Nouveaux Elémens de géometrie*, Arnauld takes the approach that both Wallis and Saccheri condemned as 'defeatist'. He believes himself capable of proving Nasīr ad-Dīn's postulate as they appear in Clavius' work. (Arnauld never mentions Nasīr ad-Dīn himself, and it is very unlikely that he knew of the Persian scholar's work: *Nouveaux Elémens* were composed between 1655 and 1667, when they were published; they were thus formulated prior to the print edition of Pocock's translation of Nasīr ad-Dīn's works from Arabic. Arnauld never even mentions Clavius, though he certainly read his work.) In the many Lemmata at the start of Book VI of *Nouveaux Elémens* (pp. 393–401), Arnauld essentially reproduces Clavius' proof of Nasīr ad-Dīn's principle. His demonstration appears in a more articulated form yet it contains the same error. In Axiom 6 of Book V (p. 361), however, he also assumes that two straight lines approaching each other will intersect (he seems to think this is equivalent to Euclid's Fifth Postulate – as Wallis and Saccheri mention here). Thus, by combining Nasīr ad-Dīn's postulate (two straight lines approach each other if a third line cutting through them forms internal angles whose sum is less than π) with his own Axiom 6, Arnauld arrives (in Corollary 2 to Theorem 11 in Book VI, p. 413) at Euclid's original Fifth Postulate. The posterity of Nasīr ad-Dīn's proof, in fact, mostly depends on the very good diffusion of Arnauld's *Nouveaux Elémens* in France. See for instance Malézieu's version of the proof, which is based on Arnauld's: N. DE MALÉZIEU, *Elemens de geometrie de Monseigneur le Duc de Bourgogne*, Paris, Boudot 1705.

[4] In Proposition 25, Saccheri proves that a hyperbolic straight line cannot approach more and more another straight line remaining however above a given finite distance from it (i. e. without intersecting or being asymptotic to it). If we assume this were the case, we destroy the acute angle hypothesis, and the only case remaining would be that of Euclidean geometry, in which converging straight lines meet at a finite distance.

[5] Wallis' 'common notion', namely his Lemma 8 ("Datae cuicunque Figurae, Similem aliam cujuscunque magnitudinis possibilem esse"), which Saccheri here quotes verbatim, together with the remark that this principle is also valid for circumferences, and the remark concerning the definition of proportional figures, are all to be found in Wallis, *Opera Mathematica*, vol. 2, p. 676. The contention that it is possible to draw circumferences of any radius is, of course, Euclid's Third Postulate. In order to understand Wallis' and Saccheri's fallacious argument concerning the similarity of circles, it suffices to see that in Euclidean geometry the length of a circumference C is a multiple of its radius r; thus similarity transformations of the first transform the second by similarity (i. e. in proportion), and *vice versa*. On the other hand, in hyperbolic geometry (of unitary curvature) the relation is $C = 2\pi \sinh(r)$, hence radius variations do not correspond to a proportional (similar) variation of the circumference.

[6] Saccheri thus rejects the validity of Wallis' proof as an absolute proof of the Fifth Postulate. I am not aware of anyone before Saccheri who made similar remarks concerning this result. What's more, twenty years or so before the publication of *Euclid Vindicated*, Saccheri had already examined and criticized Wallis' Lemma, objecting that it was proved more by metaphysics than geometry: "… quantunque il suo lemma ottavo resti da lui provato più tosto con la metafisica che con rigore geometrico, onde pare che fosse necessaria la costruzione problematica per torre ogni sospetto di petizione di principio" (from a letter Ceva wrote to Grandi on the 12th of July 1713, in which Ceva relates Saccheri's words; cf. TENCA, *Relazioni fra Gerolamo Saccheri e il suo allievo Guido Grandi*, p. 35). A similar position was later held by LAMBERT, *Theorie der Parallellinien*, §§ 79–81 (pp. 350–2).

[7] The definition of similar (rectilinear) figures as figures with equal angles and sides of proportional length, is *Elements* VI, def. 1.

[8] Saccheri's use of Proposition 13 hints at the fact that Archimedes' principle may be necessary to develop the proof (see *Note* 3 to Propositions 11, 12 and 13): this is indeed the case, and Wallis' proof, too, makes explicit use of this principle. Concerning the relationship between Wallis' proof and various continuity principles, see U. Cassina, *Sulla dimostrazione di Wallis del Postulato Quinto di Euclide*, in *Actes du 8e Congrès International d'Histoire des Sciences* (1956), Paris, Hermann&Cie 1958, pp. 33–8.

[9] Further implicit application of Pasch's Axiom (see *Note* 4 to Propositions 11, 12 and 13).

Notes to Proposition 21, Scholium 4

It is obviously unlikely that Euclid had Saccheri's Fig. 25 in mind when he formulated the Fifth Postulate. Nonetheless, this Scholium is important in the context of *Euclid Vindicated*, as it summarizes all the results obtained up to this point in the text and drafts a program outlining Saccheri's plan of attack regarding the acute angle hypothesis.

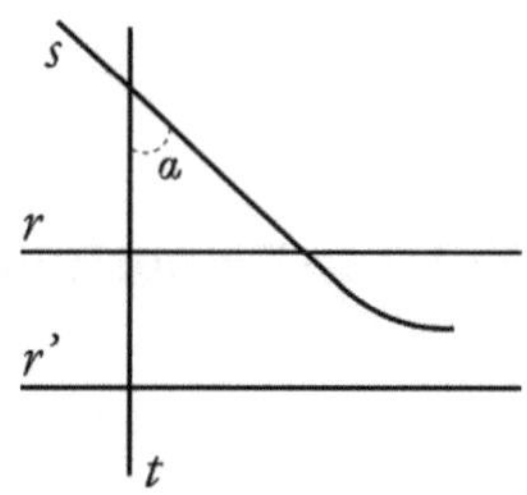

Saccheri has already shown that if a transversal line t intersects straight lines r and s forming one right angle and one acute angle α, then it is always possible to find *some* length of the segment t for which r and s meet; and in order to prove the Fifth Postulate it must be shown that intersection occurs for *any whatever* length of t. Therefore, take r (which intersects s) to move parallel along t at an increasing distance from s. If r and s always form equal angles at their intersection point, we have Nasīr ad-Dīn's assumption, and the Fifth Postulate is easy to prove. However, it may be the case that these acute angles progressively decrease (without any limit) as r moves downwards, and hence that at a certain point r' no longer intersects s. Saccheri maintains that, were this to be the case, then there would be an intermediate position of r, at which r and s are tangent or share a common segment. But this is *against the nature of the straight line*, because two straight lines cannot be tangent to one another at a point or share a common segment. This possibility is hence excluded, and the Fifth Postulate is proved.

In the following propositions, Saccheri further develops the theory of intersection of r and s in hyperbolic geometry, proving some of the most advanced theorems in *Euclid Vindicated*. He proves that the point of tangency for r and s cannot be found at any given distance, and, hence, that the two straight lines can at best be asymptotic (this is perhaps Saccheri's most important achievement in relation to of the proof strategy here outlined). Subsequently, in the Lemmata to Proposition 33, he attempts to show that the tangency and the common segment are "against the nature of the straight line" (in other words, that two straight lines cannot be tangent at one point and cannot have a segment in common), from which, by (improperly) applying these Lemmata to the point of tangency at infinity, he concludes to have thereby proved the Fifth Postulate.

The diagram exhibited in Fig. 25, in fact, is the diagram to which Saccheri (if not Euclid) will refer throughout the rest of *Euclid Vindicated.*

We may remark that this diagram and construction was in fact employed for a proof of the Fifth Postulate at least by a Medieval Hebrew mathematician (who may have been relying on Ancient sources); the demonstration is partially edited in T. LÉVY, *Gersonide, le Pseudo-Tūsī, et le postulat des parallèles. Les mathématiques en Hébreu et leurs sources arabes*, "Arabic Science and Philosophy", 2, 1992, pp. 39–82. We should also note that Abraham Kästner, who was most certainly familiar with Saccheri's work (see the *Introduction*, § 7), believed that the Fifth Postulate could not be proved, but nonetheless maintained that he could put forward an argument that would at least show its plausibility: the argument simply consists in taking Saccheri's diagram and the movement of straight line *r* perpendicular to *t*. Kästner's diagram was later used by Klügel in his dissertation. It is therefore certain that all the founders of non-Euclidean geometry were familiar with it, given the popularity enjoyed by both of these books. See G.A. KÄSTNER, *Anfangsgründe der Arithmetik, Geometrie, ebenen und sphärischen, Trigonometrie und Perspectiv*, Göttingen, Vanderhoeck 1758, Corollary 12 to Proposition 12, pp. 189–90 (and diagram 46); see KLÜGEL, *Recensio*, § 14, p. 18.

[1] Due to a misprint, the original Latin text reads YDH, YHP.

Notes to Propositions 22 and 23

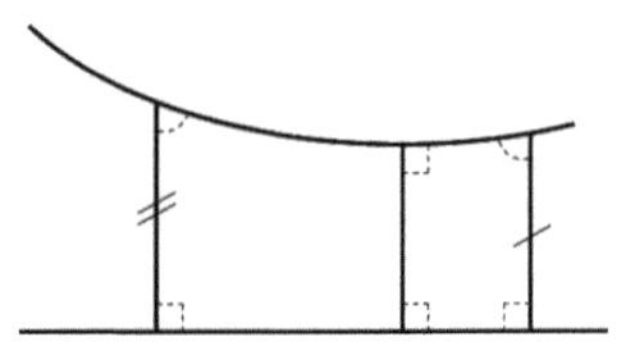

Proposition 22 is a generalization of Proposition 2. It states that for a quadrilateral with two right angles on the base and two acute angles on the opposite side (regardless of whether the other two opposite sides are equal), the side forming acute angles must be concave towards the base, and there is only one point minimizing the distance between the base and the opposite side. The reasoning is developed by means of a continuity argument, which Hilbert later showed to be unnecessary (in Appendix Three in *Grundlagen*, pp. 159–77). In any case, Proposition 22 is just a technical lemma that is useful for proving the following Proposition.

Proposition 23, on the other hand, is absolutely fundamental; it offers an outline and a starting point for a complete classification of the properties of incidence of hyperbolic straight lines (completed in the following propositions). Here, Saccheri correctly shows that given two straight lines in hyperbolic space, there are three mutually exclusive possibilities: (1) the two lines intersect at one point; (2) the two lines are non-incident and asymptotic, and share no common perpendicular (hyperbolic parallel lines; see also Proposition 25); (3) the two straight lines are both non-incident and non-asymptotic, and share one common perpendicular (ultraparallel lines). For the latter, Saccheri also shows (in the Scholium) how to construct the common perpendicular line (whose uniqueness follows from this explicit construction).

Saccheri's elegant proof of Proposition 23 has many points in common with the corresponding proof conducted by Hilbert in *Grundlagen der Geometrie* (Appendix Three, Theorem 2, pp. 164–5).

Corollary 2 proves *Elements* I, 27 and 28 in hyperbolic geometry. The two Euclidean demonstrations, which do not employ the Fifth Postulate, were of course sufficient to obtain these results. Saccheri, however, hopes to construct an alternative proof that does not immediately rely on *Elements* I, 16 and 17. Here he ultimately fails in his attempt, for he employs Proposition 3, which is also fundamentally dependent on *Elements* I, 16. But that is not the point here. Saccheri seems to be saying (as we saw in *Note* 8 to the Preface) that the purpose of applying *Elements* I, 16 and 17 solely to the case of bounded triangles (*omni ex parte circumscripti*) is not that of limiting its application to instances in which the exterior angle theorem holds good in elliptic geometry, but rather to avoid possible confusion regarding the intersection of non-incident hyperbolic lines at infinity. In fact, if we were to apply *Elements* I, 16 and 17 to unbounded triangles (where two sides are described by parallel straight lines and the (finite) base is described by a transversal line forming with the sides interior angles whose sum is π) we could be induced to think (says Saccheri) that parallel lines always meet at infinity. That this is not the case in Euclidean geometry is demonstrated, according to Saccheri, by the standard proofs of *Elements* I, 27 and 28. But the Jesuit shows that in hyperbolic geometry there are indeed cases of parallel asymptotic straight lines. He therefore advocates that we seek to strengthen the proof of *Elements* I, 27 and 28 in the hyperbolic case, in order to explicitly show that ultraparallel lines (i. e. hyperbolic straight lines that form with at least one transversal internal angles whose sum is equal to π) aren't asymptotic straight lines, i. e. that they never share a common point, even at infinity. He thus distinguishes between parallel and ultraparallel lines in the hypothesis of the acute angle.

Note, in any case, that a proof of *Elements* I, 27 that does not rely on *Elements* I, 16 (but rather *directly* employs the concept of infinite straight lines, i. e. the fundamental hypothesis of *Elements* I, 16) is found in Proclus, *In primum Euclidis*, 362–3 and can be traced back to Ptolemy. It is not clear, however, that Saccheri had any knowledge of these developments, as they are not mentioned in Clavius' commentary.

Lastly, it is worth noting that Figure 27 presents a novelty for eighteenth-century readers in that it depicts straight line AX as a curved line. This is the first instance of what is to become a very common approach to visually representing hyperbolic geometry. Practically all of Saccheri's diagrams are altered with regards to the text they illustrate; this is because most propositions in *Euclid Vindicated* are proved by *reductio*, and are developed in the obtuse and acute angle hypotheses. Most of the diagrams employed by Saccheri and contemporary authors dealing with similar topics (such as Clavius and Giordano) are non-conformal representations of the hyperbolic model, though isomorphic to this model with regards to its affine connection (that is to say: angle sizes differ from what is assumed in the text, but straight lines are represented by straight lines). However, from Fig. 27 onwards Saccheri uses (nearly) conformal diagrams, wherein geodesics are drawn as curved lines. This approach will be employed by all subsequent geometers, from Lambert to Lobachevsky, up to Poincaré and the majority of contemporary treatises. The role of diagrams in geometric proofs during Antiquity and the Modern Period is nowadays a topic of great interest to epistemologists. Good introductions to the subject, that have the dual advantage of both paying particular attention to proofs by *reductio* and of mentioning Saccheri's *Euclid Vindicated* include K. Manders, *The Euclidean Diagram* and *Diagram-Based Geometric Practice*, in *The Philosophy of Mathematical Practice*, ed. by P. Mancosu, Oxford, Oxford University Press 2008, pp. 65–133. Note, furthermore, that Fig. 27 is also manifestly 'wrong' with respect to the dimensions of segment NK (used in proof of the following Proposition 24), perhaps due to an error committed by the drawer, or perhaps in order to emphasize the remarkable abstraction of Saccheri's mathematical argument.

Notes to Propositions 24 and 25

Proposition 24 is a very important technical lemma, which is first applied in Proposition 25 and subsequently utilized in a large number of occasions. The Corollary and the Scholium to Proposition 24 are similarly technical in nature, and solve certain problems that might otherwise have shown up in Saccheri's proof.

Proposition 25, which is obtained via a repeated application (and a limiting procedure via Archimedes' Axiom; cf. Corollary 3 to Proposition 3) of Proposition 24, completes the classification of the properties of incidence introduced in Proposition 23. Indeed, in Proposition 23 Saccheri proves only that two hyperbolic lines approach each other if they are neither incident nor ultraparallel; he does not, however, show that their convergence is necessarily unbounded, i. e. that the two hyperbolic parallel lines are asymptotic. Such is the content of Proposition 25, which states (in other words) that there is no lower bound to the distance between converging parallel lines.

The two following Corollaries are correct but formulated in a somewhat convoluted fashion. The first states that if one were to prove that a converging but non-asymptotic line does in fact exist in hyperbolic geometry, the acute angle hypothesis would be refuted by *consequentia mirabilis* (but Saccheri will not attempt a proof by this line of reasoning). The second Corollary returns to Borelli's definition of parallel lines (discussed in Scholium 2 after Proposition 21) and shows that in hyperbolic geometry this definition can be applied only to the restricted class of ultraparallel lines. It is thus incorrect to think that the statement according to which if two straight lines are not parallel (according to this definition) then they intersect at finite or infinite distance, is equivalent to asserting the Fifth Postulate (since the same is true of incident straight lines and asymptotic parallel lines in hyperbolic geometry). In order to prove the Fifth Postulate one would need to show that Borelli's parallel lines (i. e. ultraparallels) must form, with *any* whatever (not only with one) transversal, interior angles whose sum is π (this last implication being a consequence of Proposition 27).

Notes to Propositions 26 and 27

Proposition 26 is yet another a technical result needed for the proof of Proposition 27.

Proposition 27 states that the Fifth Postulate is equivalent to the assumption that a straight line s, forming any acute angle α with transversal line t, must meet *all* lines r

perpendicular to *t*. In the three following Scholia, Saccheri notes that this result is the natural culmination of many previous propositions, as stated on a number of occasions: in Scholium 2 to Proposition 17; in Scholium 4 to Proposition 21; and in Corollary 2 to Proposition 25.

Proposition 27 is a complement to Proposition 17, because it shows that *given* any acute angle α ('however small') it is always possible, in hyperbolic geometry, to find a segment *t* (depending on α) whose length is such that *r* and *s* do not intersect. Were this not true, then the Fifth Postulate would hold good. As we have seen in the *Notes* to Proposition 17, this anticipates, to some extent, later (and far more mature) reflections on the angle of parallelism.

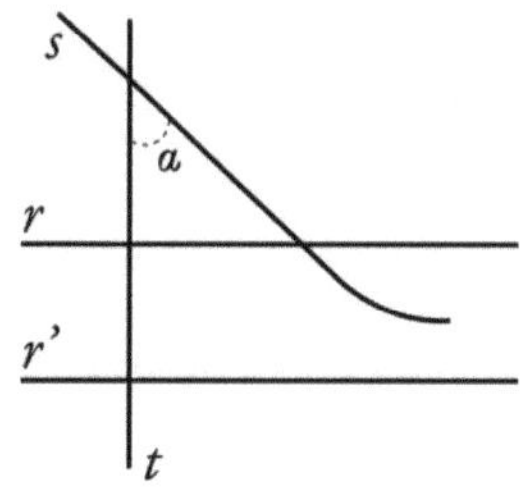

Proposition 27 was also mentioned in Scholium 4 to Proposition 21. What had there been presented in the form of a figure is, at this point in the text, expressed as a theorem: in hyperbolic geometry it must *always* be the case that, by a parallel transport of straight line *r* along *t*, there will be a point at which *r* and *s* no longer intersect. Were this not true, then the Fifth Postulate would hold good.

Finally, Proposition 27 confirms what had been said in Corollary 2 to Proposition 25: throughout the movement of *r*, not only must it be the case that at a certain point *r* and *s* no longer intersect, but it must also be the case (further on) that *r* and *s* are not asymptotic but in fact ultraparallel (in other words, they must have one common perpendicular). Were this not true, then the Fifth Postulate would hold good.

[1] We have already seen that Saccheri always accepts this principle, which was not explicit in Euclid's *Elements* but was rather an addition made by many later commentators and geometers. This principle is Axiom 14 in Clavius' work (*Euclidis*, pp. 25–6); Saccheri later develops his own demonstration of the principle in Lemma 1 to Proposition 33.

[2] There is an ambiguity regarding the meaning of this Corollary. The first sentence can be understood as referring to the infinite extension of AX or to that of BX. If we simply take it to mean that after extension at infinity (of AX, or BX, or both), AX and BX do not *intersect* but only *meet* (either as tangents, or by merging into a single straight – two hypotheses that Saccheri explicitly takes into account in Proposition 33), then Saccheri definitely seems to be incorrectly employing the notion of infinity, for he supposes that two straight lines can be extended after their meeting point at infinity. In other words, Saccheri is treating a point at infinity as if it were a point at finite distance. Such an interpretation is not absurd, for Saccheri's conclusion in Proposition 33 regarding the impossibility of hyperbolic geometry is in fact the result of a similar mistake.

Another possibility is that the first sentence refers to the infinite extension of BX, and that Saccheri is thus discussing the possibility that AX and BX meet at a point that is to be found at infinity for BX, but at finite distance from AX. Such a formulation would enable him to speak of the extension of AX after the intersection point, and conclude (*per absurdum*) that it is impossible for the meeting point to occur at infinite distance from B (although at finite distance from A). Unlike Stäckel and Engel, Halsted seems to favor this last interpretation.

[3] At this point in the text there is some confusion over quantifiers and limiting procedures – a rather common confusion persisting throughout the whole of *Euclid Vindicated*. Saccheri is most certainly saying that since length AB is (by hypothesis) however great, then the number of segments taken into consideration must also be however great; he does not seem to allow for the possibility that this number is greater than any *assignable* finite number (*plures quolibet assignabili numero finito*), i. e., that the number be actually infinite.

Notes to Proposition 28

Proposition 28 contains a rather complex proof, which provides Saccheri with the technical instruments needed in order to prove the propositions to come. The demonstration also bears a certain significance of its own, as it represents Saccheri's last word on Nasīr ad-Dīn's (and Clavius') failed proof, to which Scholium 3 to Proposition 21 of *Euclid Vindicated* is devoted. The self-evident principle that Nasīr ad-Dīn took as allowing him to prove Euclid's Fifth Postulate in fact corresponds to point (1) of Proposition 28, which Saccheri proved in Corollary 1 to Proposition 23. Yet the Persian mathematician and many of his successors never took into account the possibility that the obtuse angles towards X could decrease until they became right angles. This is in fact what comes to pass in hyperbolic geometry, as Saccheri explains and demonstrates in points (2) and (3), thus markedly exceeding all previous reflections on the topic.

The architectural autonomy of Proposition 28 is unveiled in its Corollary. Here, Saccheri shows that in hyperbolic geometry two asymptotic parallel lines share one common perpendicular at one same point – point which, however, is to be found at infinity. Note that we cannot interpret the property of having a "common perpendicular in one same point" simply as to be tangent at that point, because Saccheri also allows for another possibility, namely, that instead of being tangent to one another the two straight lines coincide at one interval. This Corollary is of great importance, as it is the only result which will be later employed in Proposition 33 to state that the acute angle hypothesis is "repugnant to the nature of the straight line". Thus Saccheri proves the following Propositions 29–32, which shed significant light on the properties of hyperbolic space, only for their geometrical significance, and not because they are needed to prove the Fifth Postulate. In fact, a 'proof' of the latter may well be developed immediately after Proposition 28.

Additionally, we note that both the structure of Saccheri's proof and the phrase *post tantum apparatum* hint at the standard Euclidean proof structure: Saccheri first develops all auxiliary constructions (κατασκευή, *constructio*), and then proceeds to devise a theoretical demonstration (ἀπόδειξις, *demonstratio, ostensio*). Note, however, that this is a *unicum* within *Euclid Vindicated*'s demonstrative procedure that does not adhere to the traditional division of mathematical reasoning as outlined by Proclus (Proclus, *In primum Euclidis*, 203), preferring instead to align itself more with seventeenth-eighteenth century methods than with Classical tradition.

[1] Should read *Elements* I, 19.
[2] Due to a misprint, the Latin text reads NC; corrected by Halsted.

Notes to Propositions 29, 30, 31 and 32

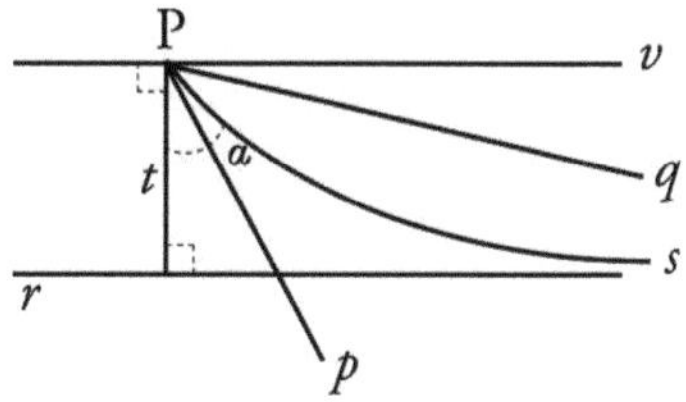

This is an intimately connected group of Propositions, which is also linked to Proposition 28, that serves as their demonstrative basis. Though not in themselves essential to the book's ultimate objective, these propositions provide important insights for hyperbolic geometry at large. Together, the four propositions demonstrate the limiting nature of the asymptotic parallel line *s*, for which all straight lines passing through P and contained within the angle of parallelism α (as a function of the length of *t*), intersect at finite distance *r*; while all those passing through P, but falling outside α, are ultraparallel to *r*.

The importance of the separation property of asymptotic parallel lines is such that both Gauss and Lobachevsky – the leading figures in the non-Euclidean revolution – adopt it as the very *definition* of parallel line. (In a recent publication however J. Tanács, *Grasping the Conceptual Difference between János Bolyai and Lobachevskii's Notions of Non-Euclidean Parallelism*, "Archive for the History of Exact Sciences", 63, 2009, pp. 537–52, claims that this is not the definition of parallel lines employed by Bolyai.)

Proposition 29 demonstrates that the asymptotic parallel line *s*, forming angle α with transversal line *t*, is the 'extrinsic limit' of all lines passing through P intersecting line *r*. In other words, all straight lines *p* forming with transversal line *t* an acute angle whose size is strictly less than α, intersect *r* at finite distance; and *vice versa*. What it means for this limit to be 'extrinsic' is that all angles for which lines *p* intersect *r*, will have angle α as an upper bound (supremum), not as a maximum (since *s* does not intersect *r* at finite distance). Clearly, Saccheri borrowed the concept of 'extrinsic limit' and the distinction between upper bound and maximum (as we would say today) from recent developments in Calculus. The distinction is interesting, as it shows that Saccheri was not so far from functional (trigonometric and analytic) considerations of the angle of parallelism after all. In fact, his faithfulness to Classical (rather than quantitative) synthetic constructions is mainly stylistic (see the *Introduction*, § 6). It is also true, however, that Saccheri employs the notion of extrinsic limit in an ambiguous way (as is clear from Proposition 32).

The first part of Proposition 30 proves that the ray *v* forming with *t* a right angle at P is the intrinsic limit of all rays *q* ultraparallel to *r*, and share with *r* a common perpendicular at two different points (the limit is intrinsic because *v* itself has a common perpendicular with *r*, namely *t*). Although Saccheri does not refer to rays, but always to straight lines (as was the custom at that time), it is obvious that he always and only has in mind half-lines originating in P. This has a certain historical relevance, given that all future non-Euclidean geometers will always define hyperbolic lines first and foremost as half-lines (rays).

The second part of Proposition 30 proves that all ultraparallel lines through P have an extrinsic limit (which, as will be shown later, is the asymptotic ray *s*). In other words, all rays forming at P an acute angle strictly greater than a certain angle (itself either smaller than or equal to a right angle; it will be shown that this is in fact angle α) do not intersect straight line *r* but share a common perpendicular with it; and *vice versa*. In this context, what it means that the limit is 'extrinsic' is that *s* does not have a common perpendicular (at finite distance) with *r*, even though it does not intersect *r*, i. e. it is not an ultraparallel line.

Proposition 31 concludes the second part of Proposition 30. It shows that the common perpendiculars of the ultraparallel lines can be chosen to be however small. In other words, all ultraparallel lines *q* passing through P approach *r* indefinitely. As a consequence, the extrinsic limit of all these lines is the asymptotic line *s* itself.

Proposition 32 brings together all previous results and explicitly proves that the extrinsic limit of all lines through P incident to *r* (Proposition 29), and the extrinsic limit of all lines through P ultraparallel to *r* (Propositions 30 and 31), coincide. This limit is therefore the asymptotic parallel line *s*.

Note that in this last Proposition the terms 'intrinsic' and 'extrinsic' are employed rather ambiguously, in a way that is no longer consistent with their application in Calculus. Saccheri states that the asymptotic parallel *s* is the *partim intrinsecum, partim extrinsecum* limit of both incident and ultraparallel lines because, like incident lines, *s* does not have a common perpendicular with *r* at a finite distance (but has one at infinity), and, like ultraparallel lines, *s* does not intersect *r* at any point at a finite distance (but does at infinity). Thus, 'intrinsic' and 'extrinsic' are but qualitative indications of the properties of an asymptotic straight line, which are difficult to grasp.

This set of propositions can also be regarded as a proof of the existence of asymptotic parallel lines in the hypothesis of the acute angle. The proof relies on an explicit continuity argument, and employs the same method later utilized by Gauss and Lobachevsky. It was János Bolyai who first provided an explicit ruler-and-compass construction of an asymptotic parallel line to a given straight line, passing through a given point (J. Bolyai, *Appendix Scientiam spatii absolute veram exhibens*, in F. Bolyai, *Tentamen juventutem studiosam in elementa matheseos purae*, Maros Vásárhelyini, Kali 1832–1833, § 34, p. 20; now also available in reprint in J. Gray, *János Bolyai, Non-Euclidean Geometry and the Nature of Space*, Cambridge, Burndy 2004, which is also a good introduction to debates on parallelism) – though Saccheri probably would not have had much difficulty devising a similar construction from his present theorems. Nevertheless also Bolyai relies on a principle of continuity in one passage of his ruler-and-compass construction. Indeed, such a principle is *almost* necessary, for W. Pejas, *Die Modelle des Hilbertschen Axiomensystems der absoluten Geometrie*, "Mathematische Annalen", 143, 1961, pp. 212–35, has shown that if we take the hyperbolic axiom to be the negation of the Euclidean axiom (rather than explicitly assuming the existence of asymptotic parallel lines) without assuming Dedekind's Continuity Axiom (but rather, assuming only Archimedes' Axiom), then we have a model without asymptotic straight lines – a model in which all parallel lines are ultraparallel. More recently, however, M.J. Greenberg, *On J. Bolyai's Parallel Construction*, "Journal of Geometry", 12,

1979, pp. 45–64, has proved that assuming Aristotle's Principle (which is weaker than the Archimedean Axiom) and Line-Circle continuity (see above, *Notes* to Propositions 20 and 21) is enough to get Bolyai's construction. As these two principles were both accepted by Saccheri, his appeal to full continuity is in fact superfluous.

At any rate, we can readily ascertain that Propositions 29–32 represent the highest point of Saccheri's entire mathematical research.

[1] I take this to be a *lapsus calami*, because Saccheri writes *conveniat* attracted by the previous sentence, despite the fact that the sense and symmetry of the phrase demand that it should read *occurrat*.

Notes to Proposition 33

Proposition 33 marks the end of the First Part of Book One, and concludes that hyperbolic geometry is impossible in virtue of the 'absolute falsity' of the acute angle hypothesis. Five long Lemmata lay the grounds for the proof's development. Saccheri goes to great foundational lengths to prove and justify the main axioms on straight lines that had been either assumed by Euclid, or added to the *Elements* by later commentators (Saccheri, of course, does not, and could not, distinguish between the two). All five Lemmata therefore contribute significantly to 'emaculate' Euclid (Savile's expression) from the errors and shortcomings that tradition attributes to him, as all five demonstrate that, whether authentic or interpolated, such axioms are but consequences of the explicit Euclidean definitions. Saccheri is thus advancing towards the goal that guides his entire work, namely, that of *proving* the Euclidean principles as a means of *showing* that they are indeed axioms, i. e. evident propositions.

In accordance with his explicit epistemological intentions concerning the nature of geometrical axioms (see the *Introduction*, § 5) Saccheri proves these five Lemmata without ever employing any prior Proposition of *Euclid Vindicated* or any theorem of the *Elements*. They could just as easily be found at the start of the treatise; indeed, Saccheri has already made both explicit and implicit use of them on many occasions throughout the 32 previous Propositions (for instance, in his proof of Proposition 26).

Also note that, despite being of central importance for studies concerning the foundations of geometry, proofs of these Lemmata do not properly contribute to the problem of parallelism, nor do they play a role in proof of Proposition 33. In other words, if one were to assume these Lemmata as unprovable principles (as sometimes happened in Saccheri's era), the proof would still be unsatisfactory, as all five Lemmata are still valid in hyperbolic geometry.

The proof of this proposition simply consists in taking the result of the Corollary to Proposition 28 (which is perhaps better understood and contextualized in the following propositions), which states that in hyperbolic geometry there are asymptotic straight lines meeting at infinity, and then *asserting* (without further demonstration) that such a meeting point at infinity "is repugnant to the nature of the straight line". We must admit, however, that Saccheri's justification for this assertion is indeed grounded in these five Lemmata: he (correctly) proves that lines with (at least) one common point at infinity should also have one common perpendicular at that point, meaning that they should either be tangent at one point or have one common segment. In other words, Saccheri proves that at that point at infinity asymptotic straight lines do not intersect in the same way straight lines normally (at finite distance) do, i. e. by forming a non-zero angle. The five Lemmata thus state (amongst other things) that two straight lines cannot be tangent to one another or have a common segment. This leads Saccheri to conclude with the impossibility of the acute angle hypothesis. Obviously, however, Saccheri is simply extending those properties holding at finite distance to points located at infinity (or, in today's terminology, at the convergence limit). In fact, nowhere in the five Lemmata does he ever refer to infinity.

In other words, Saccheri's error is of a metaphysical rather than mathematical nature, in the sense that no significant geometrical error can be identified in any of the first 32 propositions of *Euclid Vindicated*. In Proposition 33 Saccheri states that hyperbolic geometry is contradictory, without providing (rather than expounding an incorrect) proof thereof.

Yet it is clear that Saccheri had contemplated proving the Fifth Postulate with reference to the elementary properties of a straight line (i. e. by employing these five Lemmata), since at least 1713. In an excerpt from the (aforementioned) letter from Ceva to Grandi, which was written on the 12th of July of that year, we find: "Le parallele poi, pensa il P. Saccheri potersi dimostrare col solo concetto della linea retta, onde pur ricava egli, come due linee rette non possono racchiudere spazio, né avere comune segmento …" (Tenca, *Relazioni fra Gerolamo Saccheri e il suo allievo Guido Grandi*, p. 35). We recall that proving the Fifth Postulate with reference to the nature of the straight line was a popular strategy amongst geometers, most of whom, both before and after Saccheri, objected that the postulate could not be proved since there was no appropriate definition of straight line; Saccheri himself had expressed this concern in Scholium 2 after Proposition 21 (pp. 34–35). As for Classical sources, see Aristotle's statement (*Phys. B* 9, 200a2–4) that from the essence of the straight line it follows that the sum of the angles of a triangle is equal to two right-angles. As for Modern authors, I may mention d'Alembert's *Éclaircissemens sur différens endroits des Élémens de Philosophie*, xi, where he discusses the definition of parallel lines and the difficuly in proving the Postulate, drawing on his article *Parallèle* for the *Encyclopedie*, and adds that the real reason of the failure is the missing definition of a straight line: "On parviendroit peut-être plus facilement à la [a theory of parallel lines] trouver, si on avoit une bonne definition de la ligne droite; par malheur cette definition nous manque. … La definition & les proprieties de la ligne droite, ainsi que des lignes paralleles, sont donc l'écueil, & pour ainsi dire, le scandale des élemens de Géometrie" (*Mélanges de literature, d'histoire et de philosophie*, vol. 5, Amsterdam, Chatelain 1767, pp. 202 and 206–207). The same was maintained by Abraham Kästner, who had such a great influence on the German researches on the Parallel Postulate that eventually resulted in the creation of non-Euclidean geometries: "Der Grund, warum man in diesem Axiome [the Parallel Postulate] nicht die Evidenz der übrigen findet, ist … daß man von der geraden Linie nur einen klaren Begriff hat, nicht einen deutlichen" (*Ueber den mathematischen Begriff des Raums*, "Philosophisches Magazin", II 4, 1790, p. 414).

In the Scholium to Proposition 33 Saccheri adds that he is not fully satisfied with his proof. His dissatisfaction, however, does not stem from that awkward passage to infinity, but rather from the fact that he had been only able to find a contradiction between the consequences of the acute angle hypothesis and the elementary properties of the straight line (exposed in the five Lemmata). Thus he may conclude that the acute angle hypothesis is *repugnant to the nature of the straight line*. Yet Saccheri would prefer a *consequentia mirabilis* (according to him, a stronger form of reasoning) like the one he developed (or so he believed) for the obtuse angle hypothesis in Proposition 14. Such a strategy would show that the hypothesis of the acute angle is *repugnant to its own nature* without recourse to the elementary properties of a straight line. This is what he will attempt to do – says he – by means of a different demonstrative method in the Second Part of Book One. There are, however, some doubts as to whether this is the reason for which he expounds upon the second proof: see the *Introduction*, §§ 5–6.

Note that, throughout the whole book Saccheri systematically employs the Latin term *redargutio* to refer to a refutation conducted by *consequentia mirabilis*, which is thereby considered an absolute (i. e. without hypothesis) refutation of a proposition. A redargution is defined to a lesser degree in *Logica demonstrativa*: "Est autem Elenchus grece idem, ac latinè redargutio: nimirum sic definies. Elenchus est syllogismus probans alteri contradictoriam propositionis ab eo concessae, aut assertae" (ed. 1701, p. 152 [ed. 1697, p. 241]). The term is thus the Latin translation of the Greek term ἔλεγχος.

Notes to Proposition 33, Lemma 1

Saccheri's proof of the five Lemmata concerning the elementary properties of straight lines begins with a discussion of one of the most fundamental principles of Euclidean geometry, which had been the subject of much debate since Classical Antiquity. According to current reconstructions of Euclid's original text, this principle does not in fact appear among the original postulates of the *Elements*; it was discussed by Proclus and An-Nayrīzī, and it appears (as the Sixth Postulate) in the very important (tenth-century) manuscript of the *Elements* discovered by Peyrard later employed by Heiberg in his translation of Euclid's work – the best version currently available. In any case, evidence shows that it was interpolated, as early as the Classical period, in order to compensate for Euclid's implicit use of it in *Elements* I, 4. By Saccheri's time, the principle was universally considered authentic; it appears in Clavius' work as Axiom 14.

It is nevertheless important to develop a thorough understanding of this principle's content, and its various formulations. Generally, the interpolated principle is reformulated to state that through two points there is one and *only one* straight line (also phrased as: *no more than* one straight line. The existence of *at least* one straight line passing through the two points is guaranteed by Euclid's First Postulate). This last principle is not straightforwardly equivalent to the pseudo-Euclidean principle according to which two straight lines do not enclose a space. Moreover, we must also determine whether 'straight line' is best interpreted to mean a complete line, or (as was commonly believed at the time) a segment of a straight line.

The situation is thus as follows. The principle according to which no more than one complete straight line passes through two points does not hold in double elliptic geometry (on the surface of a sphere), but does hold in single elliptic geometry (on the surface of the projective place). The principle according to which no more than one segment of a straight line passes through two points is false in both spherical geometries (in this case, there are at least two segments passing through two points). The principle according to which two complete straight lines do not enclose a space is false in both geometries of spherical metrics, because two geodesics disconnect the double elliptic plane in four domains and the single elliptic plane in two domains. The principle according to which two segments of a straight line do not enclose a space is false in double elliptic geometry, but true in single elliptic geometry.

This last interpretation comes closest to the formulation explicitly discussed by Saccheri in this Lemma, and occasions a discussion of spherical metrics (i. e. the obtuse angle hypothesis) that does not immediately contradict the "nature of the straight line", i. e. the statement of Lemma 1. Of course, if this were then we would have to take Saccheri's investigations of the obtuse angle hypothesis to (implicitly) provide the foundations for single elliptic geometry in the projective plane (see *Note* 1 to Propositions 3 and 4). The first propositions of *Euclid Vindicated* and the subsequent propositions concerning the obtuse angle hypothesis, wherein Saccheri always assumes the principle of Lemma 1 to be valid, may justify such an understanding of Saccheri's work. Indeed, the Jesuit never attempts to refute the obtuse angle hypothesis by demonstrating that it is incompatible with the "nature of the straight line", however it is intended – that one only straight line passes through two points, or that two straight lines do not enclose a space. We must also note, however, that neither Saccheri nor any other geometer of his time ever explicitly acknowledged the differences between the four possible (above expounded) interpretations of this Lemma 1.

Concerning the history of Lemma 1 before Saccheri, some authors believed that the principle in question should be introduced as an axiom *per se*, while other mathematicians claimed that it may be implicitly recovered from Euclid's formulation of the First Postulate – though the most accurate recent interpreters deny that it can be found there, as it seems that Euclid could easily have explicitly emphasized that there is one unique straight line passing through two points if he had really wanted to. As we have observed, Clavius' Axiom 14 (*Euclidis*, pp. 25–6) states that two straight lines do not enclose a space; later, however, he attempts to prove the principle. The first such attempt that we know of was carried out by Proclus (*In primum Euclidis*, 239) and repeated by Simplicius in An-Nayrīzī's commentary (Tummers, *The Latin translation of Anaritius*, p. 31–32). This was also the first proof presented by Clavius, who, however, found it inconclusive and included in his own text an alternative, corrected, version (see, for instance, Heath's critique in: *The Elements*, vol. 1, pp. 195–6). Such modification did not convince Saccheri, who constructs yet another alternative demonstration.

Saccheri's 'demonstration' is just a definition (he calls it *intelligentia*, in other words 'interpretation') of straight line, both clearer and more useful than Euclid's obscure definition. For Saccheri, a straight line is defined as an axis of rotation, in other words, as the line fixed by the three-dimensional motion of a figure with two fixed points. Although this definition dates back to Antiquity and may be found in the work of Hero (*Definitiones* 4; also mentioned by PROCLUS, *In primum Euclidis,* 110), it does not seem to have circulated until the seventeenth century. Roberval was probably the first to employ it extensively; it appears in Book I, definition 26 of his 1675 *Élemens de Géometrie.* Although Roberval's book was never published, Leibniz read the manuscript and came to regard the definition of straight line therein as the best available. Indeed, he employed it extensively in his own geometrical works. The definition also probably appeared in Vitale Giordano's *Archimede*, the second volume of his *Corso di matematica,* which was also never published (cf. also Demonstration 4 of the *Annotatione* after Proposition 3 in *Euclide restituto*, p. 24). Since all of these major sources remained unpublished, and Saccheri presumably did not have any first hand knowledge of Proclus (Clavius does not refer to, or translate, this passage), he may have found this definition of straight line in Borelli's *Euclides restitutus*, where it is briefly discussed (in a very polemical manner, again in relation to Saccheri's dear topic of 'real definition') in Definition 8 of the circle. If Saccheri takes this as the central definition of his five Lemmata, it is at least partly (in the absence of other sources) a function of his propensity for foundational issues and perhaps also a function of his unspoken desire to oppose Borelli on this point.

Roberval's text is found in its entirety in V. JULLIEN, *Eléments de géométrie de G.P. de Roberval*, Paris, Vrin 1996, p. 100; on this matter, see also: V. JULLIEN, *Les étendues géométriques et la ligne droite de Roberval*, "Revue d'histoire de Sciences", 46, 1993, pp. 493–521. Concerning Roberval's influence on Leibniz (who presented his calculating machine at the Academy of Paris on the same day Roberval discussed the foundations of elementary geometry), see T. HAYASHI, *Introducing Movement into Geometry: Roberval's influence on Leibniz's Analysis Situs*, "Historia Scientiarum", 8, 1998, pp. 53–69. Leibniz's broadest discussion on the matter is probably found in a 1676 text published in G.W. LEIBNIZ, *La caractéristique géométrique*, edited by J. Echeverría and M. Parmentier, Paris, Vrin 1995, p. 66. Information concerning the definition employed by Giordano can be found in his November 1689 letters to Leibniz, published in the critical edition of Leibniz' writings: A III, 4, n. 217, p. 425. Borelli's discussion is found in *Euclides restitutus*, pp. 5–6.

Following Saccheri, the definition of straight line as an axis of rotation was employed by many scholars of non-Euclidean geometry and of the Theory of Parallels, in particular by Legendre, Gauss and Lobachevsky. See for instance G. LECHALAS, *Une définition géométrique du plan et de la ligne droite apres Leibniz et Lobatchewsky*, "Revue de metaphysique et de morale", 20, 1912, pp. 718–21; K. ZORMBALA, *Gauss and the Definition of the Plane Concept in Euclidean Elementary Geometry*, "Historia Mathematica", 23, 1996, pp. 418–36.

Most modern treatises of elementary geometry take Saccheri's Lemma 1 to be an indemonstrable principle. Indeed, when formulated as "at most one straight line can pass through two points", the lemma is Hilbert's axiom I, 2 in *Grundlagen* (p. 3).

Corollary 1, which provides some sort of proof of Euclid's First Postulate, states, in modern terms, the geodesic connection of space. Here, Saccheri assumes both metric completeness and arc connectedness. According to the epistemological principles by him espoused, the importance of this Corollary cannot be overstated. Up to this point, the Jesuit limited his discussion to an interpretation of the *nominal* definition of straight line given by Euclid – and took this nominal definition to refer to an axis of rotation. According to the methodology that he explicitly embraces in *Demonstrative Logic* and *Euclid Vindicated*, Saccheri must either postulate the existence of these straight lines as axes of rotation or go about proving it. We see, therefore, that such a proof (or at least a hint at it) is to be found in this Corollary, which should itself support the geometrical construction of the subsequent five Lemmata (not to say the whole of elementary geometry). But the demonstrative insufficiency of the Corollary, based on intuitive passages by continuity, is obvious. In order to succeed in his argument, Saccheri would have to deal with certain very abstract concepts that are far beyond the geometric capabilities of the eighteenth century.

Corollary 2 is a trivial extension of previous conclusions to plane surfaces. It also contains a sort of existential demonstration of the plane (by generation), which is not found in Euclid, and which some seventeenth century geometers were already exploring. The greatest efforts of this sort were those of Roberval in Book Two of his *Élemens de géometrie* (pp. 143–5).

[1] The original Latin text, here and below, has ABX instead of ADX.

Notes to Proposition 33, Lemma 2

Saccheri expounds on a second, certainly spurious, principle which probably did not belong to the original text of the *Elements*. Euclid makes implicit use of it in many propositions; he employs it for the first time in Book I, and then, in proof of *Elements* XI, 1, explicitly states that two straight lines cannot have a common segment – which is probably the reason why the principle later appeared as an independent axiom. Clavius represents it as Axiom 10, although he never attributes it to Euclid and readily grants that it is employed without justification throughout the *Elements*.

Note that this principle is a specific case of the broader one stating that through two points there is only one line (for which, see *Note* to Lemma 1). Indeed, Modern discussions (such as Hilbert's) that assume the latter axiom, render the former superfluous. But the principle according to which two straight lines do not enclose a space (which can be found both in Clavius and Saccheri, and whose weaker formulation depends on its application in Euclid's *Elements* I, 4) does not imply that two straight lines cannot have a common segment. Consequently, Renaissance treatises usually present them as two distinct axioms. Also note that the principle according to which two straight lines do not enclose a space is an extension of Euclid's First Postulate by means of the *uniqueness* of the straight line passing through two points. Analogously, then, the principle according to which two straight lines do not have a common segment is a refinement of Euclid's Second Postulate by means of the *uniqueness* of the extension of a straight line's segment.

The principle expressed in Lemma 2 holds in Euclidean geometry, as well as in spherical and hyperbolic geometry (in fact, it holds in any Riemannian manifold). Thus, discussions concerning this principle do not contribute to the topic of parallelism.

The traditional proof of Lemma 2 is found not only (very sketchy) in *Elements* XI, 1 but also in Proclus, *In primum Euclidis*, 215–8, who makes mention of a second proof formulated by Zeno of Sidon; and in the Medieval translation of An-Nayrīzī's commentary (see Tummers, *The Latin Translation of Anaritius*, pp. 28–29). Clavius employs Proclus' proof without providing any further commentary (*Euclidis*, p. 24). Vitale Giordano follows Clavius and Proclus (*Euclide restituto*, pp. 22–3). Note, however, that this proof has its drawbacks, as it presupposes that the diameter bisects the circle – a supposition that had not been proved by Euclid and could hardly be demonstrated without assuming the truth of this Lemma 2. Saccheri (more correctly) attempts to provide an independent proof of Lemma 2, from which he then concludes (in Lemma 4) that the diameter bisects the circle.

Saccheri's proof, which is perhaps the longest though not the most complex in *Euclid Vindicated*, does not contain any specific difficulties or errors. The proof is entirely based the definition of straight line as axis of rotation introduced in Lemma 1.

The first two conclusions mentioned in the Corollary are irrelevant, demonstrating only Saccheri's enduring confusion over the relationship between infinitesimal analysis and Classical geometry. In order to demonstrate anything of mathematical import, they would have to discuss the problem of Lemma 2 at infinity (i.e. the possibility that two straight lines converge to one straight line at infinity), which is the case that Saccheri considers in Proposition 33 to refute the acute angle hypothesis. The third conclusion evaluated in the Corollary, namely, the proof of *Elements* XI, 1, is valid in that Euclid's proof does in fact employ Lemma 2. But since this is done in an unjustified way (according to the deductive structure of the *Elements*), it was considered unsatisfactory by all interpreters. Clavius attempted to devise an independent proof of *Elements* XI, 1 (in *Euclidis*, p. 485) by explicitly appealing to the principle here described as Lemma 2, and believed himself to have done so successfully. Saccheri, however, believed otherwise. The Jesuit can thus legitimately claim to have vindicated yet another blemish that Modern interpreters often attributed to Euclid, namely, *Elements* XI, 1.

Throughout the proof (and also briefly in Corollary 2 to the previous Lemma 1) and in the foundational discourse on the nature of the straight line, Saccheri uses the expression *aequalia et similia* ("Ισα δὲ καὶ ὅμοια, cf. *Elements* XI, def. 10) to refer to congruent figures. At the time, such a definition, in which congruent figures are characterized by similarity and equality of size, was much disputed. One such definition of congruence appears, for instance, as Axiom 9 in Borelli's first edition (1658) of *Euclides restitutus*. A foundational difficulty arises, however, from the fact that equality of size can only be defined by decomposing figures into congruent parts. Thus, in the 1663 Italian version of his work (and in later editions of the Latin text), Borelli changes his mind and defines equality on the basis of congruence, abandoning the Euclidean characterization: "E quelle cose, che sono eguali potranno combaciarsi per opra dell'intelletto, trasportando, overo piegando le loro parti, se sarà di mestieri". In other words, the risk was of falling into a *petitio principii* that defined congruence on the basis of equality, and equality on the basis of congruence. Many prominent seventeenth-century mathematicians, such as Barrow, Wallis and Leibniz, honed in on this difficulty. Saccheri, who was usually very interested in foundational matters, completely ignores this matter, and remains faithful to a Classical demonstrative style that does not pay attention to 'modern' issues of functional relations (rather than geometrical objects) in elementary geometry. It is known that Euclid never gives a definition of equality of size, and then begins to employ the concept in *Elements* I, 35 without further comments.

[1] Saccheri provides further descriptions of, and historical commentary to, his own interpretation of the Euclidean definition of straight line as an axis of rotation. He states that the 'evenly' clause ("recta linea est, quae *ex aequo* sua interiacet puncta"; Clavius, *Euclidis*, p. 14) of the original definition refers to the property of congruence characterized by rotations and reflections. In fact, when restricted to a plane, Saccheri's definition implies that a straight line is the only curve between two points lacking an incongruent counterpart, i. e. a different enantiomorphic curve. This characterization of a straight line was employed extensively by Leibniz in his foundational studies on Euclidean geometry (see below *Note* to Lemma 4).

[2] The Latin text contains a misprint, as Saccheri means to refer to *Elements* XI, 1, and not to *Elements* XI, 4. Moreover the translation is problematic, since Euclid writes ἐν τῷ ὑποκείμενῳ, which some modern authors translate as 'the given plane', i. e. any plane on which there lies a straight line, and others translate as 'the underlying plane' as opposed to the next 'more elevated plane'. Saccheri employs Clavius' Latin translation (*Euclidis*, p. 485), which does not solve the interpretative problem, as it employs the Greek term alongside the correspondent Latin (and likewise ambiguous) phrase '*in subjecto plano*'. We thus do not know how Saccheri understood the Greek phrase.

Notes to Proposition 33, Lemma 3

This Lemma states that if two straight lines have one point in common, then they intersect and cannot be tangent to one another at that point. Like Lemma 2, this principle does not to belong to the original *corpus* of Euclid's work but was interpolated later. It appears in Clavius' work as Axiom 11. It is interesting to see that Clavius explicitly states (*Euclidis*, p. 24) that he counts Axioms 10 and 11 (Saccheri's Lemmata 2 and 3) among Euclid's axioms with the specific goal of proving the Fifth Postulate (i. e. Axiom 13 in his version of the *Elements*). But while Axiom 10 (Lemma 2) was certainly present in the Ancient tradition of the *Elements* (if not as a postulate then at least as a subject of discussion), Axiom 11 (Saccheri's Lemma 3) seems to appear for the first time in Clavius.

Clavius also tries to provide a demonstration of his own Axiom 11. His proof is identical to Proclus' proof of the previous Axiom 10, and is based on the fact that a circle is bisected by a diameter. Such a proof faces the same problem as the preceding Axiom: it easily falls into a *petitio principii*. Here, again, Saccheri is very skillful in reconstructing the only correct strategy of proof.

Lemma 3 plays a critical role in Saccheri's work. Indeed, a substantial part of his proof of Proposition 33 is grounded on it: in fact, asymptotic hyperbolic straight lines, whose nature Saccheri wants to show to be contradictory, can be considered *tangent* to one another at a point at infinity. Proof of Proposition 33 thus relies on an improper transfer of the validity of Lemma 3 to the converging limit of straight lines. Aside from this improper use, however, the Lemma is valid for geodesics (also in hyperbolic geometry) at any real point (i. e. not at infinity) of the plane.

Notes to Proposition 33, Lemma 4

This Proposition, which states that a diameter bisects a circle, was not characterized by Euclid as either an axiom or a postulate, nor had it been proved as a theorem. Rather, it was simply assumed to be true as part of the definition of diameter in *Elements* I, def. 17. Later, no one accepted it as an axiom, and many consequently attempted to prove it. The earliest such proof predates Euclid and is attributed to Thales. Proclus (*In primum Euclidis*, 156–8) makes brief mention of a proof, though we do not know whether it was Thales' original one, which can also be found in Clavius' work (*Euclidis*, p. 18. Note that in the Scholium to this Lemma Saccheri does not mention Proclus, of whom he probably had no first-hand knowledge; instead, he refers to Thales' proof such as it appears in Clavius). The Ancient proof does not differ much from Saccheri's: both demonstrations depend on the rotation (in three dimensions) of a semicircle about its diameter. It has been suggested (see K. von Fritz, *Die* APXAI *in der griechischen Mathematik*, "Archiv für Begriffsgeschichte", 1, 1955, pp. 13–103; K. von Fritz, *Gleichheit, Kongruenz und Ähnlichkeit in der antiken Mathematik bis auf Euklid*, "Archiv für Begriffsgeschichte", 4, 1959, pp. 7–81) that Euclid prefers to characterize this principle as a definition rather than a proposition requiring Thales' proof because the latter calls upon movement and superposition. This is supposedly the same reason that led Euclid to reject the proof of the equality of right angles (Saccheri's Lemma 5), preferring instead to postulate it at the beginning of Book I of the *Elements*. It is possible that this was not Euclid's real motivation, but there is no doubt as to the fact that Saccheri is comfortable appealing to motion in his studies of geometry, despite the degree to which this practice was criticized by his contemporaries (among them, the aforementioned Peletier and Wallis); see, however, *Note* 1 to Propositions 5, 6 and 7. Clavius was also unopposed to employing motion and superposition in geometry, as evidenced by his dispute with Peletier over the Contact Angle (*Euclidis*, pp. 117–26) and his Scholium to Axiom 8 (*Euclidis*, p. 14). Clavius' Axiom 8 is, in fact, Euclid's Common Notion 4 of the *Elements*: for a historical overview of the controversies concerning motion in geometry, see Heath's long comment to the latter (cf. *The Elements*, vol. 1, pp. 224–31). Saccheri's definition of straight line as axis of rotation clearly shows that he allows for kinematical considerations in geometry.

In any case, as we have already seen, many Ancient commentators (among them Proclus, Simplicius and Clavius) use Lemma 4 to demonstrate some important Euclidean principles. But these principles are themselves necessary for a rigorous proof of this Lemma; specifically, those which here in Saccheri appear as Lemmata 1, 2 and 3. Giordano employs a similar strategy in his *Annotatione* after Proposition 3 (cf. *Euclide restituto*, pp. 21–5). From this point of view, Saccheri is perfectly right in criticizing 'Thales' demonstration of Lemma 4. Indeed, his own proof is far more exact than the Ancient one, especially in those passages where Saccheri draws on Lemmata 1, 2 and 3 and thereby obtains maximal foundational coherence. In some senses, then, Saccheri's proof of Lemma 4 is the peak of *Euclid Vindicated*, the masterpiece of a rigorous proof of elementary definitions.

Yet another foundational problem, which Saccheri fails to identify, is associated with this Euclidean principle: it remains to be proven that the diameter (the straight line passing through the center of the circle) does in fact intersect (prior to bisecting) the circumference. In other words, Saccheri overlooked the problem of continuity. This point is salient because Borelli and Leibniz had already pinpointed the problem, and the fact that Saccheri does not touch on it may be seen as further evidence of his lack of familiarity with (and interest in) foundational problems that extend beyond elementary investigations of synthetic geometry and into the domain of Analysis. Borelli had a discussion on continuity in commenting *Elements* I, 1, and he introduced an axiom to guarantee the existence of the points of intersection of two circles (Axiom 13 of *Euclides restitutus*, in the 1658 edition only). This axiom by Borelli may well be the first instance in history of the Circle-Circle continuity principle, which is in turn equivalent to the Line-Circle continuity axiom that we have mentioned before (*Notes* to Propositions 21 and 29–32). Leibniz himself (as noted above) defines a straight line in terms of an axis of rotation, thus believing himself to have improved upon the proofs by Thales, Proclus and Clavius, by means of a very similar argument to (though not as subtle as) Saccheri's. Leibniz's most important passage on the bisection of a circle is to be found in one of his texts written in 1712, published only in 1858 by Gerhardt as *In Euclidis πρῶτα* (GM v, pp. 195–9; cf. also the important letter to Schenk written in that same period, parts of which are published in V. De Risi, *Geometry and Monadology. Leibniz's Analysis Situs and Philosophy of Space*, Basel, Birkhäuser 2007, pp. 620–1).

Notes to Proposition 33, Lemma 5

This is Euclid's Fourth Postulate and Clavius' Axiom 12. Although most commentators acknowledged the authenticity of the principle, its purpose within the *Elements* remains unclear, for which reason it generated much foundational debate. Euclid explicitly employs it only in *Elements* I, 47 (which raises doubts as to its authenticity), although he makes implicit use of it in *Elements* I, 14, 15, 28 and 46. According to some interpreters, the principle is a prerequisite of the Fifth Postulate on parallel lines (which deals with right angles, and must therefore allow for their equality in general). Such an interpretation is dubious, but has the advantage of illustrating the importance of the principle in the context of *Euclid Vindicated*.

Anachronistic interpretations take this principle to be roughly equivalent to our idea of homogeneity and isotropy of space: see Heath's comment in *The Elements*, vol. 1, p. 200; for a formal modern exposition devoid of philological aims cf. J.G. Ratcliffe, *Foundations of*

Hyperbolic Manifolds, New York, Springer 2006[2], p. 335. The first (coherently developed) formulation of this interpretation appeared in Klein's famous lectures on elementary mathematics, where he maintained that to prove this principle by means of rigid motion and superposition is comparatively inelegant and even contradictory, as this very axiom should guarantee the possibility of rigid motion. Considerations of this kind were foreign to Greek mathematics; they started gaining importance only at the time of *Euclid Vindicated*'s composition, though Saccheri does not appear to have been influenced by them.

In any case, many Ancient interpreters held that this principle could be demonstrated. Proclus, *In primum Euclidis*, 188–9, provides a proof by superposition, which was subsequently adopted by all Modern geometers (Clavius accepts it in *Euclidis*, p. 25). As was the case in Lemma 4, Saccheri follows Proclus' and Clavius' lead, reproducing the proof they had initially devised but applying previous Lemmata in order to construct a more rigorous demonstration (if one allows the use of rigid motion). In theorem 21 of *Grundlagen* (p. 23), Hilbert proves that all right angles are equal, thus providing more rigorous support (though of course in a different manner than Saccheri) to the Ancient proof presented by Proclus.

According to Thomas Heath (*The Elements*, vol. 1, p. 201), Saccheri's statement in Lemma 5, which emphasizes the importance of rectilinear angles, may have been inspired by the commentary by Proclus in *In primum Euclidis*, 189–91. Here, the Ancient scholar recalls that Pappus (whose work has since been lost) had shown that curvilinear right angles are not necessarily equal to one another (in other words, are not necessarily congruent). Although Saccheri does not seem to have had any first hand knowledge of Proclus' long argument, Heath's conjecture is quite reasonable, given that such a problem is mentioned by Clavius at the end of the Scholium to Axiom 12 in *Euclidis* (p. 25).

At the end of the Corollary, Saccheri concludes that a straight line has one unique perpendicular at one point. For his part, Euclid, as usual, provided no proof of this, perhaps because he believed that it followed from the explicit construction of perpendiculars in *Elements* I, 11.

[1] Refers to *Elements* I, def. 10; also found in Clavius' *Euclidis*, p. 16. Clavius maintains that the existence of right angles is proved by Euclid in *Elements* I, 11. His contention is important for Saccheri's general interpretation, who holds that Euclidean definitions are nominal, and therefore the possibility and existence of the objects to which they refer must either be proved or postulated.

Notes to Propositions 34, 35 and 36

These three Propositions mark the beginning of the Second Part of Book One, which aims to refute hyperbolic geometry by *consequentia mirabilis*. As we have already seen in the *Introduction* (§ 6), Saccheri here adopts a more traditional approach to proving the Fifth Postulate – an approach that considers equidistant lines. The proof he undertakes to devise is difficult in that it requires him to show that a line equidistant to a straight line is itself a straight line – a principle equivalent to the Fifth Postulate itself, for a line equidistant to a geodesic is not itself a geodesic in hyperbolic (but also in spherical) geometry. The lengthy discussion on equidistant lines in Scholium 3 to Proposition 21 reveals that Saccheri was fully aware of the difficulties associated with this proof, and well acquainted with recent attempts to proceed in this direction. His dismissal of Clavius' simplistic demonstration (*Euclidis*, pp. 50–1) of the identity between parallel straight lines and equidistant curves indicates that he was in a much more advanced position than his predecessors.

Proposition 34 can be considered a problem in the Classical sense. It gives a pointwise *construction* of a line equidistant to a straight line in the hypothesis of the acute angle (i. e. in hyperbolic geometry). It is immediately clear, that the case here is that of a line that is not straight, a curve that is always concave to the straight line to which it is equidistant. Saccheri could have confined his definition of an equidistant line to the locus of the extremes of the equal segments perpendicular to a given straight line (as did Clavius and all other geometers at the time). Instead, he lingers on a slightly more complex construction, as he wants to more precisely show the relationship (and difference) between a hyperbolic straight line and a hypercycle.

Although perfectly correct in themselves, Propositions 35 and 36 are quite complex to understand, because it is not immediately clear how they fit into Saccheri's overall strategy. Both, in a way serve as technical preparation for Proposition 37, to which they will be (very inappropriately) applied. They seem to signify that a straight (hyperbolic) line is perpendicular to the perpendicular of a given straight line if and only if it is tangent to the hypercycle at that point (i. e. to the curve equidistant from the given straight line). Proposition 35, in particular, states that if a straight line *s* is perpendicular to the perpendicular line *t* at the base *r*, then it must meet hypercycle *p* at a *point of tangency*. Proposition 36 states the inverse, namely, that if a straight line *s'* is not perpendicular to the perpendicular line *t* at the base *r* (but instead forms an acute angle at *t*), then *s'* intersects hypercycle *p*, and is therefore not tangent to it.

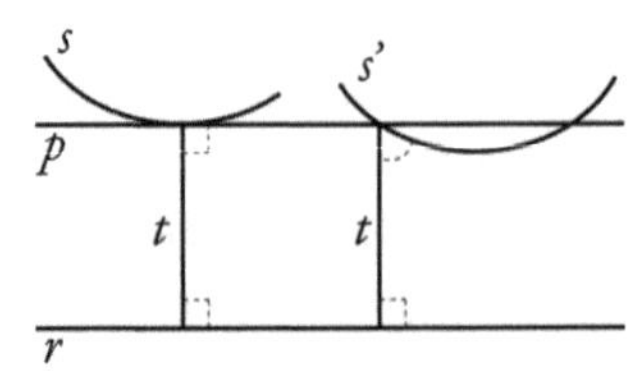

Taken together, these two propositions seem to assert the existence and uniqueness of a tangent to the hypercycle at every point on the hypercycle. The existence of the tangent is guaranteed, through Proposition 35, by the existence of a perpendicular to a given straight line (i. e. *Elements* I, 11); the uniqueness of the tangent is guaranteed, through Proposition 36, by the uniqueness of a perpendicular to a given straight line (implicitly by construction: again, in *Elements* I, 11; explicitly: in the Corollary to Lemma 5, Proposition 33). At any rate, what induced Saccheri to produce these two proofs was not the difficulty of guaranteeing the existence of the tangent to a curve (which, in the eighteenth century, was normally assumed for any continuous line), but rather the difficulty of showing that this property of continuous curves (of having one and only one tangent) holds also in the hypothesis of the acute angle, i. e. when dealing with *hyperbolic* tangent lines. Both the existence and uniqueness of a tangent to the hypercycle will in fact be used in Proposition 37 to justify the application of infinitesimal calculus to the length of the very hypercycle.

The Corollary to Proposition 36 clarifies the technical statement of Proposition 36. In the following part of *Euclid Vindicated* Saccheri will employ this Corollary rather than the Proposition itself.

[1] Note that Saccheri's statement reads 'any ordinate LF' (*quaevis ordinata LF*), which could also be interpreted universally quantified as 'every ordinate'. If we take it to have this latter meaning, however, the theorem's statement would only be trivially true, for it would then imply that straight line XF is either asymptotic or incident to AB; so when extended beyond F, no point of this straight line would ever fall "without the cavity of the curve".

The Latin word *ordinata* seems to have been employed for the first time by Leibniz in 1676.

[2] Saccheri's explanation is somewhat confused. He appears to state that there may not be a point X (of the ray *s'*) falling above the hypercycle *p*, if *s'* is asymptotic or incident to *r*, or if we are dealing with the case in which (as the size of the acute angle between *s'* and *t* decreases) *s'* and *t* coincide.

[3] This is, once again, *Elements* I, 19 (cf. *Note* 1 to Propositions 8, 9 and 10, and *Note* 1 to Proposition 28). Note that Clavius' Latin statements of *Elements* I, 18 and 19, most certainly fostered this confusion.

Notes to Proposition 37

This Proposition contains the fundamental, and essentially the sole, error in *Euclid Vindicated* that allows Saccheri to believe (or so it seems) he has succeeded in refuting the acute angle hypothesis. As we have already noted in the *Introduction* (§ 6), it is difficult to understand how such careful and skillful geometer could have failed to see such obvious and extremely gross mistake.

The error is rooted in a misinterpreted theory of infinitesimal calculus, and ends up confounding infinitesimals and indivisibles. It cannot be excused on the grounds of the immaturity of the discipline of Calculus at the time, since no mature eighteenth-century analyst, nor any geometer working with indivisibles as early as the preceding century, would have ever made such blunder.

The purpose of the Proposition is to show (always in the hypothesis of the acute angle) that a hypercycle between two perpendiculars to a given hyperbolic line is of equal length to the straight line itself. Saccheri knows this statement to be false, since the opposite statement is a very simple consequence of Proposition 3. But it is precisely because of this that he wants to prove Proposition 37: so as to arrive at a contradiction and thereby the falsity of the acute angle hypothesis.

Saccheri begins by stating a true and evident axiom found both in ordinary geometry and in infinitesimal analysis: if two magnitudes are divided into an equal number of parts, all of which are equal to one other, then the two magnitudes are equal if and only if part of the one is equal to part of the other. Saccheri additionally states that this division can be infinite: if we take this to mean that the parts can be however small, we then obtain a real principle of Calculus; if, however, we take it as referring to actual infinitesimals, then more care is required in the subsequent reasoning (as was the case, for instance, in debates on indivisibles). Saccheri does not appear to have exercised the necessary caution. In fact, in the most delicate part of the proof he simply assumes that both a segment of a straight line cut by two perpendicular lines and the corresponding segment of the curve (cut by the same perpendicular lines) count as infinitesimals. He essentially concludes that since both segments are infinitesimals then both can be considered points (that is, as indivisibles), and since all points are equal to one another, it follows, from the previous axiom, that the two lines must be of the same magnitude.

This is a very gross mistake. Were we to follow this reasoning, we could demonstrate that every curve between the same extremes are of the same length. Certainly no theorist of indivisibles would have committed such an error, yet alone any analyst. In fact, the first technical discussion of *Euclid Vindicated*, carried out by Klügel in 1763, immediately identifies this error and its paradoxical consequence: "... elementa singula curvae elementis basis aequalia esse ostendit, sed ita, ut ostendi posset quamlibet curvam, quae ordinata quadam lineae Abscissarum perpendiculari in dimidia similia dividitur, basi suae aequalem esse" (Klügel, *Recensio*, § 5, p. 10).

Yet, in order to derive the desired result, Saccheri seems to rely on some additional principles. He avails himself of the results obtained in Propositions 35 and 36, which (as we have seen) assert the existence and uniqueness of a derivative at every point on a hypercycle. Saccheri's argument is thus the following: since every point on the hypercycle has a well defined tangent at that point, then the length of an infinitesimal segment of that hypercycle can be considered equal to the length of an infinitesimal segment of its tangent; we can thus infer that Saccheri was not entirely unfamiliar with studies of infinitesimal analysis carried out at his time – which is what we would nowadays refer to as the linear approximation of a curve by means of its derivative. Saccheri's reasoning here is correct, at least *cum grano salis*. He concludes, however, that since the tangent is a straight line and the base is also a straight line, then the infinitesimal segment of the one is equal to the infinitesimal segment of the other. This reasoning represents a gross distortion of the meaning of linear approximation and analytical methods. I will not further dwell on the obvious mathematical errors of such a proof: *Notes* 1 and 4 below already exemplify how the demonstrative procedure employed in Proposition 37 conflicts with eighteenth-century geometrical common sense, and even with several other (explicit) points that appear in *Euclid Vindicated*.

The next two Scholia adopt a different demonstrative approach, based on motion. Saccheri commits yet another trivial error that is in many respects more glaring than the previous, and it is very strange that he failed to see it. Many geometers who hoped to prove the Fifth Postulate by way of the equidistance of straight lines employed a motion-based strategy, which states that the continuous translational motion of a segment with one extremity always forming a right-angle with the base, draws with the other extremity an equidistant line that is also straight. This was the strategy employed by several Medieval authors with whom Saccheri was certainly unfamiliar (such as Thābit ibn Qurra and Al-Haytham; see Rosenfeld, *A History on Non-Euclidean Geometry*, pp. 49–64), and especially by Clavius (Lemma 2 in the Scholium to Proposition 28; in *Euclidis*, p. 51) and Borelli (*Euclides restitutus*, p. 23); but whereas the Arabs and Clavius believed to have successfully proved this, Borelli held that it was to be assumed as an axiom. Saccheri discusses these attempts in

Scholium 2 to Proposition 21, and believes that he has refuted them once and for all (Clavius' and Borelli's exact formulations are found in *Note* 7 to Proposition 21, Scholium 2). In these Scholia, he therefore takes a slightly different approach: instead of immediately attempting to show that the equidistant line is straight, he first shows that its length is equal to the length of the base; yet once he has shown that the length of the one is equal to the length of the other, it follows immediately that the equidistant line is a Euclidean straight line (see the Scholium to Proposition 38).

Saccheri's argument, however, assumes that a point on the circumference of a rolling circle always describes a line of length equal to the circumference of the circle. Such an assumption was already viewed with suspicion in the Modern Period, and was regarded as a classic example of the difficulties encountered when attempting to escape from the labyrinth of the composition of the continuum. In fact, it is just a variation of the well-known 'Aristotle's wheel' paradox, namely, Problem 24 in *Mechanica*, attributed to the great Greek philosopher (855a28–856a38). The paradox was the subject innumerable commentaries throughout the seventeenth century. Amongst these, the most renowned were probably those authored by Galileo (*Discorsi e dimostrazioni matematiche intorno a due nuove scienze*, in *Opere*, Firenze, Barbera 1968, vol. 8, pp. 68 and ff.) and by the celebrated Jesuit mathematician André Tacquet (A. Tacquet, *Dissertatio physico-mathematica de circulorum volutionibus*, in *Opera Mathematica*, Antwerp, Jacob van Meurs 1669). Saccheri was surely acquainted with both of these texts. For an overview of the issues raised in the seventeenth century by Aristotle's wheel, see the classic I.E. Drabkin, *Aristotle's wheel: notes on the History of a Paradox*, "Osiris", 9, 1950, pp. 162–98.

In any case, Saccheri's proof shows only that there is a one-to-one correspondence between points on the semi-circumference DLB and the straight line AB, and between points on the semi-circumference BHD and hypercycle CD. Since there is an obvious bijection between the two semi-circumferences, Saccheri can conclude that points in AB and CD are equinumerous, though he certainly is not justified in concluding that the two segments are of equal size. Yet this is exactly what he attempts to do (in Scholium 1) when stating that all points must be considered 'exactly equal' to one another and thus must form, when taken together, two segments of equal size. In other words, he commits the typical error of confusing cardinality and magnitude that plagued a large portion of Modern mathematics (at least before the development of Analysis), and relies once again on points of infinitesimal length, employing them incorrectly. Like the similar passage in the body of the proof of Proposition 37, Saccheri's present reasoning could be used (as in the pseudo-Aristotelian paradox) to prove that the magnitude of straight line AB is equal to the magnitude of any other curve between the same extremities. In this case, too, Saccheri's error was immediately identified by interpreters; Klügel discusses it in the same place mentioned above (§ 5 of *Recensio*; p. 10).

For a brief discussion of the errors in Saccheri's proof of Proposition 37 and its Scholia, see the above-mentioned article by Dou, *Remarks on Saccheri's Geometry*, pp. 392–4. We may also remark that later writers on hyperbolic geometry have proved that the ratio between the length of the base and the corresponding hypercycle does not depend on the base itself, but only on the height of the perpendiculars: see J.-M. de Tilly, *Etudes de méchanique abstraite*, "Mémoires de l'Académie royale de Belgique", 21, 1870, pp. 1–98 (p. 12).

[1] This axiom is a specific case of Dechales' definition of equiproportionals, which is discussed by Saccheri in Book Two of *Euclid Vindicated* (see later, pp. 104 and ff.). Although Saccheri questions the suitability of such a definition, preferring instead the Euclidean one, he nonetheless accepts its validity. It is therefore unsurprising that he employs it here. What is, however, surprising, in the hypothesis that *Euclid Vindicated* was composed unitarily, is that Saccheri did not consider this dependency and connection worth noting: I take this as strong evidence supporting the claim that the two Books of *Euclid Vindicated* were composed separately. In any event, it is important to underline that Saccheri's 'infinitesimal analysis' is still completely grounded in Classical principles of the theory of proportions – as was, after all, Cavalieri's theory of indivisibles.

We can also identify a more complex problem: in the Appendix to *Euclid Vindicated* (pp. 139–42), which was most certainly written so as to establish an explicit connection between the two parts of the work – the theory of parallels and the theory of proportions – Saccheri strives to prove that *analytical* methods themselves are all based on the theory of proportions, and that this theory is itself (at least when applied to geometry) based on the Euclidean theory of parallels. Thus Saccheri runs a serious risk of committing a *petitio principii* when using Proposition 37 to prove the Fifth Postulate, at least according to the line of thought running through *Euclid Vindicated.* As a matter of fact, Saccheri's objections to the reasoning concerning proportional figures outlined in p. 141 (if one were not to employ the Fifth Postulate) are also applicable to his proof of Proposition 37, in the very same way.

In more general and abstract terms, we can say that eighteenth-century standard analytical procedures require tangent spaces with Euclidean structure. Therefore, any attempt to prove the Fifth Postulate by means of this procedure is at risk to be inconclusive, unless it is subjected to rigorous verification and sound theory of limits. But we should also observe that Saccheri's confusion as to this point is only natural: although during the first centuries attempts to lay the foundations of infinitesimal analysis sometimes relied on the theory of proportions and similarity, and although everybody knew, after Wallis' proof, that similarity is well defined only if the Euclidean axiom on parallel lines is valid, yet there are no other theoretical texts (that I know of) explicitly stating that (and to which extent) Calculus is founded upon the Fifth Postulate.

[2] In other words, quadrilateral ANFC can be translated onto quadrilateral NLFF, or otherwise reflected over quadrilateral NLFF about the FN axis. The quadrilaterals are congruent in either case. Note that Saccheri employs here that the curve is convex (although the same would also hold for a concave curve; it is only relevant that there are no points of inflection).

[3] This is the critical point, where Saccheri moves from considerations of magnitudes however small, i. e. the vertical bands ACFN and their submultiples, to actual indivisibles: by imagining the straight line segment MK to be of "infinitely small breadth". In the discussion that follows, this seems to admit that straight line MK of infinitesimal breadth can cut different segments on straight line AB and curve CKD respectively (even if both are infinitesimals, i. e. point K and point M). However, he does not provide any real argument

as to why they should be in fact equal: they are equal simply because they are both points, in other words, because they are actual indivisibles and not just 'potential' infinitesimals (i. e. of size however small).

For a philosophical discussion of the Aristotelian and Zenonist positions on the resolution of the continuum into points, see *Logica*, where Saccheri devotes many pages to exemplify the fallacies often embedded in such arguments. Cf. *Logica demonstrativa*, pp. 250–65 (ed. 1697); and also the Physical Thesis 22, here in pp. 274–5.

A harbinger of Saccheri's confusion was already to be found in his 1708 work on statics, where he takes points to be infinitesimal magnitudes, and then states that throughout the book he will always assume these points to be equal to one another unless the 'context' clearly shows that they are in fact different infinitesimals (not exactly what we would call a rigorous methodical precept): "Ubi rursum advertes, nomine puncti, & puncti intelligi à me particulas infinitesimas spatij inter se aequales, nisi tamen ex ipso contextu clarescat venire ibi duas quaslibet infinitesimas spatij, sive aequales, sive inaequales" (*Neo-Statica*, Book Three, *Admonitio* after Proposition 1; p. 85).

[4] This is another proof passage that contradicts what Saccheri explicitly claims in *Euclid Vindicated*. Dealing with infinitesimal segments, Saccheri takes the curve (the hypercycle) to be equal to its tangent – and we may even allow for such a procedure (albeit with some reservations for its vagueness and lack of rigor). But Saccheri now states that the breadth of an infinitesimal of the base (say, for instance, segment PM in Fig. 45, reduced to breadth of infinitesimal size) is equal to the breadth of the tangent straight line which, in linear approximation, joins the corresponding points FK of the hypercycle. Well: this statement contradicts not only the principles of mature infinitesimal analysis but also Corollary 3 to Proposition 3 of *Euclid Vindicated*. In fact, Corollary 3, which seems rather artificial and which Saccheri never explicitly employs throughout the rest of the work, clearly (and somewhat uselessly) states that the result obtained in Proposition 3 – that the hyperbolic straight line (here FK) joining the extremities of equal perpendiculars (PF and MK) to the base (PM) is longer than the base – holds for finite and also for infinitesimal magnitudes. In other words: the infinitesimal tangent (hyperbolic) straight line FK is longer than the corresponding base PM. So this result, which is true also according to the modern way of thinking, was one of the explicit results (which Saccheri intended to make explicit) of *Euclid Vindicated*; and it blatantly contradicts what is stated throughout Proposition 37. (Note that in the Corollary here at issue Saccheri only mentions the infinitesimal length of the perpendiculars, not that of the base and the join; but in proof of Proposition 3 he showed that the two statements are equivalent).

[5] *Elements* III, 19 does not in fact rely on the Fifth Postulate and is thus valid also in hyperbolic (and spherical) geometry. The initial propositions of Euclid's Book III represent an important example of 'absolute geometry', as the Fifth Postulate is not employed until *Elements* III, 20.

[6] In fact, as Klügel pointed out and later commentators remarked again, the points on the semi-circumference do not describe curve CD throughout the rolling motion of the circle. It seems, however, that Saccheri does not want to rely on a motion-based description of a

line in this Scholium. Rather, he attempts to construct an actual one-to-one correspondence between points of 'equal size', as can be inferred from the following paragraph.

[7] Saccheri here seems to justify the equality between base and hypercycle by the uniform motion that generates the equidistant line. He thus abandons the discussion of the simple one-to-one correspondence of points (as in Scholium 1), and consequently no longer needs (says he) to concern himself with the size of infinitesimals. So now the equality of size of two lines should be guaranteed by the simple fact that both lines are described by the uniform motion of two points in the same interval of time. But this is precisely what Saccheri should *prove*, since this property of uniform motion can be considered a good definition of a space with vanishing curvature – as demonstrated by any modern (differential and kinematic) treatment of the notion of curvature.

Notes to Propositions 38 and 39

In these last Propositions of Book One, Saccheri draws conclusions from the preceding results and then declares himself to have proved the Fifth Postulate.

Proposition 38 briefly states that, in the hypothesis of the acute angle, the hypercycle segment cut by two perpendiculars is longer than the segment of straight line from which it is equidistant. This result eluded those geometers who, before Saccheri, attempted to establish the theory of parallelism on the basis of equidistant lines; Saccheri can here easily prove it by drawing on his own previous discussions in *Euclid Vindicated.* It contradicts, however, the (erroneous!) result obtained in Proposition 37, which states that the hyperbolic straight line and the hypercycle are of the same length. Saccheri thus concludes that the acute angle hypothesis is contradictory and false. In the following Scholium, he shows that such reasoning is not applicable to the obtuse angle hypothesis – not a real problem, as he takes this to have been refuted in Proposition 14.

The statement of Proposition 39 is the very Fifth Postulate, which at this point Saccheri believes himself to have proved by refuting the obtuse angle hypothesis (in Proposition 14) and the acute angle hypothesis (in Propositions 33 and 38, according to two different methods). Here, he is in fact employing 'Saccheri's Theorem' (Propositions 5, 6 and 7), which allows him to conclude that the only true hypothesis is the right angle hypothesis; while the proof of the Fifth Postulate from the truth of the right angle hypothesis is the content of Proposition 11 (or, rather, the generalization thereof found in Proposition 13).

The last, very disorganized and almost fragmentary, Scholium deals with two distinct topics. The first concerns *consequentia mirabilis*: at the end of the First Part of Book One, Saccheri expressed his intention to embark on yet another proof of the Fifth Postulate, this time one that does not rely on the elementary properties of a straight line, but rather concludes through a logical procedure which refutes (*redargutio*, used in the technical sense) a hypothesis on the basis of its own assumption. It is not clear, however, that the refutation of the acute angle hypothesis in Proposition 38 is done by *consequentia mirabilis.* In fact, Saccheri carries out a simple proof by contradiction: it is shown that assuming the acute angle hypothesis leads to the statement that the hypercycle is both longer than and equal to the subjacent hyperbolic straight line. In order to obtain a refutation by *consequentia mirabilis*, Saccheri would have needed to show that the truth of the right angle hypothesis follows from the result of Proposition 37, in other words, from the fact that hypercycle and straight line are of the same length. But Saccheri does not, and probably cannot, provide such demonstration. He would first have to prove that the equidistant line is straight, then apply Proposition 3 in order to arrive at the right angle hypothesis. But he does not know how to prove that the equidistant line is straight (which is what Clavius' attempted to do) without previously assuming the Fifth Postulate ("but I say with us is first, that it is equal; then however, that it is straight ..."). He thus seems to concede that an absolute redargution of the acute angle hypothesis would have been possible if the path embarked upon by many other geometers (namely, to prove that the equidistant line is straight) had been successful. But since that path is still unviable, he bases his proof upon considerations of length – a strategy that does not lead to the desired redargution. The protestations expressed in this Scholium, namely, that considerations of length derive "from the very viscera" of the acute angle hypothesis, illustrate his dissatisfaction (not with the result, but rather) with the structure of the proof of the Second Part of Book One. In this connection, see BELLISSIMA, PAGLI, *Consequentia Mirabilis*, pp. 106–11.

The second half of the last Scholium addresses some points concerning the definition of parallel straight lines as equidistant lines that he had already touched upon in the Preface. Saccheri correctly notes that the assumption according to which the line equidistant to a straight line is itself a straight line is in fact a postulate (or, if you prefer, a real definition), not just a simple definition. In other words, it is an axiom in disguise. But if this is the case, we cannot expect to prove the Fifth Postulate with reference to such a definition without committing a *petitio principii*. To my knowledge, the first mathematician to explicitly recognize the *petitio* is Alfonso de Valladolid (see the aforementioned LÉVY, *Gersonide, le Pseudo-Tūsī, et le postulat des parallèles*), even though the repeated attempts of the Arab and Persian geometers to produce a proof of the Fifth Postulate not grounded on equidistance may reveal some awareness of the problem. It is clear, in any case, that the difficulty was not easily detected in the Renaissance and the Early Modern Age.

Saccheri holds, somewhat cautiously, that no author mentioned in *Euclid Vindicated* has ever committed such a mistake, and in fact Clavius, Borelli, and Wallis, whom Saccheri explicitly mentions in this work, never assume equidistance within their definition of parallel straight lines. Clavius, however, immediately discusses the matter in his commentary on the definition of parallel straight lines (*Euclidis*, pp. 21–2), and then holds that he has proved such a result in Lemma 1 to the Scholium to Proposition 28 (*Euclidis*, pp. 50–1). Borelli assumes this property instead of the Fifth Postulate; this same strategy was adopted by all of Borelli's school. For example, Angelo Marchetti, with whose writings Saccheri was certainly acquainted (cf. TENCA, *Relazioni fra Gerolamo Saccheri e il suo allievo Guido Grandi*, p. 35), makes this assumption in his theory of parallels in *Euclides reformatus*, Axiom 18, pp. 10–1. However, it is also the case that less shrewd geometers, with whom Saccheri was also familiar, merely assumed that formulating the definition of parallel straight line in terms of equidistance was unproblematic: amongst the Jesuits stand Tacquet and Dechales, whose works on elementary geometry were well known, as well as the shorter treatment of Fabri. It is significant that both Tacquet and Dechales openly acknowledged that the notion of non-incidence in the Euclidean original definition of parallelism is different from the notion of equidistance, and that one could thus advance the objection of asymptotic straight lines – hence Euclid's definition "parallelismi natura non satis explicat" (Tacquet, *Elementa Geometriae*, pp. 10–1). Both, however, resolve such objection by explicitly defining parallel lines as being non-incident and equidistant, and are thus vulnerable to criticism in that they commit a very blatant error of 'complex definition'. Both Jesuits employ, in addition, two axioms about parallels (which, given their definitions, are totally useless). Tacquet assumes an Axiom 11, which states that parallel lines have a common perpendicular, and then an Axiom 12, which states that two perpendiculars cut off equal segments from each of two parallel lines – which is Saccheri's Proposition 3 in the hypothesis of the right angle (*Elementa Geometriae*, p. 12). Armed with such overabundant definition and two axioms, Tacquet proceeds to prove the Fifth Postulate in its original form (*Elementa Geometriae*, pp. 35–6), and goes on to criticize Clavius' long-winded proof. Dechales assumes an Axiom 11, which is the Fifth Postulate itself, and then an Axiom 12, which states that all common perpendiculars to two parallel lines are equal to each other (cf. *Cursus*, p. 112). Fabri, on the other hand, defines parallel straight lines in terms of equidistance (in def. 4 of *Synopsis Geometrica*, p. 30) and then easily proves the Fifth Postulate, without even mentioning it as such, under the name of Corollary 3 to Proposition 1 (p. 83). Saccheri's irritation with blatant argumentative fallacies of this sort is thus understandable.

On the other hand, in *Nouveaux Elémens de géometrie,* Arnauld explicitly distinguishes between a negative notion of parallelism as non-incidence and a positive notion of parallelism as equidistance – though he basically takes them to be equivalent to one another in light of his assumption of the Fifth Postulate (pp. 391–2).

P.A. Cataldi, *Operetta delle linee rette equidistanti et non equidistanti*, Bologna, Rossi 1603, a booklet quite famous at that time, also bears mentioning: in its dealings with the problem of Euclidean parallelism, it refers only to equidistance (never to non-incidence). Throughout his various proofs, Pietro Antonio Cataldi (1548–1626) often employs Saccheri Quadrilaterals, although only in the trivial (i. e. Euclidean) case. In general, he does not seem worried about proving the possibility of equidistance (even though Clavius had, by then, already noted the difficulties associated with such a proof).

Moreover, this definition of equidistant parallel lines was also the one provided by Christian Wolff, whom Saccheri explicitly mentions in Book Two of *Euclid Vindicated* (pp. 133–4): C. Wolff, *Elementa Geometriae*, I, 1, § 81, in *Elementa Matheseos universae*, vol. 1, p. 103. Wolff's use of geometrical definitions, especially this one on equidistant parallels, was of much dispute in eighteenth-century German mathematical epistemology, and the issue continued up to Kant's time. Wolff's proudest opponent (in many ways akin to Saccheri) in this respect was Lambert, who, in addition to discussing the matter both privately and in his philosophical works, also referred to it in § 4 of *Theorie der Parallellinien* (pp. 142–3), where he expounds the defects of the Wolffian definition of equidistant straight lines (a '*vitium subreptionis*', in other words Saccheri's 'fallacy of complex definition'). Wolff's position had certainly several sources, and might even stem directly from Leibniz's works (with which Saccheri is not familiar). In *Nouveaux Essais*, in fact, Leibniz maintains that the definition of parallels as non-incident straight lines is only nominal because it resorts to the notion of infinity (this is the same argument that appears in the works of Borelli, and Leibniz was well acquainted with it; see the *Notes* to Scholium 2 to Proposition 21), whereas the definition of parallel lines as equidistant straight lines is, in fact, real (constructive). This stands in direct opposition to Saccheri's understanding of the issue. Leibniz' passage reads: "Peu de gens ont bien expliqué, en quoy consiste la difference de ces deux definitions [the nominal and the real], qui doit discerner aussi l'essence et la proprieté. A mon avis cette difference est, que la reelle fait voir la possibilité du defini, et la nominale ne le fait point: la definition de deux *droites paralleles*, qui dit qu'elles sont dans un même plan et ne se rencontrent point quoyqu'on les continue à l'infini, n'est que nominale, car on pourroit douter d'abord si cela est possible. Mais lorsqu'on a compris, qu'on peut mener une droite parallele dans un plan à une droite donnée, pourveu qu'on prenne garde que le pointe du Stile qui decrit la parallele, demeure tousjours egalement distante de la donnée, on voit en même temps que la chose est possible et pourquoy elles ont cette proprieté de ne se rencontrer jamais, qui en fait la definition nominale, mais qui n'est que la marque de la parallelisme, que lorsque les deux lignes sont droites, au lieu que si l'une au moins estoit courbe, elles pourroient estre de nature à ne se pouvoir jamais rencontrer, et cependant elles ne seroient point paralleles pour cela" (G.W. Leibniz, *Nouveaux Essais*, III, iii, § 18; A vi, 6, n. 2, p. 295). Note that Leibniz' *Nouveaux Essais*, written in 1702–1705, were only published in 1765 (when Wolff was already dead), but some of his ideas on the topic had already surfaced in his writings on the Calculus. Several other papers, however, remained unpublished: among them, many short essays and attempts to prove the Fifth Postulate show Leibniz' clear understanding that one has to prove that the line equidistant to a straight line is itself straight (something that Wolff was unable to grasp).

We have already seen how the assumption that a line equidistant to a straight line is itself straight is equivalent to the Fifth Postulate only if we allow for Archimedes' Axiom.

[1] It is well known that Euclid's *Elements* does not explicitly prove that a straight line is the shortest path between two points; Archimedes assumes this principle in *On the Sphere and Cylinder*, where it appears as the First Assumption. Some mathematicians from the Modern Period incorporated this property within their definition of straight line, while others characterized it as an axiom, and still others tried to prove it using alternative definitions of straight lines. Saccheri belongs to this last group. At this point in *Euclid Vindicated*, he briefly mentions the standard proof employed at the time, which is a generalization to infinity of the triangular inequality proved by Euclid in *Elements* I, 20. This is mainly the same proof that we use nowadays (without the same standards of rigor), based on the integration of the modules of the curve's derivatives.

[2] Both here and systematically in later sections of this work, Saccheri refers to a common notion or the common sense (*notio communis*). He may be referring to the common notions (i. e. the axioms) of the *Elements*, and hence to those who considered a straight line

as the shortest path between two points to be an axiom; alternatively, he may be referring to the axiom concerning the whole-part relation implicitly used in the proof of this fact. Saccheri may also be saying that all geometers accepted a straight line to be the shortest line between two points (by common sense), and that he thus considers a proof (like those that he advanced in the five Lemmata in Proposition 33) thereof quite unnecessary. Furthermore, the absence of such a proof hints at the fact that this Second Part of Book One of *Euclid Vindicated* may not have been revised as carefully as Saccheri may perhaps have deemed necessary on the basis of the maximum rigor that he wanted to apply in other parts of his work.

[3] Saccheri seems to explicitly admit that his *redargutio* is simply a *reductio ad absurdum*.

[4] The Latin text provides a literal reproduction of Clavius' translation of the Fifth Postulate, which he called 'Axiom 13' (*Euclidis*, p. 25). Note that in Proposition 39 Saccheri employs for the first time (besides the Preface) the term 'Axiom' to denote the principle that he has up to now systematically called 'Assertion'. He probably wants to emphasize that he has achieved the aim of demonstrating that a true assumption with dubious evidence (an assertion) is in fact a self-evident principle (an axiom).

[5] Proposition 13 does not appear to be very relevant here. Saccheri is probably referring to the Scholium after Proposition 13, which he mentioned shortly above.

[6] In other words, without requiring convexity.

[7] The *consequentia mirabilis* is introduced in *Logica demonstrativa*, ed. 1701, p. 82 [ed. 1697, p. 130]. See the *Introduction*, § 5.

[8] In Saccheri's *Logica*, we read: "Suppositio, seu hypothesis, & postulatum in eo differunt, quod postulatum est essentiae postulatum, & hypothesis est postulatum alicuius proprietatis seu accidentis. Definit geometra rectam lineam esse, *quae ex aequo sua interiacet puncta*: iam postulat dari, seu, possibilem esse lineam rectam à se definitam, & hoc est propriè postulatum, est enim essentiae postulatum: at ubi accidet geometram postulare, v.g. rectam lineam palmarem, illud est proprietatis, seu accidentis postulatum, ac propterea propriè dicitur hypothesis" (*Logica demonstrativa*, 1701 ed. p. 118 [ed. 1697 p. 187]; and more extensively: pp. 127–34 [pp. 200–11]).

[9] The Latin text simply reads *ratio objectiva*, which Halsted rightly translates as 'concept', even though it may also mean 'relation' (of a straight line with a line equidistant to it).

[10] In *Logica,* Saccheri finds a similar instance of a *complex definition* in the Euclidean definition of diameter. As in the case of this Euclidean definition, where the diameter is defined as a line *both* passing through the center *and* bisecting the circle (see above, Lemma 4 to Proposition 33), Saccheri claims that a line parallel to a given straight line is *both* equidistant *and* a straight line. In *Logica demonstrativa*, we read: "Ubicumque definitio *quid nominis* est complexa, non est facilè admittendum postulatum. Voco autem definitionem complexam, quae partibus constat, quarum una est sufficienter determinativa quiddiatis rei definitae. Sit exemplum: diameter est *linea recta, quae per centrum circuli transiens, & utrinque in circumferentiam terminata, bifariam dividit circulum*. Definitio est complexa quia una eius pars, quòd diameter sit *linea recta per centrum circuli transiens, & utrinque in circumferentiam terminata* sufficientèr determinat quidditatem diametri, ut constat; quòd

verò, linea eiusmodi bifariam dividat circulum, non est assumendum, sed probandum" (ed. 1701, p. 129 [ed. 1697, p. 203]).

[11] The Scholium's conclusion is obviously messy; it seems more like a personal remark or a lecture's rough draft than a book prepared for publication. Concerning the circumstance in which Saccheri authored *Euclid Vindicated*, see the *Introduction*, § 4.

Notes to the First Part of Book Two

In Book Two of *Euclid Vindicated*, Saccheri discusses the theory of proportions. In this First Part in particular, he discusses the Euclidean definition of the equality of ratios (equiproportionality) and the problem of existence of the fourth proportional.

Since this section of the book is presented in a particularly disordered way, I will provide a brief overview of its structure. According to Saccheri, those who accused Euclid of devising an obscure definition of equiproportionals (Borelli, for instance) and reproached him for not proving that it effectively applies to the equality of ratios (Dechales, for instance), have failed to grasp the meaning of Euclid's definition. They consider it a *real definition*, that is, as a principle that states a certain property of equal ratios. But Euclid surely regarded it as a *nominal definition*, that is, as a stipulation of what it means for two ratios to be mutually equal. Thus, he does not need to prove anything at all. He is only bound to apply such a definition consistently (pp. 102–4 concerning Dechales, and pp. 124–6 concerning Borelli and the theory of the scientific definition). Saccheri proceeds to consider Dechales' alternative definition of equiproportionals (pp. 104–5), which depends on Tacquet, who in turn draws from the Galilean school's research on the theory of proportions, and whose work was also resumed by Arnauld. Saccheri shows (pp. 105–8) that this alternative definition is just as valid as the Euclidean definition in the theorems of the *Elements* (and explicitly discusses its application in *Elements* VI, 1). However, if the definition is *nominal* (p. 109), i. e. an arbitrary definition different from the Euclidean one, then Dechales' definition does not improve the Ancient's, but is actually equivalent to it, while if it is *real* (pp. 109–11), i. e. an effective principle of proof, then it necessarily requires an additional axiom, namely, an axiom stating the existence of the fourth proportional to three given magnitudes. In fact, both Dechales and all his predecessors who also followed this line of reasoning accepted the axiom of the fourth proportional – and it seems (pp. 111–2) that Euclid needed it in order to lay the foundations of several proofs, particularly *Elements* V, 18 and XII, 2. But Saccheri, who wants to reduce the number of principles employed, rejects this additional axiom, and intends to 'vindicate' Euclid by showing that this axiom, which the Ancient mathematician did not explicitly assume, was indeed unnecessary to him. Saccheri thus proceeds to prove (pp. 112–4) a specific case of *Elements* V, 18 without employing the axiom of the fourth proportional; but encounters greater difficulties when dealing with the general form of the proposition and with *Elements* XII, 2. Consequently, he decides to formulate an alternative axiom: the trichotomy of ratios (p. 115, later reformulated in pp. 128–9). With this additional axiom, a couple of important Lemmata, and a strategic application of the principle of continuity (pp. 118–9), Saccheri proves *Elements* XII, 2 (pp. 119–22) and the full generalization of *Elements* V, 18 (pp. 129–31). Finally, he believes himself to have proved (pp. 115–7) his own Trichotomy Axiom via recourse to another Euclidean definition, *Elements* V, def. 8, which he discusses at length (pp. 122–3, and then again in pp. 126–8), such that his own proofs of *Elements* V, 18 and *Elements* XII, 2 turn out to be unhypothetical. He thus takes himself to have shown that Euclid's definition is equivalent to Dechales' definition (if the latter is taken as *nominal*), or even better than Dechales' (if it is taken as *real*), since it does not assume any additional hypotheses. Euclid's blemish has thereby been vindicated.

[1] Book V, Definition 5 of Heiberg's Greek edition, also presented as Definition 6 in Clavius' edition (Saccheri provides a literal transcription of *Euclidis*, pp. 208–9). The number and especially the order of definitions in *Elements* V are relevant, and in fact gave rise to different interpretations of the theory of proportions. Some scholars have proposed (Giusti, *Euclides Reformatus*, pp. 9–13) that, by placing the definition of equiproportionals in position six, Clavius may have been endorsing a different foundational path from the more traditional one followed by Commandino (whose version of the *Elements* seems to us to provide a more faithful interpretation of Euclid on this point) who instead placed it in position five (Commandino, *Euclidis*, p. 58). Clavius' placement of the definition may have prompted the Italian school's reformation of the theory of proportions (which Saccheri is here attempting to contrast). At any rate, Saccheri seems not to have read Clavius from such a perspective, since *Elements* V, def. 4 (i. e. Clavius' displaced definition) is only mentioned once in *Euclid Vindicated* (p. 132), and he strongly downplays its importance, classifying it as a merely grammatical explanation of a term that will be considered philosophically (i. e. conceptually) only through *Elements* V, def. 6. In other words, Saccheri read Clavius, but interpreted Commandino.

Euclid's definition, which most probably came from Eudoxus, is nowadays considered one of the masterpieces of Ancient mathematics; it does not, in fact, require any substantial revision. From this perspective, Saccheri certainly has a point in wanting to 'vindicate' Euclid (shortly before, Giordano referred to this definition as the great composition of Euclid, "la gran compositione d'Euclide"; *Euclide restituto*, p. 176). In modern terms, *Elements* V, def. 6 on the equality of ratios and *Elements* V, def. 8 on the inequality of ratios (always according to Clavius' numeration) can both be expressed as follows (and this is how we will interpret them in the rest of these *Notes*):

Given that four magnitudes are pairwise homogeneous (Archimedean), $\mathrm{A},\mathrm{B} \in \mathcal{P}$ and $\mathrm{C},\mathrm{D} \in \mathcal{Q}$:

$$\mathrm{A}:\mathrm{B}=\mathrm{C}:\mathrm{D} \Leftrightarrow \forall n,m \in \mathbb{N}:((n\mathrm{A}<m\mathrm{B} \wedge n\mathrm{C}<m\mathrm{D}) \vee (n\mathrm{A}=m\mathrm{B} \wedge n\mathrm{C}=m\mathrm{D}) \vee (n\mathrm{A}>m\mathrm{B} \wedge n\mathrm{C}>m\mathrm{D})),$$

$$\mathrm{A}:\mathrm{B}>\mathrm{C}:\mathrm{D} \Leftrightarrow \exists n,m \in \mathbb{N}:(n\mathrm{A}>m\mathrm{B} \wedge n\mathrm{C}\leq m\mathrm{D}),$$

$$\mathrm{A}:\mathrm{B}<\mathrm{C}:\mathrm{D} \Leftrightarrow \exists n,m \in \mathbb{N}:(n\mathrm{A}<m\mathrm{B} \wedge n\mathrm{C}\geq m\mathrm{D}).$$

In *Elements* V, def. 8, Euclid does not mention the last case of smaller ratio, since it could easily be obtained by conversion from the previous case of greater ratio. I here denote equiproportionality by the standard equality symbol instead of the usual specific symbol of identity (in this context it is generally written as A:B::C:D). The latter would probably be more reflective, as a symbolic transcription, of the original Greek theory, but most of the mathematicians of the Modern Period did not draw any distinction between the notion of identity and the notion of equality of ratio in their discussions of the matter (see, for instance, Saccheri's use of language in the proof at p. 137). In reference to Greek mathematics, one could also object to the formalization of equimultiples as magnitudes multiplied by natural numbers (see, for instance, B. Vitrac, *Euclide. Les Éléments*, Paris, PUF 1990–2001, vol. 2, p. 43). Such formalization, however, is again consistent with Saccheri's seventeenth-eighteenth century idea that equimultiples are abstract magnitudes multiplied by a 'scalar' element (see p. 117 and *Note* 30). I combined various equimultiple relationships in conjunctive form (with '$\wedge$'), whereas other interpreters prefer to formalize these same relationships in terms of implications or equivalences ("$\rightarrow$" or "$\leftrightarrow$"), drawing on the exact meaning (in so far as it can be reconstructed) of the Greek phrasing and of the ἅμα there employed. It is certainly the case, however, as is testified by translations and Latin paraphrases, that Modern mathematicians regarded that clause in its conjunctive form; Saccheri himself systematically employs *et* (a contrasting example from the period is Roberval, *Éléments de géométrie*, p. 305). In this way, moreover, *Elements* V, def. 6 can be understood by perfect analogy with *Elements* V, def. 8, which does not allow for a translation in terms of implication or equivalence. Finally, I take this to be the only possible symbolic and extensional interpretation that effectively fulfills Eudoxus' and Euclid's intentions: by employing either implication or equivalence, one may find easy counterexamples (even if we were to triplicate quantifiers in Definition 6) to the equality and inequality of ratios – and it would be rather strange to provide a similar modern interpretation and then acknowledge that Euclid's definitions were flawed. A well-articulated and philologically acute discussion on this point is found in F. Acerbi, *Euclide. Tutte le opere*, Milano, Bompiani 2007, pp. 340–60, and F. Acerbi, *Drowning by Multiples. Remarks on the Fifth Book of Euclid's Elements, with Special Emphasis on Prop. 8*, "Archive for History of Exact Sciences", 57, 2003, pp. 175–242.

[2] Saccheri refers to the Scholium to Definition 6 (*Euclidis*, pp. 209–10), where Clavius criticizes the incorrect interpretation of the criterion for the equality of ratios first put forward by Giovanni Campano da Novara (1220–1296) and later defended by Oronce Fine (1494–1555), among others. Campano in fact worked with a very deteriorated version of the *Elements* derived from Adelard of Bath's edition, which was translated from Arabic. As Campano did not want to modify the inherited Euclidean text, he openly allowed for a circular Definition, which, according to his interpretation, would serve to explain the equiproportionality of magnitudes by means of the equiproportionality of their multiples. Campano's text, which widely circulated in its manuscript form, was first published in print as *Preclarissimus liber elementorum Euclidis perspicacissimi*, Venezia, Ratdolt 1482. Although Campano's translation and interpretation sparked objections as early as the fifteenth century (by Regiomontanus, for instance), they were nevertheless employed and accepted by

many Renaissance editions in the fifteenth and sixteenth century. One of most important, and debated, during the sixteenth century was O. Fine, *In sex priores libros Geometricorum Elementorum Euclidis Megarensis demonstrationes*, Paris, Simon de Colines 1536, who accepts Campano's reasoning (pp. 97–8) in a Scholium to Definition 6. In the second half of the sixteenth century, far more philologically correct editions of the *Elements* appeared, and the theory of proportions was more widely discussed from a scientific point of view. These developments precipitated the universal rejection of Campano's (and Fine's) edition. Some such criticisms can be found, for instance, in Tartaglia (*Euclide Megarense philosopho, solo introduttore delle scientie mathematice*, Venezia, Rossinelli 1543), later in Commandino (*Euclidis*, p. 58), and finally in Clavius (see the aforementioned reference).

Concerning the definitions of Book V of the *Elements* throughout the Middle Ages and particularly in Campano da Novara, see J.E. Murdoch, *The Medieval Language of Proportions: Elements of the Interaction with Greek Foundations and the Development of New Mathematical Techniques*, in *Scientific Change*, edited by A.C. Crombie, London, Heinemann 1963. A 'defense' of Campano's position in the context of the lacunas and textual corruptions of Euclid's text in Medieval times is found in A.G. Molland, *Campanus and Eudoxus; or, Trouble with Texts and Quantifiers*, "Physis", 25, 1983, pp. 213–25.

Concerning Oronce Fine, I think Imre Tóth is worthy of mention, as he recently discovered a disc model of the Earth in which straight lines are represented by arcs of a circle perpendicular to the disc's circumference in the astronomical and geographical work of the French mathematician (O. Fine, *De Mundi Sphaera sive Cosmographia*, Paris, Simon de Colines 1542), i. e. in the same way in which hyperbolic geodesics are represented in Poincaré's disk model. I find it difficult to share Tóth's enthusiasm about his findings, as there is no evidence demonstrating that suh a model was meant to represent, by isomorphism, some geometrical property of these curves (which, for instance, in Poincaré's model are *conformal* to hyperbolic straight lines), nor is there any reason to believe that Fine ever concerned himself with the theory of parallels. Above all, there is no definitive evidence of Saccheri's direct knowledge of this or of any other work by Fine (contrary to Tóth's belief), nor is there any reason to believe that Saccheri could have even conceived of the Beltramian idea of a non-Euclidean model. In any case, it is interesting that the works of Fine and Saccheri intersect, however modestly. Cf. Tóth, *Aristotele e i fondamenti assiomatici della geometria*, pp. 503–4.

[3] These kinds of criticisms of the Euclidean definition were rather common in Renaissance and Modern mathematical literature, and were developed on the basis of two other passages of the *Elements*. First, they were articulated with reference to the same Definition 4 of Book V mentioned in the previous *Note* 1 ("proportion is the similarity of ratios"), which, if placed before the definition of equiproportionals (as it happens in Clavius), rather than being postponed (as in Commandino), may suggest that it is *itself* the definition of the equality of ratios (proportion), and hence that *Elements* V, def. 6 is only a characterization of an *already defined* relation: thus *Elements* V, def. 6 should be proved by means of the preceding definitions. But, secondly, the problem is aggravated by the presence of *another* definition of the equality of ratios in Euclid's work, that concerns numerical ratios (rather than ratios between any whatever magnitude), and is found in *Elements* VII, def. 20 (or

def. 21 in some modern editions). This last definition is, of course, only applicable to commensurable ratios and is thus different from, and in a way more specific than, the previous case; it was, however, generally considered both clearer and more intuitive than *Elements* V, def. 6, which explains why commentators at the time often attempted to reduce this latter definition to an arithmetic ratio – appropriately modified so as to include incommensurable magnitudes. This amounted to yet another proof of *Elements* V, def. 6. Finally, the introduction of the theory of anthyphairesis in the Middle Age (which was itself based on *Elements* VII, def. 20), produced some attempts to prove *Elements* V, def. 6, starting from the new definition of anthyphairesis; this was the path followed by Al-Māhānī, Al-Haytham and others (see B. Vahabzadeh, *Al-Māhānī's commentary on the concept of ratio*, "Arabic Science and Philosophy", 12, 2002, pp. 9–52).

Since Saccheri later considers Dechales' criticisms of the *Elements*, he may have Dechales, *Cursus*, p. 151 in mind. In this text, the French Jesuit laments that Euclid did not prove his definition. Tacquet articulates the same concern (*Elementa Geometriae*, pp. 128–9): he strives to prove *Elements* V, def. 6 from his own new definition of equiproportionals (in the Second Part of Book V of his *Elementa Geometriae*, pp. 154–64). A similar approach was adopted by all the members of the Italian school, starting with Galileo, who attempts to demonstrate the Euclidean definition from his own principles in the *Fifth Day* of *Discorsi* (Galilei, *Opere*, vol. 8, pp. 353–6). Viviani, Borelli, Torricelli, and many others would follow along a similar path.

Another defense of Euclid against Tacquet's accusation (rather similar to Saccheri's, although better articulated) is found in I. Barrow, *Lectiones Mathematicae XXIII in quibus Principia Matheseos generalia exponuntur*, London, Playford 1683 (lect. xxii); now in *The Mathematical Works*, edited by W. Whewell, Cambridge, Cambridge University Press 1860, vol. 1, pp. 348–57. In the following *Lecture* (pp. 361–7), Barrow also defends Euclid against the analogous (and more robust) criticisms put forward by Borelli. Here he explicitly says that he would like to *vindicate* Euclid: "Euclideam definitionem, et ab eo pendentem proportionalitatis doctrinam, utcunque vindicârim" (pp. 367–68). Barrow's *Lectiones* date 1664–1666, but were published only twenty years later.

[4] Once again, Saccheri is probably referring to Dechales, although Tacquet had already stated that proofs of these eleven propositions *in nostra methodo sunt superfluae* (*Elementa Geometriae*, p. 139). Also Torricelli, taken as a representative of the Italian school, either completely suppressed the propositions preceding *Elements* V, 14 or accepted them as axioms.

[5] Clavius, *Euclidis*, pp. 310–1. Euclid's original definition in fact only comprises the first half of this statement. The second half was added on by Clavius, who explains in the following Scholium that the additional clause is needed in order to account for all possible cases, and that the absence of it in the Greek text can be attributed to a lacuna within the manuscript tradition. At this point Clavius exemplifies what he has in mind (I will simplify a little): the ratio $12:4=9:3$ is of the first type (12 is an equimultiple of 4, as 9 is of 3); $4:12=3:9$ is of the second type (4 is a part of 12, as 3 is of 9); $6:8=9:12$ is of the third type (6 has the same number of parts of 8, namely the three quarters, as 9 has of 12); $11:5=22:10$ is of the fourth type, and is not found in Euclid (11 contains two times 5 and

its part, i. e. a fifth; likewise 22 contains two times 10, and a fifth); 12 : 5 = 24 : 10 is of the fifth and last type, which was also missing in Euclid (12 contains two times 5, and its parts, i. e. two fifths; likewise 24 with 10). Current interpreters believe that there is no textual lacuna, and that the Euclidean definition concludes without considering the two last cases (and maybe some others that Clavius here does not consider either). It may be that some of the latter cases, which Euclid wanted to, and could, handle with his own methods, can be obtained by inversion or separation of ratios, i. e. by a manipulation of the proportions found in the definitions and theorems of *Elements* V. It is not clear, though, that this could be the Euclidean solution to the problem, nor is it clear that all cases can be easily restored; see Vitrac, *Les Éléments*, vol. 2, pp. 262–7.

Note that *Elements* VII, def. 20, could be regarded as completely unrelated to *Elements* V, def. 6, as it refers to a kind of object – numbers – whose characteristics are quite different from the (continuous, or at least dense) magnitudes dealt with in Book V of the *Elements*. As evidenced by numerous Ancient sources, such a disparity in the treatment of ἀριθμός and μέγεθος has an immediate geometrical application with respect to the different way of intending a multiplication, proper to the two types of objects (see below, *Note* 30); to the definition of homogeneity (Clavius' *Elements* V, def. 5; Heiberg's *Elements* V, def. 4); and to the very notion of equimultiples. The irreducibility of these two concepts, however, tends to disappear in *Elements* X, 5 and 6, as they explicitly reconnect the concept of numerical ratio to the concept of magnitude ratio, thus making it difficult to determine whether one should here turn to the definition of equiproportionals in *Elements* V, def. 6 or to *Elements* VII, def. 20. So, if Renaissance studies on the foundations of mathematics merged the two definitions of proportions, it was, to some extent, due to reasons that were internal to the structure of the *Elements*. This merging, however, presented manifold difficulties. First and foremost that of accepting the principle of existence of the fourth proportional (which could be allowed for magnitudes, but definitely not for numbers), which Saccheri will soon deal with.

For some of these issues in the Ancient and Modern Period, see Rabouin, *Mathesis Universalis*.

[6] Euclid's theory of incommensurable magnitudes is found in *Elements* X, and goes through a very complex classification of 13 different types of lines of irrational magnitude. For a useful overview of this issue, see Vitrac, *Les Éléments*, vol. 3, pp. 51–63; and D.H. Fowler, *An invitation to read Book X of Euclid's Elements*, "Historia Mathematica", 19, 1992, pp. 233–64. Saccheri later employs the terms 'incommensurable' and 'irrational' interchangeably.

[7] I have translated *pars aliquota* as 'submultiple', as this phrase is more intelligible to a contemporary audience than the literal translation from Latin. *Submultiplex* was often used as a synonym for *pars aliquota* in Saccheri's time (it appears, for instance, in *Euclides reformatus* by Angelo Marchetti).

[8] Saccheri is referring to the Jesuit Milliet Dechales. He transcribes here the definition (and examples) given by Dechales and systematically employed by him it in his own commentary on Euclid. According to Dechales, the origin of this alternative definition is to be found in the work of another Jesuit, André Tacquet, who did not, however, employ it

extensively. Saccheri could hardly ignore Tacquet's work, which, at the time, together with Dechales' work, was the major educational text utilized by Jesuit schools. We can thus say that Saccheri is here challenging his own Order, or in any case the recent (post-Clavian) turn taken by studies about elementary mathematics conducted by the Society of Jesus. Also note that Tacquet's definition had, in turn, significant similarities with the reformed theory of proportions put forth by the Galilean school in Italy (in particular, by Borelli, as we will see later on), with which Saccheri must have been well acquainted. Tacquet's definition was also employed in the first edition (1667) of Arnauld's *Nouveaux Elémens de géometrie*, so that we find a perfect anti-Euclidean consensus among Jesuits and Jansenists. The second edition (1683) of *Nouveaux Elémens*, however, supported the more 'modern' ideas of François de Nonancourt's *Euclides Logisticus* (1652), thereby diverging even further from Euclid, and interpreting ratios as fractional numbers; Arnauld does, nevertheless, continue to employ Tacquet's definition in a theorem characterizing equiproportionality (*Nouveaux Elémens*, pp. 200–2). In any case, it is quite possible that Saccheri's criticism of Tacquet and Dechales is also aimed at Arnauld.

Dechales' passage reads: "Primam ad secundam eandem rationem habebit; ac tertia ad quartam, si prima toties contineat secundae partes aliqutas quascumque quoties tertia quartae similes partes aliquotas continet; ut si magnitudo A toties continet magnitudinis B partes centesimas, millesimas, centies millesimas, & quascumque alias aliquotas; quoties C continet magnitudinis D partes centesimas, millesimas, centies millesimas & quascumque alias aliquoties similes; ita ut nulla sit pars magnitudinis B quae pluries contineatur in magnitudine A, quàm similis pars aliquota ipsius D contineatur in C licet in irrationalibus restet semper aliqua quantitas, tunc est A ad B ut C ad D. Haec proprietas est generalis convenitque tam proportionibus aequalitatis quàm inaequalitatis, rationalibus quàm irrationalibus, assignatúrque à R.P. Tacquet licèt ea non utetur in demonstrationibus hujus libri" (*Cursus*, p. 151). Tacquet's definition was considerably more generic, and no *direct* reference to this definition is found in Saccheri's work: "Duae rationes (A ad B, & C ad F) sunt similes, aequales, eadem; cum unius antecedens (A) aequè seu eodem modo (hoc est, nec magis, nec minus) continet suum consequens (F)" (*Elementa Geometriae*, p. 132; followed by an analogous definition for the case B > A). Then, in a scholium, the Flemish Jesuit outlines a criterion, a *proportionum aequalium indicium primum & infallibile*, that is very similar to the one later considered by Dechales: "Rationes aequales sunt quando & consequentes ipsi, & consequentium similes partes aliquotae quaecunque, in antecedentibus aequali semper numero continatur" (p. 136). Tacquet subsequently proves this to be equivalent to his general definition (pp. 161–2). Arnauld's definition may be worth mentioning so as to better showcase the great similarity between the two: "Deux raisons sont appelées égales quand toutes les aliquotes pareilles des antécédents sont chacune également contenues dans chaque conséquent"; he proceeds to offer the standard example concerning the division of magnitudes in ten, hundred, and thousand parts, and finally concludes: "Mais si *x* n'est jamais précisément tant de fois dans *c*, mais toujours avec quelque résidu, il faut qu'y soit aussi autant de fois dans *g*, mais avec quelque résidu: et alors leur raison est sourde" (first edition of *Nouveaux Elémens*, Book Two, § 38; p. 163 of the cited modern reprint).

Isaac Barrow (*Lectiones mathematicae*, lect. XXIII; in *The Mathematical Works*, vol. 1, pp. 367–78) maintains, as does Saccheri, that the Euclidean definition is better than these of Tacquet and Borelli.

Concerning the relationship between Tacquet's and Borelli's respective theories, see PALLADINO, *Sulla teoria delle proporzioni nel Seicento.* Concerning Borelli's theory, see also F. PODETTI, *La teoria delle proporzioni in un testo del XVII secolo*, "Bollettino di biliografia e storia delle scienze matematiche", 15, 1913, pp. 1–8. Regarding Borelli's work in the context of the Galilean reform of proportions, see GIUSTI, *Euclides Reformatus*, pp. 115–33.

[9] This is Euclid's proof, which Saccheri copies (with only slight stylistic changes) directly from Clavius (*Euclidis*, pp. 247–8).

[10] By 'nature' of parallels, Saccheri here means equidistance. It is difficult to tell whether Saccheri expresses himself in this way because he is still committed to the notion of parallelism as equidistance (that is, if we suppose that Book Two of *Euclid Vindicated* was written before the treatise on parallels) or because he believes himself to have proved, in the final Scholium to Book One, that non-incident straight lines (i. e. parallels) must have the property of equidistance – which thus constitutes their nature. Note that, in his commentary to *Elements* I, 38 (which is the proposition here employed to prove *Elements* VI, 1), Proclus states that "to have the same height is the same thing as to be in the same parallels" (*In primum Euclidis,* 405).

[11] Saccheri reproduces Dechales' proof (*Cursus*, p. 160), once again with some insignificant variation in style and punctuation. A very similar proof (in different terminology) was devised by Tacquet (*Elementa Geometriae*, pp. 182–4). This proof was certainly a source of inspiration for Dechales. It was not, however, the only such proof, as very similar demonstrations were to be found in the Italian school, by which Tacquet may have also been inspired. Borelli, for instance, provides a similar proof in *Euclides restitutus* (pp. 146–52), although his text was published four years after Tacquets' *Elementa Geometriae*. Even more similar a proof is found in Torricelli's *De Proportionibus Liber* (1647), on which Borelli's definition probably depended, and which may have been the origin of the whole debate.

[12] Saccheri means that Euclid's definition, which he intends to defend, is only a stipulation of the meaning of 'equiproportionality', whereas Tacquet's and Dechales' definitions seek to characterize the nature of equiproportionality (assumed as already known). From a historical perspective, Saccheri's understanding of the situation is not inaccurate and in fact quite possible, as both authors strive to explain their new definitions, rather than simply assuming it. Moreover (as we have noted before), their definition looks at *Elements* V, def. 4, as well as at *Elements* VII, def. 20. Saccheri's differentiation between nominal and real definition is extensively explained in *Logica demonstrativa*; see *Note* 44.

Saccheri's position (i. e., his belief that *Elements* V, def. 6 is a nominal definition) was also advocated by Vitale Giordano (whom we know to be a possible source for Saccheri's work on parallels). Giordano, however, does not levy any explicit criticism against these Jesuit authors. Instead, he focuses his attack on Borelli. In *Euclide restituto* we read: "Devesi notare, che l'antecedente sesta definitione non serve per conoscere quando sono date quattro quantità, se quelle siano nella medesima proportione; perché il modo di conoscere, se quattro date quantità habbiano l'istessa proportione, si haverà nella 16. propos. del sesto libro, come à suo luogo si farà noto; e questa sesta definit. serve solo per spiegatione di quello, ch'intende Euclide, e che noi dovemo intendere, quando nelle cose seguenti si dirà la proportione della prima alla seconda quantità è l'istessa, che la proportione della terza alla quarta quantità ..." (*Euclide restituto*, p. 177). Nevertheless, Giordano also 'vindicates', or better 'restores', Euclid and his *Elements* V, def. 6 by providing an extensive discussion of the legitimacy of such a definition (i. e. by proving that the definition is well posed in mathematical terms) in a long series of remarks anticipating the definition itself (pp. 170–5).

[13] Tacquet's and Dechales' definition relies on the possibility of approximating any irrational magnitude to a rational magnitude to any degree of accuracy: it depends (if we read it in numerical terms) on the density of $\mathbb{Q}$ in $\mathbb{R}$. Note that similar reasoning was also employed by Borelli in his reformation of the Euclidean theory: he (I simplify a bit) takes *Elements* VII, def. 20 as the definition of equiproportionals, which holds for rational ratios, and then goes on to consider the approximation of irrational to rational ratios. See the extensive analysis by Giusti, *Euclides Reformatus*, pp. 122–30. Tacquet's and Dechales' 'approximating' definition of equiproportionals was sometimes used in later accounts of the theory of proportions and foundational studies: Heath, *The Elements*, vol. 2, p. 126, mentions G. Veronese and A. Faifofer; Palladino, *Sulla teoria delle proporzioni nel Seicento*, pp. 76–81, mentions G.C. Fagnano, G. Saladini e F. Severi.

[14] Concerning the legitimacy of accepting any nominal definition, within the limits posed by Saccheri in the following passage (and reiterated in Scholium 3 (pp. 125–6)), and as long as it is not a *complex nominal definition* (see above, *Note* 10 to Propositions 38 and 39), see Propositions 5 and 9 in Chapter 4 of *Analytica Posterior* in *Logica demonstrativa*: "omnis definitio *quid nominis* est bona" and later "definitio *quid nominis* cadere non potest in controversiam nisi purè historicam" (ed. 1701, pp. 122 and 126 [ed. 1697, pp. 195, 200]).

[15] Saccheri will later return to the issue of transforming a nominal definition into a real definition. For the time being, cf. *Logica demonstrativa*: "Hinc habes definitionem quidditativam [a kind of real definition] esse plerumque fructum post longam seriem demonstrationum de aliquo subiecto. Neque enim constare potest de genere, & differentia proximis, nisi post longum examen proprietatum, seù praedicatorum convenientium alicui subjecto. Dixi *plerumque* quia accidere potest ut definitio *quid nominis* primo loco stabilita, sit ea ipsa quidditativa: quòd tamen sit quidditativa vix ab initio constare potest, ut patet experientia. Quàre, saltem notitia reflexa, quòd aliqua definitio sit quidditativa, erit semper plurium demonstrationum fructus" (ed. 1701, p. 120 [ed. 1697, pp. 190–1]).

[16] F. Enriques, *Gli Elementi di Euclide e la critica antica e moderna*, Roma, Stock 1923, vol. 2, p. 12, notes that, in his definition Euclid avails himself of multiplication rather than division because he does not have a universal and algorithmic method for dividing any magnitude whatsoever (or at least those magnitudes dealt with in the *Elements*) into any number of parts. For instance, it was impossible for him to trisect arbitrary angles. Thus Borelli was obliged to *postulate* the principle of partition of any magnitude into any number of equal parts and then to employ this principle to provide a proof of the existence of the fourth proportional. Saccheri may find a similar difficulty here, though this is quite unlikely,

since he later accepts the universal partition principle without question, referring to a proof of it by Clavius that however only deals with segments (see p. 128).

[17] Saccheri refers here to the problem of incommensurable magnitudes – the very same problem he mentioned a few paragraphs above. If two magnitudes are incommensurable, then Tacquet's and Dechales' progressive partition method only provides a sequence of infinite rational partitions that approach the desired ratio. As Saccheri stated earlier: in the case of *Elements* VI, 1, this definition of equiproportionals proves that triangle DEF is to base EF as triangle AKC is to base CK, where point K varies at every new partition (indefinitely approaching point F). This proof by Saccheri of the uniqueness of the fourth proportional (which is entirely correct) therefore shows that, in the case of incommensurable magnitudes, Tacquet's and Dechales' definition only show that no approximating rational partition is the required ratio. If it is thus taken as a *real* definition, such that it assumes the existence of the object it describes, then the existence of the (irrational) limit of the rational sequence must be *postulated*. As no clear theory of real numbers (in Cantor's or Russell's fashion) had been developed at the time, this limit is assumed via the axiom of the existence of the fourth proportional (stretching its meaning quite a bit).

[18] All the authors mentioned so far accepted the axiom of existence of the fourth proportional, which they employed in fact as a cornerstone of their theory. Dechales, whom Saccheri quotes here, refers to it as a 'request' (*petitio*): "Datis tribus quantitatibus A, B, C petitur concedi dari posse quartam quantitatem D, ad quam C eamdem rationem habeat, quam habet et ad B" (*Cursus*, p. 153). We find it in Tacquet as the only *axioma* of Book V (*Elementa Geometriae*, p. 139), and in the work of all the mathematicians of the Italian school. Note, however, that the axiom was of varying mathematical strength. Taken as it stands – stating that, given three magnitudes A, B and C, there is always a fourth magnitude X such that $A:B=C:X$ – it establishes only that the domain of magnitudes is closed with respect to the relation of 'ratio'. The simplest examples arise from the notion of a numerical *field*: if I were to consider only integer or rational numbers, then the axiom of existence of the fourth proportional constrains the domain of the theory of proportions to the whole $\mathcal{Q}$, instead of, for instance, $\mathbb{N}$ or Z (which are not closed fields with respect to the ratios). Likewise, if I only want to deal with certain irrational magnitudes, then the axiom constrains certain extended rational fields, as for instance $\mathcal{Q}(\sqrt{2})$, or (more appropriate to the Euclidean context) the set of constructible numbers $\mathbb{E}$, or the set of algebraic numbers. In other words: the axiom of existence of the fourth proportional does not imply, in itself, the *completeness* of the considered set. But these principles were easily confounded, for completeness was tacitly assumed by all geometers of the time. Borelli, for instance, only explicitly accepts the weak version of the axiom of existence of the fourth proportional, i. e. the one concerning rational numbers, but then, when approximating irrational to rational ratios, he also seems to implicitly assume completeness (see above, *Note* 13). Indeed, Vitale Giordano criticized him (though somewhat inappropriately) precisely for this reason (*Euclide restituto*, pp. 186–92). In any event, Tacquet and Dechales seem to take this axiom of the existence of ratios as some kind of completeness axiom; at least, this is how Saccheri seems to understand the matter.

[19] In Latin, *reperiri demonstrative*, i. e. construct. This phrase should be read in conjunction with the expression *problematice demonstratur* (solving a problem through construction), which is found two paragraphs below. At this point in *Euclid Vindicated*, Saccheri seems to direct the more abstract discussion concerning domains of magnitudes back towards his primary concern, namely, geometrical objects. With reference to the previous *Note*, thus, he may mean that the axiom of existence of the fourth proportional forces the domain of magnitudes to be the whole set of constructible numbers $\mathbb{E}$ (or rather: a set of points on a plane isomorphic to $\mathbb{E} \times \mathbb{E}$), instead of the entire real field. Saccheri's (and others') objections to assuming the axiom would then come down to the fact that these geometrical magnitudes must be effectively and explicitly constructed (with ruler and compass), and not just abstractly postulated. Note that, since $\mathcal{Q}$ is dense in $\mathbb{E}$, all previously described approximation processes are still valid. From a foundational point of view, this transition from domains of abstract magnitudes to geometrical domains is very significant: according to the modern algebraic presentation of constructible numbers, they constitute a field without having to assume additional axioms – so the difficulty concerning the existence of the fourth proportional only arises in the general case of any magnitude whatsoever, and not in this specific case. Saccheri's restricted interpretation is thus only applicable to the particular case of *Elements* V, 18 which is proved in the following pages, and then also to *Elements* XII, 2 which Saccheri will later discuss extensively; not to the general case of *Elements* V, 18.

To summarize: in the case of geometrical magnitudes, there are two meanings according to which a magnitude may be said to be 'constructible'. The broad sense means constructibility by ruler and compass; the narrow sense demands an *explicit* procedure to construct a given magnitude by means of ruler and compass. The existence of the fourth proportional magnitude can be proved by construction if we allow for the broad meaning, whereas this is not necessarily the case if we consider the narrow sense. For instance (*Elements* XII, 2), given two segments d_1 and d_2 and given a circle C_1 constructed on d_1 taken as its diameter, we are to find the fourth proportional S with respect to the square of the diameters, i. e. an S for which $(d_1)^2 : (d_2)^2 = C_1 : S$. If $d_1, d_2, C_1 \in \mathbb{E}$, we can then easily see that $S = (d_2)^2 \times C_1 / (d_1)^2 \in \mathbb{E}$. Yet it is not clear (before *Elements* XII, 2 is proved) how magnitude S can be effectively constructed from three given magnitudes (the standard Euclidean procedure would construct a square of area S, but this cannot be done – squaring the circle). The principle of existence of the fourth proportional is thus a non-constructive existential principle (in both the broad and narrow sense) when it refers to abstract magnitudes, and it is non-constructive in the second sense of the term when it refers to geometrical magnitudes.

We may finally remark that Euclid himself admits that if three magnitudes are *given*, then their fourth proportional is also *given* (*Data*, 2). The existential or custructive meaning of 'givenness' in Euclid's *Data*, however, was controversial since Antiquity.

[20] Saccheri literally transcribes Clavius' axiom and commentary from *Euclidis*, p. 221. The non-constructive nature of *Elements* V, 18 is nowadays highlighted by all of Euclid's interpreters as an (not entirely isolated) oddity, for his work, at least according to an interpretation that was quite widespread at the end of the nineteenth century (initiated by Zeuthen), should be entirely constructive. Without going into the complex details of the

constructive character of Classical mathematics, we can still point out that Renaissance and Modern mathematics were hardly constructive, and that Clavius' axiom on the fourth proportional undoubtedly contributed to this state of affairs. The contemporary reader will certainly find Clavius' justification of the non-constructive existence of certain objects rather striking: he simply states that they exist since their existence does not contradict all other assumptions – something that we nowadays may want to understand, following Hilbert or Poincaré, as the statement that mathematical existence is just non-contradiction, or that every consistent system has a model. In any case, Clavius' position was popular, and Saccheri himself admits something of the kind (although not the exact same concept) in *Logica demonstrativa*: "Ubicumque definition sit incomplexa admittendum est postulatum. Sit definitio incomplexa: *linea est, quae ex aequo sua interiacet puncta.* Hanc postulat sibi concedi geometra, videlicet licitum sibi esse, à punto ad punctum rectam lineam ducere. Non est recusandum: *quia nullum est fundamentum, suspicandi impossibilitatem lineae sic definitae.* [...] Itaque habes admittendum facilè non esse postulatum, ubi definitio sit complexa, propter periculum erroris; non item ubi sit incomplexa, *cùm nulla esse possit positiva erroris suspicio*" (ed. 1701, pp. 130–1 [ed. 1697, pp. 205–6]; my emphasis). We do not know whether Saccheri regarded the definition of fourth proportional magnitude as a non-complex definition, but he certainly seems to align himself with Clavius, at least in principle. It is therefore confusing that he voices constructivist concerns in the previous paragraph of *Euclid Vindicated*. In other words, why reject Clavius' postulate? I think that Saccheri's position can be explained in at least two ways. The first is simply philological and classicist: Saccheri wants to demonstrate that Euclid did not need to rely on any implicit principle, and so hopes to 'vindicate' the Ancient geometer from what some interpreters took to be a blemish (an omission). The second reason is less explicit and more structural and intrinsic: the principle of existence of the fourth proportional precludes the possibility of unifying the theory of proportions for continuous magnitudes, found in Book V, with the theory of numerical proportions found in the arithmetic books of the *Elements*. Saccheri's propensity for a unified theory of proportions was already apparent in his discussion of *Elements* VII, def. 20 (p. 104), as he claimed that the latter definition is not different from, but rather only a restricted case of, *Elements* V, def. 6.

Whatever the case, in this passage of *Euclid Vindicated* Saccheri explicitly advances a constructivist instance against the principle of existence of the fourth proportional. We had to underline, however, that his mathematical epistemology does not explicitly exclude non-constructive procedures from geometry (indeed, *Logica demonstrativa* almost moves in the opposite direction); and we should not take for granted that a seventeenth- or eighteenth-century mathematician lacking an explicit epistemological commitment necessarily embraced constructivist positions. In fact, as we have already seen, most mathematicians had no trouble accepting this additional axiom, and many actually sought to incorporate it into their work as the cornerstone of new foundational endeavors. It is also true, in any case, that Saccheri's position, which was (so to speak) more Euclidean than Euclid's (as *Elements* V, 18 seems not to be spurious, as Clavius and Saccheri suggest), was not a complete anomaly:

various Modern mathematicians attempted to prove the proposition of the composition of ratios without employing the axiom here at issue. The first such attempt was probably put forward by Campano in his commentary on Euclid. The resulting proof, however, was deeply flawed, and earned Clavius' reproaches (*Euclidis*, p. 231). The best-known Modern proof was probably the one developed by Simson, who wrote shortly after Saccheri and did not believe (with Clavius and Saccheri) that Euclid had truly authored the proof appearing in the *Elements* (SIMSON, *The Elements of Euclid*, pp. 381–85; *Euclidis libri sex*, pp. 365–8).

In my belief, what can be considered an isolated case is rather the strategy employed by Mercator, who took *Elements* V, 18 to be an axiom and thus exempted himself from the more general discussions concerning the existence of, or possibility of constructing, the fourth proportional. Here is his somewhat vague formulation of the axiom: "Qualem rationem habent omnes partes simul sumptae ad aliquid, tale rationem habet & totum ad idem" (Axiom 14; *Euclidis Elementa Geometrica*, pp. 54–5).

We may also recall that *Elements* V, 18 acquired extraordinary importance within Renaissance geometrical studies (and thus also received considerable foundational attention), in part because it constituted the crux of all the (Cavalierian and non-Cavalierian) treatises on indivisibles.

One of the first philological debates on the axiom of the fourth proportional is O. BECKER, *Eudoxos-Studien II. Warum haben die Griechen die Existenz der vierten Proportionale angenommen?*, "Quellen und Studien zur Geschichte der Mathematik, Astronomie und Physik", 2, 1932–1933, pp. 369–87. An excellent analysis of geometric constructivism during the first half of the seventeenth century (immediately after Clavius) is contained in H. BOS, *Redefining geometrical exactness: Descartes' transformation of the early modern concept of construction*, New York, Springer 2001.

[21] That is, *resolvendo* rather than *componendo*: instead of starting with the first principles to prove its consequences, one begins with the theorem (*Elements* V, 18) to be proved and works his way towards the proof's principles. Saccheri is not utterly faithful to this promise: in the following paragraphs, he re-demonstrates (through synthesis) *Elements* VI, 2 and 12, although repeatedly relying on the classic Euclidean proofs, which he could very well have suppressed to follow his own reasoning. We should also connect Saccheri's expression regarding analysis to his premise in Book Two of *Euclid Vindicated* (above, p. 102), where he claims to employ 'reasoning only', which is to say, no geometrical intuition (in construction) – which, in fact, is not very useful in *Elements* V.

[22] These are Euclid's proofs of *Elements* VI, 2 and 12, which Saccheri simply presents without providing any additional commentary. Elements VI, 12 is in fact the principle of existence of the fourth proportional as applied to segments. Proof of *Elements* VI, 2, i. e. Thales' Theorem, explicitly employs (in Euclid) *Elements* I, 31, and hence the Fifth Postulate. As we have seen, Saccheri proved a generalized version of this theorem for the case of non-Euclidean geometries in Proposition 19 of *Euclid Vindicated* – the fact that he does not here recall this could be considered as evidence supporting the claim that the two Books were independently composed.

Note that in the proof's last lines Saccheri employs *proportio* to mean 'ratio', a terminological slip originating in Medieval translations of Euclid's work. This use was rather common in Clavius' writings, even though the terminological distinction between *ratio* and *proportio* was well established by the Modern Period.

[23] We are unaware of the exact extent of Saccheri's restriction in proving *Elements* V, 18 only for the case of segments, because we are unsure as to the class of magnitudes to which he thought the proofs in Book V could be applied. Ancient authors display no consistent use of the term μέγεθος. Aristotle, for instance, only uses it for segments, surfaces and solid bodies (*Metaph.* Δ 13; although his position is more nuanced elsewhere), and Euclid himself applies the results of the theory of proportions to this set of objects in the remaining books of the *Elements*; however, he also considers angles "given in magnitude" in *Data*, def. 1. Numbers were normally not considered magnitudes by Greek geometers, but we may register some inconsistency in the common practice, and in any case the first attempts towards the introduction of the concept of a real number in the Islamic Middle Age and the Renaissance paved the way to include them among magnitudes as well. Early Modern geometers harbored diverse opinions on this point. According to some very restricted (Aristotelian) interpretations, the magnitudes referred to in *Elements* V were only geometrical magnitudes, i. e. lengths of segments, areas of plane figures, and volumes of solids. Such magnitudes were not considered mutually homogeneous, because in these cases Archimedes' Axiom, which was taken to be the condition of a possible ratio in *Elements* V, def. 4 (Heiberg's numeration), fails to hold good: areas cannot be obtained by summing lengths, nor volumes by summing areas (I am here modernizing, since at the time (lacking real numbers) there was some uncertainty regarding the distinction between a segment and the size of a segment, a surface and the area of a surface, a solid and the volume of a solid). According to this interpretation, Book V of the *Elements* is more comprehensive than other Euclidean treatises, for it deals with three different, inhomogeneous classes of objects falling under the same general category of 'magnitude'. Such a narrow meaning of magnitude was probably grounding the important editions of the *Elements* of Commandino and Clavius (see the *Notes* to the Second Part of this Book). Other geometers offered much broader interpretations of the Euclidean concept of magnitude: some included numbers or angles; others time, velocity, musical tones and various other intensive quantities; others still took magnitude to be the most general genus (category) of quantity *tout court*. Quite obviously, the theory of proportions was developed and employed in the seventeenth century just because mathematicians were able to take a more abstract (more than geometrical) approach to the concept of quantity, which allowed them to apply it to philosophy of nature and mechanics. In fact, some of the Galilean school's major difficulties derived from the problem of inhomogeneity of magnitudes within physics, and could only be solved by a broader interpretation of the generic concept of magnitude of Euclid's Book V; and without such an interpretation, it would have been meaningless to dedicate so much effort to developing the foundations of the theory of proportions.

If we thus have in mind a narrow understanding of the concept of magnitude, such as Clavius' or Commadino's, then Saccheri's proof of *Elements* V, 18 restricted to segments, will seem to be entirely satisfactory, for the generalization of this theorem to areas and volumes does not appear to present insurmountable difficulties (as Saccheri states later in the book). If, on the other hand, we allow for a generalized interpretation of the concept of magnitude and ratio, and for their application in universal mathematics and in physics, then Saccheri's proof is rather disappointing, for it only covers a very small portion of the possible applications of the theory of proportions (because Saccheri employs theorems of *Elements* VI that require, for instance, ruler-and-compass constructions, and thus cannot be applied to a velocity or a *momentum* of an impulse). In any case, Saccheri must have had at least some of these generalizations in mind, and he states as much in p. 139. Furthermore, he expresses dissatisfaction with his particular proof of *Elements* V, 18, and attempts to generalize it in pp. 129–31.

Concerning these issues, see the *Notes* to the Second Part of Book Two, and CRAPULLI, *Mathesis Universalis.*

[24] This is one of the most debated propositions of the *Elements* in the Modern Age, the first in which Euclid employs the so-called method of exhaustion (the first person to use this term, which is never mentioned in Saccheri's work, seems to have been the Jesuit Grégoire de Saint-Vincent), and the main text of the first period of research on methods of infinite analysis. The generalization of this proposition to any curvilinear figure was perhaps the greatest challenge for the various theories on indivisibles. The proposition was much discussed also with regards to the need of proving it by *reductio*, and this yielded some epistemological difficulties (see *Introduction*, § 5). For a brief summary of the Euclidean procedure, see *Note* 35. Concerning the original meaning of Euclid's method of exhaustion, see (amongst many others) VITRAC, *Les Éléments*, vol. 3, pp. 237–59.

[25] Euclid obviously never considered circles as polygons of infinite sides; had he done so, then the whole reasoning in *Elements* XII, 2 would have been of no avail. Once again, Saccheri is here referring to misinterpreted and poorly rigorous limiting procedures, typical of seventeenth- and eighteenth-century mathematics.

[26] In Latin: *universa geometria*, which can simply mean 'the whole of geometry'. Since these problems relate to the theory of proportions, however, the phrase may in this instance refer to *mathesis universalis*, i. e. to a theory of any magnitude whatsoever, which could be considered the subject of Book V of the *Elements*. In any case, as we have seen, Saccheri does not seem to afford these modern speculations much space. The expression also appears in the subtitle of *Euclid Vindicated.*

[27] Saccheri's principle is the trichotomy of ratios: it is always possible to order two given ratios $A:B$ and $C:D$ as $A:B<C:D$, $A:B=C:D$, or $A:B>C:D$, and these orderings are mutually exclusive. The principle may also be expressed stating that ratios constitute a *totally ordered* domain. This is obvious according to our common practice, which takes Euclidean ratios to be fractions, as ratios are themselves magnitudes (numbers). But it is important to recall that in Antiquity and in most interpretations of Euclid in the Renaissance and the Early Modern Period, ratios were understood to be relations, not magnitudes. Consequently, it does not follow immediately that *Elements* V, def. 6 and 8, introducing an order relation among ratios, also introduce a total order, such as the one obtaining among the magnitudes

that appear in the ratios. Euclid developed some arguments that paved the way for a proof of this ordering principle, but he never explicitly took it into consideration. An instance of the law of trichotomy as applied to ratios seems, however, to be employed implicitly in *Elements* V, 10 (see the commentary by Heath, *The Elements*, vol. 2, pp. 156–7), and some modern interpreters have gone so far as to propose that a total order of ratios is implicit in *Elements* V, def. 3. More to the point, one could perhaps refer to Propositions 14–18 of Euclid's *Data*, which show some interest in trichotomy of ratios. In any case, note that Euclid and Saccheri always take the total order of magnitudes (employed in ratios) to be self-evident – neither of them ever mentions it. Euclid tacitly employs it in his proof of *Elements* V, 9, 10 and 13, and applies it repeatedly also in several passages outside Book V: it is, for instance, crucial (as we will presently see) in *Elements* XII, 2.

During the eighteenth century, the dissolution of the theory of proportions into a more general theory of rational and irrational numbers had as a natural consequence the law of trichotomy: as ratios were themselves becoming numbers. In fact, this path was followed by all 'modern' geometers, and firstly (for instance) by Wallis, Nonancourt and, influenced by Nonancourt, Arnauld (cf. the second edition of *Nouveaux Elémens*, Book Two, § 8, pp. 188–9).

In *Euclid Vindicated*, the explicit formulation of the principle of trichotomy for ratios is rather important, as it is telling about the logical origin of many of Saccheri's reflections on geometry. In fact, only in the nineteenth century were order relations taken into consideration in the context of foundational aspects of mathematical reasoning; in the Classical and Modern Age, they were just tacitly assumed. It was not until De Morgan (see later, *Note* 32) that the total ordering principle was applied to the theory of ratios between magnitudes. Thus, in advocating for more rigorous foundations of fundamental algebraic relations, Saccheri seems to have been ahead of his time. Before him, only Roberval, as far as I know, had expressed this need (Roberval's work, as we have already mentioned, remained unpublished): see the Scholium to Proposition 11 in *Éléments de géométrie*, p. 313.

Also note that Saccheri does not attempt to prove the principle of the existence of the fourth proportional by means of the Trichotomy Principle (an attempt that would be bound to fail), because he does not wish to assert the absolute validity of the existence of the fourth proportional. Rather, he hopes to demonstrate that trichotomy (that he accepts in general) is sufficient to prove all theorems in which Euclid makes use of the existence of a fourth proportional. After all, if Saccheri could apodictically prove the principle of existence of the fourth proportional, there would be no need to criticize those geometers (Tacquet, Dechales, Borelli) whose theories of proportion rest upon this principle – geometers who are the primary object of criticism of *Euclid Vindicated*. Saccheri thus seems to regard the total ordering as a *weaker* principle than the existence of the fourth proportional. For a theoretical development of this issue, see the appendix to Giusti, *Euclides Reformatus* (pp. 169–71): Giusti shows, within a modern mathematical context meant to represent the Galilean theory of proportions, that the Law of Trichotomy (Theorem 19) follows in fact from the principle of existence of the fourth proportional (Axiom 11).

[28] Saccheri has to assume that the magnitudes constituting a ratio are homogeneous, i. e. that A and B must be homogeneous if one is to allow the ratio A:B (this is a classic

requirement of the theory). However, Saccheri explicitly states that, if we compare ratios in the form $\mathrm{A}:\mathrm{B} \lesseqgtr \mathrm{C}:\mathrm{D}$, then C and D (which, of course, have to be homogeneous to each other) need not be homogeneous to A and B. In addition to being fully consistent with the Classical theory of proportions and with Modern developments in mechanics (which absolutely had to allow any kind of inhomogeneity between pairs of magnitudes), this is also the only reasonable requirement in the present context of *Euclid Vindicated*, because if the four magnitudes were all mutually homogeneous (not only pairwise homogeneous), then *Elements* V, 18, which sparked this whole discussion, could be demonstrated by means of Euclidean theorems and without recourse to the existence of the fourth proportional: cf. Heath, *The Elements*, vol. 2, pp. 172–4, and Vitrac, *Les Éléments*, vol. 2, pp. 108–9.

[29] This proof presents a significant problem for the interpretation of *Euclid Vindicated*, as it contains such a macroscopic mistake that we can hardly imagine it to be authored by the same mathematician who wrote the rest of the book.

The trichotomy proof (see previous *Note*) is actually very simple (almost immediate) if we allow for Euclid's definition of equality and inequality of ratios (*Elements* V, def. 6 and def. 8, according to Clavius' and Saccheri's numeration) and if we assume (as everyone certainly did in Saccheri's days): (a) the Law of Trichotomy for those magnitudes constituting the ratios, i. e. that A and B belong to a totally ordered domain of magnitudes $\mathcal{P}$, and, similarly, that C and D belong to a totally ordered domain of magnitudes $\mathcal{Q}$ (see above, *Note* 27); and (b) that the total order of the two domains is preserved in their 'scalar' multiplication with natural numbers (those needed to obtain equimultiples), i. e. given $n \in \mathbb{N}$ and $\mathrm{A} \in \mathcal{P}$, then $n\mathrm{A} \in \mathcal{P}$ (see below, *Note* 30). What is yet to be proved, in this case, is simply that Euclid's criteria for the inequality of ratios in *Elements* V, def. 8 are well posed and cannot both be satisfied by the same pair of ratios: in other words, it cannot be the case that $\mathrm{A}:\mathrm{B}<\mathrm{C}:\mathrm{D}$ for a given choice of equimultiples, and then $\mathrm{A}:\mathrm{B}>\mathrm{C}:\mathrm{D}$ for another such choice. This was proved by De Morgan (with the aid of a couple of additional assumptions that were readily accepted in the eighteenth century), and nowadays his proof can be found in Heath, *The Elements*, vol. 2, pp. 130–1. Euclid's criteria for the equality and inequality of ratios in his two Definitions, in fact, are mutually exclusive and exhaust all the possibilities. Consequently, trichotomy follows.

Yet, Saccheri here provides a flawed proof that is only applicable to commensurable magnitudes. His demonstration essentially consists in reformulating the criterion for the equality of ratios in *Euclid V*, def. 6 ("*In the first place ...*") and the criterion for the inequality of ratios in *Euclid V*, def. 8 ("*In the second place ...*") to show that they exhaust all possibilities; depending on their equimultiples, two ratios will be either equal or unequal, and thus in any case ordered (trichotomy). But this is true (if we disregard some minor difficulty noted by De Morgan) for Euclid's original definitions, which, when formulated in modern terms, are in fact mutually exhaustive (see above, *Note* 1). But Saccheri here reformulates the proof in such a way that the equality criterion is (formally) leveled onto the two inequality criteria, and the universal quantifier of *Elements* V, def. 6 becomes existential:

$$\mathrm{A}:\mathrm{B}=\mathrm{C}:\mathrm{D} \Leftrightarrow \exists n,m \in \mathbb{N}:(n\mathrm{A}=m\mathrm{B} \wedge n\mathrm{C}=m\mathrm{D}).$$

These three criteria (namely, this last one and the two analogous criteria in *Elements* V, def. 8) exhaust all possible cases only when magnitudes A, B, C and D are commensurable, i. e. not in the general case of any magnitude (which is why Euclid, or rather Eudoxus, adopted a more complex equality criterion); in the general case, the previous criterion is only a sufficient, though not a necessary, condition for equality. Saccheri's conclusion ("*I proceed therefore* ...") concerning total ordering is thus incorrect. This can be illustrated by considering, for instance, an irrational ratio (Simson's example; see below) with four magnitudes: $A=1$, $B=\sqrt{2}$, $C=2$, $D=2\sqrt{2}$. In this case, we obtain that there is no pair of natural numbers m,n satisfying equality of equimultiples:

$$\nexists\, n,m \in \mathbb{N} : \left(n = m\sqrt{2} \wedge 2n = 2m\sqrt{2}\right),$$

yet we cannot find greater or smaller equimultiples, as is required by the definition of inequality, because the two ratios are perfectly equal, and $1:\sqrt{2}=2:2\sqrt{2}$.

Such a gross error was immediately spotted by Saccheri's readers. Simson offers a lucid critique as early as 1756. In a note to the theory of proportions, he recalls the final part of Saccheri's reasoning on total order, commenting: "Not in the least; but it remains still undemonstrated; for what he says may happen, may in innumerable cases never happen, and therefor his demonstration does not hold. For example, if A be the diameter and B the side of a square; and C the diameter and D the side of another square; there can in no case be any multiple of A equal to any of B; nor any one of C equal to one of D, as is well known; and yet it can never happen that when any multiple of A is greater than or lesser than a multiple of B, the multiple of C can, upon the contrary, be lesser than or greater than the multiple of D, viz. taking equimultiples of A and C, and equimultiples of B and D. For A, B, C, D are proportionals, and so if the multiple of A be greater &c. than that of B, so must that of C be greater &c. than that of D" (Simson, *The Elements of Euclid*, p. 384; cf. *Euclidis libri sex*, p. 367). Note that Simon rightly rejects Saccheri's proof, though he does not express any doubt about the total order of ratios, i. e. about the algebraic principle here put forward by Saccheri.

It is still unclear how Saccheri could have committed such an astonishing mistake in this Second Book of *Euclid Vindicated*, whose fundamental purpose was precisely to 'vindicate' *Elements* V, def. 6 on incommensurable magnitudes – a definition that Saccheri clearly understands well enough to want to defend it from the attacks to which it was subjected. In my opinion, the only possible explanation is some extrinsic accident: in his haste to complete *Euclid Vindicated*, Saccheri could have inserted a proof that was in fact meant for some other part of the work; or perhaps a distracted pupil was charged with ordering his teacher's notes. Since the proof is correct for commensurable magnitudes, Saccheri's intention was perhaps to explicitly apply it to such magnitudes and later generalize it – just as Borelli (see above, *Note* 13) assumed an axiom of existence of the fourth proportional in the case of commensurable magnitudes only, and then generalized, by density, to incommensurable magnitudes. Yet another possibility is that Saccheri understood 'numbers' in this proof not as usual natural numbers but rather as real numbers (as when, in p. 127, he meant rational numbers; cf. *Note* 47 below). This possibility, however, seems at odds with eighteenth-century mathematical practice, and in particular with Saccheri's classicist approach. Also note that, in the following section of *Euclid Vindicated* (pp. 126–8), Saccheri returns to this whole theory and seems to reconstruct it in a more correct way. Whatever the case, we are here left with but some poor conjectures.

Finally, note that the entire proof is based solely on the definition of equiproportionality (with the exception of an explicative recourse to *Elements* V, 11), and is thus consistent with Saccheri's epistemological idea about how to prove an axiom (his own on trichotomy).

[30] This passage seems to highlight the 'scalar' character of the multiplication used in equimultiples. The product of two segments marks a transition "into a new species of being", i. e. to a surface; and likewise for higher dimensions. In the definition of equimultiples, on the other hand, a segment is multiplied by a 'pure (natural) number', thus simply yielding a longer segment. Thus it seems that Saccheri aims to justify assumption (b) which I explained in the previous *Note*, namely, that if the principle of trichotomy holds for two homogeneous magnitudes A and B, then it also holds for their multiples nA and mB. Some interpreters think to find this assumption in a strong interpretation of *Elements* V, def. 4 (the Archimedean criterion for the homogeneity of magnitudes); see VITRAC, *Les Éléments*, vol. 2, pp. 137–9. Of course in Classical theory this passage had to be justified explicitly, for there is no doubt that in Book V of the *Elements*, magnitudes are not just numbers, but (at least) lengths, areas, volumes; since these magnitudes and numbers are *inhomogeneous*, it is not obvious that they can be multiplied together. In other words, numbers must be considered both (perhaps) magnitudes in themselves, and scalar operators for the multiplication of magnitudes inhomogeneous with them (and actually, via Archimedes' Axiom, for *defining* what homogeneity is). In this regard, see the second edition of ARNAULD, *Nouveaux Elémens de géometrie*, Book Two, § 53, pp. 215–6.

[31] We said that the Euclidean criterion for greater ratio is:

$$\mathrm{A}:\mathrm{B}>\mathrm{C}:\mathrm{D}\Leftrightarrow\exists n,m\in\mathbb{N}:(n\mathrm{A}>m\mathrm{B}n\mathrm{C}\leq m\mathrm{D}).$$

This formulation, however, still seems to leave open a possibility, namely, that the first equimultiples are equal, and the second unequal. It is therefore reasonable to believe that Euclid's definition (if we want trichotomy and total order) should be completed by this last case (and the similar case of smaller ratio):

$$\mathrm{A}:\mathrm{B}>\mathrm{C}:\mathrm{D}\Leftrightarrow\exists n,m\in\mathbb{N}:(n\mathrm{A}>m\mathrm{B}\wedge n\mathrm{C}\leq m\mathrm{D})\vee(n\mathrm{A}=m\mathrm{B}\wedge n\mathrm{C}<m\mathrm{D}).$$

Together, the two cases do indeed cover all possible cases, and do not require further development. Euclid probably fails to explicitly mention this last inequality condition only for reasons of definitional economy (he also excluded an explicit definition of smaller ratio); In any case, it is rather obvious that the latter case can be brought back to the previous by means of the theorems of Book V of the *Elements*. This is exactly the strategy that Saccheri employs later on in *Euclid Vindicated* (pp. 126–8), when demonstrating that his improved version of the Euclidean definition, which is necessary to obtain a prefect trichotomy of ratios, was in fact already present in the original definition – another small 'vindication' of Euclid.

Clavius did not seem to have placed particular emphasis on this second inequality criterion, though a brief mention of something of the kind is found in the discussion of *Elements* V, def. 6 (*Euclidis*, p. 209).

[32] These two Lemmata state that the considered magnitudes possess some property of continuity, or rather density. They are very clear when expressed in the modern language of fractions: the first states that, if $a/b < c/d$ (*strictly* less than), then there is an ε such that the inequality holds also when 'perturbed' in the form $a/b < c/(d+\varepsilon)$ the second Lemma is the inverse of the first and considers the inequality $a/b > c/(d-\varepsilon)$. As the existence of ε is assumed, Saccheri understands these Lemmata as a consequence of his own Axiom, which guarantees the possibility of comparing the non-constructively obtained ratio $c/(d\pm\varepsilon)$ to a/b. In a way, then, these Lemmata take Saccheri *beyond* the simple principle of trichotomy (which is its necessary condition), for he assumes here the existence of this magnitude $(d\pm\varepsilon)$ in the form of a density principle. Nevertheless, he sees this as a weaker principle than the principle of existence of the fourth proportional – and this is certainly the case if one wants to find, in that principle, an application of completeness (as may have been the case for Tacquet and Dechales).

Many years later, Augustus De Morgan attempted to demonstrate the principle of existence of the fourth proportional by assuming these two Lemmata (cf. Heath, *The Elements*, vol. 2, p. 171), although without ever mentioning Saccheri. As we have seen, Saccheri does not attempt to use these Lemmata to obtain a proof of such generality, although his use of them in proof of *Elements* V, 18 and *Elements* XII, 2 is general enough (as he himself observes in p. 121) to be comparable to De Morgan's.

[33] Indeed, Euclid never proves the transformation laws for the proportionality of ratios of the kind $A:B=C:D \Leftrightarrow B:A=D:C$ or $A:B<C:D \Leftrightarrow B:A>D:C$, and seems to content himself with the generality of his own definitions of equality and inequality of ratios, and perhaps also for his own definition of inverse ratio (*Elements* V, def. 13). The proposition here mentioned by Saccheri (*Elements* V, 26) is not contained in the original Greek text, but was in fact added later by commentators. Clavius, who attributes it to Campano, talks about it in *Euclidis*, p. 236. Saccheri mentions it again in the following Scholium 2 (pp. 121–2).

[34] In Latin, *ab aliquo finito numero denominata*. This is the only instance of the term *denominatio* in the whole of *Euclid Vindicated*. The term – and concept – was widely accepted in the Medieval and Renaissance tradition of the theory of proportions. A 'denomination' of a ratio is a number expressing the ratio, a characterization that closely resembles the modern concept of fraction. Still, there are some important differences: first of all, most authors believed that only ratios between commensurable quantities could have a denomination (and, consequently, that denominations are rational numbers); secondly, these numbers are not systematically expressed in the modern way, but rather by two numbers – i. e. the greatest possible integer – to which we then add a fraction (*minutia*) of value less than one. For example, 15/4 is expressed as 3¾, and likewise in all other cases. Denominations are very common in Clavius, and Saccheri must have taken the concept from him (cf. *Euclidis*, pp. 176–81). Saccheri does not, however, offer an articulate theory

of denominations, and actually seems to avoid it, perhaps to propound for greater Classical Euclidean purity.

A brief overview of the theory of denomination in the Medieval period can be found in Murdoch, *The Medieval Language of Proportions*, who also shows that there had been attempts (Bradwardine) to broaden the concept of *denominatio* to include irrational magnitudes; concerning Clavius, see Rommevaux, *Clavius*.

[35] It is perhaps worth comparing Euclid's classic proof (which is the first application of the exhaustion method) with this new proof put forward by Saccheri. Given two circles C_1 and C_2 and their diameters d_1 and d_2, we want to show that $C_1 : C_2 = (d_1)^2 : (d_2)^2$.

Euclid starts with the principle of existence of the fourth proportional, and assumes that there is an S such that:

$$(1)\ (d_1)^2 : (d_2)^2 = C_1 : S.$$

He wants to prove that $S = C_2$. Suppose $S < C_2$. Now, by a repeated application of *Elements* X, 1 (a kind of Archimedes' Axiom), Euclid *constructs* a polygon P_2 inscribed in (and thus smaller than) C_2 but greater than S. We thus have $S < P_2 < C_2$.

But if we take a polygon P_1 similar to P_2, inscribed in C_1 (so $P_1 < C_1$), then by *Elements* XII, 1 the following must hold:

$$(2)\ P_1 : P_2 = (d_1)^2 : (d_2)^2$$

and, by substitution in (1) and (2), the following must also hold:

$$(3)\ C_1 : S = P_1 : P_2$$

But this is impossible, because $P_1 < C_1$, and thus, to obtain equality (3) according to the well-known *Elements* V, def. 6 on equiproportionality, it should also be the case that $P_2 < S$, whereas this is not the case and the opposite relation is true. So, hypothesis $S < C_2$ is impossible. From here, Euclid can easily demonstrate the impossibility of the opposite inequality, and hence conclude that $S = C_2$.

Saccheri's proof is very similar. He assumes *per absurdum* the inequality of ratios:

$$(4)\ C_1 : C_2 < (d_1)^2 : (d_2)^2$$

and then employs his Lemmata, which allow him to state that there is a certain magnitude ε such that the inequality of ratios holds also when C_2 is substituted by $C_2 - \varepsilon$. Instead of constructing this magnitude, Saccheri assumes that $C_2 - \varepsilon$ is a polygon P_2 inscribed in C_2 (thus $P_2 < C_2$). This assumption is very similar to the assumption appearing in Euclid's proof of the existence of a P_2 inscribed in C_2 and greater than magnitude S: both depend on the density of the set of areas of polygonal figures (inscribed in a circle) in the set of areas of smaller circles. In fact, Saccheri's formulation could be inverted to state that, *given that* polygons have such a density property (which Euclid proves by employing Archimedes' principle in *Elements* X, 1), *then* Saccheri's Lemma 2 can be applied to this specific case of *Elements* XII, 2. In any case, we now have:

$$(5)\ C_1 : P_2 < (d_1)^2 : (d_2)^2.$$

But, once again, if we take a polygon P_1 similar to P_2 and inscribed in C_1 (thus $P_1 < C_1$), then *Elements* XII, 1 is valid, and thus (2) is valid. But this is impossible, because (2) and (5) imply that $C_1 < P_1$, which contradicts what has been stated above. Thus (4) is false, and with similar reasoning we exclude the inverse inequality. Saccheri can now apply his Axiom. From the falsity of both $C_1 : C_2 < (d_1)^2 : (d_2)^2$ and $C_1 : C_2 > (d_1)^2 : (d_2)^2$ he deduces that $C_1 : C_2 = (d_1)^2 : (d_2)^2$. It is quite clear now that the reasoning which led Saccheri to develop his own proof is the following: Euclid assumes the existence of a magnitude S (a fourth proportional), then proves that it can be neither greater nor smaller than the required magnitude (C_2), and finally concludes, by means of the trichotomy of magnitudes, that they must be equal; Saccheri does not assume the existence of a magnitude but rather reasons directly in terms of ratios, and, via a procedure similar to Euclid's, proves that the first ratio is neither greater nor smaller than the second, hence concluding (by means of the trichotomy of ratios) that they must be equal. Concerning the hypotheses employed in the two demonstrations, we see that both proofs assume the density of the magnitudes in polygonal approximation (via Archimedes' Axiom); Euclid's proof also requires the existence of the fourth proportional. It is true, however, that in the case of the plane geometrical magnitudes referred to in *Elements* XII, 2, this last requirement comes down to the requirement of the existence of a *constructible* magnitude, of which no construction algorithm is given; so that Saccheri's alleged gain on Euclid may be not very big (see above, *Note* 19).

In any case, the Proposition in question is probably the most important theorem proved by Saccheri in Book Two of *Euclid Vindicated.*

[36] In fact, Clavius' proof of *Elements* V, 26 (see above, *Note* 33) to which Saccheri is here referring, relies to the existence of the fourth proportional. Note that Saccheri does not consider such proof to be a blemish on Euclid (contrary to *Elements* V, 18 and *Elements* XII, 2), as he knows that it was a later addition. In fact, it may very well be the case that Euclid himself did not explicitly prove *Elements* V, 26, because he took it to be an immediate consequence of his own definitions; in this sense, then, Saccheri's demonstration would be much more Euclidean than Clavius'.

[37] As we have already seen, Saccheri's problem at this point is that *Elements* V, def. 8 on the inequality of ratios only gives a definition of greater ratio (not of smaller ratio). But in the proof of the principle of trichotomy offered above, Saccheri has to employ a definition of smaller ratio that is shaped exactly in the same way as the original Euclidean definition of greater ratio. Clavius and Saccheri attribute such a definition to Euclid himself; and it may well be that Euclid had such a definition in mind, but never explicitly stated it due to a principle of economy of expression. But it may also be that given the explicit definition of equal ratio (*Elements* V, def. 6) and greater ratio (*Elements* V, def. 8), then the definition of smaller ratio was simply residual (i. e. all cases excluded from the previous definitions). In this case the Axiom of Trichotomy of ratios is already included in Euclid's definitional system, and one would then have to prove that the definition of smaller ratio can also be *characterized* in the manner of Clavius' and Saccheri's – which would lead us back to Saccheri's proof of the Axiom of trichotomy.

[38] Saccheri refers to Clavius' Scholium after *Elements* V, def. 8, in which smaller ratio is defined by analogy to the Euclidean definition; cf. *Euclidis*, p. 211.

[39] Modern interpreters of the *Elements* have pointed out that Euclid in fact employs it in a different way, and always assumes the multiplication factor of the first equimultiples to be different from the multiplication factors of the second equimultiples. In other words (according to the previously employed formalism) he always considers $n \neq m$. Such restriction seems useless and in some cases even harmful, and Saccheri seems to have this very problem in mind here. See Vitrac, *Les Éléments*, vol. 2, pp. 42 and 76, and Acerbi, *Drowning by Multiples*.

[40] Giovanni Alfonso Borelli authored the most important alternative to Euclid's theory of proportions. His substitute theory systematized all the work of the Galilean school, and was very similar to the theories by Tacquet and Dechales which Saccheri has previously criticized.

[41] Saccheri is quoting the sixth page of the Preface *ad lectorem geometram* in *Euclides restitutus*. Note that Borelli later provides a more detailed explanation of his criticism of the Euclidean definition of equiproportionals. In this subsequent clarification, he does not condemn the definition's obscurity so much as its uselessness: in practice, the definition requires an infinite number of checks (due to the universal quantification of equimultiples), rendering it unserviceable. From a mathematical perspective, this criticism (which was quite common, and which Saccheri does not take into consideration) is rather poor, since such infinite verifications are easily conducted by reducing them to very basic calculations, as demonstrated by factual application of this definition in later parts of the *Elements*. Borelli's critique is nevertheless significant, because he generally avoids any concept involving the infinite, regardless of whether its mathematical treatment is easy or difficult: which is why he also rejects the Euclidean definition of parallels, which is rooted in a global property (non-incidence) that has to be verified at infinite points and at an indefinite distance, hoping to substitute it with his own 'local' definition of parallelism; see above, the *Notes* to Scholium 2 to Proposition 21. Borelli evidently believed that, in a similar way, *Elements* VII, def. 20, which is the definition of equiproportionals that holds for commensurable magnitudes and that only requires an existential quantifier, had that local and finite character which is proper to mathematics. He therefore takes it as a starting point for his own theory of proportions (with both rational and irrational magnitudes).

[42] Saccheri employs the scholastic expression *quo ad nos*, usually opposed to the expressions *per se* or *secundum se*, which are commonly found in Aristotelian contexts. The issue is briefly addressed in Physical Thesis 31 at the end of *Logica demonstrativa* (ed. 1697, p. 287).

[43] *Elementa vulgaria* is a seventeenth-century scholastic term denoting the four simple Aristotelian elements.

[44] The difference between a *nominal* definition (that simply stipulates a meaning) and a *real* definition (a definition of an otherwise already known object) is explained in *Logica demonstrativa* in the same way: "Hinc demum intelliges, quòd nam sit verum discrimen inter definitionem *quid nominis*, & definitionem *quid rei*, ut a nobis intelligitur. Neque enim in eo discrepant, quod definitio *quid nominis* definiat vocem, & definitio *quid rei* definiat rem; quippè cùm utrobique definiatur res, seù possibilis illa sit, seù impossibilis. Itaque ex eo petendum est discrimen, quòd definitio *quid nominis* nullum praesupponat conceptum rei significatae per vocem, unde est, ut subiectum, de quo illa dicitur, etiam in suppositione simplici

supponat pro predicato; cùm è contrà, definitio *quid rei* anterioremnotitiam praesupponat, unde est, ut subiectum, de quo illa dicitur, non simpliciter, sed tantùm personalitèr supponat pro praedicato. Sic, dum definio *homo est animal rationale*, praecedit conceptus *hominis*, qui in suppositione simplici non est idem, atque conceptus *animali rationalis*, licet obiectivè, & in suppositione personali sint idem" (ed. 1701, pp. 122–3 [ed. 1697, pp. 194–5]). The two criteria added here, as it is clear enough, are needed to avoid that 'real' marks (non-stipulated ideas that only agree with preexistent notions) be introduced in a nominal definition. Regarding the fact that every nominal definition would otherwise be valid, see above, *Note* 14.

45 *Logica demonstrativa*, ed. 1701, pp. 166–7 [ed. 1697, p. 263]. Shortly before drawing on this example, Saccheri defines this type of fallacy (of the same nature as the one regarding complex definition; see above *Note* 10 to Propositions 38 and 39) in general terms: "Fallaciam duplicis definitionis, seu hypothesis, committunt saepissimè. Dico autem fallaciam duplicis definitionis, seu conceptus, cùm assumpta definitione *quid nominis*, non deponitur tamen omnis alius conceptus circà rem definitam" (ed. 1701, p. 166 [ed. 1697, pp. 262–3]).

46 This is the explicit (and very simple) proof of Saccheri's improved Euclidean definition of greater ratio, which he employed in his proof of the Trichotomy Axiom; see p. 116.

47 In fact, ET is not a multiple of A, but rather the multiple of a submultiple of A. The Euclidean definition, therefore, cannot be applied as such, and Saccheri is forced to speak of 'integer or fractional multiples', though these should perhaps be justified otherwise. The same holds for XH with respect to C.

48 In *Euclidis*, pp. 69–70, Clavius demonstrates in four different ways that *any* segment can be divided into equal parts: in particular, he proves that, *given* a segment divided into n equal parts, it is always the case that any other segment can also be divided in n equal parts. It is remarkable that the Fifth Postulate plays such a substantial role in all four proofs, especially since none of them employ any of the theorems of proportions in *Elements* V. In any case, Saccheri's argument is inadequate, because his proof aims to demonstrate that the definition holds for any magnitude whatsoever, not just for segments. Recall that the possibility of partition was postulated by Borelli in order to prove the principle of existence of the fourth proportional, so Saccheri's method seems not to provide any real advantage over Borelli's.

49 The "*opportunior*" assumption is that of not assuming the principle of existence of the fourth proportional, or perhaps of accepting Saccheri's Axiom of trichotomy, which the Jesuit discusses in the following paragraph. We recall that Saccheri eliminated the need for the axiom of existence of the fourth proportional in two specific cases: in *Elements* V, 18 restricted to segments, and in *Elements* XII, 2 and similar propositions. Both cases thus only concern geometrical objects. It is possible, however, that Book V of the *Elements* dealt with (or wanted to deal with) more general classes of magnitudes.

50 Since Saccheri's proof is quite involute and sometimes obscure, let us attempt to summarize it in modern terms. We begin with the hypothesis that four given magnitudes (which we here take to be strictly positive numbers) are such that $a/b = c/d$. We want to prove that

$$\frac{a+b}{b} = \frac{c+d}{d}$$

In the first part of the demonstration, Saccheri proves, with a perfectly Euclidean procedure (i. e. using *Elements* V, 8, 11 and 17), that for every (positive) magnitude x, the equality of ratios 'perturbed' by x:

$$(1)\quad \frac{a+b}{b\pm x}=\frac{c+d}{d}$$

implies (without being equivalent to it) the equality:

$$(2)\quad \frac{a\pm x}{b\mp x}=\frac{c}{d}\left(=\frac{a}{b}\right).$$

But (2) is always false for $x\neq 0$, since:

$$(3)\quad \frac{a+x}{b-x}>\frac{a}{b},$$

$$(4)\quad \frac{a-x}{b+x}<\frac{a}{b}.$$

Note that Saccheri does not deduce (4) in this form, but in one equivalent and inverse to it. In any case, it follows that for every value of $x\neq 0$, (1) leads to a contradiction and the following is therefore valid:

$$(5)\quad \frac{a+b}{b\pm x}\neq\frac{c+d}{d}\left(\forall x\neq 0\right).$$

In the following paragraph Saccheri attempts to arrive at the theorem's thesis via a continuity argument that allows him to invert the implication from (1) to (2) and thus obtain the result that, since (3) and (4) approach the equality of ratios (from the 'right' and from the 'left') when $x\rightarrow 0$, then (1) must also hold when $x=0$. This argument, which Saccheri formulates quite obscurely, is dubious (not to say flawed), and Saccheri does not rely on it. In fact, he rejects it in the next paragraph, explaining that he is considering only abstract magnitudes (not line segments) and that it is not obvious that these continuity properties hold for such magnitudes. The paragraph's conclusion is also rather confused. It seems to concern the new proof strategy that Saccheri intends to adopt to arrive at the proof's conclusion: on the basis of the previously proved Lemmata (which were incorporated into the Trichotomy Axiom shortly above), if the following were to hold:

$$(6)\quad \frac{a+b}{b}>\frac{c+d}{d},$$

then, for some value of x, the following should also hold:

$$(7)\quad \frac{a+b}{b+x}>\frac{c+d}{d}.$$

Saccheri wants to show that (7) leads to a contradiction.

Let us thus assume that (6), and consequently also (7), hold good. We hence have two possibilities: (A) formula (7), which holds good when existentially quantified for a value of x, is also universally valid for any x; or (B) there are some x for which (7) is false.

A) But the first alternative is impossible, because if $x=a$, then from (7) we have:

$$(8)\quad 1=\frac{a+b}{b+a}>\frac{c+d}{d}$$

which is absurd.

B) However, we can also show that the second alternative is false. Together, (5) and (6) show that the following equality cannot hold good for any value of x:

$$(9)\quad \frac{a+b}{b+x}=\frac{c+d}{d}$$

and since (7) is not universally quantified, then there has to be a least one value of x for which the following holds good:

$$(10)\quad \frac{a+b}{b+x}<\frac{c+d}{d}.$$

Thus inequality (7) should hold for some value of x, and inequality (10) for some other value of x, without it ever being the case that the equality of ratios (9) holds good. But this, again, goes against that minimum continuity (or better, density) requirement expressed in Saccheri's two Lemmata, namely, the monotony of strict inequality in the neighborhood of a point – which is what Saccheri expresses somewhat convolutely (but ultimately correctly) through the concept of 'extrinsic limit' (a supremum which is not a maximum), which played a role already in Book One of *Euclid Vindicated*.

From the *reductio* in (A) and (B) we thus conclude that (7) is false, and hence also (6) is false, for it implies the former. The very same reasoning applies with equal force to the case of a smaller – rather than greater – ratio in (6). We thus come to Saccheri's thesis via trichotomy of ratios.

A simpler proof of *Elements* V, 18 that also relies on the trichotomy of ratios was hatched up by Roberval in Proposition 35 of Book VI in his *Éléments de géométrie* (pp. 333–4). An even simpler proof, in certain respects more Euclidean than the one by Saccheri (although expressed in modern terms), is found in Mueller, *Philosophy of Mathematics*, p. 139.

[51] The definition of proportion in terms of similarity of ratios is the definition that was interpolated into the Renaissance editions of Euclid. It appears as *Elements* V, def. 4 in Clavius' *Euclidis*, p. 167. Saccheri briefly discusses the matter in the Second Part of this Book.

[52] In other words, with magnitudes belonging to the same species (for which Archimedes' Axiom holds good). Saccheri will be more explicit about the meaning of this expression in the Appendix to this Second Book (pp. 139–40).

[53] Due to a misprint the Latin text states AB.

Notes to the Second Part of Book Two

The Second Part of Book Two is dedicated to 'vindicating' the second Euclidean blemish that had been spotted by Henry Savile, namely, the obscurity of Euclid's definition of the composition of ratios. The reference here is to *Elements* VI, def. 5 of Clavius' edition, which is deemed spurious by almost all contemporary interpreters of Euclid. Indeed, the definition has but scarce manuscript authority; moreover, owing to the definition's general character, it would seem more appropriate for Euclid to have placed it in Book V (as was placed in some Renaissance treatises, and in Vitale Giordano's book). In *Elements* V, def. 9 and 10, Euclid defines two particular cases of the composition of ratios, which he employs throughout the *Elements*: and there are dimensional reasons for this, as the cases in question allow Euclid to pass from considerations of linear magnitudes to those of areas ('duplicated ratio') and volumes ('triplicated ratio'); nor does he allow for higher dimensions. Yet in *Elements* VI, 23 (and then again somewhat elliptically in *Elements* VIII, 5), Euclid refers to the composition of ratios in general terms. For this reason, perhaps, someone (already in Antiquity) included at the beginning of Book VI (not Book V) a more general definition of composition of ratios, which is what Saccheri here intends to analyze. For a brief discussion of the inauthenticity of *Elements* VI, def. 5, see, for example, Vitrac, *Les Éléments*, vol. 2, pp. 150–3.

In any case, despite its relative unimportance in the *Elements*, the problem of the composition of ratios did impact some important applications of the theory of proportions since Antiquity: the composition of ratio was particularly essential in the theory of music and in certain aspects of astronomy. There is no doubt, however, that the issue became more of a central concern during the scientific revolution and birth of mathematized natural philosophy – when the Euclidean theory of proportions (at least in its first seventeenth-century formulations) represented the theoretical mechanic's 'toolbox', and the notion of the composition of ratio turned out to be the most important instrument therein.

The theoretical difficulty in the definition of the composition of ratios is rooted in the specific interpretation of 'ratio' favored by Antiquity. According to this interpretation, ratios are a specific kind of relations (not magnitudes), therefore one cannot perform on them the same sorts of operations that apply to ordinary magnitudes. The introduction of an operation involving ratios themselves, such as that of composition, must therefore be separately justified. In modern terms, since a ratio is understood as a fraction, this composition is nothing but multiplication; in other words, given the ratios a/b and c/d, the compound ratio is ac/bd. In Euclidean terms, however, given the ratios A : B and C : D, one first has to apply the principle of the existence of the fourth proportional to take a magnitude X with property C : D = B : X, and then *define* the compound ratio between the two given ratios (which are now A : B e B : X) as A : X.

Perhaps the first serious Modern discussion of this issue at the end of the sixteenth century was in Giambattista Benedetti's work (1530–1590), who in some respect initiated the following century's broader discussion of the subject. In any case, in Saccheri's time, there were several different ways in which the definition of the composition of ratios was justified (given that one did in fact want to justify it – taking it, as Saccheri would say, as a 'real' definition). One way (the one favored by Clavius) was to turn to the *denomination* of two ratios, i. e. to express ratios in terms of fractions (thus moving from A : B to *a*/*b*), and then multiply them. The problem here is that a ratio cannot always be expressed in terms of rational numbers – the problem of incommensurable magnitudes. Therefore, any attempt to justify the definition of the composition of ratios turns into the (significant) foundational difficulty of devising a structured theory for the product of real numbers. This approach may have been followed since Late Antiquity, as it is witnessed by a passage of Eutocius (in Heiberg's edition of Archimedes' works, vol. 3, p. 120), which refers to older Greek views. It was quite common in the Islamic World, and was shared, for instance, by Khayyām. This kind of researches, in fact, engendered the first rules on the multiplication of irrational numbers.

Another path (which was followed, for example, by Guidobaldo dal Monte) was to consider the product between magnitudes, and to define the compound ratio as A × C : B × D. But here, nothing guarantees that, in the general case, the product of magnitudes (which are not numbers) is well defined, for magnitudes A and C (and then B and D) may not be mutually homogeneous.

The third main way was to abandon all Euclidean scruples and consider all magnitudes as mutually homogeneous: a conservative stance of this solution was the one proposed by Hobbes, who equated all magnitudes with geometrical segments; a more modern point of view straightforwardly equated magnitudes with real numbers. This Algebraic and Analytic strategy (followed by Wallis, by the aforementioned Nonancourt and Arnauld, and also by Giordano) ultimately prevailed, though it came at the cost of removing (rather than solving) the main foundational difficulties.

The references are the following. G.B. BENEDETTI, *Diversarum speculationum mathematicarum & physicarum Liber*, Torino, Bevilacqua 1585; concerning Benedetti's work see G. BODRIGA, *Giovanni Battista Benedetti filosofo e matematico veneziano del secolo XVI*, "Atti del Regio Istituto Veneto di Scienze, Lettere e Arti", 85, 1925–1926, pp. 585–754; cf. also GIUSTI, *Euclides Reformatus*, pp. 22–34. For Clavius' solution, see the long Scholium in *Euclidis*, pp. 243–7, and then the note *De Proportionum compositione*, at pp. 392–4. On Arabic sources, see B. VITRAC, *'Umar al-Khayyām et Eutocius: les antécéndents grecs du troisième chapitre du commentaire Sur certaines prémisses problématiques du livre d'Euclide*, "Farhang. Quarterly Journal of Humanities and Cultural Studies", 12, 2000, pp. 51–105; B. VAHABZADEH, *'Umar al-Khayyām and the concept of irrational number*, in *De Zénon d'Élée à Poincaré: recueil d'études en hommage à Roshdi Rashed*, ed. R. Morelon, A. Hasnawi, Paris, Peeters 2004, pp. 55–63. For Guidobaldo's solution, see his booklet *De Proportione composita* published in GIUSTI, *Euclides Reformatus*, pp. 243–75, as well as the discussion offered by Giusti himself (pp. 18–22) and P.D. NAPOLITANI, *Sull'opuscolo "De Proportione Composita" di Guidobaldo dal Monte*, in *Atti del Convegno "La storia delle matematiche in Italia"*, ed. by O. Montalto and L. Grugnetti, Cagliari, Università di Cagliari 1982, pp. 431–9. For Hobbes, see *De corpore*, II, 13 (in T. HOBBES, *Opera philosophica quae latine scripsit omnia*, ed. W. Molesworth, London, Bohn 1839–1845, vol. 1, pp. 128–53). For Wallis' arithmetical solution (I here mention Wallis rather than some other author, because of Saccheri's familiarity with his work) see for instance *De Rationum Compositione* presented as Chapter 30 in his *Mathesis Universalis* (*Opera Mathematica*, vol. 1, pp. 154–7); but also the very *De postulato quinto*, in which he attempts to remove Euclid's (Savillian) second blemish. A good description of Hobbes' and Wallis' ideas on the topic is to be found in D.M. Jesseph, *Squaring the Circle. The War between Hobbes and Wallis*, Chicago, UCP 1999, pp. 142–59. On the acceptance of the concept of real number in Britain, see K. Neal, *From Discrete to Continuous: The Broadening of the Number Concepts in Early Modern England*, Dordrecht, Kluwer 2002. On positions similar to Wallis', see also Arnauld's *Nouveaux Elémens de géometrie* (p. 255) and Giordano's *Euclide restituto* (pp. 261–71). A rather broad discussion of this issue is found in the already mentioned Fine, *In sex priores Libros*, pp. 120–4.

Saccheri's solution to this definitional difficulty is not particularly original: he reaffirms Clavius' position (though without employing the term *denomination*), which is certainly the most traditional of positions, and applies it to commensurable magnitudes. He then states that, when dealing with incommensurable magnitudes, the Euclidean definition does not require any explanatory remark, as it is a nominal definition, not a real one.

The first pages of the Second Part of Book Two do not deal with *Elements* VI, def. 5, but rather address other Euclidean definitions in Book V of the *Elements*, thereby shedding some light on Saccheri's general understanding of the theory of proportions. Rather disparate material is brought together here, material which could probably not be appropriately positioned in the previous discussion on *Elements* V, def. 6, and was placed here, by Saccheri (or an editor), for fear that it would otherwise be lost.

In any case, Saccheri does not produce any original theory nor any relevant geometrical proof in this part of his work. The significance of this section of *Euclid Vindicated* seems to be historical and cultural, rather than genuinely mathematical.

[1] Saccheri literally quotes Clavius, *Euclidis*, p. 243. Note that Clavius translates the troublesome Greek term πηλικότης as *quantitas*. The issue of translation is relevant here, for it represents the immediate (and implicit) logical connection with what follows, where Saccheri discusses the Latin translation of this same Greek word in the context of its only other definitional occurrence in Euclid, namely, in *Elements* V, def. 3, and denies that the word can be translated as *quantitas*. Saccheri then proposes another translation – *quotitas*. Given his development of this notion in the rest of the volume (which takes into account the denomination of ratios), it seems that this second translation is also appropriate to *Elements* VI, def. 5. We might almost cautiously conjecture that this connection would have been explicitly articulated in *Euclid Vindicated* if only Saccheri had had enough time to compose this organically.

[2] Once again, Saccheri is referring to Borelli, who in the Preface to *Euclides restitutus* rejects the two Euclidean definitions, denouncing them as obscure (they are both abstract and non-operative). The objection raised by Saccheri in the upcoming sentences – that *Elements* V, def. 3 and 4 should be interpreted as first sketches of the true (and operative) definitions of *Elements* V, def. 5, 6 and 8 – was of course clear to Borelli and the rest of the Italian school of geometry. But the point here is that these geometers also (or: especially) regarded Euclid's subsequent definitions as incomprehensible, and hence dismissed these. Thus Galileo and others established, on the basis of *Elements* V, def. 3 and 4, a new theory

of proportions that did not rest upon considerations of equimultiples. Borelli, who rejected also *Elements* V, def. 3 and 4, tried to develop a new theory based on *Elements* VII, def. 20. So, actually, Saccheri does not respond here to Borelli's criticisms (which focus on the ambiguity of the term *similitudo* in *Elements* V, def. 4). Rather, he simply returns the discussion to what was stated and proved in the First Part of this Book of *Euclid Vindicated*, i. e. that subsequent operative definitions, which employ equimultiples, do indeed work very well. Concerning Borelli's criticisms of these Euclidean definitions, see the brilliant treatment by Giusti, *Euclides Reformatus*, pp. 115–21; criticisms that are similar to Borelli's were also put forward by many others, for instance, by Arnauld (*La logique*, pp. 311–2). Also note that Saccheri's formulation of *Elements* V, def. 4 (with a demonstrative adjective: "a similarity of *these* ratios") immediately links this definition with the previous one – this was neither obvious nor inconsequential, given that (as we mentioned in the *Notes* to the First Part of Book Two) the interpolated *Elements* V, def. 4 was placed differently in other editions of Euclid's work and may have therefore had a different meaning.

[3] I have translated the term *quotitas* as 'multiplicity'. Saccheri employs *quotitas*, as opposed to the usually (and Clavian) adopted *quantitas*, to translate the Greek term πηλικότης (cf. the translation of Elements VI, def. 5 above). Behind the issue of the translation, in fact, one can glimpse the discussion about the acceptability of irrational numbers as denominations of ratios, and thus the possibility of compounding ratios of incommensurable magnitudes through numbers.

Saccheri's reference to the *omnimode laudandus* John Wallis is wrong – perhaps he was here quoting by heart. In fact, in Chapter 25 (p. 137) of *Mathesis Universalis* (1657), Wallis proposes that the Greek πηλικότητα in *Elements* V, def. 3 be translated as *quotientem*. Although this word is in some ways related to *quotitas*, it nevertheless expresses a very different concept that tends towards an arithmetization of the theory of ratios – a 'modern' approach which Saccheri would certainly not have advocated. Moreover, a few pages ahead, Wallis translates the same Greek term in *Elements* VI, def. 5 simply as *quantitas*, adhering to the standard translation (*Mathesis Universalis*, Chapter 30, p. 154). Wallis also composed a long polemical text entitled *De Proportionibus Dialogi a Marco Meibomio conscripti Refutatio*, which was first published that same year (1657) and later included in the first volume of his *Opera mathematica*. In this work, Wallis passionately (and also somewhat ferociously) argues against the opinions expressed in *De Proportionibus Dialogus* (1655) by Danish humanist, philologist and music historian Marcus Meibom (1630–1710 ca.), who (Meibom, not Wallis) dared to translate πηλικότης as *quotitas*. Wallis' opinion on this point: "id admodum absurde factum videtur" (*Opera Mathematica*, vol. 1, p. 260). A couple of dense pages follow this observation, and many others are found scattered here and there in the following part of the book, all of which levy mathematical and philological criticism against that 'hypercritical' Danish translator (who *hallucinatur* elsewhere too for his 'most crass' and 'astonishing' ignorance of Greek), who committed the atrocity of translating πηλικότης as *quotitas*. In any case Saccheri probably (and confusedly) had Wallis' *De Postulato Quinto* in mind (*Opera Mathematica*, vol. 2, pp. 665–6); in this text the English mathematician also mentions *quotitas* (without referencing Meibom) when briefly returning to the issue of translating πηλικότης – though he in fact proposes here to translate the term as *quantuplicitas*. We may also mention Hobbes' *Six Lessons to the Savillian professors of mathematics* (written against Wallis in 1656), where he proposes to translate πηλικότης as *tantity* or *somuchness* (cf. T. Hobbes, *The English Works*, ed. W. Molesworth, London, Bohn & Longman 1839–1845, vol. 7, pp. 192–4).

The dispute between Wallis and Meibom had some resonance in England; see, for instance, Isaac Barrow's commentaries in favor of Wallis' translation in *Lectiones mathematicae* (*The Mathematical Works*, vol. 1, pp. 291–3). Regarding seventeenth-century debates on the composition of ratios and the disagreement between Wallis and Meibom: E. Sylla, *Compounding Ratios: Bradwardine, Oresme and the first edition of Newton's Principia*, in *Transformation and Tradition in the Sciences*, ed. by E. Mendelsohn, Cambridge, Cambridge University Press 1984, pp. 11–44. Concerning the translation *quotitas* in De Morgan and others, see Heath, *The Elements*, vol. 2, pp. 116–7 (although his account of Wallis is mistaken – just as Saccheri's).

We can also note that the issue of translating and interpreting this "relation in respect of quantity" (σχέσις κατὰ πηλικότητα) of *Elements* V, def. 3, was further complicated by the Aristotelian definition of "relation in respect of number" (πρός τι κατ᾽ ἀριθμόν, in *Metaph.* Δ 15), which some authors took to be (nearly) synonymous with Euclid's expression. Moreover, the issue of the translation of πηλικότης was raised already in the Medieval Islamic world, and Thābit's Arabic translation of Euclid's *Elements* employs a word meaning 'measure', while Thābit's own treatise on the composition of ratios uses another word that means 'magnitude'. On this terminologial fluctuation and its conceptual relevance for the acceptance of irrational numbers, see P. Crozet, *Thābit ibn Qurra et la composition des rapports*, "Arabic Science and Philosophy", 14, 2004, pp. 175–211.

[4] This is the standard seventeenth-century concept of homogeneity. It arose from the (more or less explicit) fusion of the metaphysical idea of homogeneity as that which is identical in kind, and the mathematical idea of homogeneity as that which satisfies Archimedes' principle. It was this combination of notions that presented so many difficulties for the development of Modern mechanics – and induced Galileo (and many others) to reform the theory of proportions. Although the theory of proportions and the theory of homogeneity were still widely discussed and debated in the eighteenth century, especially within texts of elementary geometry, it is rather obvious that Saccheri's Euclidean (and Aristotelian) classicism lagged far behind the developments of his time. The arithmetic and algebraic approach to the theory of proportions (we have already mentioned Wallis, but we could also include the Italian school, even though it had no proclivity for an algebraic treatment of the problem) overcame the metaphysical concept of homogeneity and allowed for a structural and abstract approach to magnitudes that was unconcerned with their nature (time or line, surface or number, etc.). The development of Calculus also yielded new perspectives on the mathematical characterization of homogeneity. For Leibniz, for instance, homogeneous magnitudes are locally homeomorphic (as we would say nowadays), not just Archimedean. In any case, from a strictly mathematical foundational perspective, the restriction to Archimedean magnitudes was useless; Hilbert devoted the whole of his Third Chapter of

Grundlagen der Geometrie to developing a new foundation of the Euclidean theory of ratios – one which did not require Archimedes' Axiom (cf. also Bernays' Second Supplement).

[5] Wolff's passage (§ 127 of *Elementa Arithmeticae*) is in *Elementa Matheseos universae*, vol. 1, p. 43. The comparison with trigonometric functions is a little hard to translate into contemporary terms: Saccheri literally says that the square of the sine of an angle and the square of the sine of the complementary angle (cosine) are equal (taken together, i.e. summed) to the square of the sine of their sum (the right angle). Wolff's definitions here are, as usual, more philosophical than mathematical; in the following part of his work, however, he complies with the modern understanding of ratios and proportions (which he considers always reducible to rational or irrational numbers) without effectively questioning the homogeneity of magnitudes. From this point of view, we can certainly regard him as more modern than Saccheri.

[6] Here, again, we encounter the translation problems mentioned in the above *Notes*. In *Elements* V, def. 3 and 5 of Euclid's Latin editions, interpreters sometimes translated μέγεθος as *magnitude* – as did Zamberti in his first translation, Commandino, Clavius and Saccheri. We also see it translated as *quantitas* – for instance in Campano, Pierre de la Ramée and Vitale Giordano (while the first group of authors mentioned above take *quantitas* to mean πηλικότης). Crapulli, *Mathesis Universalis*, shows that those objects in Book V of the *Elements* translated as *magnitudo* were often limited to three geometrical magnitudes (length, area, volume), angles and (maybe) also numbers, whereas the broader and more metaphysical translation of *quantitas* also allowed for time, velocity and any other measurable entity to be considered μεγέθη. In other words, this latter translation fostered the birth of *mathesis universalis* as a science of quantity in general. We should also note, however, that the distinction is not always so clear; Clavius, for example, translates the term as *magnitudo*, but then often employs the term *quantitas* in his Scholia – as if the two terms were synonyms. Saccheri's standpoint, as we have noted above, is uncertain: he always draws on examples of geometrical quantities, and sometimes seems to refer only to these, as is the case here (where he also explicitly speaks of 'extension'); at other times, however, he refers to abstract magnitudes (p. 130) or explicitly mentions time and velocity (p. 139), though he is perhaps unable to deal with these mathematically.

The idea that quantity, taken in the metaphysical sense, should refer first and foremost to the impenetrability of bodies, was already stated by Saccheri in the Physical Thesis 25 after *Logica demonstrativa*: "Effectus formalis quantitates, est reddere suum subjectum naturaliter impenetrabile cum altero subject similiter quanto" (ed. 1697 only, p. 276).

[7] The original reads: "seu per numeros integros, seu per fractos, sive etiam per minutiam". Although there was no fixed terminology, by *numeri fracti* Saccheri seems to mean what we today call rational numbers; and by *minutiae*, a fraction of the unity (Wallis says '*unitatis fragmenta*'). These different names for fractions and fractions of the unity, which to us may seem odd, derived from the fact that rational numbers were often expressed (as we mentioned in *Note* 34 to the First Part of Book Two) as the sum of the greatest whole number plus the *minutia*: 2½, 17¾, etcetera. This was the standard practice in common calculation, and we find *minutiae*, for instance, in all treatises of the Medieval *abacus* tradition. In *Ele-*

ments IX, Clavius, provides a long *excursus* in the arithmetic operations on fractions titled *Minutiarum sive numerorum fractorum demonstrationes* (*Euclidis*, pp. 381–92), where he provides examples and rules that apply only to the case of *minutiae* of value less than 1, as he assumes that the reader is already acquainted with the arithmetic of whole numbers and that to perform calculations on rational numbers generally one can always proceed first by dealing on whole numbers, and then on small fractions by means of the rules he provides. The modern decimal expression for rational numbers (2,5 or 17,75), however, was common in the seventeenth century (it may have been initiated by Stevin), and Saccheri's reference to *minutiae* seems here quite old-fashioned.

[8] In § 126 of *Elementa Arithmeticae* (*Elementa Matheseos universae*, vol. 1, p. 43). Wolff later explains (§ 129) that, for instance, according to his definition, a fraction is a ratio because a/b is seen as a relation between two (not three) numbers. Further on, however, he holds that every ratio is reducible to a fraction (rational or irrational). So this example basically becomes the only aim of his definition – which is what Saccheri cannot accept. It is also notable that Wolff states that such a definition of ratio, which should improve upon Euclid's flawed definition, was first put forward by Leibniz. He was probably thinking of Leibniz's definition of similarity (which is often stated by Leibniz, and probably first appeared in a 1677 letter to Gallois; cf. GM I, p. 180; A III, 2, n. 79, pp. 227–8; A II, 1, n. 158, p. 380), which in fact consisted in comparing two objects without introducing a third homogeneous; Wolff himself adopted and discussed at length this Leibnizian definition (cf. § 24 of *Elementa Arithmeticae* and the Preface to *Elementa Geometriae*, in *Elementa Matheseos universae*, vol. 1, pp. 25 and 120). Nonetheless, such a definition does not seem immediately applicable to the definition of ratio, nor does Leibniz's 'phenomenological' definition seem to effectively have the meaning that Wolff and Saccheri ascribe to it here. Finally, we can note that the Wolffian definition, deprived of its reference to the third homogeneous, is the first step towards a formulation of an abstract definition of continuous function. This could very well have had Leibnizian origins – though Saccheri does not concern himself with this issue.

[9] The reference to the question of quantity, which brings us back to the previous paragraph, may allude to Wallis' criticism of Meibom, even though it was actually pointing in the opposite direction. Wallis held that (*Opera Mathematica*, vol. 1, pp. 261 e 267–9) the question about πηλικότης may be phrased as *quanta?* or *quam multa?*, while *quot magnitudo, quaerit nemo*, and hence the translation '*quotitas*' is certainly flawed (furthermore, *quot* should refer to ordinal, not cardinal, numbers).

[10] Saccheri's proof is just an explanation of the Euclidean definition in the case of commensurable magnitudes. It consists of multiplying the fractions corresponding to the given ratios (which were generally called denominations of these ratios). This was, after all, the solution proposed by Clavius, who first translated πηλικότης as '*quantitas*', and then identified the latter with the *denominatio* of a ratio.

[11] This is the hard case, because these denominations cannot be expressed in terms of rational numbers; and the difficulty in dealing with the composition of ratios, in fact, properly lies in the multiplication of irrational numbers. Instead of embarking on a foundational discussion concerning operations on real numbers, Saccheri prefers to interpret the Euclid-

ean definition as a stipulation of what it means to compound ratios – without bringing the latter back to the definition of product of numbers. Saccheri is thus not properly removing the alleged blemish, but rather getting used to its appearance.

[12] As previously mentioned, *Elements* VI, 23 is not just any proposition, but rather the only proposition in which Euclid refers to the composition of ratios, and the very proposition that induced later editors to introduce *Elements* VI, def. 5. The proof development of the following paragraph is not essential because it is similar to the one provided by Euclid and Clavius (*Euclidis*, p. 273).

[13] There is a mistake here, for the letters of the two couples are interchanged. It should read: side DC is to side CE as segment I is to segment K, and side BC is to side CG as segment K is to segment L. Perhaps Saccheri got confused when working from Clavius' edition, whose diagram has a different lettering than that in *Euclid Vindicated*; Clavius' notation is coherent with Saccheri's formulation. In any case, in the following paragraph Saccheri's notation is correct.

From a theoretical perspective it is important to note that Saccheri does not hesitate in assuming the existence of segments K and L, which are only defined as fourth proportionals of three given or constructed magnitudes (K as the fourth proportional of DC, CE and any whatever I; and L as the fourth proportional of BG, CG and K). Throughout his proof, Clavius explicitly emphasizes that he employs the Axiom of existence of the fourth proportional. But actually there is no real need for him to do so, for, as we have already seen, in *Elements* VI, 12, Euclid proves the existence of the fourth proportional in the restricted case of segments, which is the only case that is relevant here. Saccheri already acknowledged this point in pp. 112–4 of *Euclid Vindicated*. Still, the fact that he makes no reference to this here is of interest if we are to support the view that Book Two was not written organically.

Note, in any case, that proof of *Elements* VI, 23 (which is entirely correct) is sufficient to vindicate Euclid's blemish: it effectively shows that *Elements* VI, def. 5 can be assumed as a nominal definition, because in *Elements* VI, 23 Euclid only proves what is already contained in the definition (in other words, Saccheri's reasoning shows that the interpolated definition is an exact copy of the proof of this proposition). But since the existence of the fourth proportional is an *unavoidable* requirement for the Euclidean procedure of the composition of ratios (see the introductory *Note* to this section), and since Saccheri rejects its general form, then in *Euclid Vindicated* this concept of composition of ratios is in fact justified only for the case of *any* rational magnitude (with the concept of denomination, in the previous proof) and for *linear* (geometrical) irrational magnitudes. It is thus not universally justified.

[14] The Latin term *ex aequo* (as it appears here in Saccheri), *ex aequalitate* (as it appears in Clavius), or also *ex aequali*, expresses the Greek term δι'ἴσου; modern translations often leave it as such (even though Vitrac dares a '*égalité de rang*' in French, and Acerbi a '*tramite uguale*' in Italian). This expression appears in *Elements* V, def. 17, and its use is justified in *Elements* V, 22. Concerning textual problems relating to the Euclidean definition see Vitrac, *Les Éléments*, vol. 2, pp. 52–6. At any rate the meaning of the theorem is that given $2n$ magnitudes:

$$A_1, A_2, A_3, \ldots A_{n-1}, A_n,$$

$$B_1, B_2, B_3, \ldots B_{n-1}, B_n$$

such that the following proportions hold good:

$$A_1 : A_2 = B_1 : B_2,$$

$$A_2 : A_3 = B_2 : B_3,$$

$$\ldots$$

$$A_{n-1} : A_n = B_{n-1} : B_n,$$

then the following proportion between the first and last terms must also hold good:

$$A_1 : A_n = B_1 : B_n.$$

In proving *Elements* VI, 23, Saccheri (like Euclid) employs *Elements* V, 22 taking six magnitudes into account: Area(ABCD), Area(BHEC), Area(CEFG), I, K, L. Saccheri proves:

$$\text{Area(ABCD)} : \text{Area(BHCE)} = \text{I} : \text{K},$$

$$\text{Area(BHCE)} : \text{Area(CEFG)} = \text{K} : \text{L},$$

and thus concludes *ex aequo*:

$$\text{Area(ABCD)} : \text{Area(CEFG)} = \text{I} : \text{L}.$$

[15] The notions of double and triple ratio are, as we have seen, the only kind of composition of ratios explicitly (and authentically) defined by Euclid in *Elements* V, def. 9 and 10. They correspond to the ratio of a magnitude and its second or third power.

[16] Because it only appears in *Elements* VI, 23. If we interpret this as a *nominal* definition, then it is but one part of the theorem's proof process. This useless aspect of *Elements* VI, def. 5 was the subject of widespread criticism at the time. The strongest opposition, which Saccheri could well have had in mind, was in Galileo's *Discorsi* (*Opere*, vol. 8, p. 359).

[17] This is where the whole difficulty of Saccheri's treatise on the composition of ratios lies. If one compounds ratios only among (rational) numbers or linear magnitudes, as was the case in the *Elements*, then this notion of composition is rather useless (as we have seen) – and in fact, in Euclid, this definition was most certainly interpolated. Such a notion, which was rather peripheral in the Greek treatise, was subjected to so much scrutiny in the seventeenth century that Henry Savile identified it as one of Euclid's blemishes. The revival of interest in this particular point was due to the development of mechanics, which imposed (if approached by means of proportions) that ratios be compounded of areas, volumes, times, velocities, and so on. But in order to do this, the Axiom of existence of the fourth proportional in its most general form was needed. Thus, if Galileo wanted to compound the velocities A and B with the distances C and D, he would certainly be able to assume the existence of that velocity X, fourth proportional in $C : D = B : X$, allowing him to state that

the compounded ratio was A:X. But since Saccheri rejects this axiom (and perhaps wants to replace it with a principle of total order of ratios), it is hard to understand how he could have believed (in this conclusive part of *Euclid Vindicated*) that different types of ratios can in any way be reduced to homogeneity. Saccheri, who is utterly Euclidean, thus seems to have justified (in some way, though it is not clear that he needed to do so) the Ancient theory of ratios and proportions of geometrical (or at least linear) magnitudes – but he also lost sight of the very reason for which this systematization was initially required and then further developed in the seventeenth century. Saccheri was thus facing (though one and a half centuries of scientific development later) the same difficulty that impeded Guidobaldo dal Monte and his simple (markedly pre-Galilean) multiplication theory of the composition of ratios (see above, the introductory *Note* to this Part of *Euclid Vindicated*).

Notes to the Appendix of Book Two

This appendix is the only part of *Euclid Vindicated* that functions to link the two Books, i. e. the theory of parallels and the theory of proportions. Its briefness and location seem to indicate that it is a *post festum* addition that attempts to lend homogeneity to a book aimed at addressing composite demands. It may very well be the transcription of one of Saccheri's lectures, or perhaps it was purposely, and hastily, composed on the occasion of the volume's publication (since he in fact returns to the last theorem discussed in Book Two).

The Appendix's theoretical content does not present any particular novelty: it only states that the theory of proportions as applied to geometry (i. e. the theory of similarity) is fundamentally dependent on the parallel axiom. Wallis already obtained this result in his proof of the Fifth Postulate, and Saccheri has discussed it in Scholium 3 to Proposition 21.

It is noteworthy that Saccheri's polemic is aimed at 'analysis', i. e. at that purely symbolic algebraic reasoning that would not be sufficient to provide a foundational grounding for Euclidean geometry, as it is indifferent to the validity of the Fifth Postulate. What we nowadays regard as a very important quality of ordinary algebra (i. e. its generality and non-categoricity), that enables it to be profitably employed in the study of non-Euclidean geometries (and was done by Lambert and then later Gauss, Lobachevsky and Bolyai), was for Saccheri (who believed himself to have proven the Fifth Postulate) a defect of the formalism. He certainly was not the only person at that time to cultivate such conviction: Leibniz, for instance, who dreamt of a logicist foundation for geometry in which all axioms were ultimately to be proved, lamented the underdetermined nature (not: the generality) of ordinary algebra as compared with geometry. But whereas Leibniz saw the need to reform Cartesian algebra and focused on establishing a geometrical formalism expressing the properties of space (a *characteristica geometrica propria*), Saccheri responds to the new symbolic methods (which were by then already a century old) in a classicist fashion, and hopes for a revival of synthetic Euclidean geometry (inclusive of arithmetic). Concerning the relation between modern infinitesimal (not just algebraic) analysis and the Fifth Postulate, see also the first *Note* to Proposition 37.

It is also worth noting that Saccheri does not base his discussion of *Elements* VI, 23 (presented in this Appendix) on Wallis' demonstration of the Fifth Postulate (which he doesn't even mention), but rather on the distinction between a hyperbolic straight line and a hypercycle established in the Second Part of Book One (Saccheri in fact ignores both Wallis' results and the advanced theorems that he himself obtained in the First Part). This is important firstly because, for the first and only time in *Euclid Vindicated*, Saccheri actually hints at an algebraic approach to hypercycles and non-Euclidean geometry. Secondly, it is important to observe that this algebraic approach to geometry (as here advanced by Saccheri) naturally takes hypercycles, not straight lines, as its objects: because it is based upon quantitative considerations such as equidistance. Thus the problem for ordinary geometry seems to be the impossibility of discriminating between a straight line and a curve, in other words, it does not provide any quantitative determination for the 'qualitative' property of being a straight line, i. e. a geodesic – and this is in fact one of the greatest difficulties faced by Riemannian geometry in the following century.

[1] The word 'analysis' took on a vast number of meanings in the Modern Period; here Saccheri seems to be referring to general algebra, i. e. as symbolic writing. There are no references to infinitesimal analysis as a specific mathematical domain – nor, of course, to algebra as a subject exclusively devoted to non-transcendent quantities; an appropriate synonym might be *characteristics*. In any case, see *Logica demonstrativa*: "duplicitèr comparari potest Scientia de aliquo obiecto, vel resolvendo, vel componendo. Resolvit Algebra, quae proptereà vocant antonomasticè Analytica: componit Geometria" (ed. 1701, p. 114 [ed. 1697, p. 180]).

[2] Saccheri does not give any further clue as to what these higher dimensions could be. Already in the seventeenth century, it was quite common to employ the concept of dimension in a generalized way in the case of magnitudes that were not strictly geometrical, and thus sometimes more than three dimensions were taken into consideration. We do not, however, know exactly what Saccheri thought of these theoretical constructions. In any case, the point of this whole argument is that the theory of proportions is independent of the Fifth Postulate (as is shown by *Elements* V) if we deal with numerical fields or linear (or abstract) magnitudes, whereas it bears upon the Fifth Postulate (as is shown in *Elements* VI) if we are dealing with proper geometry, i. e. when more dimensions are taken into account. This statement could allude to something more profound (which Saccheri was unable to grasp): that curvature-related questions are relevant only to the Cartesian product of two or more linear domains – there is no one-dimensional Riemannian geometry.

[3] While the phrasing may seem awkward, the point of the Euclidean theorem is simply that $ac:bc=a:b$.

[4] "*Hoc opus, hic labor*", Verg. *Aen.* VI, 129.

[illegible]

Bibliography

The New Science and Jesuit Science: Seventeenth Century Perspectives, ed. by M. Feingold, Dordrecht, Kluwer 2003.

Jesuit Science and the Republic of Letters, ed. by M. Feingold, Cambridge, MIT 2003.

F. Acerbi, *Drowning by Multiples. Remarks on the Fifth Book of Euclid's Elements, with Special Emphasis on Prop. 8*, "Archive for History of Exact Sciences", 57, 2003, pp. 175–242.

F. Acerbi, *Euclide. Tutte le opere*, Milano, Bompiani 2007.

A. Agostini, *Due lettere inedite di Girolamo Saccheri*, "Memorie della Reale Accademia d'Italia. Classe di Scienze Fisiche, Matematiche e Naturali", 2, 1931, pp. 3–20.

L. Allegri, *The Mathematical Works of Girolamo Saccheri, S.J. (1667–1733)*, Ph.D. diss., Columbia University, 1960.

A.R. Amir Moéz, *Discussion of difficulties in Euclid by Omar Ibn Abrahim Al-Khayyam*, "Scripta mathematica", 24, 1959, pp. 275–303.

I. Angelelli, *Saccheri's Postulate*, "Vivarium", 33, 1995, pp. 98–111.

A. Arnauld, *Nouveaux Elémens de géometrie*, Paris, Savreux 1667[1], 1683[2]; riedizione in *Géométries de Port-Royal*, ed. by D. Descotes, Paris, Champion 2009.

A. Arnauld, P. Nicole, *La logique ou l'art de penser*, Paris, Desprez 1662[1], 1683[2]; modern edition ed. by P. Clair, F. Girbal, Paris, PUF 1965.

U. Baldini, *Giovanni Alfonso Borelli e la rivoluzione scientifica*, "Physis", 16, 1974, pp. 97–128.

U. Baldini, *Legem impone subactis. Studi su filosofia e scienza dei gesuiti in Italia, 1540–1632*, Roma, Bulzoni 1992.

U. Baldini, P.D. Napolitani, *Christoph Clavius. Corrispondenza*, Pisa, Department of Mathematics (preprint) 1992.

F. Barozzi, *Proclii Diadochi Lycii in primum Euclidis*, Padova, Percacino 1560.

I. Barrow, *Euclidis Elementorum Libri XV breviter demonstrati*, Cambridge, Nealand 1655.

I. Barrow, *Lectiones Mathematicae XXIII in quibus Principia Matheseos generalia exponuntur*, London, Playford 1683; riedizione in I. Barrow, *The Mathematical Works*, ed. by W. Whewell, Cambridge, Cambridge University Press 1860; reprint Hildesheim, Olms 1973.

O. Becker, *Eudoxos-Studien II. Warum haben die Griechen die Existenz der vierten Proportionale angenommen?*, "Quellen und Studien zur Geschichte der Mathematik, Astronomie und Physik", 2, 1932–1933, pp. 369–87.

G. Saccheri, *Euclid Vindicated from Every Blemish*, Classic Texts in the Sciences 1,
DOI 10.1007/978-3-319-05966-2,

F. Beckmann, *Neue Gesichtpunkte zum 5. Buch Euklids*, "Archive for History of Exact Sciences", 3, 1967, pp. 1–144.

F. Bellissima, P. Pagli, *Consequentia Mirabilis. Una regola logica fra matematica e filosofia*, Firenze, Olschki 1996.

E. Beltrami, *Un precursore italiano di Legendre e di Lobatschewsky*, "Rendiconti dell'Accademia dei Lincei", 5, 1889, pp. 441–48.

G.B. Benedetti, *Diversarum speculationum mathematicarum & physicarum Liber*, Torino, Bevilacqua 1585.

G. Bodriga, *Giovanni Battista Benedetti filosofo e matematico veneziano del seaolo XVI*, "Atti del Regio Istituto Veneto di Scienze, Lettere e Arti", 85, 1925–1926, pp. 585–754.

J. Bolyai, *Appendix Scientiam spatii absolute veram exhibens*, in F. Bolyai, *Tentamen juventutem studiosam in elementa matheseos purae*, Maros Vásárhelyini, Kali 1832–1833.

R. Bonola, *Sulla teoria delle parallele e sulle geometrie non-euclidee*, in *Questioni riguardanti le matematiche elementari*, ed. by F. Enriques, Bologna, Zanichelli 1924–1927 (1900[1]), Parte Prima, vol. 2, pp. 309–427.

R. Bonola, *Un teorema di Giordano Vitale da Bitonto sulle rette equidistanti*, "Bollettino di Bibliografia e Storia delle Scienze Matematiche", 8, 1905, pp. 33–36.

R. Bonola, *I teoremi del Padre Girolamo Saccheri sulla somma degli angoli di un triangolo e le ricerche di M. Dehn*, "Rendiconti dell'Istituto Lombardo", 38, 1905, pp. 651–62.

R. Bonola, *La geometria non-euclidea. Esposizione storico-critica del suo sviluppo*, Bologna, Zanichelli 1906. *Non-Euclidean Geometry*, transl. by H.S. Carslaw, Chicago, Open Court 1912.

G.A. Borelli, *Euclides Restitutus, sive priscae Geometriae Elementa brevius et facilius contexta, in quibus praecipue proportionum theoriae nova firmiorique methodo proponantur*, Pisa, Francesco Onofri 1658[1]; Roma, Mascardi 1679[3]. Italian translation as *Euclide Rinnovato*, transl. by D. Magni and revised by Borelli, Bologna, Giovan Battista Ferroni 1663.

G.A. Borelli, *Apollonii Pargei Conicorum lib. V VI VII*, Firenze, Cocchini 1661.

M.T. Borgato, *Una presentazione di opere inedite di Vitale Giordani (1633–1711)*, in *Giornate di storia della matematica*, ed. by M. Galuzzi, Cetraro, EditEl 1991, pp. 1–20.

M.T. Borgato, *Scritti inediti di Vitale Giordani*, in *Giornate di storia della matematica*, ed. by M. Galuzzi, Cetraro, EditEl 1991, pp. 21–56.

M.T. Borgato, L. Pepe, *Una memoria inedita di Lagrange sulla teoria delle parallele*, "Bollettino di Storia delle Scienze Matematiche", 8, 1988, pp. 307–335.

H. Bos, *Redefining geometrical exactness: Descartes' transformation of the early modern concept of construction*, New York, Springer 2001.

H. Bosmans, *Le géomètre Jérome Saccheri, S.J. (1667–1733)*, "Revue des Questions Scientifiques", 7, 1925, pp. 401–30.

E. Breitenberger, *Gauss's Geodesy and the Axiom of Parallels*, "Archive for History of Exact Sciences", 31, 1984, pp. 273–89.

A. Brigaglia, P. Nastasi, *Le soluzioni di Girolamo Saccheri e Giovanni Ceva al "Geometram Quaero" di Ruggero Ventimiglia: Geometria proiettiva italiana nel tardo seicento*, "Archive for History of Exact Sciences", 30, 1984, pp. 7–44.

L. Brusotti, *Gli "Elementa" di Carlo Edoardo Filippa allievo di Girolamo Saccheri*, "Atti dell'Accademia Ligure di Scienze e Lettere", 9, 1952, pp. 155–164.

F. Cajori, *George Bruce Halsted*, "The American Mathematical Monthly", 29, 1922, pp. 338–340.

J.G. Camerer, *Euclidis Elementorum libri sex priores*, Berlin, Reimer 1824.

G. Campano, *Preclarissimus liber elementorum Euclidis perspicacissimi*, Venezia, Ratdolt 1482.

F.F. Candale, *Euclidis Megarensis Mathematici clarissimi Elementa Geometrica, libris XV. ad germanam Geometriae intelligentiam è diversis lapsibus temporis iniuria contractis restituta*, Paris, Royer 1566.

P.P. Caravaggio, *Primi sei libri d'Euclide tratti in volgare*, Milano, Lodovico Monza 1671.

G. Cardano, *Opus novum de proportionibus numerorum*, Basel, Heinrich Petri 1570.

E. Caruso, *Honoré Fabri, gesuita e scienziato*, in *Miscellanea Secentesca*, Dipartimento di Filosofia dell'Università di Milano, Cisalpino 1987, pp. 85–126.

U. Cassina, *Sulla dimostrazione di Wallis del Postulato Quinto di Euclide*, in *Actes du 8e Congrès International d'Histoire des Sciences* (1956), Paris, Hermann&Cie 1958, pp. 33–38.

J. Cassinet (ed.), "Cahiers d'histoire des mathématiques de Toulouse", 9, 1986.

P.A. Cataldi, *Operetta delle linee rette equidistanti et non equidistanti*, Bologna, Rossi 1603.

G. Ceva, *De Lineis rectis se invicem secantibus statica constructio*, Milano, Lodovico Monza 1678.

A.C. Clairaut, *Éléments de géométrie*, Paris, Lambert&Durand 1741.

C. Clavius, *Euclidis Elementorum Libri XV*, Roma, Vincenzo Accolto 1574; several other edition, the last one in C. Clavius, *Opera mathematica*, Mainz, Reinhard Eltz (and others) 1611–1612; reprint ed. by E. Knobloch, Hildesheim, Olms 1999.

F. Commandino, *Euclidis Elementorum libri XV, unà cum Scolijs antiquis*, Pesaro, Camillo Franceschini 1572.

G. Cosentino, *L'insegnamento delle matematiche nei collegi gesuitici nell'Italia settentrionale*, "Physis", 13, 1971, pp. 205–217.

G. Crapulli, *Mathesis Universalis.Genesi di un'idea nel XVI secolo*, Roma, Edizioni dell'Ateneo 1969.

P. Crozet, *Thābit ibn Qurra et la composition des rapports*, "Arabic Science and Philosophy", 14, 2004 pp. 175–211.

J. D'Alembert, *Mélanges de literature, d'histoire et de philosophie*, vol. 5, Amsterdam, Chatelain 1767.

E. De Angelis, *Il metodo geometrico nella filosofia del Seicento*, Firenze, Le Monnier 1964.

C.F.M. Dechales, *Euclidis Elementorum Libri VIII*, Lyon, Coral 1660.

C.F.M. Dechales, *Cursus seu Mundus Mathematicus*, Lyon, Posuel&Rigaud 1690 (1674[1]).

M. Dehn, *Die Legendre'schen Sätze über die Winkelsumme im Dreieck*, "Mathematische Annalen", 53, 1900, pp. 405–39.

M. Dehn, *Beziehungen zwischen der Philosophie und der Grundlegung der Mathematik in Altertum*, "Quellen und Studien zur Geschichte der Mathematik, Astronomie und Physik", B4, 1938, pp. 1–28.

V. De Risi, *Geometry and Monadology. Leibniz's Analysis Situs and Philosophy of Space*, Basel, Birkhäuser 2007.

M.E. Di Stefano, M. Frasca Spada, *Logica, metodo, geometria : Saccheri, Borelli e l'equidistante da una retta*, "Epistemologia", 8, 1985, pp. 33–76.

A. Djebbar, *L'épître d'al-Khayyâm sur "l'explication des prémisses problématiques du livre d'Euclide"*, "Farhang. Quarterly Journal of Humanities and Cultural Studies", 14, 2002, pp. 83–136.

A.M. Dou, *Logical and Historical Remarks on Saccheri's Geometry*, "Notre Dame Journal of Formal Logic", 11, 1970, pp. 385–415.

I.E. Drabkin, *Aristotle's wheel: notes on the History of a Paradox*, "Osiris", 9, 1950, pp. 162–98.

P. Duhem, *Les origines de la statique*, Paris, Hermann 1906.

A. Emch, *The Logica Demonstrativa of Girolamo Saccheri*, Ph.D. diss., Harvard, 1934.

A. Emch, *The Logica Demonstrativa of Girolamo Saccheri*, "Scripta Mathematica", 3, 1935, pp. 51–60, 143–52, 221–33.

F. Engel, P. Stäckel, *Die Theorie der Parallellinien von Euklid bis auf Gauss*, Leipzig, Teubner 1895.

F. Enriques, *Gli Elementi di Euclide e la critica antica e moderna*, Roma, Stock 1923.

L. Euler, *Opera omnia*, 1907–.

H. Fabri, *Synopsis Geometrica*, Lyon, Molin 1669.

A. Favaro, *Due lettere inedite del P. Gerolamo Saccheri d.C.d.G. a Vincenzio Viviani*, "Rivista di Fisica, Matematica e Scienze Naturali", 4, 1903, pp. 426–30.

C.E. Filippa, *Euclidis priora elementa sex, auctore Carolo Eduardo Taurinense, reverendo patri Hieronymo Saccherio Societatis Iesu dictata*, Torino, Zappata 1695.

O. Fine, *In sex priores libros Geometricorum Elementorum Euclidis Megarensis demonstrationes*, Paris, Simon de Colines 1536.

O. Fine, *De Mundi Sphaera sive Cosmographia*, Paris, Simon de Colines 1542.

D.H. Fowler, *An invitation to read Book X of Euclid's Elements*, "Historia Mathematica", 19, 1992, pp. 233–64.

F.M. Franceschinis, *Teoria delle parallele rigorosamente dimostrata*, Bassano, Remondini 1787.

G. Galilei, *Opere*, Firenze, Barbera 1968 (1890–1907[1]).

V. Giordano, *Euclide Restituto, ovvero gli antichi elementi geometrici ristaurati e facilitati*, Roma, Angelo Bernabò 1686 (1680[1]).

A. Girard, *Invention nouvelle en l'algebre*, Amsterdam, Blaeuw 1629.

S. Giuntini, *Gabriele Manfredi - Guido Grandi: Carteggio (1701–1732)*, "Bollettino di storia delle scienze matematiche", 13, 1993, pp. 5–144.

E. Giusti, *Euclides Reformatus. La teoria delle proporzioni nella scuola galileiana*, Torino, Bollati Boringhieri 1993.

J.A. Goldstein, *A Matter of Great Magnitude: The Conflict over Arithmetization in 16th–, 17th–, and 18th–Century English Editions of Euclid's Elements Books I through VI (1561–1795)*, "Historia Mathematica", 27, 2000, pp. 36–53.

J.V. Grabiner, *Why Did Lagrange "Prove" the Parallel Postulate?*, "The American Mathematical Monthly", 116, 2009, p. 3–18.

G. Grandi, *Elementi geometrici piani e solidi di Euclide posti brevemente in volgare*, Firenze, Tartini 1731.

G.B. Grandi, *Thomas Reid's geometry of visibles and the parallel postulate*, "Studies in History and Philosophy of Science", 36, 2005, pp. 79–103.

J. Gray, *Non-Euclidean Geometry - A Re-interpretation*, "Historia mathematica" 6, 1979, pp. 236–58.

J. Gray, *Ideas of Space: Euclidean, Non-Euclidean, and Relativistic*, Oxford, Clarendon 1989[2].

J. Gray, *János Bolyai, Non-Euclidean Geometry and the Nature of Space*, Cambridge, Burndy 2004.

M.J. Greenberg, *On J. Bolyai's Parallel Construction*, "Journal of Geometry", 12, 1979, pp. 45–64.

M.J. Greenberg, *Aristotle's Axiom in the Foundations of Geometry*, "Journal of Geometry", 33, 1988, pp. 53–57.

M.J. Greenberg, *Euclidean and Non-Euclidean Geometries. Development and History*, New York, Freeman 2008[4].

M.J. Greenberg, *Old and New Results in the Foundations of Elementary Plane Euclidean and Non-Euclidean Geometries*, "The American Mathematical Monthly", 117, 2010, pp. 198–219.

S. Grynaeus, ... *In Euclidis Geometriae elementa Graeca. Adiecta praefatiuncula in qua de disciplinis mathematicis nonnihil*, Basel, Herwagen 1533.

G. Guarini, *Euclides Adauctus et Methodicus*, Torino, Zappata 1671.

L. Guerrini, *Matematica ed erudizione. Giovanni Alfonso Borelli e l'edizione fiorentina dei libri V, VI e VII delle Coniche di Apollonio di Perga*, "Nuncius", 14, 1999, pp. 505–568.

M. Hallet, U. Majer, *David Hilbert's Lectures on the Foundations of Geometry 1891–1902*, Berlin, Springer 2004.

R. Hartshorne, *Geometry: Euclid and Beyond*, New York, Springer 2000.

T. Hayashi *Introducing Movement into Geometry: Roberval's influence on Leibniz's Analysis Situs*, "Historia Scientiarum", 8, 1998, pp. 53–69.

T. Heath, *The Thirteen Books of the Elements*, Cambridge, Cambridge University Press 1908; reprint New York, Dover 1965.

J.C. Heilbronner, *Historia Matheseos universae*, Leipzig, Gleditsch 1742.

D. Hilbert, *Grundlagen der Geometrie*, Stuttgart, Teubner 1968 (1899[1]).

C.F. Hindenburg, *Ueber die Schwürigkeit bey der Lehre von den Parallellinien*, "Leipziger Magazin zur Naturkunde, Mathematik und Oekonomie", 1781, pp. 145–68.

C.F. Hindenburg, *Anmerkungen über das neue System der Parallellinien*, "Leipziger Magazin zur Naturkunde, Mathematik und Oekonomie", 1781, pp. 342–71.

J. Hjelmslev, *Neue Begrundung der ebenen Geometrie*, "Mathematische Annalen", 64, 1907, pp. 449–74.

T. Hobbes, *Opera philosophica quae latine scripsit omnia*, ed. W. Molesworth, London, Bohn 1839–1845.

T. Hobbes, *The English Works*, ed. W. Molesworth, London, Bohn & Longman 1839–1845.

J.J.I. Hoffmann, *Critik der Parallel-Theorie*, Jena, Erdker 1807.

C.F.A. Hoormann, *A further Examination of Saccheri's Use of the "Consequentia Mirabilis"*, "Notre Dame Journal of Formal Logic", 17, 1976, pp. 239–47.

G. de l'Hospital, *Traité analytique des sections coniques*, Paris, Montalant 1720.

K. Jaouiche, *De la fécondité mathématique: d'Omar Khayyam à G. Saccheri*, "Diogène", 57, 1967, pp. 97–113.

K. Jaouiche, *La théorie des parallèles en pays d'Islam*, Paris, Vrin 1986.

D.M. Jesseph, *Squaring the Circle. The War between Hobbes and Wallis*, Chicago, UCP 1999.

V. Jullien, *Les étendues géométriques et la ligne droite de Roberval*, "Revue d'histoire de Sciences", 46, 1993, pp. 493–521.

V. Jullien, *Eléments de géométrie de G.P. de Roberval*, Paris, Vrin 1996.

G.A. Kästner, *Anfangsgründe der Arithmetik, Geometrie, ebenen und sphärischen, Trigonometrie und Perspectiv*, Göttingen, Vanderhoeck 1758.

G.A. Kästner, *Ueber den mathematischen Begriff des Raums*, "Philosophisches Magazin", II 4, 1790, pp. 403–19.

G.S. Klügel, *Conatuum praecipuorum theoriam parallelarum demonstrandi recensio*, Göttingen, Schultz 1763.

W. Kneale, *Aristotle and the Consequentia Mirabilis*, "The Journal of Hellenic Studies", 77, 1957, pp. 62–6.

E. Knobloch, *Sur la vie et l'oeuvre de Christophore Clavius (1538–1612)*, "Revue d'Histoire des Sciences", 41, 1988, pp. 331–356.

E. Knobloch, *La connaissance des mathématiques arabes par Clavius*, "Arabic Science and Philosophy", 12, 2002, pp. 257–84.

J.H. Lambert, *Anlage zur Architectonic*, Riga, Hartknoch 1771.

J.H. Lambert, *Theorie der Parallellinien*, "Magazin für reine und angewandte Mathematik", 1786, pp. 13–64, 325–58.

G. Lechalas, *Une définition géométrique du plan et de la ligne droite apres Leibniz et Lobatchewsky*, "Revue de metaphysique et de morale", 20, 1912, pp. 718–21.

A.-M. Legendre, *Éléments de géometrie*, Paris, Didot 1794; numerose edizioni successive.

A.-M. Legendre, *Réflexions sur différentes manières de démontrer la théorie des parallèles ou le théorème sur la somme des trois angles du triangle*, "Mémoires de l'Académie Royale des Sciences de l'Institut de France", 12, 1833, pp. 367–410.

G.W. Leibniz, *Mathematische Schriften*, ed. by C.J. Gerhardt, Berlin/Halle 1849–1863. [GM]

G.W. Leibniz, *Die philosophischen Schriften*, ed. by C.J. Gerhardt, Berlin 1875–1890. [GP]

G.W. Leibniz, *Sämliche Schriften und Briefe*, Darmstadt/Leipzig/Berlin 1923–. [A]

G.W. Leibniz, *La caractéristique géométrique*, ed. by J. Echeverría, M. Parmentier, Paris, Vrin 1995.

T. Lévy, *Gersonide, le Pseudo-Ṭūsī, et le postulat des parallèles. Les mathématiques en Hébreu et leurs sources arabes*, "Arabic Science and Philosophy", 2, 1992, pp. 39–82.

N.I. Lobachevsky, *Pangeometry*, ed. A. Papadopoulos, Zürich, EMS 2010.

G. Lolli, *Saccheri e le definizioni "filiae plurium demonstrationum"*, in G. Lolli, *Le ragioni fisiche e le dimostrazioni matematiche*, Bologna, Il Mulino 1985.

E. Mach, *Erkenntnis und Irrtum. Skizzen zur Psychologie der Forschung*, Leipzig, Barth 1905.

L. Maierù, *Il quinto postulato euclideo in Cristoforo Clavio*, "Physis", 20, 1978, pp. 191–212.

L. Maierù, *L'influsso del Narbonense sui commentatori euclidei del Seicento italiano circa il problema delle parallele*, in *Atti del Convegno "La Storia delle matematiche in Italia"*, Cagliari, Università di Cagliari 1982, pp. 341–49.

L. Maierù, *Il Quinto Postulato Euclideo da C. Clavio [1589] a G. Saccheri [1733]*, "Archive for History of Exact Sciences", 27, 1982, pp. 297–334.

L. Maierù, *Il "meraviglioso problema" in Oronce Finé, Girolamo Cardano e Jacques Peletier*, "Bollettino di Storia delle Scienze Matematiche", 4, 1984, pp. 141–70.

L. Maierù, *John Wallis. Una vita per un progetto,* Catanzaro, Rubbettino 2007.

N. Malebranche, *Oeuvres complètes*, Paris, Vrin 1958–1978[1].

A. Malet, *Changing notions of proportionality in pre-modern mathematics*, "Asclepio", 1, 1990, pp. 183–211.

A. Malet, *Renaissance notions of number and magnitude*, "Historia Mathematica", 33, 2006, pp. 63–81.

A. Malet, *Euclid's Swan Song: Euclid's Elements in Early Modern Europe*, in *Greek Science in the Long Run*, ed. P. Olmos, Newcastle, Cambridge Publishing 2012, pp. 205-34.

N. de Malézieu, *Elemens de geometrie de Monseigneur le Duc de Bourgogne*, Paris, Boudot 1705.

P. Mancosu, *On the Status of Proofs by Contradiction in the Seventeenth Century*, "Synthese", 88, 1991, pp. 15–41.

K. Manders, *The Euclidean Diagram*, in *The Philosophy of Mathemtical Practice*, ed. by P. Mancosu, Oxford, Oxford University Press 2008, pp. 80–133.

K. Manders, *Diagram-Based Geometric Practice*, in *The Philosophy of Mathemtical Practice*, ed. by P. Mancosu, Oxford, Oxford University Press 2008, pp. 65–79.

M.P. Mansion, *Analyse des recherches du R.P. Saccheri, S.J. sur le postulatum d'Euclide*, "Annales de la société scientifique de Bruxelles", 14.2, 1889–1890, pp. 46–59.

A. Marchetti, *La natura della proporzione e della proporzionalità con nuovo, facile, e sicuro metodo*, Pistoia, Gatti 1695.

A. Marchetti, *Euclides Reformatus*, Livorno, Celsi 1709.

M. Meibom, *De Proportionibus Dialogus*, Copenhagen, Martzan 1655.

N. Mercator, *Euclidis Elementa Geometrica novo Ordine ac Methodo fere demonstrata*, London, Martyn 1678.

A.I. Miller,*The Myth of Gauss' Experiment on the Euclidean Nature of Physical Space*, "Isis", 63, 1972, pp. 345–8.

A.G. Molland, *Campanus and Eudoxus; or, Trouble with Texts and Quantifiers*, "Physis", 25, 1983, pp. 213–25.

E. Montucla, *L'Histoire des Mathématiques*, Paris, Jombert 1758.

I. Mueller, *Philosophy of Mathematics and Deductive Structure in Euclid's Elements*, Cambridge, MIT Press 1981.

I. Mueller, *Sur les principes des mathématiques chez Aristote et Euclide*, in *Mathématique et Philosophie de l'antiquité à l'âge classique*, ed. by R. Rashed, Paris, CNRS 1991, pp. 101–13.

I. Mueller, *On the Notion of a Mathematical Starting Point in Plato, Aristotle, and Euclid*, in *Science and Philosophy in Classical Greece*, ed. A.C. Bowen, London, Garland 1991, pp. 59–97.

I. Mueller, *Remarks on Euclid's Elements I, 32 and the Parallel Postulate*, "Science in Context", 16, 2003, pp. 287–97.

J.E. Murdoch, *The Medieval Language of Proportions: Elements of the Interaction with Greek Foundations and the Development of New Mathematical Techniques*, in *Scientific Change*, ed. by A.C. Crombie, London, Heinemann 1963.

P.D. Napolitani, *Sull'opuscolo "De Proportione Composita" di Guidobaldo dal Monte*, in *Atti del Convegno "La storia delle matematiche in Italia"*, ed. by O. Montalto, L. Grugnetti, Cagliari, Università di Cagliari 1982, pp. 431–39.

Nasīr ad-Dīn at-Tūsī: *Euclidis elementorum geometricorum libri tridecim ex tratittione doctissimi Nasiridini Tusini*, Roma, Typographia Medicea 1594.

K. Neal, *From Discrete to Continuous: The Broadening of the Number Concepts in Early Modern England*, Dordrecht, Kluwer 2002.

F. de Nonancourt, *Euclides Logisticus sive de ratione euclidea*, Louvain, Bouvet 1652; riedizione in *Géométries de Port-Royal*, ed. by D. Descotes, Paris, Champion 2009.

G. Nuchelmans, *A 17th-Century Debate on the Consequentia Mirabilis*, "History and Philosophy of Logic", 13, 1992, pp. 43–58.

P. Pagli, *Two unnoticed Editions of Saccheri's Logica Demonstrativa*, "History and Philosophy of Logic", 30, 2009, pp. 331–40.

G.M. Pagnini, *Theoria rectarum parallelarum ab omni scrupolo vindicata*, Parma, Rossi&Ubaldi 1783.

G.M. Pagnini, *Epistola ad praestantissimum virum Hieronymum canonicum Saladinum*, Parma, Rossi&Ubaldi 1794.

F. Palladino, *Sulla teoria delle proporzioni nel Seicento. Due "macchinazioni" notevoli: Le sezioni dei razionali del galileiano G.A. Borelli; Le classi di misura del gesuita A. Tacquet*, "Nuncius", 6, 1991, pp. 38–81.

P. Palmieri, *The Obscurity of the Equimultiples. Clavius' and Galileo's Foundational Studies of Euclid's Theory of Proportions*, "Archive for History of Exact Sciences", 55, 2001, pp. 555–97.

V. Pambuccian, *Zum Stufenaufbau des Parallelenaxioms*, "Journal of Geometry", 51, 1994, pp. 79–88.

V. Pambuccian, *Axiomatizations of Hyperbolic and Absolute Geometries*, in *Non-Euclidean Geometries*, ed. by A. Prékopa, E. Molnár, New York, Springer 2006, pp. 119–53.

V. Pambuccian, *Lambert or Saccheri quadrilaterals as single primitive notions for plane hyperbolic geometry*, "Journal of Mathematical Analysis and Applications", 346, 2008, pp. 531–2.

V. Pambuccian, *On the Equivalence of Lagrange's Axiom to the Lotschnittaxiom*, "Journal of Geometry", 95, 2009, pp. 165–71.

A. Pascal, *Girolamo Saccheri nella vita e nelle opere*, "Giornale di Matematiche di Battaglini", 52, 1914, pp. 229–51.

M. Pasch, *Vorlesungen über die neuere Geometrie*, Berlin, Springer 1926 (Leipzig, Teubner 1882[1]).

W. Pejas, *Die Modelle des Hilbertschen Axiomensystems der absoluten Geometrie*, "Mathematische Annalen", 143, 1961, pp. 212–35.

J. Peletier, *Demonstrationum in Euclidis elementa geometrica libri sex*, Lyon, Tornes 1557.

O. Perron, *Nichteuklidische Elementargeometrie der Ebene*, Stuttgart, Teubner 1962.

W.S. Peters, *Das Parallelenproblem bei A. G. Kästner*, "Archive for History of Exact Sciences", 1, 1962, pp. 480–87.

C.F. Pfleiderer, *Deduction der Euclidischen Definitionen 3,4,5,7 des fünftes Buchs der Elemente*, "Archiv für reine und angewandte Mathematik", 2, 1797–1798, pp. 257–87, 440–47.

E.B. Plooij, *Euclid's Conception of Ratio and his Definition of Proportional Magnitudes as Criticised by Arabian Commentators*, Rotterdam, Hengel 1950.

F. Podetti, *La teoria delle proporzioni in un testo del XVII secolo*, "Bollettino di biliografia e storia delle scienze matematiche", 15, 1913, pp. 1–8.

J.-C. Pont, *L'aventure des parallèles. Historie de la géométrie non euclidienne: précurseurs et attardés*, Berne, Lang 1986.

Proclus Diadochus, *In Primum Euclidis*, ed. by G. Friedlein, Leipzig, Teubner 1873.

D. Rabouin, *Mathesis Universalis. L'idée de " mathématique universelle " d'Aristote à Descartes*, Paris, PUF 2009.

R. Rashed, B. Vahabzadeh, *Al-Khayyâm mathématicien,* Paris, Blanchard 1999.

R. Rashed, C. Houzel, *Thābit ibn Qurra et la théorie des parallèles*, "Arabic Science and Philosophy", 15, 2005, pp. 9–55.

J.G. Ratcliffe, *Foundations of Hyperbolic Manifolds*, New York, Springer 2006[2].

T. Reid, *Philosophical Works. With notes and supplementary dissertations by sir William Hamilton*, Edinburgh, Maclachlan&Stewart 1895[8].

P. Riccardi, *Saggio di una Bibliografia Euclidea*, Hildesheim, Olms 1974 (1887–1892[1]).

B. Riemann, *Über die Hypothesen, welche der Geometrie zu Grunde liegen*, ed. by Jürgen Jost, Berlin, Springer 2013.

A. Robinet, *Leibniz et les mathématiciens italiens*, "Symposia Mathematica", 27, 1986, pp. 101–21.

A. Romano, *La contre-réforme mathématique. Constitution et diffusion d'une culture mathématique jésuite à la Renaissance (1540–1640)*, Roma, École Française 1999.

S. Rommevaux, *Clavius, une clé pour Euclide au XVIe siècle*, Paris, Vrin 2005.

B.A. Rosenfeld, *A History on Non-Euclidean Geometry. Evolution of the Concept of a Geometric Space*, transl. from Russian by A. Shenitzer, New York, Springer 1988.

A.I. Sabra, *Thābit Ibn Qurra on Euclid's Parallels Postulate*, "Journal of the Warburg and Courtauld Institutes", 31, 1968, pp. 12–32.

A.I. Sabra, *Simplicius's Proof of Euclid's Parallels Postulate*, "Journal of the Warburg and Courtauld Institutes", 32, 1969, pp. 1–24.

G. Saccheri, *Quaesita Geometrica a Comite Rugerio de Vigintimillis omnibus proposita*, Milano, Malatesta 1693.

G. Saccheri, *Logica Demonstrativa*, Torino, Paolino 1696–1697; several other editions, the last one posthumous in 1735; reprint ed. by W. Risse, Hildesheim, Olms 1980; Italian edition ed. by M. Mugnai, M. Girondino, Pisa, Edizioni della Normale 2012.

G. Saccheri, *Neo-Statica*, Milano, Malatesta 1708.

G. Saccheri, *Confermazione teologica*, senza indicazione dell'editore, 1729.

G. Saccheri, *L'Euclide emendato del p. Gerolamo Saccheri*, ed. by G. Boccardini, Milano, Hoepli 1904.

G. Saccheri, *Euclides Vindicatus*, ed. by G.B. Halsted, Chicago, Open Court 1920.

G. Saccheri, *Euclide liberato da ogni macchia*, ed. by P. Frigerio, with an introduction by I. Tóth and E. Cattanei, Milano, Bompiani 2001.

G. Saccheri, *Euclide vendicato da ogni neo*, ed. by V. De Risi, Pisa, Edizioni della Normale 2011.

G. Saccheri, *Logica dimostrativa*, ed. by M. Mugnai and M. Girondino, Pisa, Edizioni della Normale 2012.

G. Saladini, *Nuovo trattato delle parallele*, Bologna, S. Tommaso 1805.

C. Sasaki, *The Acceptance of the Theory of Proportion in the Sixteenth and Seventeenth Centuries. Barrow's Reaction to the Analytic Mathematics*, "Historia Scientiarum", 29, 1985, pp. 83–116.

H. Savile, *Praelectiones tresdecim in principium elementorum Euclidis*, Oxford, Lichfield&Short 1621.

F. Schur, *Über die Grundlagen der Geometrie*, "Mathematische Annalen", 55, 1902, pp. 265–92.

C. Segre, *Congetture intorno all'influenza di Girolamo Saccheri sulla formazione della geometria non euclidea*, "Atti della Regia Accademia di Scienze di Torino", 38, 1902–1903, pp. 351–63.

A. Seidenberg, *Did Euclid's Elements, Book I, Develop Geometry Axiomatically?*, "Archive for History of Exact Sciences", 10, 1975, pp. 263–295.

L. Simonutti, *Guido Grandi, scienziato e polemista, e la sua controversia con Tommaso Ceva*, "Annali della Scuola Normale Superiore di Pisa. Classe di Lettere", III 19, 1989, pp. 1001–26.

R. Simson, *Euclidis Elementorum libri priores sex item undecimus et duodecimus ex versione latina Federici Commandini*, Glasgow, Foulis 1756. English version as *The Elements of Euclid, viz. the first Six Books together with the Eleventh and Twelfth. In this Edition, the Errors, by which Theon, or others, have long ago Vitiated these Books, are Corrected, and some of Euclid's Demonstrations are Restored*, Glasgow, Foulis 1756.

C.E. Sjostedt, *Le axiome de paralleles de Euclide a Hilbert: un probleme cardinal in le evolution del geometrie*, Uppsala, Interlingue–Fundation 1968.

D.E. Smith, *Euclid, Omar Khayyâm, and Saccheri*, "Scripta Mathematica", 3, 1935, pp. 5–10.

D.C. Smolarski, S.J., *The Jesuit Ratio Studiorum, Christopher Clavius, and the Study of Mathematical Sciences in Universities*, "Science in Context", 15, 2002, pp. 447–64.

D.M.Y. Sommerville, *Bibliography of Non–Euclidean Geometry including the Theory of Parallels, the Foundations of Geometry, and Space of n Dimensions*, London, Harrison 1911.

S. Stevin, *Oeuvres mathematiques*, ed. A. Girard, Leyden 1634.

E. Sylla, *Compounding Ratios: Bradwardine, Oresme and the first edition of Newton's Principia*, in *Transformation and Tradition in the Sciences*, ed. by E. Mendelsohn, Cambridge, Cambridge University Press 1984, pp. 11–44.

A. Tacquet, *Elementa Geometriae planae ac solidae*, Antwerp, Jacob van Meurs 1654.

A. Tacquet, *Opera Mathematica*, Antwerp, Jacob van Meurs 1669.

J. Tanács, *Grasping the Conceptual Difference between János Bolyai and Lobachevskii's Notions of Non-Euclidean Parallelism*, "Archive for History of Exact Sciences", 63, 2009, pp. 537–52.

N. Tartaglia, *Euclide Megarense philosopho, solo introduttore delle scientie mathematice*, Venezia, Rossinelli 1543.

L. Tenca, *Relazioni fra Gerolamo Saccheri e il suo allievo Guido Grandi*, "Studia Ghisleriana", 1, 1952, pp. 19–45.

J.-M. de Tilly, *Etudes de méchanique abstraite*, "Mémoires de l'Académie royale de Belgique", 21, 1870, pp. 1–98.

M.M. Toeppell, *Über die Entstehung von David Hilberts „Grundlagen der Geometrie"*, Göttingen, Vanderhoeck 1986.

E. Torricelli, *Opere*, ed. by G. Loria, G. Vassura, Faenza, Montanari 1919–1944.

M. Torrini, *Dopo Galileo. Una polemica scientifica (1684–1711)*, Firenze, Olschki 1979.

I. Tóth, *Das Parallelenproblem im Corpus Aristotelicum*, "Archive for History of Exact Sciences", 3, 1967, pp. 249–422.

I. Tóth, *Aristotele e i fondamenti assiomatici della geometria. Prolegomeni alla comprensione dei frammenti non-euclidei nel "Corpus Aristotelicum" nel loro contesto matematico e filosofico*, Milano, Vita e Pensiero 1997.

I. Tóth, *Fragmente und Spuren nicheuklidischer Geometrie bei Aristoteles*, Berlin, DeGruyter 2010.

P.M. Tummers, *The Latin Translation of Anaritius' Commentary on Euclid's Elements of Geometry, Books I–IV*, Nijmegen, Ingenium 1994.

B. Vahabzadeh, *Al-Khayyām's conception of ratio and proportionality*, "Arabic Science and Philosophy", 7, 1997, pp. 247–63.

B. Vahabzadeh, *Al-Māhānī's commentary on the concept of ratio*, "Arabic Science and Philosophy", 12, 2002, pp. 9–52.

B. Vahabzadeh, *'Umar al-Khayyām and the concept of irrational number*, in *De Zénon d'Élée à Poincaré: recueil d'études en hommage à Roshdi Rashed*, ed. R. Morelon, A. Hasnawi, Paris, Peeters 2004, pp. 55–63.

G. Vailati, *Di un'opera dimenticata del P. Girolamo Saccheri ("Logica Demonstrativa" 1697)*, "Rivista Filosofica", 6, 1903.

G. Vailati, *Sur une classe remarquable de raisonnements par réduction à l'absurde*, "Révue de Métaphysique et de Morale", 12, 1904, pp. 799–809.

G. Vailati, *Sulla teoria delle proporzioni*, in *Questioni riguardanti le matematiche elementari*, ed. by F. Enriques, Bologna, Zanichelli 1924–1927 (1900[1]), Parte Prima, vol. 1, pp. 143–91.

P. Valéry, *Oeuvres*, ed. by J. Hytier, Paris, Gallimard 1957.

C. Vasoli, *Fondamento e metodo logico della geometria nell'Euclides Restitutus del Borelli*, "Physis", 11, 1969, pp. 571–98.

G. Venturi, *Memoria intorno alle linee parallele*, Modena, Società Tipografica 1784.

G. Veronese, *Fondamenti di Geometria*, Padova, Seminario 1891.

B. Vitrac, *Euclide. Les Éléments*, Paris, PUF 1990–2001.

B. Vitrac, *'Umar al-Khayyām et Eutocius: les antécéndents grecs du troisième chapitre du commentaire Sur certaines prémisses problématiques du livre d'Euclide*, "Farhang. Quarterly Journal of Humanities and Cultural Studies", 12, 2000, pp. 51–105.

B. Vitrac, *Umar al Khayyam et l'anthyphérèse : Étude du deuxième Livre de son commentaire "Sur certaines prémisses problématiques du Livre d'Euclide"*, "Farhang. Quarterly Journal of Humanities and Cultural Studies", 14, 2002, pp. 137–92.

B. Vitrac, *Les classifications des sciences mathématiques en Grèce ancienne*, "Archives de philosophie", 68, 2005, pp. 269–301.

V. Viviani, *Quinto Libro degli Elementi di Euclide, ovvero Scienza Universale delle Proporzioni spiegata colla dottrina del Galileo*, Firenze, Condotta 1674.

V. Viviani, *Elementi piani e solidi d'Euclide*, Firenze, Carlieri 1690.

J.-D. Voelke, *Renaissance de la géométrie non euclidienne entre 1860 et 1900*, Bern, Lang 2005.

K. Volkert, *Das Undenkbare denken. Die Rezeption der nichteuklidischen Geometrie im deutschsprachigen Raum (1860–1900)*, Berlin, Springer 2013.

K. von Fritz, *Die* APXAI *in der griechischen Mathematik*, "Archiv für Begriffsgeschichte", 1, 1955, pp. 13–103.

K. von Fritz, *Gleichheit, Kongruenz und Ähnlichkeit in der antiken Mathematik bis auf Euklid*, "Archiv für Begriffsgeschichte", 4, 1959, pp. 7–81.

J. Wallis, *Opera mathematica*, Oxford, Theatrum Sheldonianum 1693–1699; reprint ed. by C.J. Scriba, Hildesheim, Olms 1972.

C. Wolff, *Elementa Matheseos universae*, Halle, 1713–1741; reprint Hildesheim, Olms 1968.

K. Zormbala, *Gauss and the Definition of the Plane Concept in Euclidean Elementary Geometry*, "Historia Mathematica", 23, 1996, pp. 418–36.

Appendix 1

Propositions of Euclid's *Elements* used by Saccheri in *Euclid Vindicated*

(Latin text by Clavius; English text by T. Heath)

I, 4

Si duo triangula duo latera duobus lateribus aequalia habeant, utrumque utrique; habeant vero & angulum angulo aequalem sub aequalis rectis lineis contentum: Et basim basi aequaliem habebunt: eritque triangulum triangulo aequale; ac reliqui anguli reliquis angulis aequales erunt, uterque utrique, sub quibus aequalia latera subtenduntur.

If two triangles have the two sides equal to two sides respectively, and have the angles contained by the equal straight lines equal, they will also have the base equal to the base, the triangle will be equal to the triangle, and the remaining angles will be equal to the remaining angles respectively, namely those which the equal sides subtend.

I, 5

Isoscelium triangulorum, qui ad basim sunt, anguli inter se sunt aequales: Et productis aequalibus rectis lineis, qui sub basi sunt, anguli inter se aequales erunt.

In isosceles triangles the angles at the base are equal to one another; and, if the equal straight lines be produced further, the angles under the base will be equal to one another.

I, 8

Si duo triangula duo latera habuerint duobus lateribus, utrumque utrique, aequalia, habuerint vero & basim basi aequalem: Angulum quoque sub aequalibus rectis lineis contentum angulo aequalem habebunt.

If two triangles have the two sides equal to two sides respectively, and have also the base equal to the base, they will have the angles equal which are contained by the equal straight lines.

I, 11

Data recta linea, à puncto in ea dato, rectam lineam ad angulos rectos excitare.

To draw a straight line at right angles to a given straight line from a given point on it.

I, 12

Super datam rectam lineam infinitam, à dato puncto, quod in ea non est, perpendicularem rectam deducere.

To a given infinite straight line, from a given point which is not on it, to draw a perpendicular straight line.

I, 13

Cum recta linea super rectam consistens lineam angulos facit, Aut duos rectos, aut duobus rectis aequales efficiet.

If a straight line set up an a straight line make angles, it will make either two right angles or angles equal to two right angles.

I, 14

Si ad aliquam rectam lineam, atque ad eius punctum, duae rectae lineae non ad easdem partes ductae eos, qui sunt deinceps, angulos duobus rectis aequales fecerint; in directum erunt inter se ipsae rectae lineae.

If with any straight line, and at a point on it, two straight lines not lying on the same side make the adjacent angles equal to two right angles, the two straight lines will be in a straight line with one another.

I, 15

Si duo rectae lineae se mutuo secuerint, angulos ad verticem aequales inter se efficient.

If two straight lines cut one another, they make the vertical angles equal to one another.

I, 16

Cuiuscunque trianguli uno latere producto, externus angulus utrolibet interno, & opposito, maior est.

In any triangle, if one of the sides be produced, the exterior angle is greater than either of the interior and opposite angles.

I, 17

Cuiuscunque trianguli duo anguli duobus rectis sunt minores, omnifariam sumpti.

In any triangle two angles taken together in any manner are less than two right angles.

I, 18

Omnis trianguli maius latus maiorem angulum subtendit.

In any triangle the greater side subtends the greater angle.

I, 19

Omnis trianguli maior angulus maiori lateri subtenditur.

In any triangle the greater angle is subtended by the greater side.

I, 20

Omnis trianguli duo latera reliquo sunt maiora, quomodocunque assumpta.

In any triangle two sides taken together in any manner are greater than the remaining one.

I, 21

Si super trianguli uno latere, ab extremitatibus duae rectae interius constitutae fuerint, hae constitutae reliquis trianguli duobus lateribus minores quidem erunt, maiorem vero angulum continebunt.

If on one of the sides of a triangle, from its extremities, there be constructed two straight lines meeting within the triangle, the straight lines so constructed will be less than the remaining two sides of the triangle, but will contain a greater angle.

I, 23

Ad datam rectam lineam, datumque in ea punctum, dato angulo rectilineo aequalem angulum rectilineum constituere.

On a given straight line and at a point on it to construct a rectilineal angle equal to a given rectilineal angle.

I, 24

Si duo triangula duo latera duobus lateribus aequalia habuerint, utrumque utrique, angulum vero angulo maiorem sub aequalibus rectis lineis contentum: Et basin basi maiorem habebunt.

If two triangles have the two sides equal to two sides respectively, but have the one of the angles contained by the equal straight lines greater than the other, they will also have the base greater than the base.

I, 25

Si duo triangula duo latera duobus lateribus aequalia habuerint, utrumque utrique, basin vero basi maiorem: Et angulo sub aequalibus rectis lineis contentum angulo maiorem habebunt.

If two triangles have the two sides equal to two sides respectively, but have the base greater than the base, they will also have the one of the angles contained by the equal straight lines greater than the other.

I, 26

Si duo triangula duos angulos duobus angulis aequales habuerint, utrumque utrique, unumque latus uni lateri aequale, sive quod aequalibus adiacet angulis, seu quod uni aequalium angulorum subtenditur: & reliqua latera reliquis lateribus aequalia, utrumque utrique, & reliquum angulum reliquo angulo aequalem habebunt.

If two triangles have the two angles equal to two angles respectively, and one side equal to one side, namely, either the side adjoining the equal angles, or that subtending one of the equal angles, they will also have the remaining sides equal to the remaining sides and the remaining angle to the remaining angle.

I, 27

Si in duas rectas lineas recta incidens linea alternatim angulos aequales inter se fecerit: parallelae erunt inter se illae rectae lineae.

If a straight line falling on two straight lines make the alternate angles equal to one another, the straight lines will be parallel to one another.

I, 28

Si in duas rectas lineas recta incidens linea externum angulum interno, & opposito, & ad easdem partes, aequalem fecerit; Aut internos, & ad easdem partes duobus rectis aequales: Parallelae erunt inter se ipsae rectae lineae.

If a straight line falling on two straight lines make the exterior angle equal to the interior and opposite angle on the same side, or the interior angles on the same side equal to two right angles, the straight lines will be parallel to one another.

I, 29

In parallelas rectas lineas recta incidens linea; Et alternatim angulos inter se aequales efficit; & externum interno, & opposito, & ad easdem partes aequalem; & internos, & ad easdem partes, duobus rectis aequales facit.

A straight line falling on parallel straight lines makes the alternate angles equal to one another, the exterior angle equal to the interior and opposite angle, and the interior angles on the same side equal to two right angles.

I, 31

A Dato puncto, datae rectae lineae parallelam rectam lineam ducere.

Through a given point to draw a straight line parallel to a given straight line.

I, 32

Cuiuscunque trianguli uno latere producto: Externus angulus duobus internis, & oppositis est aequalis. Et trianguli tres interni anguli duobus sunt rectis aequales.

In any triangle, if one of the sides be produced, the exterior angle is equal to the two interior and opposite angles, and the three interior angles of the triangle are equal to two right angles.

I, 37

Triangula super eadem basi constituta, & in eisdem parallelis, inter se sunt aequalia.

Triangles which are on the same base and in the same parallels are equal to one another.

I, 38

Triangula super aequalibus basibus constituta, & in eisdem parallelis, inter se sunt aequalia.

Triangles which are on equal bases and in the same parallels are equal to one another.

I, 46

A data recta linea quadratum describere.

On a given straight line to describe a square.

III, 7

Si in diametro circuli quodpiam sumatur punctum, quod circuli centrum non sit, ab eoque puncto in circulum quaedam rectae lineae cadant: Maxima quidem erit ea, in qua centrum, minima vero reliqua; aliarum vero propinquior illi quae per centrum ducitur, remotiore semper maior est: Duae autem solum rectae lineae aequales ab eodem puncto in circulum cadunt, ad utrasque partes minimae, vel maximae.

If on the diameter of a circle a point be taken which is not the centre of the circle, and from the point straight lines fall upon the circle, that will be greatest on which the centre is, the remainder of the same diameter will be least, and of the rest the nearer to the straight line through the centre is always greater than the more remote, and only two equal straight lines will fall from the point on the circle, one on each side of the least straight line.

III, 19

Si circulum tetigerit recta quaepiam linea, à contactu autem recta linea ad angulos rectos ipsi tangenti excitetur: In excitata erit centrum circuli.

If a straight line touch a circle and from the point of contact a straight line be drawn at right angles to the tangent, the centre of the circle will be on the straight line so drawn.

V, 7

Aequales ad eandem, eandem habent rationem; Et eadem ad aequales.

Equal magnitudes have to the same the same ratio, as also has the same to equal magnitudes.

[if $A = B$ then $A:C = B:C$]

V, 8

Inaequalium magnitudinum maior ad eandem, maiorem rationem habet, quam minor: Et eadem ad minorem, maiorem rationem habet, quam ad maiorem.

Of unequal magnitudes, the greater has to the same a greater ratio than the less has; and the same has to the less a greater ratio than it has to the greater.

[if $A > B$ then $A:C > B:C$]

V, 11

Quae eisdem sunt eaedem rationes, & inter se sunt eaedem.

Ratios which are the same with the same ratio are also the same with one another.

V, 13

Si prima ad secundam eandem habuerit rationem, quam tertia ad quartam; tertia vero ad quartam maiorem rationem habuerit, quam quinta ad sextam: Prima quoque ad secundam maiorem rationem habebit, quam quinta ad sextam.

If a first magnitude have to a second the same ratio as a third to a fourth, and the third have to the fourth a greater ratio than a fifth has to a sixth, the first will also have to the second a greater ratio than the fifth to the sixth.

[if $A:B=C:D$ and $C:D>E:F$ then $A:B>E:F$]

V, 15

Partes cum pariter multiplicibus in eadem sunt ratione, si prout sibi mutuo respondent, ita sumantur.

Parts have the same ratio as the same multiples of them taken in corresponding order.

V, 17

Si compositae magnitudines proportionales fuerint, hae quoque divisae proportionales erunt.

If magnitudes be proportional *componendo*, they will also be proportional *separando*.

[if $A:B=C:D$ then $(A-B):B=(C-D):D$]

V, 18

Si divisae magnitudines sint proportionales, hae quoque compositae proportionales erunt.

If magnitudes be proportional *separando*, they will also be proportional *componendo*.

[if $A:B=C:D$ then $(A+B):B=(C+D):D$]

V, 22

Si sint quotcunque magnitudines, & aliae ipsis aequales numero, quae binae in eadem ratione sumantur: Et ex aequalitate in eadem ratione erunt.

If there be any number of magnitudes whatever, and others equal to them in multitude, which taken two and two together are in the same ratio, they will also be in the same ratio *ex aequali*.

V, 26

Si prima ad secundam habuerit maiorem proportionem, quam tertia ad quartam: habebit convertendo secunda ad primam minorem proportionem, quam quarta ad tertiam.

[If a first magnitude has to a second a greater ratio than a third to a fourth, then by conversion the second will have to the first a lesser ratio than the fourth to the third.]

[if $A:B>C:D$ then $B:A<D:C$]

VI, 1

Triangula & parallelogramma, quorum eadem fuerit altitudo, ita se habent inter se, ut bases.

Triangles and parallelograms which are under the same height, are to one another as their bases.

VI, 2

Si ad unum trianguli latus parallela ducta fuerit recta quaedam linea, haec proportionaliter secabit ipsius trianguli latera. Et si trianguli latera proportionaliter secta fuerint, quae ad sectiones adiuncta fuerit recta linea, erit ad reliquam ipsius trianguli latus parallela.

If a straight line be drawn parallel to one of the sides of a triangle, it will cut the sides of the triangle proportionally; and if the sides of the triangle be cut proportionally, the line joining the points of section will be parallel to the remaining side of the triangle.

VI, 12

Tribus datis rectis lineis, quartam proportionalem invenire.

To three given straight lines to find a fourth proportional.

VI, 19

Similia triangula inter se sunt in duplicata rationem laterum homologorum.

Similar triangles are to one another in the duplicate ratio of the corresponding sides.

VI, 20

Similia polygona in similia triangula dividuntur, & numero aequalia, & homologa totis: Et polygona duplicatam habent eam inter se rationem, quam latus homologum ad homologum latus.

Similar polygons are divided into similar triangles, and into triangles equal in multitude and in the same ratio as the whole, and the polygon has to the polygon a ratio duplicate of that which the corresponding side has to the corresponding side.

VI, 23

Aequiangula parallelogramma inter se rationem habent eam, quae ex lateribus componitur.

Equiangular parallelograms have to one another the ratio compounded of the ratios of their sides.

VII, 18

Si duo numeri numerum quempiam multiplicantes secerint aliquos: Geniti ex ipsis eandem rationem habebunt, quam multiplicantes.

If two numbers by multiplying any number make certain numbers, the numbers so produced will have the same ratio as the multipliers.

[that is, $ac:bc=a:b$]

VII, 19

Si quatuor numeri proportionales fuerint, qui ex primo, & quarto fit, numerus, aequalis erit ei, qui ex secundo, & tertio fit, numero. Etsi, qui ex primo, & quarto, fit, numerus, aequalis fuerit ei, qui ex secundo, & tertio fit, numero; ipsi quatuor numeri proportionales erunt.

If four numbers be proportional, the number produced from the first and fourth will be equal to the number produced from the second and third; and, if the number produced from, the first and fourth be equal to that produced from the second and third, the four numbers will be proportional.

XI, 1

Rectae lineae pars quaedam non est in subiecto plano, quaedam vero in sublimi.

A part of a straight line cannot be in the plane of reference and a part in a plane more elevated.

XI, 33

Similia solida parallelepipeda, inter se sunt in triplicata ratione homologorum laterum.

Similar parallelepipedal solids are to one another in the triplicate ratio of their corresponding sides.

XII, 1

Quae in circulis polygona similia: inter se sunt, ut à diametris quadrata.

Similar polygons inscribed in circles are to one another as the squares on the diameters.

XII, 2

Circuli inter se sunt, quemadmodum à diametris quadrata.

Circles are to one another as the squares on the diameters.

Appendix 2

Original diagrams from *Euclides Vindicatus* (1733).

Reproduced with kind permission from Niedersächsische Staats- und Universitätsbibliothek Göttingen, Germany.

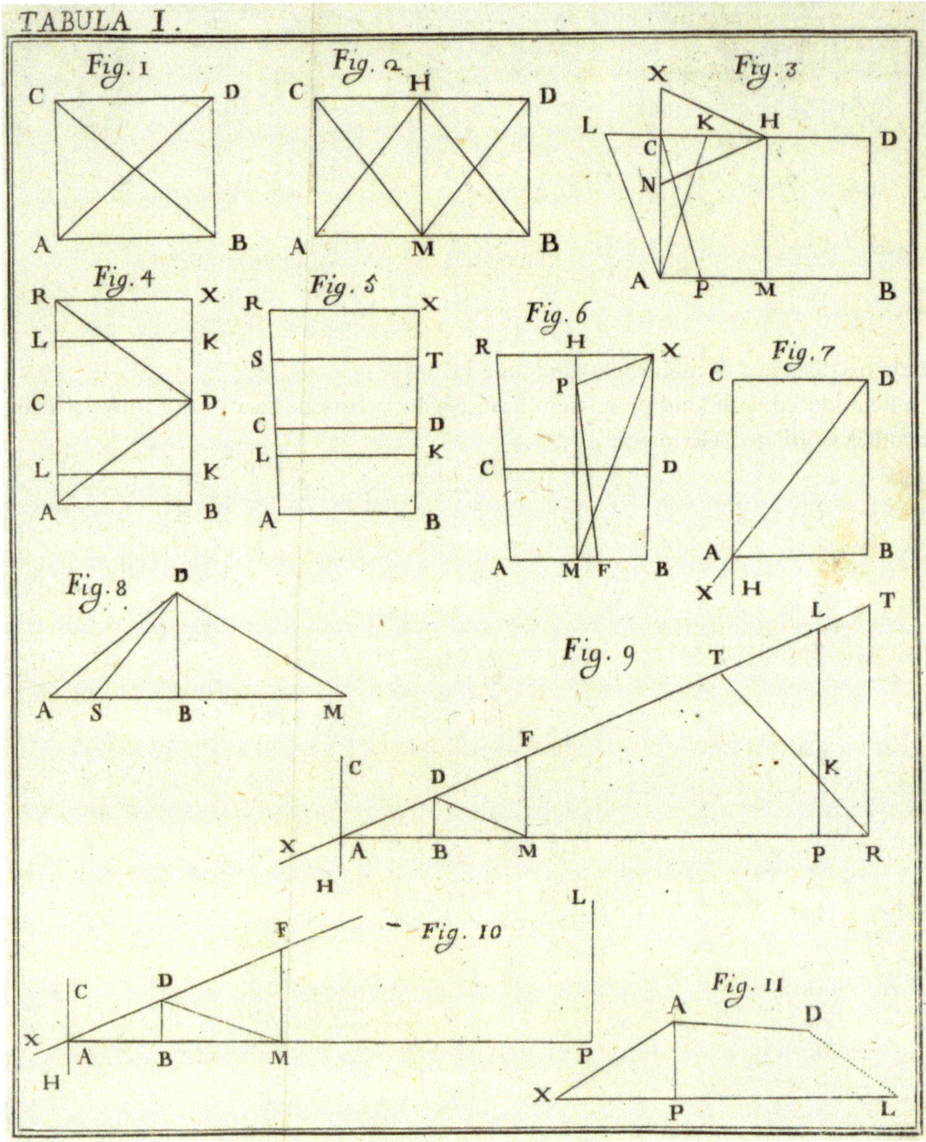
TABULA I.
Fig. 1
C D A B
Fig. 2
C H D A M B
Fig. 3
X L K H D C N A P M B
Fig. 4
R X L K C D L K A B
Fig. 5
R X S T C D L K A B
Fig. 6
R H X P C D A M F B
Fig. 7
C D A B X H
Fig. 8
D A S B M
Fig. 9
L T T F K C D X A B M P R H
Fig. 10
L F C D X A B M P H
Fig. 11
A D X P L

TABULA II.

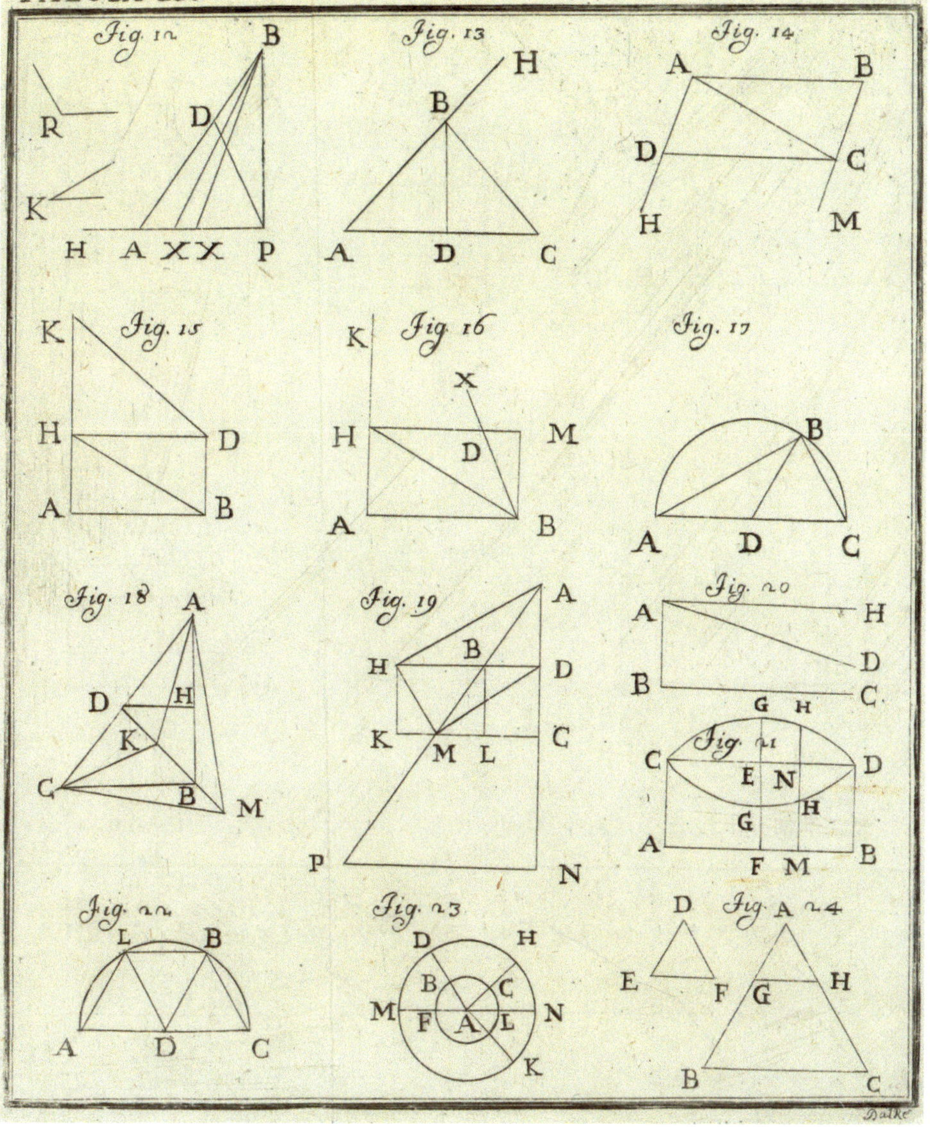
Fig. 12
Fig. 13
Fig. 14
Fig. 15
Fig. 16
Fig. 17
Fig. 18
Fig. 19
Fig. 20
Fig. 21
Fig. 22
Fig. 23
Fig. 24

TABULA III.

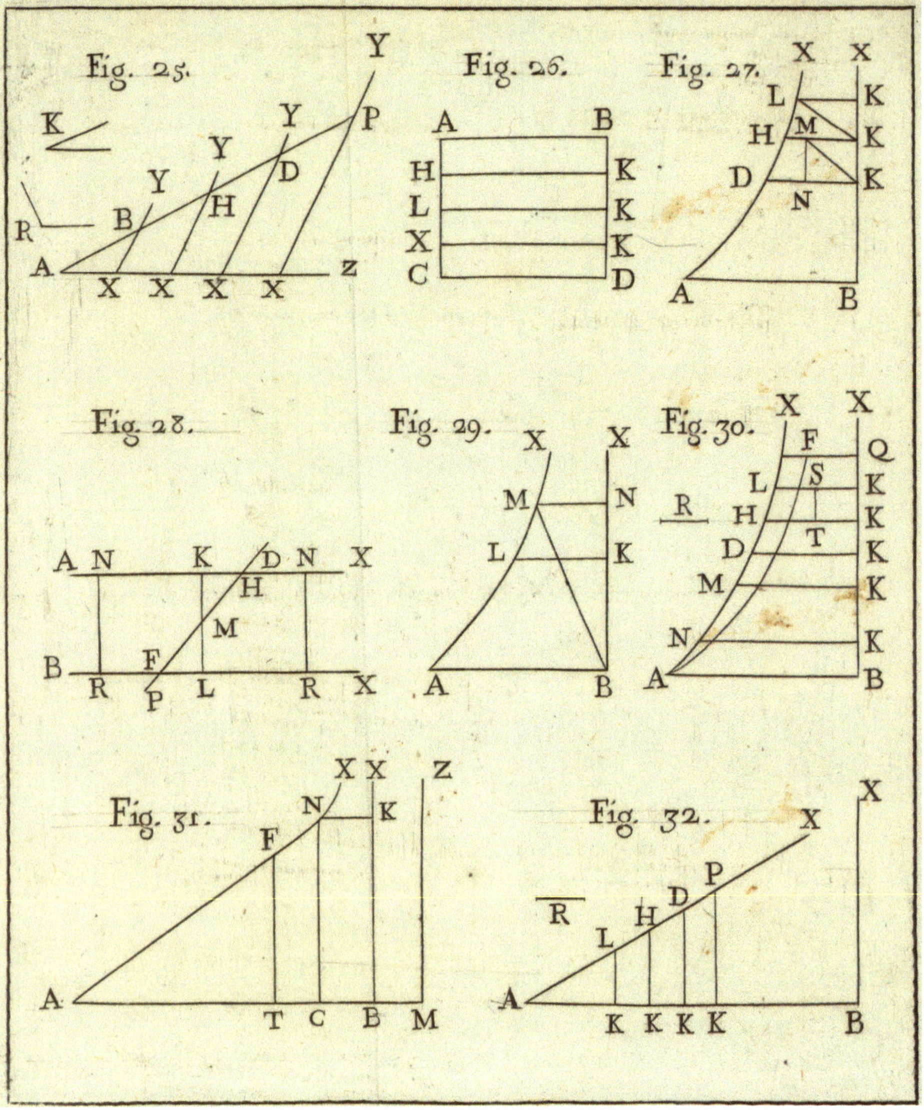

TABULA IV.

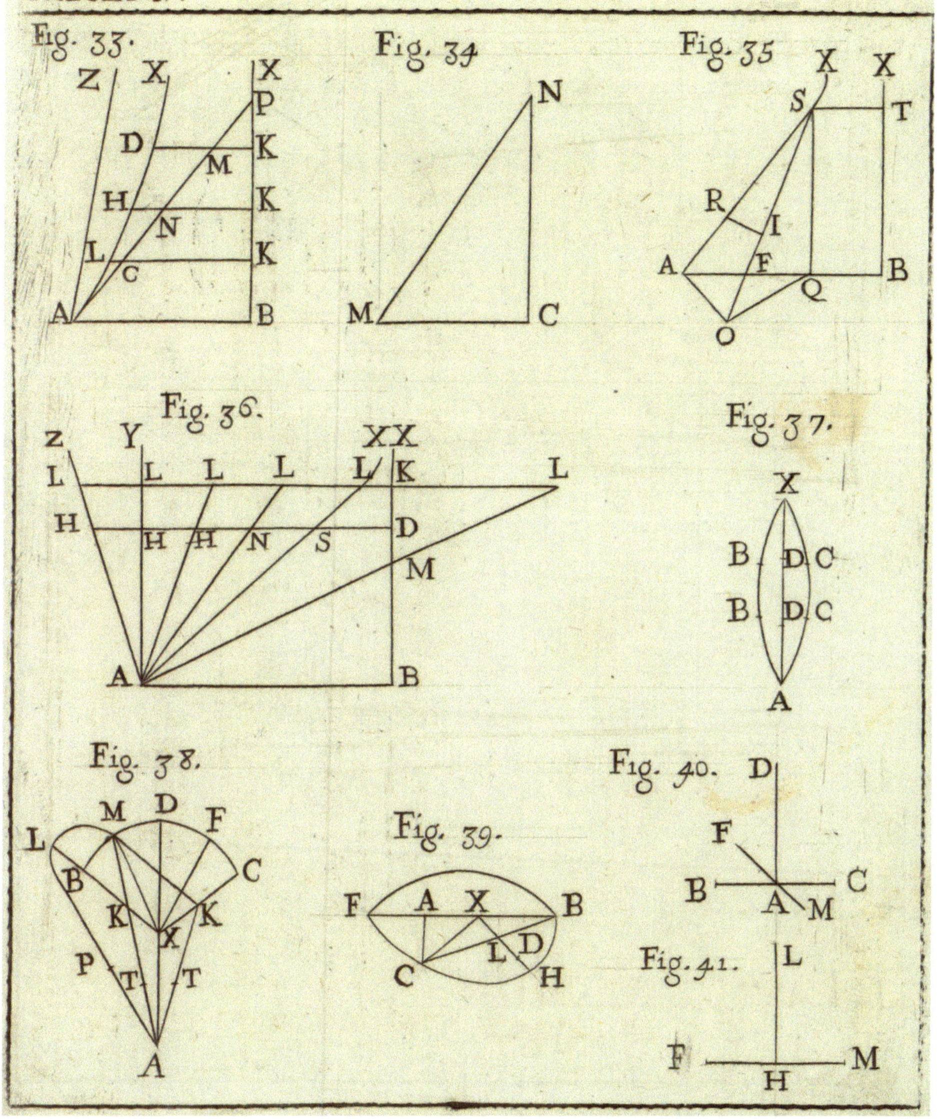

Fig. 33.
Z X X P D K M H K N L K C A B
Fig. 34
N M C
Fig. 35
X X S T R I A F B Q O
Fig. 36.
Z Y XX L L L L L K L H D H H N S M A B
Fig. 37.
X B D C B D C A
Fig. 38.
M D F L C B K K X P T T A
Fig. 39.
F A X B C L D H
Fig. 40.
D F B C A M
Fig. 41.
L F M H

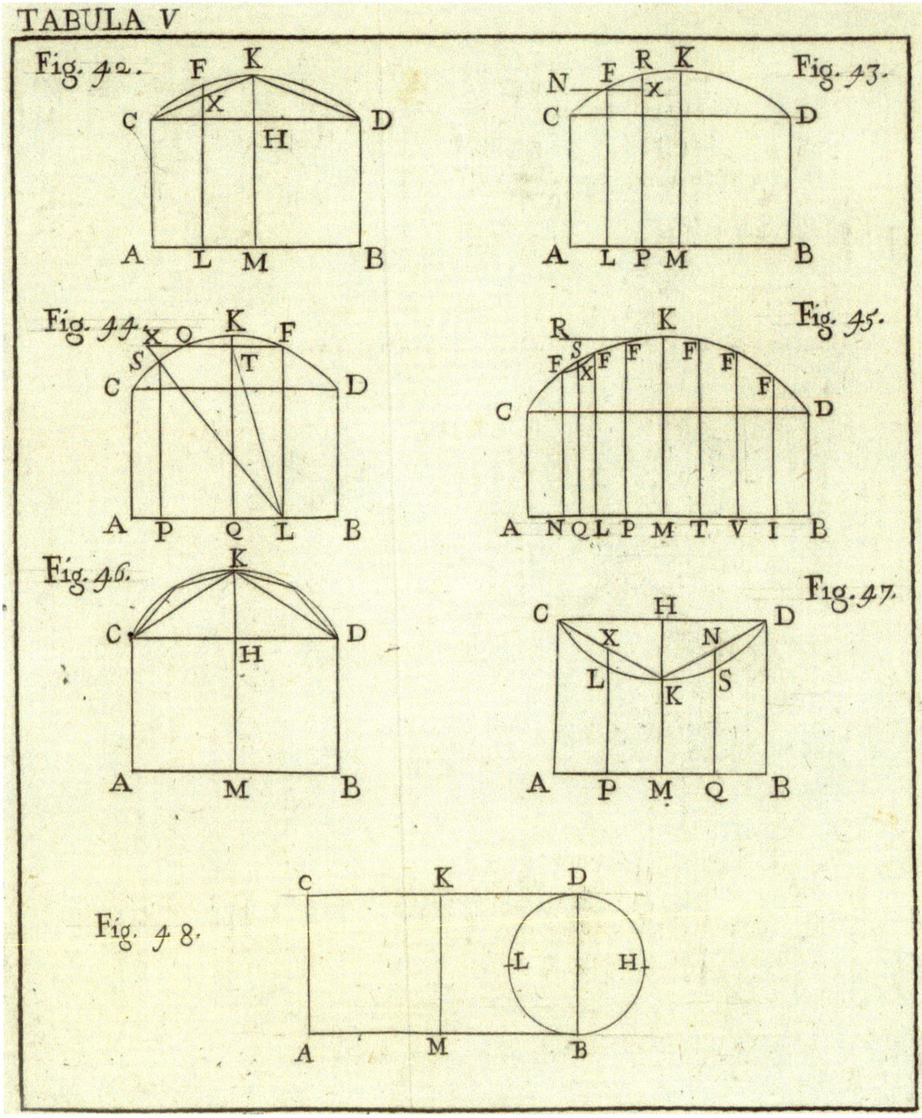
TABULA V
Fig. 42.
F K X C D H A L M B
Fig. 43.
N F R K X C D A L P M B
Fig. 44.
X O K F S T C D A P Q L B
Fig. 45.
R K S F X F F F F F C D A N Q L P M T V I B
Fig. 46.
K C D H A M B
Fig. 47.
C H D X N L S K A P M Q B
Fig. 48.
C K D L H A M B

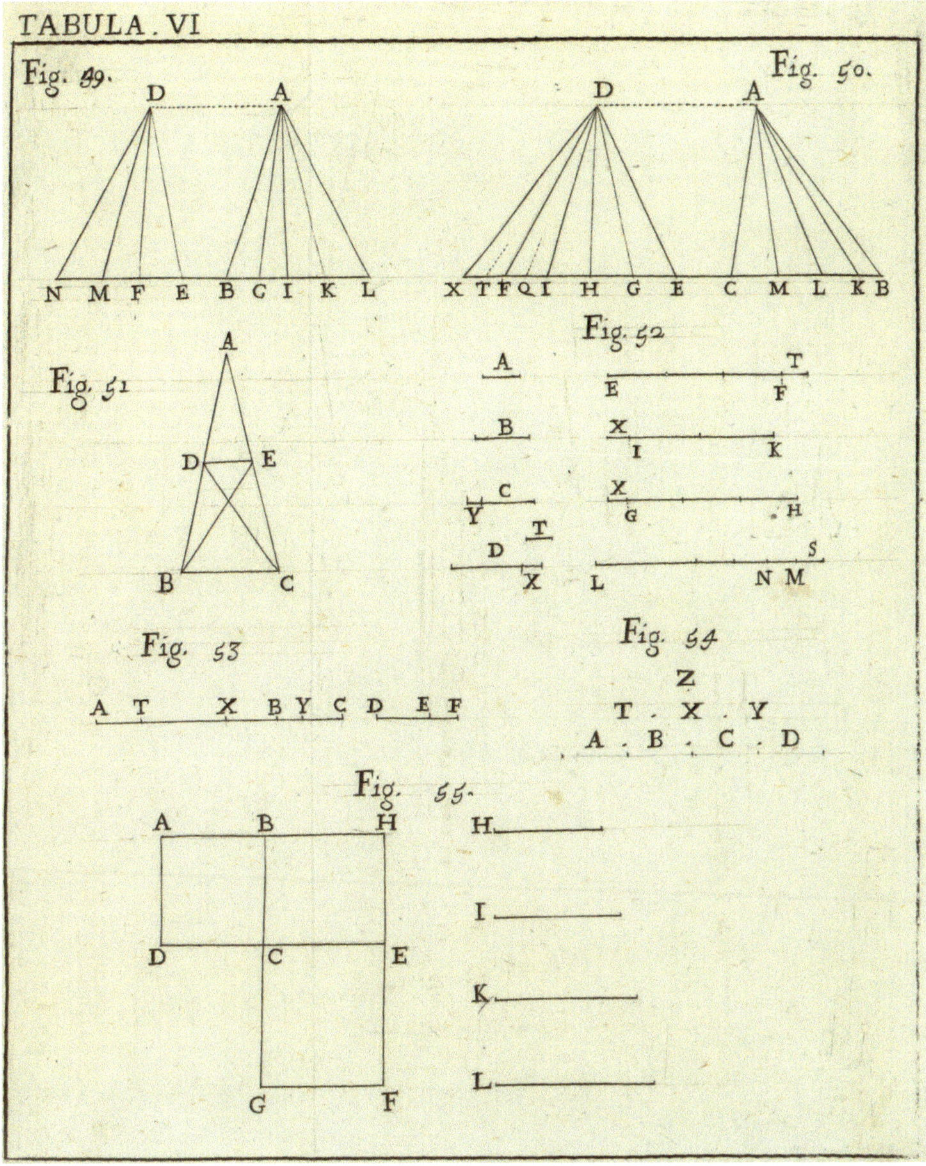
TABULA . VI
Fig. 49.
D A
N M F E B G I K L
Fig. 50.
D A
X T F Q I H G E C M L K B
Fig. 51
A
D E
B C
Fig. 52
A
B
C
Y
T
D
X
T
E F
X
I K
X
G H
S
L N M
Fig. 53
A T X B Y C D E F
Fig. 54
Z
T . X . Y
A . B . C . D
Fig. 55.
A B H
D C E
G F
H
I
K
L

Index

Z

MIX
Papier aus verantwortungsvollen Quellen
Paper from responsible sources
FSC® C105338

If you have any concerns about our products,
you can contact us on
ProductSafety@springernature.com

In case Publisher is established outside the EU,
the EU authorized representative is:
Springer Nature Customer Service Center GmbH
Europaplatz 3, 69115 Heidelberg, Germany

Printed by Libri Plureos GmbH
in Hamburg, Germany